Scattering Theory in Mathematical Physics

NATO ADVANCED STUDY INSTITUTES SERIES

Proceedings of the Advanced Study Institute Programme, which aims at the dissemination of advanced knowledge and the formation of contacts among scientists from different countries

The series is published by an international board of publishers in conjunction with NATO Scientific Affairs Division

A	Life Sciences	Plenum Publishing Corporation
B	Physics	London and New York
C	Mathematical and Physical Sciences	D. Reidel Publishing Company Dordrecht and Boston
D	Behavioral and Social Sciences	Sijthoff International Publishing Company Leiden
E	Applied Sciences	Noordhoff International Publishing Leiden

Series C – Mathematical and Physical Sciences

Volume 9 – Scattering Theory in Mathematical Physics

Scattering Theory in Mathematical Physics

Proceedings of the NATO Advanced Study Institute held at Denver, Colo., U.S.A., June 11–29, 1973

edited by

J. A. LAVITA and J.-P. MARCHAND

Dept. of Mathematics, University of Denver, Denver, Colo., U.S.A.

D. Reidel Publishing Company

Dordrecht-Holland / Boston-U.S.A.

Published in cooperation with NATO Scientific Affairs Division

Library of Congress Catalog Card Number 73–91205

ISBN-13:978-94-010-2149-4 e-ISBN-13:978-94-010-2147-0
DOI: 10.1007/978-94-010-2147-0

Published by D. Reidel Publishing Company
P.O. Box 17, Dordrecht, Holland

Sold and distributed in the U.S.A., Canada, and Mexico
by D. Reidel Publishing Company, Inc.
306 Dartmouth Street, Boston, Mass. 02116, U.S.A.

CONTENTS

PREFACE

These proceedings contain lectures given at the N.A.T.O. Advanced Study Institute entitled "Scattering Theory in Mathematics and Physics" held in Denver, Colorado, June 11-29, 1973. We have assembled the main series of lectures and some presented by other participants that seemed naturally to complement them. Unfortunately the size of this volume does not allow for a full account of all the contributions made at the Conference; however, all present were pleased by the number and breadth of those topics covered in the informal afternoon sessions.

The purpose of the meeting, as reflected in its title, was to examine the single topic of scattering theory in as many of its manifestations as possible, i.e. as a hub of concepts and techniques from both mathematics and physics. The format of all the topics presented here is mathematical. The physical content embraces classical and quantum mechanical scattering, N-body systems and quantum field theoretical models. Left out are such subjects as the so-called analytic S-matrix theory and phenomenological models for high energy scattering.

We would like to thank the main lecturers for their excellent presentations and written summaries. They provided a focus for the exceptionally strong interaction among the participants and we hope that some of the coherence achieved is reflected in these published notes. We have made no attempt to unify notation. Thus the presentations are consistent with each author's previous work rather than with the rest of this volume. Still, we feel the essential unity of concept and technique shows through.

We appreciate the contributions of all the attendees and the patience shown with any fault of organization of which we are guilty.

We thank N.A.T.O. for their generous financial support of this humanistic, peaceful enterprise.

Thanks are also due to the University of Denver and its Conference Coordinating Office.

We owe much to the people who helped in the organization of the whole meeting, in particular to L.P. Horwitz, who gave the initial stimulus, to J.M. Jauch, Calvin Wilcox and A. Barut who served on various advisory committees and to Herbert Greenberg, Chairman of the Mathematics Department at the University of Denver, for his hospitality. Finally, special thanks are due to Loretta Norris and Anita Johnson, on whom the burden of typing the manuscript fell, and to Mark Balas and Dexter Strawther for technical assistance in the preparation of the manuscript.

J.A. LaVita

J.-P. Marchand

NATO Scattering Theory Institute

Participants

S. Albeverio
U. of Napoli

W. Amrein
U. of Geneva

Victor Barcilon
U. of Chicago

James R. Beale
Stanford University

Désiré Bollé
U. of Leuven, Belgium

John Chadam
Indiana University

Colston Chandler
U. of New Mexico

Kuang-Ho Chen
Louisiana State U.
in New Orleans

F. Coester
Argonne National Lab.

Jean Michel Combes
U. of Toulon, France

James H. Cooke
U. of Texas
at Arlington

John V. Corbett
Indiana University

John A. DeSanto
Naval Research Lab.
Washington D.C.

Charles L. Dolph
U. of Michigan

Hans Ekstein
Argonne National Lab.

Archie G. Gibson
U. of New Mexico

Charles Goldstein
Brookhaven National Lab.

Jerome A. Goldstein
Tulane University

William L. Goodhue
U. of Notre Dame

J.C. Guillot
U. of Dijon

Karl Gustafson
U. of Colorado

Mazhar Hasan
Northern Illinois University

F. Joanne Helton
Dowling College

J. William Helton
SUNY, Stonybrook

James R. Klein
Iona College

Richard B. Lavine
U. of Rochester

James A. LaVita
U. of Denver

Peter D. Lax
Courant Institute

Chul H. Lee
U. of Notre Dame

Charles T. McAbee
U. of New Mexico

Andrew J. Majda
Stanford University

Jean-Paul Marchand
U. of Denver

Philippe Martin
U. of Colorado

Gustavo P. Menzala
Brown University

Roger G. Newton
Indiana University

Participants cont.

Tuong D. Nguyen
U. of Ottawa, CANADA

T. A. Osborn
Brooklyn College

John Palmer
Stanford University

Ralph Phillips
Stanford University

Edward Prugovecki
U. of Toronto

Peter A. Rejto
U. of Minnesota

Albert W. Saenz
NRI, Washington, D.C.

James T. Sandefur
Tulane University

Norman Shenk
U. of California
at San Diego

Ronald Shonkwiler
Georgia Institute of Technology

Lawrence Horwitz
Tel Aviv University

James Howland
University of Virginia

Mark Balas
U. of Denver

S. R. Singh
York University, Toronto

Walter A. Strauss
Brown University

R. F. Streater
Bedford College, London

Kwong-Tin Tang
Pacific Lutheran University

Lawrence E. Thomas
U. of Geneva, Switzerland

Clasine van Winter
University of Kentucky

Michael J. Westwater
U. of Washington

Calvin H. Wilcox
U. of Utah

Woodford W. Zachary
Naval Research Lab.
Washington, D.C.

Khelifa Zizi
U. of Paris-Nord

S. Steinberg

D. Strawther
U. of Denver

List of Afternoon Talks

J.T. Beale, "Scattering Frequencies of Resonators"

J. Chadham, "Coupled Maxwell-Dirac Equation"

C. Chandler, "Multichannel Time-Interdependent Scattering Theory for Changed Particles"

K.H. Chen, "Semmetric Hyperbolic System"

F. Coester, "Unstable Particles"

J.V. Corbett, "Unbounded Motion and Scattering Theory"

J.A. DeSanto, "Diagram Methods for Electromagnetic Scattering from Random Rough Surfaces"

H. Ekstein, "Scattering Theory Without the S Operator"

C. Goldstein, "Scattering Theory in Wave Guides"

J. Goldstein, "Temporally Inhomogeneous Scattering Theory"

J.C. Guillot & K. Zizi "Perturbation of the Laplacian by Coulomb-type Potentials"

K. Gustafson, "H_{ac}"

L. Horwitz, "Relativistic Dynamics"

J. Howland, "Spectral Concentration"

R.B. Lavine, "Commutator Methods in Scattering"

P. Martin, "The Kato-Rosemblum Lemma"

T.A. Osborn, "Primary Singularities in Three-Body Wave Functions"

E. Prugovecki, "Scattering Theory for Asymptotically Coulomb Potentials"

P. Rejto, "The Titchmarsh-Weidmann Theorem on Absolute Continuity"

S. Steinberg, "Analytic Continuation of the Resolvent; Local Time Decay"

L.E. Thomas, "Algebraic Theory of Scattering from Impurities in Crystals"

C. Van winter, "Complex Dynamic Variables and Asymptotic Completeness in Multichannel Scattering"

M.J. Westwater, "A Convergence Theorem for the Kohn Variational Method"

W.W. Zachary, "Comment on a Paper of Amrein, Martin and Misra"

SCATTERING THEORY FOR DISSIPATIVE HYPERBOLIC SYSTEMS

P. D. Lax and R. S. Phillips†

Abstract. An abstract theory of scattering is developed for dissipative hyperbolic systems, a typical example of which is the wave equation: $u_{tt} = \Delta u$ in an exterior domain with lossy boundary conditions: $u_n + \alpha u_t = 0$, $\alpha \geq 0$. In this theory, as in an earlier theory developed by the authors for conservative systems, a central role is played by two distinguished subspaces of data common to both the perturbed and unperturbed problems. Associated with each subspace is a translation representation of the unperturbed system. When these representations coincide they provide a convenient tool for extending the data so as to include a large class of generalized eigenfunctions for both the perturbed and unperturbed generators. The scattering matrix is characterized in terms of these generalized eigenfunctions; it is shown to be meromorphic in the whole complex plane and holomorphic in the lower half plane. The zeros and poles of the scattering matrix correspond, respectively, to incoming and outgoing generalized eigenfunctions of the perturbed generator.

Introduction. Most of scattering theory to date has dealt with energy or mass conserving systems and the corresponding groups of unitary operators. However a perfectly conservative system is an idealization which can only be approximated in nature. It is therefore important to extend the theory to dissipative systems. Several authors [1,2,5,10,11] have already explored this problem for quantum mechanical systems for which the perturbed system still generates what is basically a group of operators. The perturbation problem for

†This work was supported in part by the United States Atomic Energy Commission under contract AT(30-1)-1480, by the National Science Foundation under grant GP-8867 and by the United States Air Force under contract AF-F44620-71-C-0037.

J. A. LaVita and J.-P. Marchand (eds.), Scattering Theory in Mathematical Physics, 1–34.

semigroups of operators was first treated by S. C. Lin [9] in 1969. Our purpose in this paper is to present a scattering theory for dissipative hyperbolic systems based on the ideas introduced by the authors in [6] and [7] which can also handle perturbed systems which do not generate groups of operators.

We suppose that there is an unperturbed system, described by a one-parameter group of unitary operators $U_0(t)$, acting on a Hilbert space H_0; we denote by A_0 the generator of U_0. We suppose that there are two distinguished closed subspaced D_- and D_+, with the following properties:

$$\begin{aligned} &\text{(i)} && U_0(t)D_- \subset D_- \text{ for } t \leq 0, \\ & && U_0(t)D_+ \subset D_+ \text{ for } t \geq 0, \\ &\text{(ii)} && \wedge\, U_0(t)D_- = \{0\}, \quad \wedge\, U_0(t)D_+ = \{0\} \\ &\text{(iii)} && \overline{\vee\, U_0(t)D_+} = H_0, \quad \overline{\vee\, U_0(t)D_-} = H_0. \end{aligned} \tag{1}$$

According to the translation representation theorem (cf. Chapter 2 of [6]) there correspond to D_+ and D_- unitary representations $\mathcal{J}_\pm$ of H_0 as vector valued functions on the real axis $\mathbb{R}$:

$$\begin{aligned} &\mathcal{J}_+ : H_0 \to L_2(\mathbb{R}, N) \\ &\mathcal{J}_- : H_0 \to L_2(\mathbb{R}, N) \end{aligned} \tag{2}$$

such that

$$\begin{aligned} &\mathcal{J}_+ : D_+ \to L_2(\mathbb{R}_+, N) \\ &\mathcal{J}_- : D_- \to L_2(\mathbb{R}_-, N), \end{aligned} \tag{3}$$

where N denotes an auxiliary Hilbert space; and

$$\mathcal{J} U_0(t) = \mathcal{T}(t)\mathcal{J}; \tag{4}$$

here $\mathcal{T}(t)$ denotes right translation by t. We call $\mathcal{J}_-$ the *incoming*, $\mathcal{J}_+$ the *outgoing representation.*

We note that Fourier transformation

$$(2\pi)^{-1/2} \int_{-\infty}^{\infty} k(s)\, e^{i\sigma s}\, ds$$

changes translation representation into spectral representation; when applied to $\mathcal{J}_+$ and $\mathcal{J}_-$ the resulting spectral representations take D_+ into the space of vector valued functions analytic in the upper half plane Im $\sigma > 0$, D_- into the space of vector valued functions analytic in the lower half plane, respectively. It is the translates of D_+ and D_-, rather than D_+ and D_- themselves that enter into the theory. We shall denote these as follows:

$$D_+^a = U_0(a)D_+, \quad D_-^a = U_0(-a)D_-, \quad a \geq 0. \tag{5}$$

We turn now to the perturbed system; we suppose it to be described by a semigroup of contraction operators $T(t)$, $t \geq 0$, acting on a Hilbert space H. The relation between U_0, H_0 and T, H is that for some $\rho > 0$, H contains D_+^ρ and D_-^ρ as subspaces and that $U_0(t)$ and $T(t)$ coincide on D_+^ρ whereas $U_0(-t)$ and $T^*(t)$ coincide on D_-^ρ:

$$T(t)g = U_0(t)g \quad \text{for all } g \text{ in } D_+^\rho \tag{6}$$

$$T^*(t)f = U_0(-t)f \quad \text{for all } f \text{ in } D_-^\rho. \tag{6*}$$

Since $T(t)$ and $T^*(t)$ agree with $U_0(t)$ and $U_0(-t)$ on D_+^ρ and D_-^ρ respectively, it follows that they inherit some of the properties of U_0 on these subspaces. The relations (1)i and (1)ii are such properties, and when combined with (6) and (6)* they yield

$$\begin{aligned} &\text{(i)} \quad T^*(t)D_-^\rho \subset D_-^\rho, \quad T(t)D_+^\rho \subset D_+^\rho. \\ &\text{(ii)} \quad \wedge\, T^*(t)D_-^\rho = \{0\}, \quad \wedge\, T(t)D_+^\rho = \{0\}. \end{aligned} \tag{7}$$

Property (1)iii refers to actions of U_0 which carry $D_\pm^\rho$ outside of themselves; therefore (1)iii implies nothing about the action of T and T^*. Instead we have to assume an analogous property for T and T^*.

To state the needed property we introduce two projections P_+^ρ and P_-^ρ, defined as orthogonal projections onto the orthogonal complements of D_+^ρ and D_-^ρ, respectively. That is, P_+^ρ removes the D_+^ρ component, P_-^ρ removes the D_-^ρ component. The analogue of (1)iii is then:

For every f in H

$$\begin{aligned} &\lim_{t\to\infty} P_-^\rho T^*(t)f = 0, \\ &\lim_{t\to\infty} P_+^\rho T(t)f = 0. \end{aligned} \tag{8}$$

Property (8) expresses the fact that under the action of T or T^*, data eventually end up in D_+^ρ, respectively D_-^ρ.

The following two examples will illustrate the kind of systems we are concerned with:

Example 1. A very transparent model for our theory is: $H_0 = L_2(R)$, $[U_0(t)f](s) = f(s-t)$, $D_-^\rho = L_2((-\infty,-\rho))$, $D_+^\rho = L_2((\rho,\infty))$; $H = L_2(R)$ and $[T(t)f](s) = f(s-t)$ if $s \leq 0$ or $s > t$ and $= 0$ otherwise.

Example 2. A more interesting example is provided by the acoustic equation. Here the unperturbed system satisfies the equation

$$u_{tt} = \Delta u \quad \text{in } \mathbb{R}^n, \tag{9}$$

the Hilbert space H_0 is the set of all initial data $d = (f_1,f_2)$:

$$u(x,0) = f_1(x), \quad u_t(x,0) = f_2(x)$$

with energy norm:

$$\|d\|_E^2 = \int [|\partial_x f_1|^2 + |f_2|^2]\, dx. \tag{10}$$

The subspace D_-^ρ $[D_+^\rho]$ consists of the set of all initial data for solutions which vanish in the truncated backward [forward] cone: $|x| \leq \rho-t$, $t \leq 0$ $[|x| \leq \rho+t,\ t \geq 0]$; here ρ is chosen so that the scattering obstacle is contained in the ball $\{|x| < \rho\}$. The perturbed system satisfies the acoustic equation (9) in the exterior G of the obstacle and dissipative boundary conditions:

$$u_n + \alpha u_t = 0 \text{ on } \partial G,\ \alpha \geq 0; \tag{11}$$

here n denotes the outward normal to G on ∂G. In this example property (8) is a consequence of local energy decay.

In Section 1 we define the scattering operator S as

$$S = \operatorname*{st.lim}_{t\to\infty} U_0(-t)J^*T(2t)JU_0(-t), \tag{12}$$

where J maps H_0 into H. We show that if conditions (i)-(iii) in (1), (6), (i)-(ii) in (7) and (8) hold, the limit above exists. The operator S commutes with U_0, from which it follows that in a translation representation for U_0 the operator S acts as convolution with an operator valued distribution which we denote as S(s):

$$\mathcal{J}S = S * \mathcal{J}. \tag{13}$$

In the corresponding spectral representation S acts as multiplication by the Fourier transform $\mathcal{S}$ of S;

$$\mathcal{S}(\sigma) = \int S(s)\, e^{is\sigma}\, ds. \tag{14}$$

$\mathcal{S}(\sigma)$ is called the *scattering matrix*.

In many important applications the two translation representations $\mathcal{J}_-$ and $\mathcal{J}_+$ discussed earlier coincide; when this occurs one can show that in this representation S is *causal* in the sense that it maps $L_2(\mathbb{R}_- - \rho, N)$ into $L_2(\mathbb{R}_- + \rho, N)$. Consequently in this case the function S(s) has its support in $\mathbb{R}_- + 2\rho$; and its Fourier transform, the scattering matrix $\mathcal{S}(\sigma)$, is then analytic in the lower half plane $\operatorname{Im} \sigma < 0$ and is of exponential growth. Even when S is not causal, it is often possible to continue $\mathcal{S}(\sigma)$ into the lower half plane from $\mathbb{R}_+$ or $\mathbb{R}_-$, but usually not from both simultaneously; a general result of this kind is given at the end of Section 1.

If we convolve S with an exponential function $e = e^{-izs}n$, n in N, we get, using (14),

$$S * (e^{-izs}n) = \int S(r)\, e^{-iz(s-r)}n\, dr = e^{-izs} \int S(r)e^{izr}n\, dr = e^{-izs}\mathcal{S}(z)n. \quad (15)$$

This suggests that the scattering matrix $\mathcal{S}(z)$ can be studied by observing the action of the scattering operator S on data whose translation representation is exponential. Since exponential functions are not in L_2 over $\mathbb{R}$, it follows that they do not represent elements in H_0. In order to be able to talk of such elements, we extend the spaces H_0 and H in Section 2 and then define suitable extensions for the operators U_0, T, $P_\pm^a$, A_0 and A to the extended spaces. In Section 3 we use this modest machinery to legitimize formula (15).

In Section 4 we introduce the associated semi-group:

$$Z(t) = P_+^\rho T(t) P_-^\rho \text{ on } K = H \ominus (D_-^\rho \oplus D_+^\rho). \quad (16)$$

This turns out to be a useful tool in the study of the scattering matrix. In fact if the resolvent of the generator B of Z is meromorphic then the set of poles of $\mathcal{S}(z)$ coincide with $-i$ times the spectrum of B. In Section 5 we show that the poles of $\mathcal{S}(z)$ correspond to outgoing (orthogonal to D_-^ρ) generalized eigenfunctions of A: $Aa = iza$ and the zeros of $\mathcal{S}(z)$ to the incoming (orthogonal to D_+^ρ) generalized eigenfunctions of A. The zeros of $\mathcal{S}(z)$ in the upper half plane correspond to ordinary eigenfunctions of A in H.

The spectrum of A always contains a continuous part filling out the imaginary axis. In Section 6 we present a criterion for the rest of the spectrum of A to be point spectra and hence determined by the zeros of $\mathcal{S}(z)$.

In an expanded version of this paper which is being published in the Journal of Functional Analysis the abstract theory is applied to Example 2 above, that is to the wave equation in the exterior of an obstacle subject to dissipative boundary conditions.

1. <u>The Wave and Scattering Operators</u>. Throughout this part we assume that

(a) $U_0(t)$ is a group of unitary operators satisfying conditions (i)-(iii) in (1);

(b) T(t) is a semi-group of contractions satisfying conditions (i)-(ii) in (7) and (8);

(c) U_0 and T are related through conditions (6) and $(6)^*$ which can be restated as:

$$U_0(-t)T(t) = I \text{ on } D_+^\rho \tag{1.1}$$
$$U_0(t)T^*(t) = I \text{ on } D_-^\rho \, . \tag{1.1*}$$

We start by deducing from (1.1) and (1.1)* a further useful relation between U_0 and T:

$$T(t)U_0(-t) = I \text{ on } D_-^\rho \tag{1.2}$$
$$T^*(t)U_0(t) = I \text{ on } D_+^\rho \tag{1.2*}$$

The proof relies on the following general proposition about Hilbert space operators:

Lemma 1.1. Let W be an operator mapping one Hilbert space H_0 into another H, satisfying the following conditions:

(a) W is contractive, i.e., $\|W\| \leq 1$;

(b) W is the identity on some subspace D common to both H_0 and H. Then

(c) W maps the orthogonal complement of D in H_0 into the orthogonal complement of D in H;

(d) W^* is the identity on D.

Proof. Let x be any element of D, y any element of $D^\perp$ in H_0. By property (b),

$$W(x + \varepsilon y) = x + \varepsilon Wy.$$

By property (a)

$$\|x + \varepsilon Wy\|^2 \leq \|x + \varepsilon y\|^2.$$

Since y is orthogonal to x, the coefficient of ε on the right in the inequality above is zero. Therefore it must be zero also on the left, which proves part (c). This shows that D and $D^\perp$ completely reduce W, from which part (d) follows.

We should like to apply Lemma 1.1 to $W = U_0(-t)T(t)$. However this operator is not well defined because U_0 and T act on different spaces which have in common only D_-^ρ and D_+^ρ (not necessarily orthogonal). We therefore need linear mappings that take us from H_0 to H and back. Let J_0 and J be such maps:

$$J_0 : H_0 \to H,$$
$$J : H \to H_0.$$

We only require these maps to be linear and continuous, and

that they act like the identity on the common part of H_0 and H, which is $D_+^\rho + D_-^\rho$. For instance we could choose J_0 and J to be orthogonal projections of their domain spaces onto $D_+^\rho + D_-^\rho$. Note that for this choice J is the adjoint of J_0; such a choice is convenient for some calculations, and we shall use it whenever we need it.

We can now apply Lemma 1.1 to $W = U_0(-t)JT(t)$ and $D = D_+^\rho$. Clearly $\|W\| \leq 1$ and according to (1.1) condition (b) is fulfilled. Hence conclusion (d) follows so that $W^* = T^*(t)J_0U_0(t)$ acts like the identity on D_+^ρ. This proves $(1.2)^*$ and (1.2) follows similarly from $(1.1)^*$.

In the important special case when D_-^ρ and D_+^ρ are orthogonal, we can decompose H into three orthogonal components:

$$H = D_-^\rho \oplus K^\rho \oplus D_+^\rho,$$

where $K^\rho = H \ominus (D_-^\rho \oplus D_+^\rho)$. It is instructive to observe the action of $T(t)$ on these three components. From properties (1.1) and (1.2) and the orthogonality of the three components one can easily deduce:

(i) $T(t)$ maps D_-^ρ into all three components, but the $D_+^{\rho+t}$ component of D_+^ρ is uneffected.

(ii) $T(t)$ maps K^ρ into K^ρ and D_+^ρ, and again the $D_+^{\rho+t}$ component of D_+^ρ is uneffected.

(iii) $T(t)$ maps D_+^ρ into itself.

It is useful to think of D_-^ρ, K^ρ and D_+^ρ as past, present and future and to measure time in the past and future by the parameter s in the incoming and outgoing translation representations. If we think of the action of $T(t)$ as the flow of time, then the above observations on the action of $T(t)$ can be described picturesquely as follows:

The past influences the present and the future, and the present influences the present and the future. No instant of time t_0 can have any influence on a later time t_1 in less time than $t_1 - t_0$.

Such action can be described as *causal, with signal speed* ≤ 1.

Next we turn to studying the asymptotic relation between the actions of U_0 and T.

<u>Theorem 1.2</u>. Define the operators $W_1(t)$ and $W_2(t)$ by

$$W_1(t) = T(t)J_0U_0(-t) \tag*{$(1.3)_1$}$$

$$W_2(t) = U_0(-t)JT(t) \tag{1.3$_2$}$$

then the limits

$$W_1 = \text{strong} \lim_{t\to\infty} W_1(t) \tag{1.4$_1$}$$

and

$$W_2 = \text{strong} \lim_{t\to\infty} W_2(t) \tag{1.4$_2$}$$

exist for any admissible J_0 and J, and are independent of the choices of J_0 and J. W_1 and W_2 are called the wave operators.

Proof. Let f be of the form

$$f = U_0(s)k, \quad k \text{ in } D_-^\rho \tag{1.5}$$

then

$$W_1(t)f = T(t)J_0U_0(-t)f = T(t)J_0U_0(s-t)k. \tag{1.6}$$

It follows from (1.2) (with t-s in place of t) that for $t > s$, (1.6) equals $T(s)k$. Being independent of t for $t > s$, the limit of (1.6) as t tends to ∞ exists for every f of form (1.5). According to condition (iii) in (1) elements of the form (1.5) are dense in H_0; since the operators $W_1(t)$ are uniformly bounded for all t, convergence on a dense set implies convergence everywhere. It is clear from the above that this limit does not depend on the choice of J_0.

To prove the existence of W_2, let f be any element of H, and decompose $T(t)f$ as a sum of elements of D_+^ρ and its orthogonal complement:

$$T(t)f = d(t) + e(t), \tag{1.7}$$

where

$$d(t) \in D_+^\rho \quad \text{and} \quad e(t) \perp D_+^\rho.$$

Using (1.7) and the fact that d is in D_+^ρ and that J is the identity on D_+^ρ, we get

$$W_2(t)f = U_0(-t)JT(t)f = U_0(-t)d(t) + U_0(-t)Je(t). \tag{1.8}$$

Similarly, for $s > 0$,

$$\begin{aligned} W_2(t+s)f &= U_0(-t-s)JT(t+s)f \\ &= U_0(-t-s)T(s)d(t) + U_0(-t-s)JT(s)e(t). \end{aligned}$$

According to (1.1), T and U_0 agree on D_+^ρ, so we get

$$W_2(t+s)f = U_0(-t)d(t) + U_0(-t-s)JT(s)e(t). \tag{1.9}$$

Subtract (1.8) from (1.9):

$$W_2(t+s)f - W_2(t)f = U_0(-t-s)JT(s)e(t) - U_0(-t)Je(t). \tag{1.10}$$

By assumption (8), $\|e(t)\|$ tends to zero as t tends to ∞; this shows that the right side of (1.10) tends to 0, which proves the convergence of (1.10) to a limit. The independence of this limit from the choice of J follows from formula (1.8).

We now list some properties of the wave operators which will be used subsequently:

<u>Theorem 1.3</u>. (a) U_0 and T are intertwined by both W_1 and W_2, i.e.

$$W_1U_0(s) = T(s)W_1 \tag{1.11}$$

and

$$W_2T(s) = U_0(s)W_2; \tag{1.12}$$

(b)

$$W_1 = I \text{ on } D_-^\rho \tag{1.13}$$

and

$$W_2 = I \text{ on } D_+^\rho; \tag{1.14}$$

(c) For any $a \geq \rho$, W_1 maps $(D_-^a)^\perp$ in H_0 into $(D_-^a)^\perp$ in H and W_2 maps $(D_+^a)^\perp$ in H into $(D_+^a)^\perp$ in H_0;

(d) Suppose D_+^ρ and D_-^ρ are orthogonal; then for any $a \geq 0$, W_1 maps $U_0(a)D_-^\rho$ into the orthogonal complement of $D_+^{a+\rho}$ into W_2 maps $D_-^{a+\rho}$ the orthogonal complement of $U_0(-a)D_+^\rho$.

<u>Proof</u>. From the definition (1.3) of $W_1(t)$, $W_2(t)$, we see that

$$W_1(t)U_0(s) = T(s)W_1(t-s),$$

and

$$W_2(t)T(s) = U_0(s)W_2(t+s).$$

Hence, letting t tend to ∞ we deduce the intertwining relations (1.11) and (1.12).

Again from the definition of $W_1(t)$, $W_2(t)$ and the relations (1.1) and (1.2) we see that

$$W_2(t) = I \text{ on } D_+^\rho$$
$$W_1(t) = I \text{ on } D_-^\rho.$$

Letting t tend to ∞ we deduce (1.13) and (1.14).

Relations (c) follow from part(c) of Lemma 1.1 on using part (b) of the theorem proved above.

To prove the first part of (d) we write $D_+^{a+\rho}$ as $U_0(a)D_+^\rho$ and state the result to be proved in the following slightly symbolic form:

$$(W_1 U_0(a) D_-^\rho,\ U_0(a) D_+^\rho) = 0.$$

Using the intertwining property (1.11) we can replace $W_1 U_0(a)$ by $T(a)W_1$. Since by (1.13), $W_1 = I$ on D_-^ρ, we can rewrite the above as

$$(T(a) D_-^\rho,\ U_0(a) D_+^\rho) = 0$$

which is the same as

$$(D_-^\rho,\ T^*(a) U(a) D_+^\rho) = 0.$$

By $(1.2)^*$, $T^*(a)U_0(a) = I$ on D_+^ρ, so the above relation merely expresses the orthogonality of D_-^ρ and D_+^ρ; but that was assumed to be true, so the proof of the first part of (d) is complete.

The second part of (d) is handled similarly, $D_-^{a+\rho}$ is written as $U_0(-a)D_-^\rho$, and the result to be proved is written as

$$(W_2 U_0(-a) D_-^\rho,\ U_0(-a) D_+^\rho) = 0.$$

Moving $U_0(-a)$ across the inner product and using the second intertwining property (1.12) we get

$$(W_2 T(a) U_0(-a) D_-^\rho,\ D_+^\rho) = 0.$$

By (1.2), $T(a)U_0(-a) = I$ on D_-^ρ, so after shifting W_2 we can rewrite the above as

$$(D_-^\rho,\ W_2^* D_+^\rho) = 0.$$

It follows from (1.14) upon the application of part (d) of Lemma 1.1 that $W_2^* = I$ on D_+^ρ; so the above relation expresses the orthogonality of D_+^ρ and D_-^ρ. This completes the proof of Theorem 1.3.

We turn now to the *scattering operator*, defined by

$$S = W_2W_1. \tag{1.15}$$

Some of its properties are summarized in

Theorem 1.4. (a) S commutes with U_0:

$$U_0(s)S = SU_0(s) \tag{1.16}$$

(b) S-I maps D_-^ρ into the orthogonal complement of D_+^ρ

(c) If D_-^ρ and D_+^ρ are orthogonal, S maps D_-^ρ into the orthogonal complement D_+^ρ.

Proof. Part (a) for $s \geq 0$ is an immediate consequence of the intertwining relations (a) of Theorem 1.3 and the definition (1.15) of S. On multiplying both sides of (1.16) by $U_0(-s)$ we see that this relation holds for all real s. To prove part (b), take f in D_-^ρ; then according to (1.13) of Theorem 1.3,

$$W_1f = f$$

we conclude, using the definition (1.15) of S that

$$Sf = W_2f.$$

Next decompose f into orthogonal parts with respect to D_+^ρ:

$$f = d + e, \quad d \text{ in } D_+^\rho, \quad e \perp D_+^\rho \tag{1.17}$$

then

$$Sf = W_2d + W_2e$$

and subtracting (1.17) from this given

$$Sf - f = W_2d - d + W_2e - e.$$

According to parts (b) and (c) of Theorem 1.3, $W_2d = d$ and W_2e is orthogonal to D_+^ρ; this shows that

$$Sf - f = W_2e - e,$$

is orthogonal to D_+^ρ, as asserted in part (b). Part (c) is a corollary of part (b).

Since S commutes with U_0, it follows that in a spectral representation for U_0 the operator S acts as multiplication by some function $\mathcal{S}(\sigma)$. This is true even if we employ two different spectral representations of H_0 as the domain and as the range of S. We shall in fact take as the domain of S the *incoming* and as the range the *outgoing spectral representations* of U_0, defined in the introduction, and define the *scattering matrix* to be the multiplying factor $\mathcal{S}(\sigma)$ describing the action of S in these particular representations. Since these representations employ vector valued functions, $\mathcal{S}(\sigma)$ is an operator valued function.

Since the identity also commutes with U_0, it follows that when the above representations are employed for domain and range, the identity acts as multiplication by some factor which we denote by $\mathcal{K}(\sigma)$.

We noted in the introduction that in the incoming spectral representation D_-^ρ is represented by analytic functions in the lower half plane Im $\sigma < 0$, bounded by $e^{\rho \text{ Im } \sigma}$, and in the outgoing spectral representation the orthogonal complement of D_+^ρ is represented by functions analytic in the lower half plane, bounded by $e^{-\rho \text{ Im } \sigma}$. According to part (b) of Theorem 1.4, S-I maps D_-^ρ into the complement of D_+^ρ; consequently we conclude from the foregoing that multiplication by $\mathcal{S}(\sigma) - \mathcal{K}(\sigma)$ maps analytic functions in the lower half plane, bounded by $e^{\rho \text{ Im } \sigma}$, into analytic function in the lower half plane, bounded by $e^{-\rho \text{ Im } \sigma}$. It follows, see [6], that $\mathcal{S}(\sigma) - \mathcal{K}(\sigma)$ is *itself analytic in the lower half plane, and bounded by* $e^{-2\rho \text{ Im } \sigma}$.

Suppose the incoming and outgoing representations coincide; then $\mathcal{K}(\sigma) \equiv 1$ and we conclude that $\mathcal{S}(\sigma)$ itself can be continued analytically into the lower half plane where it is of exponential growth. In Section 3 we shall rederive this result.

In one of the applications studied in Part II, the factor $\mathcal{K}(\sigma)$ is $\not\equiv 1$ but can be continued analytically into the lower half plane from either the positive or the negative real axis. It then follows from the analyticity of $\mathcal{S} - \mathcal{K}$ that the same is true for $\mathcal{S}(\sigma)$.

2. <u>Extensions of the Spaces H_0 and H</u>. Throughout the remainder of Part I we assume that *the incoming and outgoing translation representations for* U_0 *described in the introduction are the same.* We shall denote this representation by the unsubscripted symbol $\mathcal{J}$. Note that D_-^ρ and D_+^ρ are automatically orthogonal in this case.

We define the space $\overline{H_0}$ *to consist of all* N-*valued locally* L_2 *functions on* $\mathbb{R}$. $\overline{H_0}$ is an extension of $L_2(\mathbb{R},N)$; identifying $L_2(\mathbb{R},N)$ with H_0 via the representation $\mathcal{J}$ makes $\overline{H_0}$ an extension of H_0.

We show now how to extend the operators U_0, A_0 and $P_\pm^a$ to $\overline{H_0}$:

We define $U_0(t)$ as right translation. We define A_0 to be $-\partial/\partial s$, its domain consisting of all functions whose distribution derivatives are locally L_2.

We define P_+^a and P_-^a on $\overline{H_0}$ as multiplication by the characteristic function of $\mathbb{R}_- + a$ and $\mathbb{R}_+ - a$ respectively.

We shall mainly use the one-sided extension of H_0 defined as consisting of all N-valued functions which are L_2 over any interval of the form (a,∞) with a arbitrary. We denote this extension by $\overline{H_0'}$. We shall say that f_n converges to f in $\overline{H_0'}$ if $f_n \to f$ in $L_2((a,\infty),N)$ for each a. Clearly U_0, A_0 and $P_\pm$ can also be extended as mappings of $\overline{H_0'}$ into itself.

We define similarly an extension of the space H; first we decompose H into three orthogonal parts:

$$H = D_-^a \oplus K^a \oplus D_+^a, \tag{2.1}$$

where $a > \rho$ and K^a is the orthogonal complement in H of $D_+^a \oplus D_-^a$. We then define $\overline{D_+^a}$ and $\overline{D_-^a}$ to consist of all N-valued locally L_2 functions on $\mathbb{R}_+ + a$ and $\mathbb{R}_- - a$, respectively. Identifying D_+^a with $L_2(R_+ + a,N)$ and D_-^a with $L_2(R_- - a,N)$ via the representation $\mathcal{J}$ makes $\overline{D_+^a}$ and $\overline{D_-^a}$ an extension of D_+^a and D_-^a, respectively. We define $\overline{H}$ as

$$\overline{H} = \overline{D_-^a} \oplus K^a \oplus \overline{D_+^a} \tag{2.2}$$

and we define the one-sided extension $\overline{H}'$ as

$$\overline{H}' = D_-^a \oplus K^a \oplus \overline{D_+^a}.$$

Note that $\overline{H}$ and $\overline{H}'$ do not actually depend on the value of a used in their definition.

We define P_+^a as annihilating the $\overline{D_+^a}$ component and being the identity on the K^a and $\overline{D_-^a}$ components; P_-^a is defined analogously.

On $\overline{D_+^a}$ we define T(t) as right translation. On $\overline{D_-^a}$ we define T(t) as right translation, provided that $a \geq \rho+t$; since the choice of a is arbitrary this can always be achieved. The action of T(t) on K^a is the same as before carrying K^a into $K_a \oplus (D_+^a \ominus D_+^{a+t})$. $T^*(t)$ is defined similarly.

To define Af for f in $\overline{H}$ we first decompose f as in (2.2)

$$f = f_- + k + f_+. \tag{2.3}$$

Then we choose two smooth cutoff functions ζ_1 and ζ_+, defined on $(-\infty,-a)$ and (a,∞) respectively, having the property that $\zeta_\pm$ is equal to 0 near $\pm a$, equal to 1 near $\pm\infty$ and smooth in between. We write f as

$$f = \zeta_- f_- + [(1-\zeta_-)f_- + k + (1-\zeta_+)f_+] + \zeta_+ f_+ \quad . \tag{2.4}$$

Note that the element in the square brackets belongs to H, whereas the other two terms lie in D_-^a and D_+^a where the actions of A and A_0 coincide. We say that f belongs to the domain of A if both f_- and f_+ have locally L_2 derivatives, and if the element in the square brackets in (2.3) belongs to the domain of A in H. Af is defined using the decomposition (2.3), with A acting as $-\partial/\partial s$ on the two outer components. Clearly, this definition doesn't depend on the choice of the cutoff functions and extends the action of A on H.

We turn now to the operator W_1. According to relation (1.13) of Theorem 1.3, W_1 is the identity on D_-^a; we extend W_1 by taking it to be the identity on $\overline{D}_-^a$. This defines a one-sided extension of W_1 as a mapping from $\overline{H}_0'$ to $\overline{H}'$; it is clear that W_1 is a continuous transformation.

It turns out that the operator W_2 has a two-sided extension, although we shall only use its one sided extension. First of all, according to (1.14) W_2 is the identity on D_+^a and can be extended as the identity to $\overline{D}_+^a$. To extend W_2 on the negative side we make use of part (d) of Theorem 1.3, according to which W_2 maps $D_-^{b+\rho}$ into $[U_0(-b)D_+^\rho]^\perp$. In our situation, where the incoming and outgoing representations are the same, $D_-^{b+\rho}$ is represented by functions supported on $[-\infty,-b-\rho]$, and $U_0(-b)D_+^\rho$ by functions supported on $[-b+\rho,\infty]$. The orthogonal complement of $U_0(-b)D_+^\rho$ in H_0 is therefore represented by functions supported on $[-\infty,-b+\rho]$. This analysis shows that in the present situation property d) of W_2 amounts to a kind of causality:

For $b \geq 2\rho$ W_2 *maps* $D_-^{b+\rho}$ *into* $D_-^{b-\rho}$.

We can now extend W_2 to any f in D_-^a; decompose f as a sum of functions,

$$f = \sum f_j$$

where f_j is L_2 and supported in $[-j-1,-j+1]$. Set

$$W_2 f = \sum W_2 f_j .$$

According to the foregoing the support of $W_2 f_j$ is confined to $s \leq -j+1+2\rho$; so the above sum defining $W_2 f$ is locally finite and defines a locally L_2 function. For the same reason the value of $W_2 f$ does not depend on the particular decomposition used for f. In this fashion we can define W_2 to map H' into H_0' and it is clear from this construction that W_2 is continuous.

The scattering operator S is the product $W_2 \underline{W_1}$; it can therefore be extended as an operator mapping $\bar{H}_0'$ into $\bar{H}_0'$ continuously.

It is easy to show that these extensions are independent of the value of a, and that the extended operators retain most of their previous properties. In particular *the intertwining properties* (1.11) *and* (1.12) *hold for the extended operators* W_1, W_2, U_0 *and* T. It follows that S *extended commutes with* U_0 *extended* and that S is *causal:*

The values of f *at* $s \leq a$ *do not influence the values of* Sf *at* $s > a + 2\rho$.

Remark. A less direct but more systematic method of obtaining these extensions is to regard $\bar{H}_0$ $[\bar{H}]$ as the dual of the subspace $\underline{H}_0$ $[\underline{H}]$, consisting of the elements in H_0 [H] whose incoming and outgoing components have compact support. The relevant operators then be defined as the duals of their formal adjoints on $\underline{H}_0$ or $\underline{H}$.

3. The Scattering Matrix.

We have shown in Section 2 that the scattering operator S can be extended as an operator from $\bar{H}_0'$ into $\bar{H}_0'$. Consider now the element e_0 of $\bar{H}_0'$ defined by

$$\mathcal{J} e_0 = e^{-izs} n , \qquad \text{Im } z < 0 , \tag{3.1}$$

where n is some element of N. Then e_0 belongs to $\bar{H}_0'$ and hence $e_1 = Se_0$ is defined.

Being represented by an exponential function, e_0 is an eigenfunction of the group U_0:

$$U_0(t) e_0 = e^{izt} e_0. \tag{3.2}$$

Since S commutes with U_0, it follows that $e_1 = Se_0$ also satisfies

$$U_0(t) e_1 = e^{izt} e_1,$$

from which it follows that e_1 is also represented by an exponential function

$$\mathcal{J} Se_0 = \mathcal{J} e_1 = e^{-izs} m. \tag{3.3}$$

Here m is an element of N, determined by z and n; clearly m depends linearly on n and it is consistent with (15) to write

$$m = \mathcal{S}(z)\, n. \tag{3.3$'$}$$

$\mathcal{S}(z)$ *is called the scattering matrix;* we summarize its properties in

Theorem 3.1. (a) *For any* z *in* Im z < 0, $\mathcal{S}(z)$ *is a bounded operator on* N:

$$\|\mathcal{S}(z)\| \leq e^{2\rho|\text{Im } z|} \tag{3.4}$$

(b) $\mathcal{S}(z)$ *depends analytically on* z *in the lower half plane.*

(c) *For any element* f *in* H_0 *which is represented by a supported on* $\mathbb{R}_-$,

$$Sf = g \tag{3.5}$$

is represented by a function supported on $\mathbb{R}_- + 2\rho$. *Therefore both* f *and* g *have spectral representations which can be extended to be analytic in the lower half plane. The relation between these extended spectral representations is*

$$\mathcal{S}(z)\tilde{f}(z) = \tilde{g}(z) \ . \tag{3.6}$$

Proof. Denote as usual by P_-^0 and $P_-^{2\rho}$ the operators which remove the D_-^0 and $D_-^{2\rho}$ components respectively. By causality, stated near the end of Section 2, it follows that $\mathcal{R}SP_-^{2\rho}e_0$ coincides with $\mathcal{R}Se_0 = \mathcal{R}e_1$ on $\mathbb{R}_+$. It follows therefore that

$$\|P_-^0 e_1\| \leq \|SP_-^{2\rho} e_0\| \ .$$

Since S is a contraction, we deduce that

$$\|P_-^0 e_1\| \leq \|P_-^{2\rho} e_0\| \ .$$

These norms are easily computed explicitly:

$$\left\|P_-^0 \ 1\right\| = \frac{\|m\|}{\sqrt{2|\text{Im } z|}} \ ,$$

$$\left\|P_-^{2\rho} e_0\right\| = \frac{\|n\|}{\sqrt{2|\text{Im } z|}} e^{|\text{Im } z|2\rho} \ ;$$

and inequality (3.4) follows.

To prove the analyticity of $\mathcal{S}(z)$, define h in H_0 to be

$$h = p(s)k$$

where p(s) is any continuous real valued function with compact support contained in $\mathbb{R}_+$ and k is an element of N. Arguing as before we have

$$(SP_-^{2\rho}e_0,h) = (e_1,h) = (\mathcal{S}(z)n,k)\tilde{p}(-z);$$

here $\tilde{p}$ denotes the Fourier transform of p. The left side can be re-written as

$$(P_-^{2\rho}e_0,S^*h),$$

this clearly depends analytically on z. But then so does the right side; dividing by $\tilde{p}(-z)$ we conclude that $\mathcal{S}(z)$ is weakly and hence strongly analytic, as expected.

To prove part (c) we make use of the fact that S and U_0 commute:

$$SU_0(r)f = U_0(r)g.$$

It is convenient to work with the translation representation of U_0 for which $[U_0(r)g](s) = g(s - r)$. Multiply both sides by e^{-izr} and integrate with respect to r from -R to R; the right side is

$$\int_{-R}^{R} g(s-r)e^{-izr}\,dr = e^{-izs}\int_{s-R}^{s+R} g(t)e^{izt}\,dt .$$

The limit of this as R tends to ∞ is

$$e^{-izs}\,\tilde{g}(z) . \tag{3.7}$$

On the left side we bring the factor e^{-izr} under S and perform the integration on the argument of S; the resulting argument of S is

$$e^{-izs}\int_{s-R}^{s+R} f(t)\,e^{izt}\,dt .$$

As R tends to ∞, this tends to

$$e^{-izs}\tilde{f}(z)$$

Futhermore the convergence holds in the L_2 sense over any s-interval $[a,\infty]$. It follows therefore from the fact that S is continuous in $\bar{H}_0'$ that the left side tends to

$$S(e^{-izs}\,\tilde{f}(z)) \tag{3.8}$$

in the L_2 sense over any s-interval $[a,\infty]$, Using the definition (3.3), (3.3)' of $\mathcal{S}$, with $n = \tilde{f}(z)$, we see that (3.8) can be written as

$$e^{-izs}\,\mathcal{S}(z)\,\tilde{f}(z) . \tag{3.9}$$

Finally the expressions (3.7) and (3.9) yield the desired relation (3.6). This completes the proof of Theorem 3.1.

Remark. We shall show in Section 4 that $S(z)$ can be continued analytically to be meromorphic in the entire complex plane, and to be holomorphic on the real axis. Continuing (3.6) to the real axis gives

$$\mathcal{S}(\sigma)\tilde{f}(\sigma) = \tilde{g}(\sigma) \qquad \sigma \in R \tag{3.6)'$$

for all f in D_-^a ; however because of $(1)_{iii}$ and (1.16) we see that (3.6)' provides the spectral representation of S for all f in H_0 and that $\mathcal{S}(\sigma)$ is in fact what is usually called the scattering matrix.

We return now to the eigenfunction e_0 of U_0 defined by (3.1). For $\operatorname{Im} z < 0$, e_0 belongs to H_0' and this lies in the domain of W_1 extended. We define e by

$$e = W_1 e_0. \tag{3.10}$$

Theorem 3.2. *Let* e *denote the element of* $\overline{H_0'}$ *defined by* (3.1), (3.10). *Then*

(a) e *is an eigenvector of* T:

$$T(t)e = e^{izt}e . \tag{3.11}$$

(b) *The* D_-^ρ *component of* e *is of the form*

$$\mathcal{R}e_- = e^{-izs}n \tag{3.12}$$

whereas the D_+^ρ *component of* e *is of the form*

$$\mathcal{R}e_+ = e^{-izs}\,\mathcal{S}(z)\,n. \tag{3.13}$$

Proof. Apply W_1 to the eigenvalue relation (3.2) satisfied by e_0; using the first intertwining relation (1.11) we get (3.11); this completes the proof of part (a).

To prove part (b) we use the fact that T(t) acts as translation on the D_+^a component as well as on the $\overline{D_-^a}$ component. This implies that both $e_-(s)$ and $e_+(s)$ are represented by exponential functions;

$$\mathcal{R}e_-(s) = e^{-izs}n_- \;, \quad \mathcal{R}e_+(s) = e^{-izs}n_+ \;. \tag{3.14}$$

Since W_1 was defined to be the identity on $\overline{D_-^a}$ it follows that $e_-(s) = e_0(s)$ for $s < -a$, so that $n_- = n$; this proves (3.12). Now apply W_2 to e; using the definition of S as W_2W_1 and relations (3.3), (3.3)' we get

$$\mathcal{J} W_2 e = \mathcal{J} e_1 = e^{-izs} \mathcal{S}(z) n \quad .$$

Since by (1.14), $W_2 = I$ on D_+^a we conclude that $\mathcal{J} e_+(s) = \mathcal{J} e_1(s)$ for s a, so that $n_+ = S(z)n$; substituting this into (3.14) we get (3.13); this completes the demonstration of Theorem 3.2.

Being an eigenfunction of the semigroup T, e is also an eigenfunction of the extended A:

$$Ae = ize \quad . \tag{3.15}$$

This is easy to show on the basis of the decomposition (2.3) used in Section 2 to define the domain and action of A and T in $\bar{H}$.

The next remark will be useful in Section 10 when we obtain an explicit representation for the scattering matrix. Let J_0 be one of the mappings introduced in Section 1 to take us from the space H_0 to H, so chosen that J_o carries the domain of A_o into the domain of A. Since J_o acts like the identity on D_-^ρ and D_+^ρ components, it can be extended without further ado as a mapping from $\bar{H}_o$ to $\bar{H}$. Consider now

$$v = e - J_o e_o \tag{3.16}$$

v is sometimes called the outgoing *scattered wave*. According to (3.12) and (3.13), the D_-^ρ compoment of v is zero and the D_+^ρ component of v is represented by $[S(z)n - n]e^{-izs}$. Since $\text{Im} z < 0$, we see that v belongs to H.

We now derive a useful equation for v: Apply (A-iz) to (3.16); then since e itself satisfies (3.15) we get

$$(A - iz)v = (iz - A)J_0 e_0 . \tag{3.17}$$

Observe that the right side belongs to the subspace K^ρ. Since the right side is a known quantity, equation (3.17) can be used to determine v for $\text{Im} z < 0$; the D_+^ρ component of v is represented by a function of the form

$$\mathcal{J} v_+ = e^{-izs} m \tag{3.18}$$

where m is determined by e_0 , i.e. by z and n:

$$m = \mathcal{K}'(z)n \quad . \tag{3.19}$$

This gives the formula

$$S(z) = I + \mathcal{K}'(z) \tag{3.20}$$

for the scattering matrix.

4. The Associated Semi-group of Operators and the Analytic Continuation of the Scattering Matrix.

We now turn our attention to another tool for the study of the scattering operator, a semi-group of operators desctibing the local behavior of $T(t)$:

$$Z(t) = P_+^a T(t) P_-^a \ , \tag{4.1}$$

where P_+^a and P_-^a are the orthogonal projections on the orthogonal complements of D_+^a and D_-^a, respectively, We choose $a > \rho$.

Theorem 4.1. *The operators* $\{Z(t)\}$ *form a strongly continuous semi-group of contraction operators on* $K^a = H \ominus (D_-^a \oplus D_+^a)$. $Z(t)$ *annihilates* $D_-^a \oplus D_+^a$.

Proof. For any f in H we decompose $T(t)f$ into orthogonal parts:

$$T(t)f = d(t) + e(t) \ ,$$

where $d(t) \varepsilon D_+^a$ and $e(t) = P_+^a T(t) f \perp D_+^a$. According to $(7)_i$, $T(s)d(t)$ belongs to D_+^a and hence $P_+^a T(s)\, d(t) = 0$. It follows that

$$P_+^a T(s + t) f = P_+^a T(s) P_+^a T(t) f \ ; \tag{4.2}$$

this is the semi-group property for the operators $\{P_+^a T(t)\}$. Next we show that when $f \perp D_-^a$ so is $P_+^a T(t) f$. Now for $f \perp D_-^a$ and g in D_-^a

$$(P_+^a T(t) f, g) = (T(t) f, P_+^a g) = (T(t) f, g)$$

since we have assumed that $D_-^a \perp D_+^a$. Finally because of $(7)_i$, $T^*(t) g$ lies in D_-^a and hence

$$(P_+^a T(t) f,\ g) = (f, T^*(t) g) = 0.$$

That $Z(t)$ annihilates D_-^a is obvious; that it annihilates D_+^a follows from the orthogonality of D_+^a to D_-^a and the fact that T maps D_+^a into itself. This completes the proof of Theorem 4.1.

Denote by B the infinitesimal generaror of Z. Since each $Z(t)$ is a contraction, the spectrum of B lies in the left half plane:

$$\operatorname{Re} \sigma(B) \leq 0 \ .$$

According to property (8) in the introduction

$$\lim P_+^a T(t)\ f = 0$$

for every f in H; this implies that

$$\lim Z(t)f = 0 \ . \tag{4.3}$$

It follows from this that the point spectrum of B contains no-purely imaginary values.

It will be helpful to analyze the action of B a little more precisely. Let $f_\pm$ denote the translation representers of the $D_\pm^\rho$ components of f in H or $(\bar{H})$. We recall that the action of A on th these components is giben by

$$\begin{aligned} \mathcal{J}(Af)_+ &= -\partial_s \mathcal{J} f_+ \qquad \text{for} \quad s > \rho \\ \mathcal{J}(Af)_- &= -\partial_s \mathcal{J} f_- \qquad \text{for} \quad s < -\rho \ . \end{aligned} \tag{4.4}$$

For f in K^a, $Z(t)f = P_+^a T(t)f$. The effect of P_+^a is simply to chop off $\mathcal{J}f_+$ at s = a and consequently

$$\mathcal{J}(Bf)_+ = \begin{cases} -\partial_s \mathcal{J} f_+ & \rho < s < a \\ 0 & a < s. \end{cases} \tag{4.5}$$

Finally if g lies in the domain of A and is orthogonal to D_-^a then $P_+^a g$ lies in K^a and $Z(t)P_+^a g = P_+^a T(t)g$ as in (4.2). Differentiating with respect to t at t = o, we see that for such g

$$BP_+^a g = P_+^a Ag \ . \tag{4.6}$$

The next theorem reveals an interesting connection between the semi-group Z and the scattering matrix $\mathcal{S}(z)$:

Theorem 4.2. *Let* $\rho_0(B)$ *denote the principal component of the resolvent set of* B, i.e. *that component of the resolvent set containing the right half plane. Then* $\mathcal{S}(z)$ *can be continued analytically from the lower half plane into* $-i\rho_0(B)$.

Proof. For any z in the lower half plane and any n in N let e be the element of $\bar{H}'$ defined by (3.10); according to Theorem 3.2, e is an eigenvector of A. We now introduce two cut-off functions, one on $\mathbb{R}_- - \rho$ and the other on $\mathbb{R}_+ + \rho$:

$$\xi_-(s) = \begin{cases} 0 & \text{for } -\rho-\delta < s < -\rho \\ & \text{smooth in between} \\ 1 & \text{for} \quad s < -a \end{cases}$$

$$\xi_+(s) = \begin{cases} 0 & \text{for} \quad \rho < s < a \\ 1 & \text{for} \quad a < s \ ; \end{cases}$$

here $0 < \delta < a-\rho$. We define the element $f = f(z)$ by

$$f = e - \mathcal{J}^{-1}\xi_- \mathcal{J}e_- - \mathcal{J}^{-1}\xi_+ \mathcal{J}e_+; \tag{4.7}$$

where as before e_- and e_+ are the $\dot{D}^{\rho}_{-}$' and D^{ρ}_{+} components of e. For convenience we will write this as

$$f = e - \xi_- e_- - \xi_+ e_+ \tag{4.7)'$$

and trust that the reader will be able to keep in mind the role of the translation representation. According to (3.12) and (3.13)

$$e_- = e^{-izs}n \quad \text{for} \quad s < -\rho \quad \text{and} \quad e_+ = e^{-izs}\mathcal{S}(z)n \text{ for } s>\rho. \tag{4.8}$$

Making use of (4.4)and (4.7) we obtain

$$A\xi_- e_- = \xi_-' e_- + iz\xi_- e_- \ . \tag{4.9}$$

Now f belongs to K^a and $f + \xi_+ e_+$ lies in the domain of A so that (4.6) holds with g replaced by $f + \xi_+ e_+$. Combining (4.6),(4.7) and (4.9) with the relation $Ae = ize$ we get

$$Bf = P^a_+ A(f + \xi_+ e_+) = P^a_+ A(e - \xi_- e_-) = P^a_+ (ize - iz\xi_- e_- - \xi_-' e_-).$$

Subtracting iz times $P^a_+ \xi_+ e_+ = 0$ from the right hand side gives

$$Bf = izf - \xi_-' e_- \ . \tag{4.10}$$

Since Re $iz > 0$, we have $iz \in \rho(B)$ and the relation (4.10) can be solved for f:

$$f = (iz - B)^{-1}(\xi_-' e_-). \tag{4.11}$$

This relation holds for Im $z < 0$ and shows that f(z) can be analytically continued into a set in the upper half plane for which iz remains in the resolvent set of B, that is into $-i\rho_0(B)$.

According to (4.8) e_+ is of the form $e^{-izs}\mathcal{S}(z)n$. It follows from the definition (4.7)' of f that f_+ , the D^{ρ}_{+} component of f is

$$f_+(s,z) = e^{-izs}\mathcal{S}(z)n \qquad \text{for } \rho < s < a. \tag{4.12}$$

We have already shown that f(z) can be continued analytically into $-i\rho_0(B)$, the same is true of its component f_+ and therefore the same is true for $\mathcal{S}(z)n$. This shows that $\mathcal{S}(z)$ has a strong analytic continuation into $-i\rho_0(B)$, as asserted in Theorem 4.2.

The proof of Theorem 4.2 can be run backwards, as follows:

Suppose z belongs to $-i\rho_0(B)$, and n is any element of N. Take e_- to be

$$e_- = e^{-izs}\, n\ , \qquad s < -\rho\ ,$$

and $\xi_\pm$ as defined previously. Define f to be the solution (4.11). Define e_+ to be

$$e_+ = e^{-izs}\, \mathcal{S}(z)n$$

and solve for e from (4.7)' :

$$e = f + \xi_- e_- + \xi_+ e_+ \tag{4.13}$$

Since $\xi_+ e_+$ is a smooth continuation of f, it is clear that e belongs to the domain of A. Moreover it follows from the proof of Theorem 4.2 that for Im z < 0 the element e is an eigenvalue of A, i.e.

$$Ae - ize \tag{4.14}$$

is zero. On the other hand, by construction, the $\overline{D_-^a}$ and $\overline{D_+^a}$ components of e are exponential and hence it follows that the $\overline{D_-^a}$ and the $\overline{D_+^a}$ component of (4.14) are zero.

Denote the element (4.14) by r(z); then as observed above, r(z) belongs to H and is zero when Im z < 0. The explicit form of e shows that r depends analytically on z; it follows therefore that r(z) has to be zero everywhere. We formulate what we have proved as

Theorem 4.3. *For any* z *in* $-i\rho_0(B)$ *and any* n *in* N *there is an eigenvector* e *of* A *in* $\overline{H}$:

$$Ae = ize \tag{4.15}$$

whose $\overline{D_-^\rho}$ *and* $\overline{D_+^\rho}$ *components are given by*

$$e_- = e^{-izs} n\ , \qquad e_+ = e^{-izs}\, \mathcal{S}(z)n\ . \tag{4.16}$$

Remark. Under the additional assumptions made in Section 5, e is uniquely determined by (4.15) and (4.16).

A related result which will be needed in Section 10 is the following:

Theorem 4.4. *For* g *in* K^ρ *and any* z *in* $-i\rho(B)$ *there exists a solution in* $\overline{H}$ *to*

$$(iz - B)h = g \tag{4.17}$$

which is outgoing in the sense of having zero $\overline{D_-^a}$ *component.*

Remark. Under the additional assumptions made in Section 5, this solution will be unique.

Proof. Let f be the solution to

$$(iz - B)f = g \,, \tag{4.18}$$

and let f_+ denote the D_+^ρ component of f as before. Since f is in K^a it is automatically orthogonal to D_-^a. Since $g_+ = 0$ we see (4.5) that

$$(iz + \partial_s)f_+ = 0 \qquad \text{for} \quad \rho < s < a \,.$$

Consequently f_+ is of the form

$$f_+ = \begin{cases} e^{-izs}\, m & \text{for} \quad \rho < s < a \\ 0 & \text{for} \quad a < s \,. \end{cases}$$

We now define

$$h_+ = e^{-izs}\, m \qquad \text{for} \quad s > \rho \,.$$

We shall show that

$$h = f - f_+ + h_+$$

is a solution of (4.17); it is clear that h is orthogonal to D_-^a. We begin by constructing a smooth function ξ:

$$\xi(s) = \begin{cases} 0 & \text{for} \quad s < \rho + \delta \\ 1 & \text{for} \quad s > a - \delta \,, \end{cases}$$

where $\delta < (a - \rho)/4$. In this case we can write

$$h = f - \xi f_+ + \xi h_+ \,. \tag{4.19}$$

Moreover since ξf_+ obviously lies in the domain of B so does $f - \xi f_+$. Now $f - \xi f_+ = 0$ for $s > a - \delta$ so that

$$T(t)(f - \xi f_+) = Z(t)(f - \xi f_+)$$

for $t < \delta$. differentiating at $t = 0$ we get

$$A(f - \xi f_+) = B(f - \xi f_+) \,.$$

Combining this with (4.19) we obtain

$$A(h - \xi h_+) = B(f - \xi f_+) \ . \tag{4.20}$$

According to (4.4) and (4.5) we can write

$$B(\xi f_+) = \xi' f_+ + \xi \partial_s f_+ = \xi' f_+ + iz\xi f_+$$

$$A(\xi h_+) = \xi' h_+ + \xi \partial_s h_+ = \xi' h_+ + iz\xi h_+ \ .$$

Since the support of ξ' lies in $\rho + \delta < s < a - \delta$ where f_+ and h_+ are the same, we have $\xi' f_+ = \xi' h_+$. Thus

$$A\xi h_+ - B\xi f_+ = iz(\xi h_+ - \xi f_+) = iz(h - f)$$

by (4.19). Combining this with (4.18) and (4.20) we finally get

$$g = izf - Bf = izf - B(f - \xi f_+) - B(\xi f_+)$$

$$= izf - A(h - \xi h_+) - A(\xi h_+) + iz(h - f) \ ,$$

so that

$$g = (iz - A)h \ .$$

This completes the proof of Theorem 4.4.

Remark. If g belongs to K^ρ then it is not difficult to show that f is orthogonal to D^ρ_- so that the h constructed in Theorem 4.4 is actually orthogonal to D^ρ_- .

We conclude this section with a brief discussion of the spectral mapping theorem for Z and one of its consequences. Let m(dt) be any complex measure on Borel subsets of the non-negative reals with finite total measure. The Laplace transform of m:

$$g(\mu) = \int_0^\infty e^{\mu t}\, m(dt) \qquad \mathrm{Re}\ \mu \le 0 \tag{4.21}$$

will be analytic in the open half-plane and continuous in the closed half-plane. For functions g of this kind we define g(B) as

$$g(B) = \int_0^\infty Z(t)m(dt) \ . \tag{4.22}$$

The spectral mapping theorem for this set-up is well known (see Theorem 16.6.1 of [3]).

Theorem 4.5. $\sigma(g(B)) \supset g(\sigma(B))$.

Theorem 4.6. *Suppose that* g *is a function of the form* (4.18) *and that* g(B) *is compact. If* $g(\mu)$ *has no zero on the spectrum of* B, *then* B *has a pure point spectrum and the resolvent of* B, *in symbols* $R_\mu(B)$, *is meromorphic in the whole complex plane.*

As we have previously noted the point spectrum of B contains no point on the imaginary axis and therefore if B satisfies the hypothesis of this theorem we have

Corollary 4.7. Re $\sigma(B) < 0$.

Proof of Theorem 4.6. According to Theorem 4.5 the spectrum of B is contained in the set of roots of $\lambda = g(\mu)$ for λ in $\sigma(g(B))$. Since 0 is not in $\sigma(g(B)$, it suffices to consider $\lambda \neq 0$ in $\sigma(g(B))$. For such a λ we define

$$E_\lambda = \frac{1}{2\pi i} \oint R_\zeta(g(B))\, d\zeta$$

where the path of integration is a small circle about λ containing no other spectral point of g(B). According to Lemma 5.7.2 of [3] the range L of E_λ will be finite dimensional; let M denote the range of $I - E_\lambda$. According to Theorem 5.9.3 of [3], $R_\zeta(g(B))E_\lambda = R_\zeta(g(B)E_\lambda)$ on L can be extended to be regular except at $\zeta = \lambda$.whereas $R_\zeta(g(B))(I - E_\lambda)$ can be extended to be regular at $\zeta = \lambda$. It is easy to see that E_λ commutes with Z(t) and it follows from this that Z(t) leaves L and M invariant.

Because L if finite dimensional $R_\zeta(B)E_\lambda = R_\zeta(E_\lambda B)E_\lambda$ is meromorphic with a finite set of poles at say $\mu_1, \mu_2, \ldots, \mu_k$. These points are obviously in the point spectra of B and BE_λ. Note also that $g(B)E_\lambda = g(BE_\lambda)$ on L has λ as its only spectral point. Applying Theorem 4.5 to Z(t) restricted to L we see that $g(\mu i) = \lambda$ for $i = 1,2, \ldots ,k$. Since g(B) restricted to M does not have λ in its spectrum, if we apply Theorem 4.5 to Z(t) restricted to M we see that $g(\mu) = \lambda$ has no roots on the spectrum of B restricted to M. In other words $R_\mu(B)(I - E_\lambda)$ can be extended to be regular at all of the roots of $g(\mu) = \lambda$. Combining these two results we see that

$$R_\mu(B) = R_\mu(B)E_\lambda + R_\mu(B)(I - E_\lambda)$$

is regular at all but a finite number of roots of $g(\mu) = \lambda$ and there its singularities are poles. Finally if a sequence μ_j belonging to $\sigma(B)$ converged to μ_0 then μ_0 lies in $\sigma(B)$ and since g is analytic on the spectrum of B by Corollary 4.7 (we have already established that B has a pure point spectrum), an infinite subset of the $g(\mu_j)$ are distinct; thus $g(\mu_0)$ is a point of accumulation of the spectrum of g(B) and since g(B) is compact this implies that $g(\mu_0) = 0$; contrary to our assumption on g.

Remark. It follows from the proof of Theorem 4.5 that the

generalized eigenspace of B at $\mu \in \sigma(B)$ is of finite dimension.

Corollary 4.8. *If* $Z(T)R_{\kappa}(B)$ *is compact for some* $\kappa > 0$, *then* $R_{\mu}(B)$ *is meromorphic in the whole complex plane.*

Proof. We apply the above theorem to

$$m(dt) = e^{\kappa(T-t)}dt \qquad \text{for} \quad t \geq T \text{ and } = 0 \text{ for } 0 < t < T.$$

In this case $g(\mu) = e^{\mu T}(\kappa - \mu)^{-1}$ and $g(B) = Z(T)R_{\kappa}(B)$.

5. The Poles of the Scattering Matrix and Outgoing Eigenvectors.

In this section we assume that *the resolvent of* B *is meromorphic in the whole plane*. In this case, the *spectrum of* B *will be pure point spectrum.* We note that at the end of Section 4 we have given a sufficient condition for this to be so; we shall show in part II that for the application considered there this condition is satisfied.

Under the above assumption $\rho_0(B)$ is all of the resolvent set of B, i.e. the whole plane minus the eigenvalues of B. According to Theorem 4.2, $\mathcal{S}(z)$ can be continued analytically into $-i\rho(B)$; this shows that the singularities of $\mathcal{S}(z)$ are isolated, actually from the proof of Theorem 4.2 a little more follows: *The singularities of* $\mathcal{S}$ *are poles.* For as iz approaches an eigenvalue of B, $(iz - B)^{-1}$ increases as some power of the distance of iz to that eigenvalue; it follows from (4.11) that the same is true of f, and therefore of f_+, in terms of which $\mathcal{S}(z)$ is defined.

According to Theorem 4.3, for iz not in $\overline{\sigma(B)}$ and $\underline{\text{any}}$ n in N, the operator A in $\bar{H}$ has an eigenvector e whose $\overline{D^{\rho}_{-}}$ and $\overline{D^{\rho}_{+}}$ components are given by (4.14). Suppose z_0 is a zero of $\mathcal{S}(z)$, i.e. a value where the equation

$$\mathcal{S}(z_0)n = 0$$

has a nontrivial solution. For such a z_0 and n the $\overline{D^{\rho}_{+}}$ component of the eigenvector e is zero. An eigenvector of this kind is called an *incoming eigenvector*, since its outgoing (i.e. D^{ρ}_{+}) component is zero. An *outgoing eigenvector* is defined analogously as one whose D^{ρ}_{-} component is zero. We shall show that just as the incoming eigenvectors are associated with zeros of $\mathcal{S}$, so the outgoing ones are associated with the poles of $\mathcal{S}$. First a few lemmas:

Lemma 5.1. *Suppose* b *is an eigenvector of* B; *then* b_-, *the* D^{ρ}_{-} *component of* b, *is zero.*

Proof. An eigenvector of B is also an eigenvector of Z:

$$Z(t)b = e^{izt}b \quad . \tag{5.1}$$

Let b_- denote the D_-^ρ component of b; form the scalar product of (5.1) with b_-:

$$e^{izt}(b,b_-) = (Z(t)b,b_-) = (T(t)b,b_-) = (b,T^*(t)b_-) ; \tag{5.2}$$

In the third step we have used the fact that since b_- is orthogonal to D_+^a , $P_+^a b_- = b_-$.

Since b_- belongs to D_-^ρ, $T^*(t)b_-$ belongs to $D^{\rho+t}$; for $t > a-\rho$ this space is orthogonal to K^a. Since b belongs to K^a, (5.2) is zero for $t > a-\rho$, and in particular

$$0 = (b,b_-) = (b_-,b_-) ,$$

so that $b_- = 0$, as asserted.

Lemma 5.2. *If* iz_0 *is an eigenvalue of* B, *then* A *has an outgoing eigenvector with eigenvalue* iz_0 *and conversely.*

Proof. Let b be the eigenvector of B; it follows that D_+^ρ component b_+ of b is of the form

$$b_+ = \begin{cases} e^{-izs}\, m & \rho < s < a \\ 0 & a < s \end{cases}$$

Here we continue the practice of not distinguishing between data in $D_\pm^\rho$ and their translation representers. We define b_{++} to be

$$b_{++} = \begin{cases} 0 & \rho < s < a \\ e^{-izs}\, m & a < s \end{cases}$$

We claim that

$$a = b + b_{++}$$

is an eigenvector of A; we leave the easy verification of this fact to the reader. Clearly, $a_- = b_-$ and, according to Lemma 5.1, $b_- = 0$; this shows that a is indeed outgoing.

To prove the converse we construct b out of a by reversing (5.3), i.e. setting

$$b = a - \zeta_+ a_+ = P_+^a\, a;$$

recall that the D_-^ρ component of a is zero. It is easy to show (see (4.6)) that b is an eigenvector of B with eigenvalue iz_0. This completes the proof of Lemma 5.2.

It follows from formula (5.3) relating outgoing eigenvectors of

A and eigenvectors of B that if an eigenvector b of B has zero D_+^ρ component, then b itself is an eigenvector of A, and futhermore b is both outgoing and incoming. We now impose the additional hypothesis (happily satisfied in part II) that this cannot happen:

Neither A *nor* A^* *have eigenvectors which are both incoming and outgoing.* This assumption has the desired consequence:

Lemma 5.3. *No eigenvector* b *of* B *has a zero* D_+^ρ *component, and no eigenvector* b^* *of* B^* *has zero* D_-^ρ *component.*

We saw earlier in Theorem 4.2 that every point in $-i\rho(B)$ is a regular point for $\mathcal{S}$. Under the additional assumption just introduced we can prove the converse.

Lemma 5.4. *If* iz_0 *is an eigenvalue of* B, *then* z_0 *is a pole of* $\mathcal{S}$.

Proof. We will reverse the proof of Theorem 4.2; the thrust of that argument was: If iz_0 is not an eigenvalue of B then f, defined by (4.7)', is analytic in z near z_0, having at most a pole at z_0. From this it followed that $f_+ = e^{izs}\mathcal{S}(z)m$ for $\rho < s < a$ is also an analytic function of z. Our first step now is to show that if iz_0 is an eigenvalue of B then f, defined by (4.7)', has a pole at z_0.

By assumption the resolvent of B has a pole at iz_0 and it follows from this that the resolvent of B^* has a pole at $-i\bar{z}_0$. As a consequence $-iz_0$ is an eigenvalue of B^*; let b^* be a corresponding eigenvector:

$$(B - iz_0)^* b^* = 0. \tag{5.4}$$

According to Lemma 5.3, the D_-^ρ component b_-^* of b^* is a non-zero exponential on $-a < s < -\rho$; from this we deduce easily that there exists a vector n and a cutoff function ξ_- such that

$$(\xi_-'(s)e^{-iz_0 s}\, n,\, b_-^*) \neq 0. \tag{5.5}$$

Let z be near z_0 and $\neq z_0$. Denote by e(z) the eigenvector of A, described in Theorem 4.3, obtained by analytic continuation of $W_1 e_0(z)$ from the lower half plane. Let's define $f = f(z)$ by formula (4.7)', then following the steps of the proof of Theorem 4.2, we find that f satisfies (4.10);

$$(iz - B)f = \xi_-'\, e_-(z). \tag{5.6}$$

Form the scalar product with b^* and use the fact that $(B-iz_0)^* b^* = 0$; we get

$$i(z - z_0)(f, b^*) = (\xi_-'\, e_-(z), b^*) \tag{5.7}$$

according to (4.16),

$$e_-(z) = e^{-izs}n ;$$

since e_- belongs to D^ρ_-, it follows from (5.5) that the right side of (5.7) is $\neq 0$. But this implies that $\|f(z)\|$ tends to ∞ as z tends to z_0. This proves that f has a pole at z_0, say of order $k_0 > 0$.

Suppose now that $\mathcal{S}(z)n$ were holomorphic at z_0 and define the Laurent co-efficients:

$$f_k = \frac{1}{2\pi i} \oint f(z)(z - z_0)^{k-1}\, dz, \qquad k > 0, \tag{5.8}$$

where the path of integration is a sufficiently small circle about z_0. Then

$$Bf_k = \frac{1}{2\pi i} \oint Bf(z)(z - z_0)^{k-1}\, dz$$

and making use of (4.10): $Bf = izf - \xi_-' e_-$, we get

$$Bf_k = iz_0 f_k + if_{k+1} \,. \tag{5.9}$$

Note that the $\xi_-' e_-$ term is holomorphic at z_0 and hence does not contribute to (5.9). Since z_0 is a pole of order k_0 we see that $f_{k_0+1} = 0$ so that f_{k_0} is an eigenvector of B. Moreover $f_+(z) = e^{-izs}\mathcal{S}(z)n$ for $\rho < s < a$ and hence if $\mathcal{S}(z)n$ were holomorphic at z_0 we could deduce from (5.8) that $(f_{k_0})_+ = 0$. However this would contradict Lemma 5.3; this proves that $\mathcal{S}(z)n$ is singular at z_0 and completes our proof of Lemma 5.4.

Remark. The above argument shows that f(z) has a removable singularity at z_0 whenever $\mathcal{S}(z)n$ is holomorphic at z_0. Defining $f(z_0)$ by analytic continuation, we see that $Bf(z_0) = iz_0 f(z_0) + \xi_-' e_-$, and hence we can repeat the proof of Theorem 4.3 to obtain an eigenvector e of A satisfying (4.16) for $z = z_0$ and this particular n.

We can now state the main result of this section:

Theorem 5.5. *Under the additional hypotheses imposed in this section:*

(I) *The resolvent of* B *is meromorphic in the whole complex plane,*

(II) *Neither* A *nor* A^* *have eigenvectors which are both incoming and outgoing, the following conditions are equivalent:*

(1) *The scattering matrix* $\mathcal{S}(z)$ *has a pole at* z_0 ;

(2) iz_0 *belongs to the spectrum of* B ;

(3) A *has an outgoing eigenvector with eigenvalue* iz_0 .

Proof: The equivalence of (2) and (3) is Lemma 5.2. That (1) implies (2) is Theorem 4.2, and that (2) implies (1) is Lemma 5.4. This completes the proof of Theorem 5.5.

We turn now to the incoming eigenvectors of A. If iz is not an eigenvalue of B then Theorem 4.3 describes a class of eigenvectors e(z) of A with eigenvalue iz; they are the ones obtained by analytic continuation of $W_1 e_0$ from the lower half plane; recall that $e_0 = e^{-izs}n$. We claim that these are all the eigenvectors for such an eigenvalue. For suppose there were another one g; being an eigenvector the D^ρ_- component of g is of the form $e_-^{-izs}n$. According to Theorem 4.3, there is an eigenvector e whose D^ρ_- component has the same form. Therefore g - e would be an outgoing eigenvector; but according to Lemma 5.2 this cannot happen when iz is not an eigenvalue of B.

An eigenvector e(z) of A described by Theorem 4.3 has D^ρ_+ component $e^{-izs}\mathcal{S}(z)n$; therefore e(z) *is incoming if and only if* $\mathcal{S}(z)n = 0$. This characterizes all incoming eigenvectors of A with eigenvalue iz, where z is a regular point of $\mathcal{S}$. There may however be additional incoming eigenvectors at poles of $\mathcal{S}$. For instance, if for a particular n the function $\mathcal{S}(z)n$ happens to be regular at a pole z_0 of $\mathcal{S}$, then we claim that the eigenfunction e(z) also is regular at z_0. This follows from the remark preceding Theorem 5.5.

If $\mathcal{S}(z)n$ is not only regular at z_0 but vanishes there, then the eigenfunction $e(z_0)$ is incoming. However we were not able to show that these are all the incoming eigenfunctions of A associated with poles of $\mathcal{S}$.

Our interest in incoming eigenfunctions for Im z > 0 is that they are the only ones which belong to H; that is, they are eigenfunctions not only of the extended operators T(t) and A but of the original operators over H:

Theorem 5.6. *The point spectrum of* A *over* H *is of the form* iz, *where* Im z > 0 *and* z *is a zero of* $\mathcal{S}(z)$ *or possibly a pole.*

6. The Spectrum of A.

In the previous section we characterized the point spectrum of A; it is contained in the left half plane Re $\lambda < 0$ and is in one-to-one correspondence with the zeros of the scattering matrix

$\mathcal{S}(z)$ in the upper half plane and possibly with some of the poles of $\mathcal{S}$, the correspondence being given by $\lambda = iz$. How completely does this characterize the spectrum of A? We shall investigate this question in the present section.

Theorem 6.1. *The imaginary axis is contained in the continuous part of the spectrum of* A.

Proof. The restriction of T(t) to D_+^{ρ} corresponds in the translation representation to translation on a half-line. It is well known that the spectrum of the of such a semi-group of operators fills out the entire half-plane Re $\lambda \leq 0$. Any extension of this generator such as A will have in its spectrum at least the boundary of this set (see Theorem 4.11.2 of [3]) - that is the imaginary axis. Neither A nor A^* can have purely imaginary eigenvalues for the corresponding eigenvector would have to be both incoming and outgoing and therefore the points on the imaginary axis must be continuous spectral points for A.

We now give a criterion for the rest of the spectrum of A to be point spectrum.

Theorem 6.2. *Suppose that there is a skew-selfadjoint operator* A_1 *on* H *such that*

$$R_{\lambda_0}(A) - R_{\lambda_0}(A_1)$$

is compact for some λ_0. *Then that portion of the spectrum of* A *in the half plane* Re $\lambda < 0$ *is entirely point spectrum and is either discrete in this half plane or fills out the entire half plane.*

Remark Both of these possibilities can occur. In fact the hypothesis of Theorem 6.2 is satisfied by both of the examples given in the introduction. In example 1 the point spectrum fills out the entire half plane whereas in Example 2 the point spectrum is discrete.

Proof. We recall that when λ_0 belongs to $\rho(A)$, then $\lambda \in \rho(A)$ if and only if

$$I + (\lambda - \lambda_0)R_{\lambda_0}(A)$$

is one-to-one and has a bounded inverse (see p.187 of [3]). Since A_1 is skewselfadjoint any λ with Re $\lambda \neq 0$ lies in $\rho(A_1)$ so that for fixed λ_0 with Re $\lambda_0 > 0$

$$M_\lambda \equiv I + (\lambda - \lambda_0)R_{\lambda_0}(A_1)$$

has a bounded holomorphic inverse for all such λ. We now write

$$I + (\lambda-\lambda_0)R_{\lambda_0}(A) = I + (\lambda-\lambda_0)R_{\lambda_0}(A_1) + (\lambda-\lambda_0)[R_{\lambda_0}(A) - R_{\lambda_0}(A_1)]$$

and multiply through on the right by M_λ^{-1} . This gives

$$M_\lambda^{-1}[I + (\lambda-\lambda_0)R_{\lambda_0}(A)] = I + (- {}_0)M_\lambda^{-1}[R_{\lambda_0}(A) - R_{\lambda_0}(A_1)].$$

By assumption the expression on the right is of the form I plus a compact operator-valued function holomorphic in the left half plane Re $\lambda < 0$. It is well known (see[12]) that this expression has a meromorphic inverse if it is invertible at one point in this half-plane. At those λ for which the inverse exists it is clear that $I + (\lambda-\lambda_0)R_{\lambda_0}(A)$ is also invertible so that $\lambda \in \rho(A)$. At those λ for which the inverse does not exist the compactness assumption assures us of nontrivial null vectors for $I + (\lambda-\lambda_0)R_{\lambda_0}(A)$ and it is easy to show that such a null vector is an eigenvector for A at λ.

Remark. It often happens that $\mathcal{S}(z)$ is of the form:

$$\mathcal{S}(z) = I + \mathcal{K}'(z)$$

where $\mathcal{K}'$ is a compact operator on N. In this case if $\mathcal{S}(z)$ is invertible for some z_0 then $\mathcal{S}^{-1}(z)$ will be meromorphic on $-i\rho(B)$. In particular $\mathcal{S}(z)$ will be invertible for some z in the upper half plane and we may conclude from Theorems 5.6 and 6.2 that the point spectrum of A will be discrete in the upper half plane provided of course that the hypothesis of Theorem 6.2 is satisfied.

References.

1. C. Goldstein, Peturbation of non-selfadjoint operators, I and II, Arch. Rat. Mech. Anal. 37, 268-296 (1970) and 42, 380-402 (1971).
2. C. Goldstein, The scattering matrix associated with non-self-adjoint differential operators, Mathematics Research Center Report 1152.
3. E. Hille and R. S. Phillips, Functional Analysis and semigroups (rev. ed.) Amer. Math. Soc. Colloquium Publ., Vol.31, 1957.
4. N. Iwasaki, Local decay of solutions for symmetric hyperbolic systems with dissipative and coercive boundary conditions in exterior domains, RIMS, Kyoto Univ. 5, 193-218 (1969).
5. T. Kato, Wave operators and similarity for some non-selfadjoint operators, Math. Ann. 162, 258-279 (1966).
6. P. D. Lax and R. S. Phillips, *Scattering theory*, Academic Press, New York, 1967.
7. P. D. Lax and R. S. Phillips, Scattering theory for the acoustic equation in an even number of space dimensions, Indian Univ. Math. Jr., 22, 101-134 (1972).

8. P. D. Lax and R. S. Phillips, On the scattering frequencies of the Laplace operator for exterior domains, Comm. Pure Appl. Math., 22 (1972).
9. S. L. Lin, Wave operators and similarity for generators of semi-groups in Banach spaces, Trans. Amer. Math. Soc., 139, 469-494 (1969).
10. K. Mochizuki, On the large perturbation by a class of non-self-adjoint operators, J. Math. Soc. Japan, 19, 123-158 (1967).
11. K. Mochizuki, Eigenfunction expansions associated with the Schrödinger operator with a complex potential and the scattering theory, RIMS, Kyoto Univ. 4, 419-466 (1968).
12. S. Steinberg, Meromorphic families of compact operators, Arch. Rat. Mech. Anal., 31, 372-380 (1968).
13. B. Sz. -Nagy and C. Foias, *Harmonic analysis of operators on Hilbert space*, Amer. Elsevier Publ. Co., New York, 1970.

SCATTERING THEORY IN WAVEGUIDES

Charles Goldstein

1. Introduction.

It is the purpose of this report to discuss the scattering of waves in a uniform cylindrical waveguide in which a perturbation has been introduced.

In Section 2, a sketch of the physics involved in propagating high-frequency electromagnetic waves along a waveguide will be given. In particular, the modal expansion of the field in a uniform guide will be discussed, as well as the necessity for introducing discontinuities (i.e. perturbations) in the guide. Futhermore, the physical significance of the scattering matrix will be pointed out.

As a result of the modal expansion for the electromagnet fields in a waveguide, the mathematical problem may be reduced to the consideration of two self-adjoint operators, A_0 and A, given by $-\Delta$ associated with zero, Dirichlet or Neumann boundary conditions in a uniform cylinder, S, and a perturbed cylinder, Ω, respectively. Particularly important quantities associated with these operators from both the mathematical and physical point of view are the "cut-off wave numbers", ν_n. The $\{\nu_n\}$ are the eigenvalues (assumed in increasing order) of the operator A_ℓ, defined analogously to A_0 with S replaced by the bounded cross section, ℓ, of S. The physical significance of the $\{\nu_n\}$ is that they determine the number of propagating waves for a given frequency.

In Section 3, the main mathematical results will be described without proof. For proofs of these results, see [1]-[3]. For the sake of generality, S will be either an infinite or semi-infinite

J. A. LaVita and J.-P. Marchand (eds.), Scattering Theory in Mathematical Physics, 35–51.

cylinder contained in N-dimensional Euclidean space, R^N $(N \geq 2)$, with an arbitrary smooth, bounded cross section, ℓ. The perturbation will be obtained by either perturbing $\dot{S}$,* the coefficients of the operator, or the boundary conditions. In any case, it will be assumed that the perturbation has bounded support and is sufficiently smooth. It will be pointed out at the end of the section how some of these conditions may be relaxed.

The first important result establishes the unitary equivalence of A_o and A^C (the continuous part of A). It should be noted that A_o differs from the corresponding operator acting in all of R^N in that its spectral multiplicity is non-uniform. Each cutoff wave number, ν_n, corresponds to a jump in the multiplicity. Hence additional channels are added to the spectrum at each such point. The unitary equivalence of A_o and A^C is established by constructing spectral mappings for these operators. These mappings are in turn employed to define the wave operators and scattering operator. This is a stationary formulation of scattering theory. A time-dependent formulation will also be given.

An explicit expression will be obtained for the $\mathcal{S}$-matrix. A means for calculating the $\mathcal{S}$-matrix will be described for the case of small perturbations. It will also be demonstrated that there are a countably infinite number of $\mathcal{S}$-matrices of different order and that each of these may be meromorphically continued off the real axis onto a Riemann surface consisting of an infinite number of sheets with branch points occurring at each ν_n. Again, this differs from the usual quantum mechanical scattering problems and is a consequence of the presence of the $\{\nu_n\}$. The limiting absorption principle and the spectral theory of self-adjoint operators are the main analytical tools employed in obtaining the results of Section 3.

2. Wave Propagation in a Waveguide.

A waveguide is a device employed for the transmission of high frequency electromagnetic waves (microwaves) from a transmitter to a receiver.† In this section, we will present a brief discussion of the electromagnetic field in an ideal waveguide, as well as some of the problems arising when a discontinuity (perturbation) is present in the guide.†† We will culminate the section with a discuss-

*The boundary of an arbitrary domain, D, will be denoted by $\dot{D}$.
†Waveguides may also be employed for the transmission of sound waves. Hence our mathematical results apply equally well to acoustics as to microwave theory.
††For a more detailed discussion of these matters, see [4].

ion of the S-matrix.

Uniform Waveguides. We begin by considering an "ideal waveguide" By an ideal waveguide, S, we shall mean an infinite cylindrical tube in R^3, made of perfectly conducting material, having perfectly cylindrical walls and filled with a medium that is homogeneous, isotropic, non-dissipative and linear. This usually represents a good approximation for ordinary waveguides.

Our second assumption will be that the field quantities vary harmonically with time, t, at frequencies $\omega/2\pi$. This time dependence is represented by the suppressed exponential factor $e^{i\omega t}$. Let $\vec{E} = (\vec{E}_t, E_z)$ and $\vec{H} = (\vec{H}_t, H_z)$ denote the electric and magnetic field vectors, where $\vec{E}_t(\vec{H}_t)$ and $E_z(H_z)$ denote components of $\vec{E}(\vec{H})$ in the transverse direction, i.e. the direction parallel to the cross section, ℓ, and the direction of propagation, i.e. the z-direction, respectively. The field vectors $\vec{E}$ and $\vec{H}$ now satisfy Maxwell's equation in the form

$$\text{curl } \vec{E} = \nabla \times \vec{E} = -i\,\omega\,\mu\,\vec{H} \quad \text{and} \qquad (2.1)$$
$$\text{curl } \vec{H} = \nabla \times \vec{H} = i\,\omega\,\varepsilon\,\vec{E},$$

where the positive real scalars μ and ε denote the permeability and permittivity, respectively. The field is also subject to the boundary condition, $E_z = 0$ and $\partial H_z/\partial n = 0$ on $\dot{S}$, where n denotes the outward directed normal to $\dot{S}$.

It is readily seen from Maxwells' equations that $\vec{E}_t$ and $\vec{H}_t$ may be conveniently expressed as infinite series of elementary particular solutions as follows:

$$\vec{E}_t = \sum_{n=1}^{\infty} V_n^E(z)\vec{e}_{n,t}^{\,E} + \sum_{n=1}^{\infty} V_n^M(z)\,\vec{e}_{n,t}^{\,M} \quad \text{and} \qquad (2.2)$$
$$\vec{H}_t = \sum_{n=1}^{\infty} I_n^E(z)\,\vec{h}_{n,t}^{\,E} + \sum_{n=1}^{\infty} I_n^M(z)\,\vec{h}_{n,t}^{\,M}$$

The solutions

$$\vec{E}_n^E = V_n^E(z)\,\vec{e}_{n,t}^{\,E} \qquad (2.3)$$
$$\vec{H}_n^E = I_n^E(z)\,\vec{h}_{n,t}^{\,E}$$

are known as transverse electric (TE) modes since they have no component of $\vec{E}_{n,t}$ in the z direction. They are also called H-modes. The solutions

$$\vec{E}^M_{n,t} = V^M_n(z)\,\vec{e}^M_{n,t}$$
$$\vec{H}^E_{n,t} = I^M_n(z)\,\vec{h}_{n,t} \qquad (2.4)$$

are known as TM or E-modes and have no component of $H_{n,t}$ in the z-direction. The vectors $\vec{e}^E_{n,t}$, $\vec{h}^E_{n,t}$, etc. are functions of the transverse coordinates only.

The vectors $\vec{e}^M_{n,t}$ and $\vec{h}^M_{n,t}$ are given by $\vec{e}^M_{n,t} = -\nabla_t\phi^M_n$, $\vec{h}^M_{n,t} = z_o \times \vec{e}^M_{n,t}$, where $\vec{z}_o$ denotes a unit vector in the positive z-direction, ∇_t denotes the gradient operator transverse to the z-axis (e.g., $\nabla_t = \vec{x}_o \frac{\partial}{\partial x} + \vec{y}_o \frac{\partial}{\partial y}$ when ℓ is defined by rectangular (x,y) - coordinate system and $\vec{x}_o, \vec{y}_o$ denote unit vectors in the positive x and y directions). The scalar functions ϕ^M_n satisfy the boundary value problem $(\Delta_t + K^{M^2}_{C_n})\phi^M_n = 0$ in ℓ, $\phi_n = 0$ on $\dot{\ell}$ when $K_{C_n} > 0$, and $\frac{\partial\phi_n}{\partial s} = 0$ on $\dot{\ell}$ when $K_{C_n} = 0$, where Δ_t denotes the Laplacian transverse to the z-axis and s denotes arc length on $\dot{\ell}$.

Similarly, we have

$$\vec{e}^E_{n,t} = \vec{z}_o \times \nabla_t\psi_n \quad , \quad \vec{h}^E_{n,t} = z_o \times \vec{e}^E_{n,t}$$

where $(\Delta_t + K^{E^2}_{C_n})\psi_n = 0$, $\frac{\partial}{\partial\psi}\psi_n = 0$ on $\dot{\ell}$, ψ denoting the outward directed normal to $\dot{\ell}$, the $K^{M^2}_{C_n}$ and $K^{E^2}_{C_n}$ are assumed to be monotonic increasing and are referred to as cutoff wave numbers. The K_{C_n} and $K^{E^2}_{C_n}$ correspond to the ν_n of Section 1. When $K_{C_n} = 0$, the corresponding mode is referred to as a TEM mode and has no z component of either $\vec{E}_{n,t}$ or $\vec{H}_{n,t}$. When ℓ is simply connected, we clearly have no non-trivial TEM mode. An example in which ℓ is not simply connected is a coaxial waveguide (the space between two cylindrical waveguides whose cross sections are circles with different radii). The functions $\vec{e}_{n,t}$ and $\vec{h}_{n,t}$ are assumed to be orthonormal. Several specific examples may be found in [4].

Again employing Maxwells' equations, we may obtain the following equations for the scalar functions $V_n(z)$, $I_n(z)$† appearing in (2.3) and (2.4):

† We delete the superscript since the equations are of the same form regardless of the type of mode.

$$\frac{dV_n(z)}{dz} = -i\,\kappa_n\, Z_n\, I_n$$
$$\frac{dI_n(z)}{dz} = -i\,\kappa_n\, Y_n\, V_n, \qquad n = 1,2,\ldots \tag{2.5}$$

For TE modes, we have $\kappa_n^E = \sqrt{\kappa^2 - K_{C_n}^{E^2}}$, $Z_n^E = \sqrt{\frac{\mu}{\varepsilon}}\frac{\kappa}{\kappa_n^E} = \frac{\omega\mu}{\kappa_n^E}$, where $\kappa = \omega\sqrt{\mu\varepsilon} = \frac{\omega}{C} = \frac{2\pi}{\lambda}$ is the free space wave number and λ is the wavelength of field excitation . For TM modes, we have

$$\kappa_n^M = \sqrt{\kappa^2 - K_{C_n}^{M^2}}\ , \quad Z_n^M = \sqrt{\frac{\mu}{\varepsilon}}\frac{\kappa_n^M}{\kappa} = \frac{\kappa_n^M}{\omega\varepsilon}\ .$$

Equations (2.5) are of transmission line form. For this reason, the amplitudes $V_n(z)$ and $I_n(z)$ are referred to as mode voltages and mode currents, respectively, and κ_n, Z_n ($Y_n = Z_n^{-1}$) as mode propagating constant and mode characteristic impedance (admittance), respectively. Thus each mode is completely characterized by the corresponding mode voltage and current . It follows readily from Maxwell's equations that

$$E_z = -\frac{i}{\kappa}\sqrt{\frac{\mu}{\varepsilon}}\sum_{n=1}^{\infty} I_n^M(z)\, K_{C_n}^{M^2}\, \varphi_n^M$$

and (2.6)

$$H_z = -\frac{i}{\kappa}\sqrt{\frac{\varepsilon}{\mu}}\sum_{n=1}^{\infty} V_n^E(z)\, K_{C_n}^{E^2}\, \psi_n^E$$

Conversely, if we know E_z and H_z, we can determine $\vec{E}_t$ and $\vec{H}_t$.

The general solution (2.5) is

$$V_n(z) = A_n\, e^{-i\kappa_n z} + B_n\, e^{i\kappa_n z}$$
$$I_n(z) = Z_n^{-1}(A_n\, e^{-i\kappa_n z} - B_n\, e^{i\kappa_n z})\ \dagger \tag{2.7}$$

Substituting (2.7) in (2.2) and (2.6) we thus have the general expression for the field. The field problem is thus seen to be equivalent to an infinite number of uncoupled transmission lines.

We now observe from (2.3), (2.4) and (2.7) that the nth TE(TM) mode is propagating without attenuation only if κ_n^E (κ_n^m) is real,

† When $B_n = 0$ in (2.7) we have a travelling wave moving toward the right and conversely when $A_n = 0$.

or equivalently if $\kappa^2 > K_{C_n}^{E^2}$ $(\kappa^2 > K_{C_n}^{m^2})$. Otherwise the mode is exponentially damped and is called an evanescent mode. Suppose that ℓ is simply connected, so that the TEM mode does not exist. Every TE(TM) mode is damped if $\kappa^2 < K_{C_1}^{E^2}$ $(\kappa^2 < K_{C_1}^{M^2})$. Hence there is propagation only if the frequency is high enough or the wavelength small enough. If a TEM mode does exist, then it propagates for all frequencies, ω.

B. Discontinuities in Waveguides. Up to now, we have been discussing uniform waveguides, which are essentially uniform hollow tubes with simple cross sections, such as rectangles or circles. Hence the electromagnetic fields may be easily obtained using the modal expansions we have described. However, while the mathematics of this situation is quite straightforward, such uniform guides do not lead to physically reasonable problems.

For one thing, we see that the phase velocity, $v_p = \frac{\omega}{\kappa_n} = \frac{K\ C}{\sqrt{\kappa^2 - K_{C_n}^2}} > C$. This represents the velocity of progagation of the wave and is usually the same as that of electrons travelling in the tube. Hence the geometry must be changed to "slow down" the waves. The second main reason for introducing discontinuities into the guide is that the waveguide is only part of a microwave circuit and hence must be connected to cavities, antennas, loops, etc., through which electrons enter and leave. Obviously, these discontinuities complicate the mathematics considerably.

The typical situation may be illustrated by the typical waveguide junction illustrated in Figure 1.

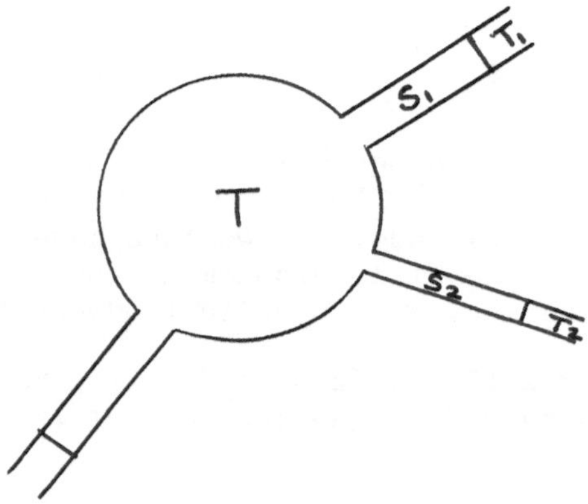

Figure 1

The junction consists of T and a number of ideal waveguide leads. A waveguide lead is a portion of a waveguide, closed off by a terminal plane perpendicular to the axis of the guide. An L-port consists of L waveguides leads, $S_1,\ldots,S_L$ with terminal planes denoted by $T_1,\ldots,T_L$. We assume that the medium inside the junction is linear and passive. We also assume that the terminal planes are located sufficiently far from T that any damped modes that are present at T are essentially zero at each T_j , $j = 1,\ldots,L$. (Usually a distance equal to a few times the cross-sectional dimensions will suffice.)

Without loss of generality, we may consider a two-port. We also suppose that there are m propagating modes. We define the $\mathcal{S}$-matrix for this two-port as follows. By an incident wave in S_j , $j = 1,2$, we shall mean one approaching T from T_j and by a reflected wave, we shall mean one approaching T_j from T. We see from our previous discussion that at a large distance from T in S_j, we have

$$\vec{E}_t^j = \sum_{n=1}^{m} (a_n^j + b_n^j)\, \vec{e}_{t,n}^{\,j}$$
$$\vec{H}_t = \sum_{n=1}^{m} \frac{(a_n^j - b_n^j)}{Z_{n,j}}\, \vec{h}_{t,n}^{\,j} \quad , \tag{2.8}$$

where a_n^j and b_n^j refer to the amplitude of the incident and reflected waves, respectively, and Z_{n_j} is the characteristic impedance of the nth mode in the jth waveguide lead.

We define the $\mathcal{S}$-matrix, $\mathcal{S} = \mathcal{S}(\kappa^2)$ as that matrix of order 2m such that

$$\mathcal{S}A = B, \tag{2.9}$$

where the column vectors A and B are defined by $A = (a_1^1,\ldots,a_m^1,a_1^2,\ldots,a_m^2)^t$ and $B = (b_1^1,\ldots,b_m^1,b_1^2,\ldots,b_m^2)^t$. $\mathcal{S}$ is a symmetric matrix. Futhermore, when the junction is lossless, $\mathcal{S}$ is unitary. In the next section, we shall obtain $\mathcal{S}(\kappa^2)$ from a different approach and discuss it at greater length. We also note that there are other important matrices, such as the impedance (admittance) matrix, $Z(Y = Z^{-1})$, which may be readily obtained from $\mathcal{S}$.

The matrices $\mathcal{S}$,Z, etc. are extremely useful in that the relevent quantities associated with the electromagnetic fields may be expressed in terms of them. However, the quantitative determination of these matrices involves the solution of a (usually rather complicated) boundary value problem. There are several methods available for specific problems, such as variational techniques, matching procedures across a discontinuity and integral equation methods. Most of these yield numerical approximations to the answer, since the fields are rarely obtainable in closed form when discontinuities are present. For illustrations of these methods, see [4] and [5].

3. Scattering Theory.

In this section, we shall consider the operator $A_o = -\Delta$ associated with zero Dirichlet or Neumann boundary conditions in an ideal waveguide, S, and establish a rigorous stationary scattering theory when A_o is perturbed by the addition of a discontinuity to the guide. We shall then obtain explicit formulas for the S-matrix from the abstract theory and show that this corresponds to the S-matrix defined in Section 2. In addition, we shall meromorphically continue the S-matrix onto an infinitely sheeted Riemann surface and obtain a means of calculating the S-matrix for small perturbations, employing the limiting absorption principle. We shall also represent the wave and scattering operators in a time-dependent framework.

It is not our intention here to give rigorous proofs of these results. Instead we shall refer to the appropriate papers where they may be found and indicate how they may be generalized. At the end of the section, some unsolved problems will be mentioned. As is to be expected, our results imply the same concepts as those used in Section 2 to describe the modal expansion of the electromagnetic field. For instance, our radiation conditions are given in terms of the travelling waves of Section 2, rather than the classical Sommerfeld radiation condition. Furthermore, the cutoff wave numbers, $K^2_{C_n}$, of Section 2 will be seen here to yield some rather interesting mathematical results concerning spectral multliplicity and meromorphic continuations.

A. Stationary Theory. For the sake of definitions (and simplicity), we shall discuss the following specific problem and later indicate how the results may be generalized. Let A_o denote the self-adjoint operator given by $-\Delta$ associated with zero Dirichlet boundary conditions acting in a semi-infinite cylinder, $S \subset R^N$ $(N \geq 2)$ with arbitrary, bounded, sufficiently smooth cross section, $\ell \subset R^{N-1}$. We assume (again for simplicity) that ℓ is simply connected. Let A_ℓ denote the analogous operator defined on ℓ with eigenvalues $\{\nu_n\}$ and the corresponding complete set of orthonormal eigenfunctions, $\{\eta_n(\overline{x})\}$, where we define an arbitrary point $x = (x_1,\ldots x_N) = (\overline{x},x_N) \varepsilon S$ by $0 \leq x_N < \infty$, $\overline{x} \varepsilon \ell$. Thus ν_n and $\eta_n(\overline{x})$ correspond to the $K^2_{C_n}$ and Φ_n of Section 2.

Let A denote the perturbed operator, obtained from A_o by perturbing a finite portion of $\dot{S}$ and deleting a finite number of bounded subsets of S. Hence $\Omega = S$ for $x_N \geq \dot{x}_N$ for some $\dot{x}_N > 0$. We also assume that $\dot{\Omega}$ is sufficiently smooth (say $\dot{\Omega}$ is of class C^{N+2}, although we shall observe later that the smoothness condition may be relaxed). We shall first show that A^c, that part of A orthogonal to its eigenvalues, is unitarily equivalent to A_o.

We may easily define a spectral mapping for A_o as follows. For each $\lambda \in [\nu_m, \nu_{m+1})$, m a fixed positive integer, set $\dot{\eta}_n(x;\lambda) = (2/\pi)^{1/2} \sin\sqrt{\lambda - \nu_n}\, x_N\, \eta_n(\bar{x})$, $n = 1,\ldots,m$. Define the Hilbert spaces $H_o = L_2(S)$, $H = L_2(\Omega)$ and

$\mathcal{H} = \bigoplus_{n=1}^{\infty} L_2\left((\nu_n,\infty);\ \frac{d\lambda}{2\sqrt{\lambda-\nu_n}}\right)$, where $\bigoplus_{n=1}^{\infty}$ denotes the direct sum. Suppose

$\Phi(\lambda) \in C_o^{\infty}(S)$ and set $(T^o\Phi)_n(\lambda) = \hat{u}_n^o(\lambda) = \begin{cases} \int_S \Phi(\lambda)\eta_n^o(x;\lambda)\,dx & \text{for } \nu_n \leq \lambda \\ 0 & \text{for } \lambda < \nu_n \end{cases}$

and

$$T^o u(\lambda) = \hat{u}^o(\lambda) = (\hat{u}_1^o(\lambda), \hat{u}_2^o(\lambda), \ldots) \qquad (3.1)$$

The following result was given in [1].

<u>Theorem 1.</u> The mapping T^o defined on $C_o^{\infty}(S)$ by (3.1) has a unitary extension mapping H_o onto $\mathcal{H}$. Futhermore $T^o(A_o u)(\lambda) = \lambda T^o u(\lambda)$ for each $u \in H_o$, $\lambda \in \sigma(A_o) = [\nu_1,\infty)$.

It follows from Theorem 1 that the points ν_n have the following mathematical significance. If $m_o(\lambda)$ denotes the spectral multiplicity of A_o at λ, then $m_o(\lambda) = m$ for $\nu_m \leq \lambda < \nu_{m+1}$. Hence the cut-off wave numbers, ν_n, also represent multiplicity jump points for A_o. This phenomenon of an operator with non-uniform spectral multiplicity in the continuous spectrum does not occur in the usual quantum mechanical scattering problems, where the multiplicity is infinite throughout the continuous spectrum.

We next consider the operator A with the objective of constructing a pair of spectral mappings, $T^{\pm}$, for A^c, mapping H^c (the orthogonal complement in H of the eigenfunctions of A) onto $\mathcal{H}$. First we recall a few results concerning the point spectrum of A. D.S. Jones, [6], has shown that there are at most a finite number of eigenvalues of A of finite multiplicity on any bounded interval. In addition, he proved that there exist eigenvalues of A below ν_1 and he estimated their number in terms of the size of the perturbation. Rellich, [7], proved that A has no eigenvalues if Ω is convex. Whether or not there actually can exist eigenvalues embedded in the continuous spectrum is still an open question in general.

The construction of spectral mappings, $T^{\pm}_{+}$, hinges on the construction of generalized eigenfunctions, $w_n^{\pm}(x;\lambda)$, for the operator A. We thus wish to obtain solutions $w_n^{\pm}(x;\lambda) = w_n^{o}(x;\lambda) + v_n^{\pm}(x;\lambda)$, of the equation

$$\begin{cases} (\Delta + \lambda)w_n^{\pm}(x;\lambda) = 0 \quad \text{in} \quad \Omega, \\ w_n^{\pm}(x;\lambda) = 0 \quad \text{on} \quad \dot{\Omega}, \end{cases} \tag{3.2}$$

such that v_n^+ and v_n^- satisfy appropriate outgoing and incoming radiation conditions, respectively, for $x_N \geq \dot{x}_N$. By outgoing we shall mean that

$$v_n^+(x;\lambda) = \sum_{n'=1}^{m} C_{n'}^{n^+}(\lambda)e^{+i\sqrt{\lambda-\nu_{n'}}\, x_N}\, \eta_{n'}(\bar{x}) + \sum_{n'=m+1}^{\infty} C_{n'}^{n^+}(\lambda)e^{-\sqrt{\lambda-\nu_{n'}}\, x_N}\, \eta_{n'}(\bar{x}), \tag{3.3}$$

where $x_N \geq \dot{x}_N$, $\nu_m < \lambda < \nu_{m+1}$ and $\sqrt{}$ denotes the positive square root. The incoming radiation condition is defined analogously with $i\sqrt{}$ replaced by $-i\sqrt{}$ and $C_{n'}^{n+}(\lambda)$ by $C_{n'}^{n-}(\lambda)$. We shall see that the constants, $C_{n'}^{n\pm}(\lambda)$, $n',n = 1,\ldots,m$, play the key role in calculating the S-matrix.

It was shown in [1] that the construction of functions, $w_n^{\pm}(x;\lambda)$, may be reduced to solving the boundary value problem

$$(-\Delta - \lambda)u^{\pm}(x;\lambda) = F(x) \tag{3.4}$$

in Ω, $u(x) = 0$ on $\dot{\Omega}$ for outgoing (incoming) functions $u^+(x;\lambda)(u^-(x;\lambda))$ where $F(x)$ has bounded support and $\lambda \in (\nu_m,\nu_{m+1})-\Lambda$, Λ consisting of the eigenvalues of A. The proof of the existence and uniqueness of the functions $u^{\pm}(x;\lambda)$, follows from the limiting absorption principle, originally proved by Eidus in [8] for semi-infinite cylinders and later extended by the author to other domains with infinite boundaries, as well as more general perturbations (see [1],[9] and [10]).

<u>Theorem 2.</u> ["Limiting Absorption Principle"]. There exist unique outgoing and incoming solutions, $u^{\pm}(x;\lambda)$, satisfying (3.4) and obtained from the equations

$$u^{\pm}(x;\lambda) = \lim_{\varepsilon \downarrow 0} u^{\varepsilon^{\pm}}(x;\lambda).$$

The convergence is uniform with respect to $\lambda \in [a,b] \subset (\nu_m,\nu_{m+1})-\Lambda$ and x in bounded subsets of Ω (provided $F(x)$ is sufficiently smooth).

Having obtained the generalized eigenfunctions, $w_n^{\pm}(x;\lambda)$, as described above, we define

$$T^{\pm}u(\lambda) = \hat{u}^{\pm}(\lambda) = (\hat{u}_1^{\pm}(\lambda), \hat{u}_2^{\pm}(\lambda), \ldots),$$
$$(T^{\pm}u)_n(\lambda) = \hat{u}_n^{\pm}(\lambda) = \begin{cases} \int u(x) w_n^{\pm}(x;\lambda)\, dx & \text{for} \quad n \le m \\ 0 & \text{for} \quad n > m \end{cases} \tag{3.5}$$

$\lambda \in (\nu_m, \nu_{m+1}) - \Lambda$ and $u(x) \in C_o^{\infty}(\Omega)$.

Theorem 3. The mappings $T^{\pm}$ defined by (3.5) have unitary extensions mapping H^c onto $\mathcal{H}$. Futhermore, we have $T^{\pm}(Au)(\lambda) = \lambda T^{\pm}u(\lambda)$ for each $u \in H$ and $\lambda \in \sigma(A_o) - \Lambda - \{\nu_n\}$.

The proofs of Theorem 2 and 3 appear in [1]. Theorem 2 follows from the theory of elliptic equations and Theorem 3 is proved employing Theorem 2 as well as results from spectral theory.

Theorem 3 implies that $T^{\pm}$ yield spectral mappings for A^c and that A^c is unitarily equivalent to A_o. (Hence A^c has an absolutely continuous spectrum). We may define wave operators, $W^{\pm}$, and the scattering operator, S, by the equations

$$W^{\pm} = T^{\pm^{-1}} T^o$$
$$S = W^{+^{-1}} W^- \tag{3.6}$$

It is clear that $W^{\pm}$ maps H_o unitarily onto H^c and S maps H_o unitarily onto itself. Futhermore, the wave operators intertwine the operators A_o and A^c.

B. The S-matrix. It was proved in [2] that the scattering operator, S, defined by (3.6) may be conveniently represented in terms of the S-matrix as follows. Suppose that $\lambda \in (\nu_m, \nu_{m+1}) - \Lambda$ and define the matrices $S_m(\lambda)$ and $T_m(\lambda)$ of order m by

$$S_m(\lambda) = I_m + T_m(\lambda),$$
$$T_m(\lambda) = (t_{n,n'}(\lambda)), \quad t_{n,n'}(\lambda) = -(2\pi)^{\frac{1}{2}}\, i C_{n'}^{n^-}(\lambda), \tag{3.7}$$

where $C_{n'}^{n^-}(\lambda)$ is defined by (3.3) (with + replaced by −) and I_m denotes the mth order identity matrix. The following result was proved in [2].

Theorem 4. For almost all $\lambda \in (\nu_m, \nu_{m+1})$ and each $u(x) \in H$, we have $\widehat{(Su)}_n^o(\lambda) = \hat{u}_n^o(\lambda) + \sum_{\nu_{n'} < \lambda} t_{n,n'}\, \hat{u}_{n'}^o(\lambda)$.

$S_m(\lambda)$ is a unitary mth order matrix for $\lambda \in (\nu_m, \nu_{m+1}) - \Lambda$. We thus observe that there are an infinite number of S-matrices,

$S_m(\lambda)$, of different order corresponding to the various intervals (v_m, v_{m+1}). This peculiar property is also a consequence of the existence of cutoff wave numbers.

We next consider the relationship between $S_m(\lambda)$ and the S-matrix defined in Section 2. Suppose that $U(x;\lambda)$ denotes an arbitrary solution of the boundary value problem: $(\Delta+\lambda)U(x;\lambda) = 0$ in Ω, where $\lambda \in (v_m, v_{m+1})-\Lambda$. For $x_N \geq \dot{x}_N$, $U(x;\lambda)$ has the Fourier expansion:

$$U(x;\lambda) = \sum_{n=1}^{m} C_n^I(\lambda)\, e^{i\sqrt{\lambda-\nu_n}\, x_N}\, \eta_n(\bar{x}) + \sum_{n=1}^{m} C_n^R(\lambda) e^{-i\sqrt{\lambda-\nu_n}\, x_N}\, \eta_n(\bar{x})$$
$$+ \sum_{n=m+1}^{\infty} C_n(\lambda)\, e^{-\sqrt{\nu_n-\lambda}\, x_N}\, \eta_n(\bar{x})$$

It may be readily shown, using the functions $w_n^-(x;\lambda)$, that $C^R(\lambda) = S(\lambda)C^I(\lambda)$ where $C^R(\lambda) = (C_1^R(\lambda),\ldots,C_M^R(\lambda))$ and $C^I(\lambda) = (C_1^I(\lambda),\ldots,C_m^I(\lambda))$. Note that $C_n^I(\lambda)\, e^{i\sqrt{\lambda-\nu_n}x_N}\, \eta_n(\bar{x})$ represents a wave incident from some distant hyperplane, $x_N >> \dot{x}_N$, and $C_n^R(\lambda)\, e^{-i\sqrt{\lambda-\nu_n}\, x_N}\, \eta_n(\bar{x})$ represents a reflected wave travelling toward the terminal hyperplane. Employing equation (2.6) for E_z as well as the boundary conditions $E_z = 0$ and on $\dot{\Omega}$, we conclude that our two definitions of the S-matrix are equivalent.

We next demonstrate a means of calculating the S-matrix in the case of small perturbations. Let ε denote a small positive number, let $h_j(x)$ denote real-valued functions such that $h_j(x) \in C^2(\dot{S})$ and suppose $h_j(x) \equiv 0$ for $x_N \geq \dot{x}_N$, $j = 1,\ldots,N$. We assume that S may be mapped onto domains Ω^ε, by means of the transformation $x^\varepsilon = T^\varepsilon x$, where $x^\varepsilon = (x_1^\varepsilon,\ldots,x_N^\varepsilon) \in \Omega^\varepsilon$, $x = (x_1,\ldots x_N) \in S$ and $x_j^\varepsilon = x_j + \varepsilon h_j(x)$, $j = 1,\ldots,N$. (Hence we are implicitly assuming that the perturbation is bounded away from the corners of S). The operator $-\Delta$ acting in Ω^ε is mapped into the operator

$$A_\varepsilon = -\Delta+\varepsilon\Big(2 \sum_{j=1}^{N} h_j \frac{\partial^2}{\partial x_j^2} + 2 \sum_{j<k} \Big(\frac{\partial h_j}{\partial x_k} + \frac{\partial h_k}{\partial x_j}\Big) \frac{\partial^2}{\partial x_j\, \partial x_k}$$
$$+ \sum_{j=1}^{N} \Delta h \frac{\partial}{\partial x_j}\Big) = -\Delta + \varepsilon\bar{V}. \qquad (3.8)$$

We now extend definition (3.8) to small complex values of ε. (Note that $\bar{V}$ has compact support).

The functions $v_n^{\varepsilon-}(x;\lambda) = w_n^{\varepsilon-}(x;\lambda) - w_n^o(x;\lambda)$ corresponding to the operators A^ε may now be expressed in terms of the unique incoming solutions, $u^{\varepsilon-}(x;\lambda)$, of

$$(-\Delta - \lambda + \varepsilon\bar{V})u_n^{\varepsilon-}(x;\lambda) = -\varepsilon\bar{V}\, w_n^o(x;\lambda) \text{ in } S,\ u_n^{\varepsilon-}(x;\lambda)=0 \text{ on } \dot{S}. \qquad (3.9)$$

The existence of $u^{\varepsilon^-}(x;\lambda)$ follows from Theorem 2.

Theorem 5. If $\lambda \in (\nu_m,\nu_{m+1})-\Lambda$, then $u^{\varepsilon}(\cdot;\lambda)$ is analytic in the topology given by $L_2^{loc}(S)$. Futhermore, if the $h_j(x)$ are sufficiently smooth functions of x, then the power series expansion for $u^{\varepsilon}(x;\lambda)$ converges uniformly for each x in bounded subsets of Ω and the coefficients of this expansion may be explicitly calculated.

The proof of Theorem 5 was given in [2] and is a variant of the proof of the limiting absorption principle. It follows by showing that $\psi_1^o(x;\lambda) = \lim_{\varepsilon \downarrow 0} \dfrac{u^{\varepsilon}(x;\lambda) - u^o(x;\lambda)}{\varepsilon}$ exists in the appropriate topology for all complex values of ε sufficiently small. The arguments also show that for $h_j(x)$ sufficiently smooth, we have $(-\Delta - \lambda)\psi_1^o = -\overline{V}w_1^o$ in S, $\psi_1^o = 0$ on $\dot{S}$, where ψ_1^o is incoming.

Since the (incoming) Green's function for the operator $A_o-\lambda$ is known to be the integral operator with kernel

$$G_o^-(x,y;\lambda) = \begin{cases} \displaystyle\sum_{n=1}^{\infty} \frac{\eta_n(\overline{x})\eta_n(\overline{y})\left(\sin\sqrt{\lambda-\nu_n}\, x_N\right)e^{-i\sqrt{\lambda-\nu_n}\, y_N}}{\sqrt{\nu_n - \lambda}} & \text{for } x_N<y_N \\ \displaystyle\sum_{n=1}^{\infty} \frac{\eta_n(\overline{x})\eta_n(\overline{y})\left(e^{-i\sqrt{\lambda-\nu_n}\, x_N}\right)\sin\sqrt{\lambda-\nu_n}\, y_N}{\sqrt{\nu_n - \lambda}} & \text{for } x_N>y_N \end{cases} \quad (3.10)$$

we can thus explicitly calculate the first coefficient, $\psi_1^o(x;\lambda)$, in the series expansion

$$u^{\varepsilon} = \sum_{n=1}^{\infty} \psi_n^o(x;\lambda)\varepsilon^n \quad (3.11)$$

from the equation $\psi_1^o(x;\lambda) = -\int_S G_o^-(x,y;\lambda)\overline{V}w_1^o(y,\lambda)\,dy$. The remaining coefficients, $\psi_n^o(x;\lambda)$ in (3.11) may be calculated similarly.†

†The same coefficients may be obtained formally by expanding $-(I + \varepsilon(A_o- \lambda+io)^{-1}\overline{V})^{-1}$ in a Neumann series for small $|\varepsilon|$ and then applying each term of the series to $-\varepsilon(A_o-\lambda+io)^{-1}\overline{V}w_n^o(x;\lambda)$; we can thus explicitly calculate the first coefficient, ψ_1^o, in the series expansion.

In view of (3.7) and the equation

$$C^{n^-}_{n'}(\lambda) = \int_{\ell} e^{i\sqrt{\lambda-\nu_n}\, x_N}\, \eta_{n'}(\bar{x}) v^-_n(\bar{x},x_N;\lambda)\, d\bar{x} \quad ,$$

we may calculate $S_m(\lambda)$ from the previous results. While convergence of the series has only been proved for ε small, the formal power series may still yield reasonable approximations for larger values of ε. Futhermore, the radius of convergence may be estimated in terms of $G^-_o(x,y;\lambda)$.

Finally, we shall briefly describe how to meromorphically continue $S_m(\lambda)$ from the interval (ν_m,ν_{m+1}) to an infinitely sheeted Riemann surface, R^∞. In view of either (3.3) or (3.10) we define R^∞ as the Riemann surface obtained by letting each point ν_n, $n = 1,2,\ldots,$ be a branch point of order one. Let $\Gamma_{n_1,\ldots,n_k}$ denote the open sheet of R^∞, consisting of those points, λ, satisfying $0 < \arg(\lambda-\nu_n) < 2\pi, n = n_1,\ldots,n_k$, and $-2\pi < \arg(\lambda-\nu_n) < 0$ for all remaining ν_n, where $n_1,\ldots,n_k$ are positive integers. The closed sheet, $\bar{\Gamma}_{n_1,\ldots,n_k}$, is defined analogously with $0 < \arg(\lambda-\nu_n) < 2\pi$ replaced by $0 \leq \arg(\lambda-\nu_n) < 2\pi$, etc. .

Thus we may proceed from the sheet $\Gamma_{n_1,\ldots,n_{k-1}}$ to $\Gamma_{n_1,\ldots,n_k}$ by traversing the point ν_{n_k} an odd numbers of times. We shall refer to the sheet Γ_o, consisting of those λ satisfying $-2\pi < \arg(\lambda-\nu_n) < 0$, $n = 1,2,\ldots,$ as the "physical sheet" of R_∞. Hence there exists an infinite number of "non-physical" sheets. By R_m, we shall mean that part of R_∞ consisting of $\bar{\Gamma}_o$ and $\bar{\Gamma}_{n_1,\ldots,n_k}$, where m is a fixed positive integer and $1 \leq n_1,\ldots,n_k \leq m$. Now, set $\kappa = \sqrt{\lambda-\nu_1}$ and $S_m(\kappa) = S_m(\lambda)$.

<u>Theorem 6.</u> $S_m(\lambda)$ has a meromorphic continuation from $(\nu_m,\nu_{m+1})-\Lambda$ onto $\bar{\Gamma}_o$. The poles of $S_m(\lambda)$ on Γ_o consist of the eigenvalues of A below ν_1. $S_m(\lambda)$ has a meromorphic continuation from Γ_o to Γ_1 across that part of the branch cut contained in $(\nu_m,\nu_{m+1})-\Lambda$ in the sense that $\bar{S}_m(\kappa)$ is meromorphic on the set $\{\mathrm{Im}\kappa<0\}\cup\{\mathrm{Im}\kappa>0\}\cup[\sqrt{a-\nu_1},\sqrt{b-\nu_1}\,]$ for each interval $[a,b] \subset (\nu_m,\nu_{m+1})-\Lambda$. Analogously, $S_m(\lambda)$ may be meromorphically continued onto all of R_m.

Remark. We have also characterized the non-real poles, λ_o, of $S_m(\lambda)$ in terms of points at which there exist exponentially blowing up solutions, $w(x;\lambda_o)$, of $(\Delta+\lambda_o)w(x;\lambda_o) = 0$ in Ω, $w(x;\lambda_o) = 0$ on $\dot{\Omega}$. Hence there exists a "resonant state" at λ_o. The details will appear in [3].

C. Time-Dependent Formulation. We also have a time-dependent formulation of the wave and scattering operators. We may assume without loss of generality that $\Omega \supset S$. Let $j(x)$ denote a function defined in Ω such that $j(x) \equiv 0$ for $x_N \le x_N^o$, $j(x) \equiv 1$ for $x_N \ge x_N^1 > \dot{x}_N$ and $j(x) \in C^\infty(\bar{S})$. For each $u(x) \in C_o^\infty(S)$, set $Ju(x) \equiv j(x)u(x)$ for $x \in S$ and zero for $x \in \Omega - S$. We extend J by continuity as a mapping from $L_2(S)$ into $L_2(\Omega)$.

Suppose that $\Phi(\lambda)$ is a real-valued function defined on $[\nu_1,\infty)$ such that $\Phi(\lambda) \in C^3([\nu_1,\infty))$. We now have the following version of the Kato invariance principle for the operators A_o and A^c.

Theorem 7. Suppose $u(x) \in L_2(S)$. Then

(a) $W_o^\pm = \lim_{t\to\pm\infty} e^{-i\Phi(A_o)t} e^{i\Phi(A^c)t}$ exists and

(b) if $\Phi'(\lambda) > 0 (\Phi'(\lambda) < 0)$ on $[\nu_1,\infty)$, then
$(\overline{W}^\pm u)_n(\lambda) = \hat{u}_n^o(\lambda)$ $((\overline{W}^\mp u)_n(\lambda) = \hat{u}_n^o(\lambda))$ a.e. on $[\nu_1,\infty)$.

This result has been proved for the wave equation in cylinders in [1]. In that case, $\Phi(\lambda) = \sqrt{\lambda}$. A more general version of Theorem 7 (in which $\Phi(\lambda)$ is only piecewise differentiable) has been proved in [10] for wedge-shaped regions. The arguments and results of [10] readily carry over to the present case. The proof is based on stationary methods.

D. Miscellaneous Remarks. We close by indicating ways of extending the previous results as well as posing some unanswered questions.

Remark 1. The cylinder, S, may be infinite instead of semi-infinite and the results still hold using the same arguments. Observe, however, that in this case the spectral multiplicity of A_o at each point is twice that for the semi-infinite cylinder. Note also that our results hold for a wider class of domains, [10], the main requirement being that we may separate variables in the unperturbed domain.

Remark 2. For many practical problems, the discontinuity or perturbation need not satisfy our smoothness conditions, but instead may have corners (such as in the case of an iris for instance). In such cases, our only difficulty is that the limiting absorption principle requires the use of elliptic estimates that may not hold in the neighborhood of such edges. In any case, all of our constructions go through, except possibly near these finite number of singular points. With more refined elliptic estimates, our constructions may go through even near these points. This will be carried out in a future publication. The numerical techniques for computing the electromagnetic fields often do not yield reasonable results near edges.

Remark 3. Our results hold for more general perturbations than those given here, such as the addition of a second order perturbation with bounded support (possibly corresponding to a local change in the medium inside the guide) or to a perturbation of the boundary condition again assuming bounded support. We have also considered non-selfadjoing perturbations in [9]. This could correspond to dissipation in part of the medium or on the surface of the guide.

Remark 4. We shall study waveguide perturbations with unbounded support in a future publication.

Remark 5. The existence of bound states in the continuous spectrum is still an open question for general boundary perturbations with compact support. It might also be of interest to look at the physical significance of such possible states as well as the poles of the S-matrix.

References.

1. C. Goldstein, Eigenfunction expansions associated with the Laplacian for certain domains with infinite boundaries, I and II. Amer. Math. Soc. Trans. 135, 1-50 (1969).
2. C. Goldstein, Analytic perturbations of the operator $-\Delta$. J. Math. Anal. Appl. 25, No. 1, 128-148 (1969).
3. C. Goldstein, Meromorphic continuation of the S-matrix for the operator $-\Delta$ acting in a cylinder. Amer. Math. Soc. Proc. (to appear).
4. N. Marcurvitz, *Waveguide Handbook*. M.I.T. Rad. Lab. Series. McGraw-Hill, (1951).
5. R. E. Collin, *Field Theory of Guided Waves*. McGraw-Hill (1960).
6. D. S. Jones, The eigenvalues of $\Delta u + \lambda u = 0$ when the boundary conditions are given on semi-infinite domains. Proc. Cambridge Phil. Soc. 49, 668-684 (1953).
7. F. Rellich, Uber das asymptotishe Verhaltung der Lösungen von $\Delta u + \lambda u = 0$ in unendlichen Gebieten. Jber. Deutsch Math. - Verein. 53, 157-165 (1943).
8. D. M. Eidus, The principle of limiting absorption. Amer. Math. Soc. Transl. 47, 157-192 (1965).

9. C. Goldstein, Perturbation of non-selfadjoint operators, I. Arch. Rat. Mech. Anal., 37, No. 4, 268-296 (1970).
10. C. Goldstein, Eigenfunction expansions associated with the Laplacian for certain domains with infinite boundaries, III. Amer. Math. Soc. Trans. 143, 283-301 (1969).

NONLINEAR SCATTERING THEORY

W. A. Strauss

Brown University

1. Introduction. Scattering theory compares the behavior in the distant future and past of a system evolving in time. It is called nonlinear if the system evolves in a nonlinear fashion. Consider a one-parameter group of operators on some linear space X:

$$U(t)U(s) = U(t + s); \qquad U(0) = I$$

$-\infty < t, s < +\infty$. We think of $U(t)f$ as representing the state at time t beginning with a state $f \in X$ at time zero. We are interested in the behavior of $U(t)f$ as $t \to \pm\infty$ and in the relationship between the behavior at $+\infty$ and at $-\infty$.

The typical features of scattering theory are the following.

(a) $U(t)$ preserves some functional E; that is, $E[U(t)f] = E[f]$ for all $f \in X$ and real t. E is frequently the square of a norm in X and is called the *energy*.

(b) There is a much simpler group of operators $U_0(t)$, called the *free* group, with the property that for all f there exists f_+ and f_- called asymptotic states such that

$$E[U(t)f - U_0(t)f_\pm] \to 0 \qquad \text{as} \qquad t \to \pm\infty \tag{1}$$

The *scattering operator* S is then defined as the operator $f_- \to f_+$. Sometimes, S is thought of as the operator $U_0(\ )f_- \to U_0(\ )f_+$

This research was supported in part by grants from the John Simon Guggenheim Foundation and the National Science Foundation.

J. A. LaVita and J.-P. Marchand (eds.), Scattering Theory in Mathematical Physics, 53–78.

acting between pairs of free states. The *wave operators* W_+ and W_- are the operators $f_\pm \to f$, or $U_0(\ )f_\pm \to U(\ )f$, acting from free states to interacting states.

In addition to the questions above, we are interested in the determination of U from U_o and S. Even when U is not unique, what part of it can be constructed from U_o and S? This is the *inverse scattering problem*, the determination of the dynamics of a system from its asymptotic behav

The plan of this paper is as follows. Part 1 consists of introductory remarks and examples. In part 2 a fairly general situation is presented in which $W_\pm$ is constructed. In addition S is constructed on incoming states which are sufficiently small. This is applied to several examples. Part 3 is concerned with the inverse scattering problem and with the non-existence of S. In Part 4 the existence and properties of S for the nonlinear Klein-Gordon equation (NLKG) are discussed.

Returning to the formal framework given above, we may modify it as follows.

(c) $U_0(\ )$ may preserve a slightly different functional E_0. We will always assume that E_0 is the square of a norm $|\ |_2$ defined on X. We will always use E_0 to define the asymptotics, so that instead of (1) our basic condition will be

$$|U(t)f - U_0(t)f_\pm|_2 \to 0 \qquad \text{as} \qquad t \to \pm\infty \tag{2}$$

We will denote by $\overline{X}$ the completion of X with respect to this norm. Note that f_+ and f_- are unique if $U_0(t)$ are linear and uniformly continuous.

(d) $U_0(t)$ may be defined on a different space X_0 with an identification map $X \to X_0$ to be placed in front of the second term in (2).

(e) Instead of groups U(t), one may consider propagators U(t,s).

The most interesting results in nonlinear scattering theory exist for systems defined by particular partial differential equations. In contrast to the linear theory, the equations may be invariant under translations. In such a context we will use the notation

$$U(t)f = u(t) = u(x,t); \qquad U_0(t)f_\pm = u_\pm(t) = u_\pm(x,t) \tag{3}$$

where x are the space variables. We will call U(t)f a perturbed solution and $U_0(t)g$ a free solution.

Example 1.1. NLKG equation which occurs as a model in quantum field theory and in elasticity. We define X to consist of pairs of functions $f = [f_1, f_2]$ in R^3 which are smooth and small at infinity. $U(t)f$ is defined by

$$u_{tt} - \Delta u + u + |u|^2 u = 0 \quad ,$$

$x = (x_1, x_2, x_3) \in R^3$, $u(x,0) = f_1(x)$, $u_t(x,0) = f_2(x)$. Thus f_1, f_2 are the Cauchy data for a solution.

$$E[f] = \frac{1}{2} \int (|f_2|^2 + |\nabla f_1|^2 + |f_1|^2 + |f_1|^4)\, dx \quad .$$

$U_0(t)$ is defined by the linear equation obtained by dropping the term $|u|^2 u$, and $E_0[f]$ the quadratic functional obtained by dropping the $|f_1|^4$ term. Then (2) is valid. Futhermore, S is a smooth nonlinear operator on a certain Banach space containing X. Details are given in Part 4.

Example 1.2. Korteweg-de Vries equation (KdV), which occurs as an approximation in many problems of fluid mechanics. Let X consist of the continuous functions of one variable whose first derivatives are square-integrable. Let E be the square of the L^2-norm. Let $U(t)$ be defined by

$$u_t + u_{xxx} + uu_x = 0 \qquad (-\infty < x < \infty) \; .$$

This equation possesses the particular "solitary wave" solutions

$$g(x-c_1 t; c_1, c_2) = 3c_1 \operatorname{sech}^2[\tfrac{1}{2} \sqrt{c_1}\, (x-c_1 t-c_2)],$$

$c_1 > 0$. The general solution (vanishing as $|x| \to \infty$) of KdV "contains" solitary waves in the sense that there is a finite number of parameters $\gamma_j^{\pm} = (c_{j1}, c_{j2})$ such that

$$u(x + c_{j1} t, t) \to g(x, \gamma_j^{\pm})$$

uniformly on bounded sets as $t \to \pm\infty$.

In fact, the speeds c_{j1} are determined as follows. Of course, they are invariants of the motion. The linear Schrödinger operators $-\partial^2/\partial x^2 + u(x,t)/6$, with $u(x,t)$ playing the role of a potential at each time t, turn out to be uniformly equivalent! Call the discrete eigenvalues λ_j. Then $c_{j1} = 4\lambda_j$. Moreover, KdV can be solved explicitly via the inverse scattering theory for these linear operators. See Gardner et al. [1] and Lax [1] for details.

Can a modification of condition (1) still be valid for KdV? If the free group $U_0(t)$ is defined by the equation $v_t + v_{xxx} = 0$, the solitary waves do not look asymptotically free because they do not decay to zero uniformly. But it is possible that the remainder does. Do there exist $f_\pm$ such that

$$E[U(t)f - \sum_j U(t)g(\gamma_j^\pm) - U_0(t)f_\pm] \to 0 \tag{4}$$

as $t \to \pm \infty$?

In linear scattering theory we are used to subtracting off the bound states. Consider, for instance, the linear Schrödinger equation with a potential $q(x)$ small at infinity. The operator $-\Delta + q(x)$ has some eigenvalues λ_j and eigenfunctions $\psi_j(x)$. Then the usual scattering theory asks that for each $f \in L^2$ there exist $f_\pm \in L^2$ so that

$$E[U(t)f - \sum_j U(t)(\gamma_j\psi_j) - U_o(t)f_\pm] \to 0$$

where $\gamma_j = (f,\psi_j)$.

The general nonlinear wave does not look asymptotically linear. However, it may do so after certain terms of its asymptotic expansion are subtracted off. The choice of a free equation is never unique, of course, but must be chosen partly on the basis of simplicity.

Example 1.3. Nonlinear Schrödinger equation (NLS), which occurs in the optics of nonlinear media. Let $U(t)$ be defined by

$$iu_t + u_{xx} + |u|^2u = 0 \qquad (-\infty < x < \infty).$$

The L^2 norm is preserved and there are solitary waves of the form

$$u(x,t) = \phi(x-c_1t)\exp i\psi(x-c_2t)$$

where ϕ and ψ are real. The situation is similar to Example 1.2. See Zakharov and Shabat [1].

We now present the intertwining properties of $W_\pm$ and S. By definition, $U(t)W_+f_+ \sim U_o(t)f_+$ as $t \to +\infty$. Replacing t by $t + T$ and using the group property, $U(t)U(T)W_+f_+ \sim U_o(t)U_o(T)f_+$. That is

$$W_+U_o(T) = U(T)W_+ \tag{5}$$

The same is true for W_-. Similarly,

$$SU_o(T) = U_o(T)S. \tag{6}$$

Equation (5) says that W_+ intertwines the free and perturbed systems, but if W_+ is not linear it is not clear whether any useful conclusion can be drawn from it. On the other hand, (6) must imply a great deal about the structure of S if the free group is unitary, say. In the notation of vector-valued differentials, (6) can be written as

$$A_o Sg = dS[g]A_o g.$$

We shall now develop the fundamental criterion for the existence of asymptotic states. Operating formally at first, let U( ) and U_o( ) have infinitesimal generators A and A_o, respectively. Let $P = A - A_o$, the perturbation. Assume that $U_o(t)$ is linear. We denote $u(t) = U(t)f$ and $\overline{u}(t) = U_o(-t)u(t)$. Then

$$\frac{d\overline{u}}{dt} = U_o(-t)Au(t) - U_o(-t)A_o u(t)$$
$$= U_o(-t)Pu(t) = U_o(-t)PU_o(t)\overline{u}(t).$$

Note that the generators A_o and A are explicitly absent from the last expression.

Lemma 1.1. Let $U_o(t)$ be a group of linear isometric operators on $\overline{X}$. Let P be an operator from a domain $D(P) \subset \overline{X}$ into $\overline{X}$. Let u(t) be a function with values in D(P) for which Pu(t) is continuous in X and which satisfies the equation

$$\frac{d}{dt}\,[U_o(-t)u(t)] = U_o(-t)Pu(t). \qquad (7)$$

If

$$\int_{-\infty}^{\infty} |Pu(t)|_2 dt < \infty \qquad (8)$$

then there exists $f_\pm \in \overline{X}$ such that

$$|u(t) - U_o(t)f_\pm|_2 \to 0 \qquad \text{as} \qquad t \to \pm\infty \qquad (9_\pm)$$

Proof. By (7), the vector

$$f_+ = U_o(-t)u(t) + \int_t^{+\infty} U_o(-s)Pu(s)\, ds$$

is independent of t. Multiplying it by $U_o(t)$, and taking norms, we have

$$|U_o(t)f_+ - u(t)|_2 \leq \int_t^{+\infty} |U_o(-s)Pu(s)|_2\, ds = \int_t^{\infty} |Pu(s)|_2\, ds.$$

By (8) this expression tends to zero as $t \to +\infty$. Similarly at $-\infty$.

For future purposes, it is convenient to write

$$u(t) = U_o(t-T)u(T) + \int_T^t U_o(t-s)Pu(s)\, ds\, . \qquad (10)$$

Letting $T \to \pm\infty$, we also have

$$u_\pm(t) = u(t) + \int_t^{\pm\infty} U_o(t-s)Pu(s)\, ds. \tag{$11_\pm$}$$

Subtracting,

$$u_+(t) - u_-(t) = \int_{-\infty}^{\infty} U_o(t-s)Pu(s)\, ds. \tag{12}$$

It should also be remarked that if the solutions of (7) are unique, they form a group of operators. This follows from the translation invariance of (7).

2. Low Energy Scattering.

In each of the three examples presented above, a battle is going on between the nonlinear effects which tend to exaggerate the size of the solutions and the dispersive effects which tend to spread the solutions out in space. If we start with a small disturbance, the dispersive effects might dominate and the solution might look asymptotically free. This idea is realized in Theorem 2.1. The next theorem constructs the wave operators by a natural iteration process. This is a pure consequence of the dispersiveness of the free system and the solutions do not have to be small. The third theorem constructs the scattering operator on small solutions. None of these results requires *a priori* information on the nonlinear system, not even that there is a conserved functional E. The assumptions are only that the free system decays in some norm $|\ |_3$ (different from $|\ |_2$, of course) and that the nonlinear interaction is of high enough degree.

<u>Hypotheses.</u> (i) X is a linear space on which are defined norms $|\ |_1$, $|\ |_2$ and $|\ |_3$ which satisfy $|f|_3 \leq c|f|_2$, for all $f \in X$. We allow $|f|_1$ to be $+\infty$ for certain $f \in X$.

(ii) $U_o(t)$ is a group of linear operators on X such that

$$|U_o(t)f|_2 = |f|_2$$

$$|U_o(t)f|_3 \leq c|t|^{-d}|f|_1 \qquad \text{if} \qquad |f|_1 < \infty$$

where c and d are positive constants.

(iii) There exist $b > 0$, $\delta > 0$ and $q \geq 1$, $dq > 1$ such that

$$|Pf|_1 + |Pf|_2 \leq b|f|_3^q \qquad \text{if} \qquad |f|_2 \leq \delta.$$

[In case $q = 1$ we assume $b = b(|f|_2) \to 0$ as $|f|_2 \to 0$.]

Theorem 2.1. Let $u(t)$ be a solution of (7). If $|u(0)|_1 + |u(0)|_2$ is sufficiently small, then there exist asymptotic states $f_\pm$ in $\overline{X}$ satisfying (9).

Proof. For convenience we denote various positive constants by c. Denote $f = u(0)$, and

$$m(t) = \sup_{0\leq s\leq t} [(1 + s)^d |u(s)|_3 + |u(s)|_2].$$

We claim that $m(t) < \delta$ for all t. By assumption, this is true when $t = 0$, and hence for small t. In any interval in which $m(t) < \delta$, we have the following estimates. From (10) with $T = 0$,

$$\begin{aligned} |u(t)|_2 &\leq |f|_2 + \int_0^t |Pu(s)|_2\, ds \\ &\leq |f|_2 + \int_0^t b|u(s)|_3^q\, ds \\ &\leq |f|_2 + b\int_0^t (1 + s)^{-dq} m(s)^q\, ds \\ &\leq |f|_2 + cb\; m(t)^q \end{aligned}$$

We note that, for any g, $|U_o(t)g|_3 \leq c|U_o(t)g|_2 = c|g|_2$, so that

$$|U_o(t)g|_3 \leq \min\{c|g|_2,\ ct^{-d}|g|_1\} \leq c(1 + t)^{-d}(|g|_1 + |g|_2) .$$

Again from (10),

$$\begin{aligned} |u(t)|_3 &\leq |U_o(t)f|_3 + \int_0^t |U_o(t-s)Pu(s)|_3\, ds \\ &\leq c(1+t)^{-d}(|f|_1 + |f|_2) + \int_0^t c(1+t-s)^{-d}(|Pu(s)|_1 \\ &\quad + |Pu(s)|_2)ds \end{aligned}$$

The integral is less than

$$cb\; m(t)^q \int_0^t (1 + t - s)^{-d}(1 + s)^{-dq}\, ds,$$

which in turn is less than $c(1+t)^{-d}$ because $dq > 1$ and $dq \geq d$.

Putting these estimates together, we obtain the inequality

$$m(t) \leq c_1 + cb\; m(t)^q ,$$

where $m(0) \leq c(|f|_1 + |f|_2) = c_1$. This is of the form $m \leq c_1 + \varepsilon(m)m$, where $\varepsilon(m) \to 0$ as $m \to 0$. Thus $m \leq 2c_1$ for all time if $2\varepsilon(2c_1) \leq 1$, say. This proves the claim.

As we showed above, $|Pu(t)|_2 \leq bm(t)^q(1+t)^{-dq}$. Since $m(t)$ is bounded and $dq > 1$, $|Pu(t)|_2$ is integrable as $t \to +\infty$, so that Lemma 1.1 applies.

Definitions. Let $v(t)$ be an X-valued continuous function of t.

$$N_T(v) = \sup_{-\infty<t\leq T} \{|v(t)|_2 + (1+|t|)^d |v(t)|_3\},$$

Y_T = the completion of $\{v | N_T(v) < \infty\}$ with respect to the norm N_T. $N(v) = N_{+\infty}(v)$, $Y = Y_{+\infty}$. Note that $N(v) \leq c(|v(0)|_2 + |v(0)|_1)$ if $v(t)$ is a free solution.

Hypothesis (iii*). There exists $q \geq 1$, $dq > 1$ and there exists b depending boundedly on $|f|_2 + |g|_2$ so that

$$|Pf - Pg|_1 + |Pf - Pg|_2 \leq b(|f|_3 + |g|_3)^{q-1}|f - g|_3$$

for all $f, g \in X$. (In case $q = 1$, assume $b \to 0$ as $|f|_2 + |g|_2 \to 0$.)

Theorem 2.2. Assume (i), (ii), (iii*). If $U_o(\ )f_- \in Y$, then there exists a finite time T and a unique $u \in Y_T$ which satisfies (7) and (9_-).

Proof. We may write (11_-) as

$$u = U_o(\ )f + \mathscr{P}u, \qquad \mathscr{P}u(t) = \int_{-\infty}^t U_o(t-s)Pu(s)\, ds$$

We claim that

$$N_T(\mathscr{P}u - \mathscr{P}v) \leq \varepsilon(T)\beta N_T(u - v) \tag{22}$$

for all $u, v \in Y_T$, where $\varepsilon(T) \to 0$ as $T \to -\infty$, and β depends boundedly on $N_T(u) + N_T(v)$. Assuming (22) for the moment, it follows that, for each $k > 0$, there exists T such that $\mathscr{P}$ is a contraction mapping on $\{u \in Y_T | N_T(u) \leq k\}$. We define $u_o = U_o(\ )f$, $u_{n+1} = u_o + \mathscr{P}u_n$. Then there is a T depending on $N_T(u_o)$ so that $\{u_n\}$ is a Cauchy sequence in Y_T. The limit is our desired solution. It is unique by (22).

The proof of (22) is similar to that of Theorem 2.1. For any $u \in Y_T$ and $t \leq T$,

$$|\mathscr{P}u(t)|_2 \leq \int_{-\infty}^t b|u(s)|_3^q\, ds \leq cb\, N_T(u)^q (1 + |T|)^{1-dq},$$

$$|\mathscr{P}u(t)|_3 \leq bN_T(u)^q \int_{-\infty}^t (1 + |t-s|)^{-d}(1+|s|)^{-dq}\, ds.$$

The last integral is $o(|t|^{-d})$ as $t \to -\infty$ because $dq > \max\{d,1\}$. So

$$N_T(\mathscr{P}u) \leq \varepsilon(T) b\, N_T(u)^q.$$

(22) follows from the analogous estimates applied to differences $Pu - Pv$.

Theorem 2.3. Assume (i), (ii), (iii*). If $U_o(\ )f_- \in Y$ and $N(U_o(\ )f_-) \le \eta$ with η sufficiently small, then there is a unique $u \in Y$ satisfying (7) and (9_-) and $N(u) \le 2N(U_o(\ )f_-)$. In addition, there is an asymptotic state f_+ so that (9_+) holds. The scattering operator $f_- \to f_+$ is one-one and continuous from $\{g \in \overline{X} | U_o(\ )g \in Y, N(U_o(\ )g) \le \eta\}$ onto itself.

Proof. We define $\mathcal{P}$ exactly as in the preceding proof except we allow arbitrary times $-\infty < t < \infty$. In exactly the same way, we have

$$N(\mathcal{P}u - \mathcal{P}v) \le \beta N(u - v)$$

where $\beta = cb(N(u) + N(v))^{q-1}$ depends on $N(u) + N(v)$ boundedly and goes to zero with it. Hence $\beta \le 1/2$, say, if $N(u) + N(v)$ is sufficiently small. Defining $u_{n+1} = u_o + \mathcal{P}u_n$, we have

$$N(u_{n+1}) \le N(u_o) + \frac{1}{2} N(u_n)$$

and hence $N(u_n) \le 2N(u_o)$, so long as $N(u_{n-1})$ is small enough. Thus for $N(u_o)$ small enough,

$$N(u_{n+1} - u_n) \le 2^{-n} N(u_1 - u_o).$$

So $\{u_n\}$ is a Cauchy sequence in Y. The limit is our desired solution. The properties of S are easy to verify.

In the following examples we use the notation $x = (x_1, \ldots, x_n) \in R^n$, $D^\alpha = (\partial/\partial x_1)^{\alpha_1} \ldots (\partial/\partial x_n)^{\alpha_n}$, $\alpha = (\alpha_1, \ldots, \alpha_n)$, $|\alpha| = \alpha_1 + \ldots + \alpha_n$. For a function $f(x)$,

$$\|f\|_p^p = \int |f(x)|^p \, dx, \qquad \|f\|_\infty = \sup_x |f(x)|$$

$$\|f\|_{k,p}^p = \sum_{|\alpha| \le k} \|D^\alpha f\|_p^p \quad .$$

Sobolev's inequality states that $\|f\|_\infty \le c\|f\|_{k,p}$ if $k > n/p$. In each example, the condition $|f|_3 \le c|f|_2$ is a special case of this inequality. In each example, also, the free group is defined by a linear differential equation, so that $U_o(t)$ is an integral operator with a Green's kernel which can be explicitly calculated.

Example 2.1. The NLS equation

$$iu_t + u_{xx} + \kappa u^p = 0 \qquad (-\infty < x < \infty)$$

with κ a real constant. Take $|f|_1 = \|f\|_1$, $|f|_2 = \|f\|_2 + \|f_x\|_2$,

$|f|_3 = \|f\|_\infty$, X = completion of C_o^∞ under $|\ |_2$. The solution of the free equation ($\kappa = 0$) satisfies

$$\|u(t)\|_\infty \leq (4\pi|t|)^{-\frac{1}{2}}\|u(0)\|_1 \quad ,$$

so that d = 1/2. We have $Pf = i\kappa f^p$,

$$|f^p|_1 \leq |f|_3^{p-2}|f|_2^2 \ , \qquad |f^p|_2 \leq |f|_3^{p-1}|f|_2,$$

so that we may take q = p-2. The restriction dq > 1 requires p > 4. The smallness condition in Theorem 2.3 is satisfied if $\kappa(|f_-|_1 + |f_-|_2)^{p-3}$ is sufficiently small.

Example 2.2. The generalized KdV equation

$$u_t + [u_{xx} + \kappa u^p]_x = 0$$

The norms and X are taken exactly as in the preceding example. In this case, d = 1/3 and $Pf = -\kappa f^{p-1}f_x$. So

$$|Pf|_1 \leq \kappa\|f\|_\infty^{p-2}\|f\|_2\|f_x\|_2 \leq \kappa|f|_3^{p-2}|f|_2^2$$

Thus we may take q = p-2 and we require p > 5. However, $\|Pf\|_2$ involves second derivatives and therefore fails. Theorem 2.1 can be salvaged by observing that the estimate of $|Pf|_2$ is required solely to obtain a bound on $|u(t)|_2$. We can obtain this bound directly by multiplying the equation by $u_{xx} + \kappa u^p$, by integrating to obtain $\int [\frac{1}{2} u_x^2 - \frac{\kappa}{p+1} u^{p+1}]dx$ = constant, and by using the smallness assumption.

The two preceding examples both possess solitary waves for all p. (Of course they do not satisfy the smallness condition.) But the asymptotic behavior of the general solution is not known.

Example 2.3. The NLKG equation

$$\phi_{tt} - \Delta\phi + m^2\phi + F(\phi) = 0 \qquad (x \in R^3)$$

where $F(\phi) = g\phi^p$, g and m are constants and m > 0. If we put $u = [u_1,u_2] = [\phi,\phi_t]$, the equation may be written as du/dt = Au, $A = A_o + P$ where

$$A_o = \begin{bmatrix} 0 & 1 \\ \Delta - m^2 & 0 \end{bmatrix} \ , \qquad P = \begin{bmatrix} 0 & 0 \\ -F & 0 \end{bmatrix}$$

Let $|f|_3 = \|f_1\|_\infty$, $|f|_2 = \|f_1\|_{2,2} + \|f_2\|_{1,2}$, $|f|_1 = \|f_1\|_{3,1} + \|f_2\|_{2,1}$.

We may define either $X = C_o^\infty$ or X = the completion of C_o^∞ with respect to $|\ |_2$. Then (ii) holds with $d = 3/2$ because the solution of the free KG equation satisfies

$$\|\phi(t)\|_\infty \leq c|t|^{-3/2}[\ \|\phi(0)\|_{3,1} + \|\phi_t(0)\|_{2,1}].$$

In order to verify (iii) and (iii*), we must show that

$$\|F(\phi)\|_{2,1} + \|F(\phi)\|_{1,2} \leq b\|\phi\|_\infty^q \ .$$

The highest-order terms on the left hand side are of the form

$$\|\phi^{p-1}\phi_{xx}\|_1,\ \|\phi^{p-2}\phi_x^2\|_1,\ \|\phi^{p-1}\phi_x\|_2 \ .$$

They are, respectively, less than

$$\|\phi^{p-2}_\infty\|\ \|\phi\|_2\ \|\phi_{xx}\|_2,\ \|\phi\ {}^{p-2}_\infty\|\ \phi_x\|_2^2,\ \|\phi\|_\infty^{p-1}\|\phi_x\|_2 \ .$$

Thus (iii) and (iii*) are satisfied with $q = p-2$ and $b = c\|\phi\|_{2,2}^2$. We must have $q \geq 1$, which means $p \geq 3$.

Example 2.4. Nonlinear symmetric hyperbolic systems

$$u_t = \sum_{j=1}^{n} A_j u_{x_j} + F(u)$$

where $x \in R^n$, $u(x,t) \in R^m$, the A_j are real symmetric $m \times m$ matrices and $F(u)$ is a $C^{3\ell}$ vector function vanishing to an integer order p at $u = 0$. Here ℓ is a fixed integer $\geq 1/2\ [(n+2)/2]$ where $[\alpha]$ means the integral part of α. We take $|f|_3 = \|f\|_{\ell,\infty}$, $|f|_2 = \|f\|_{3\ell,2}$, $|f|_1 = \|f\|_{3\ell,1}$, and $X = (C_o^\infty)^m$ or its completion in $|\ |_2$. We have $|f|_3 \leq c|f|_2$ since $3\ell-\ell > n/2$. The characteristic speeds of the free system $(F=0)$ are defined as the eigenvalues $\lambda_1(\xi),\ldots,\lambda_m(\xi)$ of the matrix $\sum A_j\xi_j$ where $\xi = (\xi_1,\ldots,\xi_n) \in R^n$. If none of these speeds vanishes and if each of the surfaces $\{\xi|\lambda_k(\xi) = 1\}$ is strictly convex, it is known that hypothesis (ii) holds with $d = (n-1)/2$. Thus we require $n > 1$.

Now for $\|f\|_\infty$ small, $F(f)$ looks like a polynomial with no term of degree $< p$. Let us denote by D^k any space derivative of order k. Then $|Pf|_1 = \|F(f)\|_{3\ell,1}$ is bounded by a sum of terms of the type

$$c\|(D^{k_1}f)\ \ldots\ (D^{k_p}f)\|_1$$

where $k_1 + \ldots + k_p \leq 3\ell$. If $k_1 \geq k_2 \geq \ldots \geq k_p$, we estimate this term by the L^2 norms of the two factors of highest order, and the L^∞

norms of the other (p-2) factors. These (p-2) other factors are derivatives of f of order $\leq \ell$. Therefore, $|Pf|_1 \leq c|f|_2^2|f|_3^{p-2}$ for $|f|_2$ sufficiently small. Similarly, $|Pf|_2$ contains terms like

$$c\|(D^{k_1}f) \ldots (D^{k_p}f)\|_2 \leq c\|D^{k_1}fD^{k_2}f\|_2 \|D^{k_3}f\|_\infty \ldots \|D^{k_p}f\|_\infty$$

If $k_2 \leq \ell$, we bound the first factor on the right by $\|D^{k_1}f\|_2\|D^{k_2}f\|_\infty$. If $k_2 > \ell$, then $k_1 < 2\ell$, and we bound it by

$$\|D^{k_1}f\|_4\|D^{k_2}f\|_4 \leq |f|_2^2$$

by Sobolev's inequality since $\ell \geq n/4$. Thus we have (iii) and (iii*) again with $b(s) = cs^2$ and $q = p-2$. We require $n > 1$ and $p \geq 3$. (We require $p > 3$ if $n = 3$, $p > 4$ if $n = 2$).

The theorems of this part and Example 2.3 are simplified versions of Segal [2][3]. Examples 2.1 and 2.2 are presented in Strauss [4]. Generalizations of Example 2.3 to nonlinear terms $F(x,t,u,u_x,u_t)$ are given by Chadam [1] and von Wahl [1]. The uniform decay estimate for free solutions in Example 2.4 is due to Costa [1]. The coupled Maxwell-Dirac equations are excluded from this example because p=2. Chadam [2] considers interaction terms of higher degree, but the question of the Maxwell-Dirac equations themselves remains open.

3. The Inverse Scattering Problem.

From knowledge of the scattering operator S and the free dynamics $U_0(t)$, we want to recover the perturbed dynamics $U(t)$. We present two examples when this can be done. The method is based on a functional which the free dynamics leaves invariant but S does not leave invariant. It is local in the sense that only a neighborhood of the zero state is required, and so we can work in the context of Part 2. It is perhaps surprising that this problem is much easier than in the corresponding linear one.

We assume (i), (ii) and (iii*) of Part 2. We consider solutions of the basic equation (7). Thus S is defined locally by Theorem 2.3. In addition, we assume

(iv) $P(\varepsilon f) = \varepsilon^p P(f)$ for $|f|_2 < \delta$, $|\varepsilon f|_2 < \delta$, $\varepsilon > 0$,

where $p \geq q$, $p > 1$. Together with (iii*), this implies

(v) $|Pf|_1 \leq c|f|_2^{p-q}|f|_3^q$,

by a homogeneity argument.

We postulate a bounded bilinear form B(f,g) on $\overline{X}$ invariant under the free group; that is,

$$|B(f,g)| \leq c|f|_2|g|_2$$

$$B(U_o(t)f,U_o(t)g) = B(f,g)$$

Theorem 3.1. Let $\phi(\ ) = U_o(\ )h$ and $\psi(\ ) = U_o(\ )k$ belong to Y. Then

$$\lim_{\varepsilon\to 0} \varepsilon^{-p-1}\{B(S(\varepsilon h),S(\varepsilon k))-B(\varepsilon h,\varepsilon g)\}$$

$$= \int_{-\infty}^{\infty} \{B(P\phi,\psi) + B(\phi,P\psi)\}\, dt \quad .$$

In particular, S determines this integral, for arbitrary free solutions.

Proof. Let u(t) and v(t) be solutions of (7). Let $\overline{u}(t) = U_o(-t)u(t)$, $\overline{v}(t) = U_o(-t)v(t)$. Assume that $f_\pm = \lim \overline{u}(t)$ and $g_\pm = \lim\ \overline{v}(t)$ exist as $t \to \pm\infty$. Then

$$\frac{d}{dt} B(\overline{u},\overline{v}) = B(\frac{d\overline{u}}{dt},\overline{v}) + B(\overline{u},\frac{d\overline{v}}{dt})$$

$$= B(Pu,v) + B(u,Pv).$$

Integrating this,

$$B(f_+,g_+) - B(f_-,g_-) = \int_{-\infty}^{\infty} \{B(Pu,v) + B(u,Pv)\}\, dt \quad .$$

(The right side converges because of the left side.)

Now let $h \in \overline{X}$ with $U_o(\ )h \in Y$. We choose $f_- = \varepsilon h$ and $u_\varepsilon(t) = U(t)W_-f_-$. For small enough ε, Theorem 2.3 applies, so that $N(u_\varepsilon) \leq 2N(U_o(\ )\varepsilon h) = O(\varepsilon)$ and

$$N(u_\varepsilon-\varepsilon U_o(\ )h) = N(\mathscr{P}u_\varepsilon) \leq bN(u_\varepsilon)^q \leq cN(u_\varepsilon)^p = O(\varepsilon^p)$$

by the estimates in the proof of Theorem 2.3 and by (v). In addition, by (iii*), for each t,

$$|Pu_\varepsilon - P(\varepsilon U_o h)|_2 \leq b(|u_\varepsilon|_3 + |\varepsilon U_o h|_3)^{q-1}|\mathscr{P}u_\varepsilon|_3$$

and so as $\varepsilon \to 0$,

$$\sup_t (1 + |t|)^{dq}|Pu_\varepsilon(t) - P(\varepsilon U_o(t)h)|_2 = o(\varepsilon^p).$$

We also choose $k \in \overline{X}$ with $U_o(\ )k \in Y$. The same estimates hold for $v_\varepsilon(t) = U\ (t)W_-(\varepsilon k)$. Using these estimates, the expression

$$\sup_t (1 + |t|)^{dq}[B(Pu_\varepsilon, v_\varepsilon) - B(P(\varepsilon U_o h), \varepsilon U_o k)]$$

is $o(\varepsilon^{p+1})$. Thus

$$\int_{-\infty}^{\infty} B(Pu_\varepsilon, v_\varepsilon)dt = \varepsilon^{p+1} \int_{-\infty}^{\infty} B(PU_o h, U_o k)dt + o(\varepsilon^{p+1}).$$

This completes the proof.

Remark. If P is the sum of homogeneous operators of various degrees, S determines a similar expression for each term.

Corollary. If for some ϕ, ψ the integral in Theorem 3.1 does not vanish, then S is not linear.

Proof. Otherwise, the limit could not exist since $p > 1$.

Example 3.1. NLS equation

$$u_t = i(-\Delta u + V(x)|u|^{p-1}u) \qquad (x \in R^n)$$

where p is an integer ≥ 3 ($p > 3$ if $n = 2$, $p > 4$ if $n = 1$) and where $V(x)$ is a real "potential" whose derivatives up to order ℓ are bounded, $\ell > 3n/4$. Then the conditions of Part 2 are met with $|f|_1 = \|f\|_{\ell,1}$, $|f|_2 = \|f\|_{3\ell,2}$, $|f|_3 = \|f\|_{\ell,\infty}$, $d = n/2$ and $q = p-2$. We choose

$$B(f,g) = \text{Im} \int f(x)\overline{g(x)}\, dx \,. \tag{31}$$

By Theorem 3.1, S determines the integral

$$\iint V[|\phi|^{p-1} - |\psi|^{p-1}]\phi\overline{\psi} \;\; dx\, dt$$

over all space-time, for all free solutions ϕ,ψ. If we take $\psi = 2\phi$, say, then

$$I[\phi] = \iint V|\phi|^{p+1}\, dx\, dt$$

is determined.

By using Fourier transformation, $V(x)$ can be determined provided it is integrable. On the other hand, the following method depends only on its continuity and boundedness. We choose the free solution ϕ_λ with the initial datum $\phi_\lambda(x,0) = \gamma(\lambda(x-x_o))$ where $\lambda > 0$, x_o is a point, and γ is any nice function. If we change variables $x' = \lambda(x-x_o)$, $t' = \lambda^2 t$ and $\phi'(x',t') = \phi_\lambda(x,t)$, then ϕ' is the free solution in the new variables with initial datum $\phi'(x',0) = \gamma(x')$. Thus

$$\lambda^{n+2} I[\phi_\lambda] = \iint V(x_o + x'/\lambda)|\phi'(x',t')|^{p+1}\, dx'\, dt'$$

$$\to V(x_o) \iint |\phi'|^{p+1}\, dx'\, dt'$$

as $\lambda \to \infty$. So $V(x_o)$ is determined by S.

Example 3.2. NLKG equation

$$\phi_{tt} - \Delta\phi + m^2\phi + V(x)|\phi|^{p-1}\phi = 0 \qquad (x \in R^3)$$

where $V(x)$ has bounded second derivatives and $p \geq 3$. S exists in the sense of Part 2. We choose the fundamental skew form

$$B(f,g) = \int (f_2 g_1 - f_1 g_2) dx \tag{32}$$

where $f = [f_1, f_2]$, $g = [g_1, g_2]$. By Theorem 3.1, S determines

$$\iint V(x)(|\phi|^{p-1} - |\psi|^{p-1})\phi\psi \quad dx\, dt$$

for all pairs of free solutions. As in the preceding example, it can be shown that S determines $V(x)$. This example can be generalized to an interaction term

$$V(x,\phi) = \sum_{j=3}^{\infty} V_j(x)\phi^j$$

analytic in a neighborhood of $\phi = 0$; $V_j(x)$ having bounded second derivatives. In this case, the coefficients are determined successively.

The method above, as applied to NLKG, is from Morawetz and Strauss [2,3]. For the corresponding linear problem, see Newton [1].

The problem one would like to solve is to characterize the properties of S which ensure that it comes from some V.

The rest of this part is devoted to a situation where there do *not* exist asymptotic free states.

Let $B(f,g)$ be as above. Assume

$$|B(Pf - Pg, g)| \leq b|g|_3^{p-1}|f-g|_2 \tag{33}$$

where $p > 1$; $|\ |_3$ is some norm; and b depends boundedly on $|f|_2 + |g|_2$.

Let $h \in \overline{X}$, $d > 0$ and $d(p-1) \leq 1$ such that as $t \to +\infty$,

$$|U_o(t)h|_3 = O(t^{-d}) \tag{34}$$

$$B(PU_o(t)h, U_o(t)h) \geq c_o t^{-d(p-1)} \tag{35}$$

where $c_o > 0$.

<u>Theorem 3.2.</u> If $u(t)$ is a solution of (7), then $|u(t)-U_o(t)h|_2$ does not go to zero as $t \to +\infty$.

<u>Proof.</u> Suppose $|u(t)-U_o(t)h|_2 \to 0$. We have

$$\frac{d}{dt} B(U_o(-t)u(t),h) = B(U_o(-t)Pu(t),h)$$
$$= B(Pu(t),U_o(t)h).$$

Therefore

$$\int_0^T B(Pu(t),U_o(t)h)\, dt$$

has a limit as $T \to \infty$. On the other hand, we have

$$|B(Pu(t)-PU_o(t)h,U_o(t)h| \leq b|U_o(t)h|_3^{p-1}|u(t)-U_o(t)h|_2$$

by (33). Since $|u(t)|_2 \to |h|_2$, b is bounded. The last factor goes to zero as $t \to \infty$. By (34) the last expression is $o(t^{-d(p-1)})$. Combining this with (35), we see that

$$B(Pu(t),U_o(t)h \geq \frac{1}{2} c_o t^{-d(p-1)} .$$

Since this is not integrable, we have a contradiction.

<u>Example 3.3.</u> Take the NLS equation

$$u_t = i(-\Delta u + V|u|^{p-1}u) \qquad (x \in R^n)$$

with V a nonzero real constant and

$$1 < p \leq 1 + 2/n \quad \text{for} \quad n \geq 2 \quad (1 < p \leq 2 \quad \text{for} \quad n = 1). \tag{36}$$

Then *the only solution with an asymptotic free state is identically zero.* ("Asymptotic" is taken in the sense of $L^2(R^n)$, the solution is assumed to have initial datum in L^2 and the asymptotic state to have initial datum in $L^1 \cap L^2$.)

To show this, we apply Theorem 3.2 with B given by (31), $|f|_2 = \|f\|_2$ and $|f|_3 = \|f\|_\infty$. Taking $V = 1$, say, we have $Pf = i|f|^{p-1}f$ and

$$B(Pf - Pg,g) = \text{Re} \int (|f|^{p-1}f-|g|^{p-1}g)\bar{g}\, dx.$$

By Schwarz's inequality, (33) holds with

$$b = (|f|_2 + |g|_2)^{p-1}|g|_2^{2-p} .$$

For $h \in L^1$, (34) holds with $d = n/2$. By assumption, $d(p-1) \leq 1$. So it remains to show (35). Let us denote $\phi(t) = U_0(t)h$. We already know that

$$B(P\phi,\phi) = \int |\phi|^{p+1}\, dx = O(t^{-(p-1)n/2}).$$

Lemma. $t^{(p-1)n/2} \int |\phi|^{p+1}\, dx$ is bounded away from zero for large t, if $1 \leq p < \infty$ and ϕ is a non-trivial free solution.

Proof. By Hölder's inequality

$$\{\int_{|x|<kt} |\phi|^2\, dx\}^{(p+1)/2} \leq \{\omega_n (kt)^n\}^{(p-1)/2} \int |\phi|^{p+1}\, dx .$$

So it suffices to bound the left side away from zero. Substituting

$$\phi(x,t) = (4\pi it)^{n/2} \int \exp(|x-y|^2/4it)\phi(y,0)\, dy ,$$

an explicit computation shows

$$\lim_{t\to\infty} \int_{|x|<kt} |\phi|^2\, dx = \int_{|\xi|<k/2} |\hat{\phi}(\xi,0)|^2\, d\xi ,$$

where $\hat{\phi}(\xi,0)$ is the Fourier transform of the initial datum. If ϕ is non-trivial, there is a k for which the limit does not vanish. This completes the proof.

Example 3.4. Take the NLKG equation in any dimension with V a non-zero constant and assume (36). Then the only solution with an asymptotic free state is identically zero. ("Asymptotic" is taken in the energy norm [the L^2 norm of the solution and its first derivatives], the solution has finite energy, and the asymptotic state decays in L^∞ like $t^{-n/2}$.)

We take B given by (32), $|f|_3 = \|f_1\|_\infty$ and $|f|_2 = \|f_1\|_{1,2} + \|f_2\|_{0,2}$. Since

$$B(Pf,g) = \int |f_1|^{p-1} f_1 g_1\, dx,$$

we are in a situation similar to the preceding example. Again $d = n/2$. It remains to verify (35). As before, it reduces to a

Lemma. Let $k > 1$ and let $\phi(x,t)$ be a free solution. Then

$$\lim_{t\to\infty} \int_{|x|<kt} |\phi|^2 dx = \tfrac{1}{2}\int [\,|\phi(x,0)|^2 + |(m^2-\Delta)^{-1/2}\phi_t(x,0)|^2]\, dx$$

if the right-hand side is finite.

Proof. We may assume ϕ has smooth data of compact support; otherwise, it may be so approximated in energy norm. Since the wave speed is 1, ϕ vanishes outside a certain cone and the integral of $|\phi|^2$ over $|x| > kt$ goes to zero. If we write the Fourier transform of ϕ explicitly, the limit of $\int |\hat{\phi}|^2 d\xi$ is easily computed. This completes the proof.

The asymptotic nature of the solutions in Example 3.3 and 3.4 remains unknown. It is conceivable that they are asymptotically free in a weaker norm than $|\ |_2$. It is possible that an analogue of (4) holds or that a different free equation should be used. Also the gap between the values of p allowed here (see (36)) and in Part 2 ($p \geq 3$ in general) should be noted. Example 1.3 falls into this gap.

These results, as applied to NLKG, are from Glassey [2].

4. Nonlinear Klein-Gordon Equation.

From now on, we restrict ourselves primarily to the real solutions of

$$u_{tt} - \Delta u + m^2 u + F(u) = 0, \tag{41}$$

$x \in R^n$, $m > 0$, $F(0) = F'(0) = 0$. Throughout this part it will be understood that only solutions whose initial data are sufficiently smooth and small at infinity will be considered. We denote $G(u) = \int_0^u F$,

$$E_R[u(t)] = \int_{|x|<R} [\tfrac{1}{2} u_t^2 + \tfrac{1}{2}|\nabla u|^2 + \tfrac{1}{2} m^2u^2 + G(u)]\, dx,$$

$$\|u(t)\|_e^2 = \int_{|x|<\infty} (u_t^2 + |\nabla u|^2 + m^2u^2)\, dx,$$

$E = E_\infty$. It is easy to construct solutions which blow up in a finite time, if G is sufficiently negative. Therefore it is reassuring to know

Theorem 4.1. Let F be any continuous function such that

$$m^2u^2 + uF(u) \geq 0 \qquad (u \text{ real}) \tag{42}$$

For any initial data of finite energy $E[u(0)]$, there exists a global solution of (41) in the sense of distributions such that $E[u(t)] \leq E[u(0)]$ almost everywhere.

Uniqueness of solutions is a major unsolved problem. (The three-dimensional Navier-Stokes equation present a similar difficulty.) Uniformly bounded solutions (which are not known to exist for all time) are easily shown to be unique. If the growth of F at infinity is severely restricted, one can show uniform boundedness and hence uniqueness and regularity. For $n = 1$, no growth condition is necessary; for $n = 2$, F should have polynomial growth; for $n = 3$, F should grow slower than u^5; as n increases, the restriction becomes more severe. The precise condition when $n = 3$ is: $G(u) \geq 0$ and

$$|F(u)| = O(G(u)^{-\varepsilon+2/3}|u|^{1-\varepsilon}) \qquad \text{as} \qquad |u| \to \infty .$$

If we assume (42) but G is somewhere negative, then there can be solutions like solitary waves which do not decay uniformly as $t \to \infty$. Suppose $n = 1$ and put $u(x,t) = \phi(x-ct)$ with $c < 1$. We get $(c^2-1)\phi'' + m^2\phi + F(\phi) = 0$, hence an equation of the form $(\phi')^2 = H(\phi)$. If H has a multiple zero at 0, a simple zero at $a > 0$ and is positive in between, there is a positive solution $\phi(x)$ which goes to zero exponentially as $t \to \pm\infty$. An analogue of (4) is possible.

Even if $G \geq 0$, there may not be free asymptotic states; see Example 3.4.

Local asymptotic behavior.

(a) Theorem 4.2. If $n \geq 3$ and

$$uF(u) \geq (2+\delta)G(u) \qquad \text{for all } u \tag{43}$$

for some $\delta > 0$, then every solution satisfies for all R:

$$\int_{-\infty}^{\infty} E_R[u(t)]\, dt < \infty , \qquad E_R[u(t)] \to 0 \quad \text{as} \quad t \to \infty$$

This theorem is based on multiplying the equation (41) by $u_t + \lambda(u_r + [(n-1)/2r]u)$ where $\lambda < 1$ and $r = |x|$, essentially a timelike derivative, and integrating over all space and time. After some integrations by parts, one obtains the sum of several positive terms bounded by a multiple of the initial energy. Among these positive terms is the integral of $r^{-1}(|\nabla u|^2 - u_r^2)$, which implies part of the conclusion. Bounds on the local L^2 norms of u_r and u_t can be derived from this.

(b) A function is called *outgoing* if it vanishes in the cone $|x| < t-t_0$ for some t_0. The ordinary wave equation ($m = 0$, $F = 0$) has a dense class of outgoing solutions if $n \neq 1$. But the KG equation for $m > 0$ has no outgoing solutions of finite energy at all. So we assume $m = 0$ for the rest of the discussion of local behavior.

If W_+ exists, it takes outgoing free solutions into outgoing perturbed solutions. Indeed for any $t \geq s \geq 0$, we have by the energy identity

$$E_s[u(t_o+s)] \leq E_t[u(t_o+t)] = E_t[u(t_o+t) - v(t_o+t)]$$

if $v(x,t_o+t)$ vanishes in $|x| < t$. But if $W_+[v(0)] = u(0)$, the right hand side tends to zero as $t \to +\infty$. Hence $E_s[u(t_o+s)] = 0$, so that $u(x,t)$ vanishes in the same cone as $v(x,t)$.

(c) If W_+ has dense range (in the space $\mathcal{E}$ of finite energy solutions of (41)), then the outgoing solutions of (41) are dense in $\mathcal{E}$. Indeed the outgoing free solutions are dense in the space Y (cf. Part 2) and W_+ is continuous on Y into $\mathcal{E}$ by Theorem 2.2.

(d) If the outgoing solutions of (41) are dense in $\mathcal{E}$, then the local energy of any solution in $\mathcal{E}$ tends to zero. Indeed let $u \in \mathcal{E}$. Given $\varepsilon > 0$, find an outgoing solution u_ε so that $\sup_t \| u(t) - u_\varepsilon(t)\|_e < \varepsilon$. In particular, the energy of u inside the cone where u_ε vanishes is less than ε. Thus the energy of u in any fixed sphere goes to zero at $t \to \infty$.

(e) We now specialixe to

$$u_{tt} - \Delta u + F(x,u) = 0 \tag{44}$$

where F vanishes outside a fixed sphere $|x| < \rho$. (This is analogous to all the linear scattering theories where the perturbation vanishes as $|x| \to \infty$.) Then we have a converse of (c). If the outgoing solutions of (44) are dense, then W_+ has dense range. Indeed, if u is outgoing, the spatial L^2 norm of $F(\ ,u(\ ,t))$ vanishes for large t, so that (8) is true. Of course, some rate of local decay of u would also suffice.

(f) Let n be odd and $F(x,u)$ have compact support. Then we have the following generalization of Huygens' Principle. The local energy of any solution with initial data of compact support decays exponentially, provided that it decays uniformly with respect to the data. This result holds for a fairly arbitrary abstract perturbation of compact support. Theorem 4.2, suitably generalized, gives a sufficient condition that the local energy does in fact decay uniformly and hence at an exponential rate. For $F(x,u) = V(r)|u|^{p-1}u$, this condition requires $p > 1$ and $V \geq 0$, $V' \leq 0$. In case there is a bounded reflecting obstacle, the condition requires it to be star-shaped.

(g) The local energy of such solutions may decay much more rapidly than exponentially. Take, for example, $F = V(r)u^3$, V positive and decreasing, $n = 3$. Then there exists $\alpha > 0$ so that, for solutions whose data have compact support and for sufficiently large t,

$$E_r[u(t)] \leq \exp(-\alpha^t).$$

This can be derived from the basic integral equation (10), which in this case (with T = 0) takes the form

$$u(x,t) = u_o(x,t) - \int_o^t R(x,y,t-s)F(y,u(y,s))\, ds, \qquad (45)$$

where $R(x,y,t-s)$ is the free solution with data $R(x,y,0) = 0$, $R_t(x,y,0) = \delta(x-y)$. This Green's or Riemann's function has support on the cone surface $|x-y| = |t-s|$. Let F and the intitial data of $u(x,t)$ have support in $|x| < \rho$. Then $u_o(x,t)$, which has the same data, vanishes in $|x| < t-\rho$. We take the local energy of (45). The result is

$$E_r[u(t)] \leq c \int_{t-2\rho}^t \{E_\rho[u(s)]\}^3\, ds$$

for $t > r+\rho$, where r is any positive number. We already know from (f) that $E_r[u(t)]$ decays exponentially. Let $\theta(t)$ be the maximum of $E_r[u(s)]$ for $s \geq t$. For $r \geq \rho$, $\theta(t) \leq 2\rho c\theta^3(t-2\rho)$. If we take $t = T + 2\rho n$ where n is an integer, we have

$$\theta(T+2\rho n) \leq (\sqrt{2\rho c})^{3^n-1}\theta(T)^{3^n} .$$

Since $\theta(T)$ may be chosen arbitrarily small, the estimate follows.

<u>Existence of S.</u> Let F(u) be independent of x. By Part 2, S exists on small solutions. Only in three dimensions is its existence known on large solutions.

<u>Theorem 4.3.</u> Let $n = 3$, $m > 0$, $F(u) = u^3$. For any solution of (41), $|t^{3/2}u(x,t)|$ is uniformly bounded. Therefore by Lemma 1.1 asymptotic free states exist.

This theorem is valid for $F(u) = g|u|^{p-1}u$ with $g > 0$, $p \geq 3$, for all uniformly bounded solutions. (The main condition to be satisfied for general F(u) is (43).) In the following discussion, however, let $p = 3$.

The proof is based on explicit use of the basic integral equation (10) = (45). In this case with $m > 0$, the Green's function $R(x-y,t-s)$ lives on the backward cone $|x-y| \leq t-s$. On the surface of the cone it is the singular function

$$(4\pi(t-s))^{-1}\delta(|x-y| - t+s).$$

Inside the cone it is

$$(4\pi\mu)^{-1} m J_1(m\mu) \qquad \text{where} \quad \mu^2 = (t-s)^2 - |x-y|^2 \quad .$$

This is used in conjunction with energy estimates. The standard one is that $E[u(t)]$ is constant. Integration of the standard identity over a characteristic cone K gives the boundedness of the integral of $u^2 + u^4$ over K. The key energy estimate is the same one used in Theorem 4.2. One of the other positive terms referred to there is the integral over space and time of $r^{-1}G(u)$. If $u(x,t)$ lives on $r < t+k$, we can replace r^{-1} by $(t+k)^{-1}$. Thus

$$\int^{\infty} f(t)\, dt/t < \infty \qquad \text{where} \quad f(t) = \int G(u)\, dx.$$

Fortunately t^{-1} is not integrable. So $\int_I f(t)\, dt$ is arbitrarily small on arbitrarily long time intervals I. The proof continues from this level like a jacking-up process. The succeeding steps are that: $u(x,t)$ is arbitrarily small on arbitrarily long time intervals; $u(x,t) \to 0$ uniformly as $t \to \infty$; $\sup_x |u(x,t)|^2$ is integrable; and finally $|u(x,t)| = O(t^{-3/2})$.

The most interesting step is the uniform convergence to zero. Let ε be a positive number. Let $T = T(\varepsilon)$ be sufficiently large. By the preceding step, $|u(x,t)| < \varepsilon$ on some time interval $[t^*-T,t^*]$. Let

$$t^{**} = \sup\{s \mid |u| < \varepsilon \quad \text{in} \quad [t^*-T,s]\} \quad .$$

If $t^{**} = \infty$, there is nothing to prove. Suppose $t^{**} < \infty$. Take a time t slightly later than t^{**}; namely $t^{**} \le t \le t^{**} + \delta$. Break up the right side of (45) into four parts. Since $t \ge t^{**} \ge T$ is large enough, $|u_0| < \varepsilon/4$. The integral over $[t^{**},t]$, the tip of the cone, is less than $\varepsilon/4$ if δ is chosen small enough. In the interval $[t-T,t^{**}]$, we have $|u(x,t)| < \varepsilon$. Since u appears in (45) to the third power and ε is small, we can arrange the integral over $[t-T,t^{**}]$ to be less than $\varepsilon/4$, no matter how large T is. The fourth part is over the large base of the cone $[0,t-T]$, where we do not know that u is small. However $t-s > T$ in that interval and so $R(x-y,t-s)$ is small in some sense. The kernel is actually constant on the hyperboloids μ = constant, but they bunch together very closely and contribute little to the integral. Altogether, we obtain $|u(x,t)| < 4(\varepsilon/4) = \varepsilon$, which contradicts the definition of t^{**} and completes the proof.

Define the space $\mathcal{F}$ as the limits of the free solutions with smooth data of compact support in the norm

$$\| v\|_F^2 = \sup \| v(t)\|_e^2 + \sup_t \sup_x |v(x,t)|^2 + \int_{-\infty}^{\infty} \sup_x |v(x,t)|^2\, dt$$

This space contains the free solutions in Y (cf. Part 2).

Theorem 4.4. S takes $\mathcal{F}$ into $\mathcal{F}$.

The proof is based in part on a variation of Theorem 2.2 with Y replaced by $\mathcal{F}$. Thus, if $u_-(x,t)$ is a free solution at $-\infty$, we construct $u(x,t)$ for $t \leq T$. Thinking of $t = T$ as the initial time, we use some estimates related to the preceding theorem to obtain the behavior as $t \to +\infty$.

If $m = 0$, similar results are valid. In that case the equation is invariant under the transformation

$$u(x) \to \frac{1}{x\cdot x}\, u\left(\frac{x}{x\cdot x}\right)$$

where x means space and time and $x\cdot x$ means the relativistic inner product. This transformation takes the standard energy identity into another one, from which one finds $\int u^4\, dx = O(t^{-2})$. Therefore the existence of asymptotic states turns out to be considerably easier than the $m > 0$ case. But it should be noted that an energy estimate is the key in both cases.

Properties of S.

(a) S maps $\mathcal{F}$ one-one onto $\mathcal{F}$.

(b) S is a diffeomorphism on $\mathcal{F}$.

(c) S is Lorentz-invariant.

(d) S commutes with the free group.

(e) S is odd.

(f) $\|Sf\|_e = \|f\|_e$.

(g) The skew form (32) is invariant under S.

(h) S is not a linear operator.

(a) To show that S is one-one, suppose $u_+ = v_+$. It is implicit in Theorem 2.2 that $u = v$, but here is an independent argument for the space $\mathcal{F}$. Subtracting equations (45) for u and for v,

$$u(t)-v(t) = \int_t^\infty R(t-s) * [u^3(s)-v^3(s)]\, ds$$

$$\sup_{t>T} \|u(t)-v(t)\|_e \leq \sup_{s>T} \|u(s)-v(s)\|_2 \int_T^\infty \sup_x (|u|+|v|)^2\, ds$$

If T is sufficiently large, the left side must vanish. Since the Cauchy problem has a unique solution (backwards), $u = v$ everywhere. So $u_- = v_-$, as remarked in Part 1. Since time is reversible in this problem, S is onto.

(b) S is once-differentiable means that

$$dS[u_-]v_- = \lim_{\varepsilon\to 0} \frac{1}{\varepsilon} [S(u_- + \varepsilon v_-) - S(u_-)]$$

exists in $\mathcal{F}$ and is a continuous linear function of v_-, where u_- and $v_- \in \mathcal{F}$. Let $u = W_-(u_-)$ as usual. Let w be the solution of the linearized equation

$$w_{tt} - \Delta w + m^2 w + 3u^2 w = 0$$

with asymptotic state v_- at $-\infty$. Then $dS[u_-]v_-$ is the asymptotic state of w at $+\infty$. A similar description holds for the higher differentials of S.

(c) and (g) are not difficult. For (d), which is a special case of (c), see equation (6). (e) is a consequence of the oddness of $u \to u^3$. (f) follows immediately from the constancy of $E[u(t)]$ and the fact that $\int G(u)dx \to 0$ as $t \to \pm\infty$. (h) was shown in Part 3.

Solutions which blow up in a finite time are studied by Glassey [1] and Levine [1], generalizing Keller [1]. Theorem 4.1 is from Strauss [2], generalizing Segal [1]. Existence and uniqueness under a growth condition for $n = 3$ is from Jörgens [1]. Waves which do not decay uniformly are considered by Roffman [1]. Theorem 4.2 is from Morawetz [2]. Outgoing free solutions for $m = 0$ are discussed by Lax and Phillips [1][2]. Exponential decay is from Lax and Phillips [1] and Morawetz [1] in the linear case and from Strauss [3] in the nonlinear case. The discussion of S for $m > 0$ is from Morawetz and Strauss [1][2], and for $m = 0$ from Strauss [1].

A question which remains open is: what is the maximal domain of S ? Can S be defined on a relativistic class of free solutions? Another problem is to determine the asymptotic behavior of (41) for any dimension other than three. It is not even known if $n = 1$. Such a generalization would probably be based on a new kind of *a priori* estimate.

References.

1. J. M. Chadam [1] Asymptotics for $\Box u = m^2 u + G(x,t,u,u_x,u_t)$, Ann. Scuola Norm. Sup. Pisa, 26, (1972) 33. [2] Asymptotic behavior of equations arising in quantum field theory, J. Applicable Anal., to be published.
2. D. G. Costa [1] The uniform behavior of solutions of linear hyperbolic systems for large times, Ph.D. thesis, Brown Univ., 1973.
3. C. S. Gardner, J. M. Greene, M. D. Kruskal and R. M. Miura [1] Method for solving the Korteweg-de Vries equation, Phys. Rev. Letters 19, 1095-7 (1967).
4. R. M. Glassey [1] Blow-up theorems for nonlinear wave equations, Math. Zeit., to be published. [2] On the asymptotic behavior of non-linear wave equations, Trans. A.M.S., to be published.
5. K. Jörgens [1] Das Anfangswertproblem im Grossen für eine Klasse nichtlinearer Wellengleichungen, Math. Zeit.77, 295-308 (1961).
6. J. B. Keller [1] On solutions of nonlinear wave equations, Comm. Pure Appl. Math. 10, 523-530 (1957).
7. P. D. Lax [1] Integrals of nonlinear equations of evolution and solitary waves, Comm. Pure Appl. Math 21, 467-490 (1968).
8. P. D. Lax and R. S. Phillips [1] *Scattering Theory* , Academic Press, New York, 1967. [2] Scattering theory for the acoustic equation in an even number of space dimensions, Indiana Math. J. 22, 101-134 (1972).
9. H. A. Levine [1] Instability and nonexistence of global solutions to nonlinear wave equations of the form $Pu_{tt} = -Au + F(u)$, to be published.
10. C. S. Morawetz [1] Exponential decay of solutions of the wave equation, Comm. Pure Appl. Math 19, 439-444 (1966). [2] Time decay for the nonlinear Klein-Gordon equation, Proc. Roy. Soc. A306, 291-6 (1968).
11. C. S. Morawetz and W. A. Strauss [1] Decay and scattering of solutions of a nonlinear relativistic wave equation, Comm. Pure Appl. Math. 25, 1-31 (1972). [2] On a nonlinear scattering operator, Comm. Pure Appl. Math. 26, 47-54 (1973). [3] to be published.
12. R. C. Newton [1] The inverse scattering problem, in this volume.
13. E. H. Roffman [1] Localized solutions of nonlinear wave equations, Bull. A.M.S. 76, 70-71 (1970).
14. I. E. Segal [1] The global Cauchy problem for a relativistic scalar field with power interaction, Bull Soc. Math. France 91, 129-135 (1963). [2] Quantization and dispersion for nonlinear relativistic equations, Proc. Conf. Math. Th. El. Particles, MIT Press, 79-108, 1966. [3] Dispersion for nonlinear relativistic equations II, Ann. Sci. Ecole Norm. Sup. (4) 1, 459-497 (1968).

15. W. A. Strauss [1] Decay and asymptotics for $\square u = F(u)$, J. Funct. Anal. 2, 409-457 (1968). [2] On weak solutions of semi-linear hyperbolic equations, Anais Acad. Brasil. Ciências 42, 645-651 (1970). [3] Decay of solutions of hyperbolic equations with localized nonlinear terms, Symposia Mathematica VII, Ist. Naz. Alta Mat., Rome, 339-355, 1971. [4] Dispersion of low-energy waves for two conservative equations, to be published.
16. W. von Wahl [1] Uber die klassische Lösbarkeit des Cauchy problems fur nichtlineare Wellengleichungen bei kleinen Anfangswerten und das asymptotische Verhalten der Lösungen, Math. Zeit. 114, 281-299 (1970).
17. V. E. Zakharov and A. B. Shabat [1] Exact theory of two-dimensional self-focusing and one-dimensional self-modulation of waves in nonlinear media, J. Exp. Th. Phys. 61, 118-134 (1971).

SCATTERING IN CLASSICAL MECHANICS

W. Hunziker

Eidgenössische Technische Hochschule

Zürich

1. Introduction. In 1668 - about 20 years before Newtons Principia appeared - John Wallis, Christopher Wren and Christian Huyghens independently formulated the "laws of impact" in response to a request by the Royal Society.

From the middle of the 19th century up to now, classical collisions have been studied extensively as the dynamical basis of the kinetic theory of heat. It was Rudolf Clausius [1] who coined the term "Wirkungssphäre "(sphere of action) from which Wirkungsquerschnitt (cross section) is derived. The virial theorem - which has recently proved useful in the quantum theory of scattering [14] - is also due to Clausius [2]. Maxwell [3] and Boltzmann [4] used general properties of collisions like reciprocity and conservation of phase volumina to derive the statistics of thermal equilibrium and to prove approach to equilibrium in the case of dilute gases. The subsequent theory of transport phenomena then related the temperature - dependence of transport coefficients to the differential cross-section for binary collisions, and a kind of inverse scattering problem had to be solved to find the intermolecular potentials [5]. All attempts to extend this theory to dense gases are still beset with the difficulties of multiple collisions between 3 and more particles [6].

However, the work on classical scattering [7] - [11] which I describe in these lectures, has its roots in the quantum theory of scattering, more precisely in the nonrelativistic N-body problem. In particular, asymptotic completeness is still an intriguing question which is incompletely solved. A study of the problem in

J. A. LaVita and J.-P. Marchand (eds.), Scattering Theory in Mathematical Physics, 79–96.

classical mechanics should reveal the dynamical origin of this universally expected property and may also provide our intuition with some classical pictures which we could try to translate into quantum mechanics. An example of this is the theory of scattering by long-range potentials.

Apart from many aspects common to the classical and the quantum case, each field has of course its own specific problems. In the classical case, we are confronted with the difficult question of stability of bound states, which eventually prevents us to go beyond the case of finite-range forces. In this respect (as in many others), quantum mechanics proves to be a blessing!

It is a pleasure to thank professors J. P. Marchand and J. La Vita for their kind invitation to Denver.

2. Scattering by a potential of finite range. Consider a masspoint of mass 1 in R^n under the influence of a potential $V \in C_0^2(R^n)$. Let $z = (x,p)$ be its position and momentum, $z \in R^{2n} = T$ = phase space. The equation of motion

$$\dot{z} = (\dot{x},\dot{p}) = (p,-\text{grad } V)$$

has a unique global solution $z_t, -\infty<t<+\infty$, for any $z_0 \in T$, and the maps

$$S^t : z_0 \to z_t$$

form a 1-parameter group of canonical transformations. Let S_0^t be the corresponding group in the case $V = 0$;

$$S_0^t : (x,p) \to (x + pt, p),$$

and let

$$T' = \{(x,p) : p \neq 0\}.$$

Theorem 1. The wave operators

$$\omega^{\pm} = \lim_{t\to\pm\infty} S^{-t}S_0^t$$

exist on T'. They are canonical from T' into T and satisfy

$$S^t\omega^{\pm} = \omega^{\pm}S^t.$$

Hence the ranges $r^{\pm}$ of $\omega^{\pm}$ are open, S^t - invariant subsets of T and the parts of S^t on $r^{\pm}$ are canonically equivalent to S_0^t on T'.

The proof is simple, since, on any compact $\subset T'$, the canonical map $S^{-t}S_0^t$ is independent of t for $\pm t$ sufficiently large.

Definitions.

$r = r^+ \cap r^-$ (scattering state)

$b_n = \{ z : |S^t z| \leq n \quad \text{for} \quad 0 \leq \pm t < \infty\}$

$$b^\pm = \bigcup_{n=1,2,\dots} b_n^\pm$$

$b = b^+ \cap b^-$

Since $S^t z$ is continuous in z, $b_n^\pm$ are compact, hence $b^\pm$ are measurable (and S^t - invariant)

Lemma 1.

$b^\pm \cap r^\pm = \phi; \quad b^\pm \cup r^\pm = T$

Proof. For any orbit in r^+ we have $|x_t| \to \infty$, hence $r^+ \cap b^+ = \phi$. Conversely, $z_0 \varepsilon b^+$ implies $|z_t| \to \infty$ for $t \to \infty$. But p_t is bounded by energy - conservation since V is bounded below, therefore $|x_t| \to \infty$, i.e. the particle moves like a free particle with nonzero momentum for large t. Therefore $z_0 \varepsilon r^+$.

Let A and B be subsets of T. Then we write $A \approx B$ for: A = B up to a set of measure zero.

Theorem 2.

$r \cap b = \phi$

$r \cup b \approx T$ (asymptotic completeness)

Figure 1.

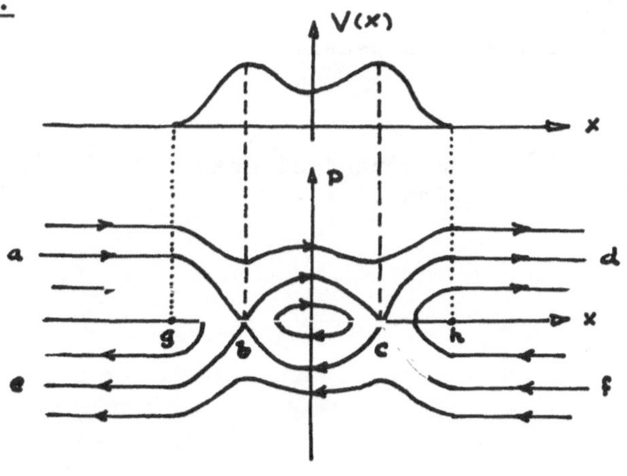

(text on following page.)

Example of a flow S^t in $T = R^2$ for $n = 1$. We have

b = closed set bounded by the loop b c b plus the x-axis without the open intervals (g,b) (c,h)

b^+ = b plus the orbits (a,b) and (f,c)

b^- = b plus the orbits (b,c) and (c,d)

$r^\pm$ = complements of $b^\pm$.

<u>Proof.</u> The first statement is obvious from Lemma 1. To prove the second, it suffices to show that $b^+ \approx b$ (By time reflection, $b^- \approx b \approx b^+$ hence, by taking complements, $r^+ \approx r^- \approx r$, and $T = r^+ \cup b^+ \approx r \cup b$). Observe that

$$b^+ - b \subset \bigcup_{n=1,2\ldots} (b_n^+ - (b_n^+ \cap b_n^-)),$$

so we only need to show that $\mu(b_n^+) = \mu(b_n^+ \cap b_n^-)$, ($\mu$ = Lebesgue measure) Let C_n be the ball of radius n in T. Then by definition,

$$b_n^+ = \bigcap_{0 \leq \tau < \infty} S^{-\tau} C_n \quad ; \qquad S^t b_n^+ = \bigcap_{-t \leq \tau < \infty} S^{-\tau} C_n \quad .$$

The set $S^t b_n^+$ is therefore shrinking for increasing t, but has constant measure, since S^t is canonical. Therefore

$$\mu(b_n^+) = \lim_{t=1,2\ldots} \mu(S^t b_n^+) = \mu\left(\bigcap_{t=1,2\ldots} S^t b_n^+ \right)$$

$$= \mu\left(\bigcap_{-\infty < t < +\infty} S^{-t} C_n \right) = \mu(b_n^+ \cap b_n^-).$$

The S-matrix is constructed as usual:

$$S = (\omega^+)^{-1} \omega^- .$$

It is defined on $(\omega^-)^{-1} r$ and canonical onto $(\omega^+)^{-1} r$, both sets being open $\approx T$ and S_0^t - invariant.

Figure 2.

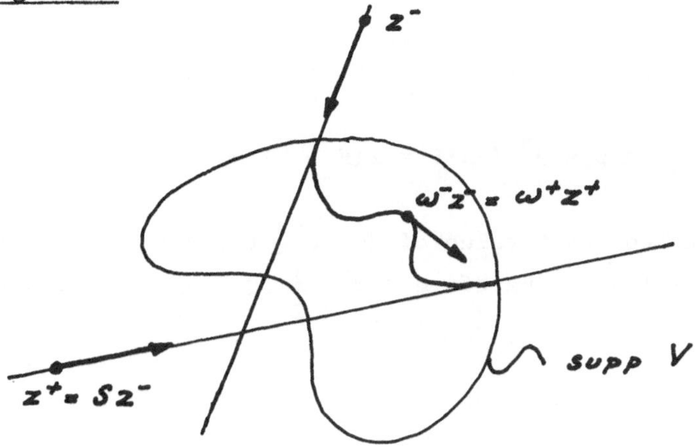

This picture shows the meaning of the map S in configuration space R^n, where a state $z = (x,p)$ is represented by an arrow p at the point x.

Notes. The results of this section are those of Cook [7] who used the language of Hilbertspace. The proof of Theorem 2 is found in Siegel [13] and obviously applies to much more general situations - the essential ingredients being the group property of S^t and the S^t - invariance of the measure.

3. Potentials of non-finite range. First, we make a typical assumption which settles all questions of existence and uniqueness at finite times:*

A_0: V(x) is continuous from R^n to $[b,+\infty]$ with $-\infty < b$, and for any $W < \infty$, V(x) has continuous, bounded derivatives up to order 2 on $\{x : V(x) < W\}$

Then $T = \{z \in R^{2n} : V(x) < \infty\}$ is the phase space and dynamics is again described by a 1-parameter canonical group S^t. In particular we still have $b^+ \approx b^- \approx b$. Our later assumptions will always imply

$$\lim_{|x|\to\infty} V(x) = \lim_{|x|\to\infty} |x||F(x)| = 0,$$

where $F = -\text{grad } V$ is the force. Let $H(z) = \frac{1}{2}p^2 + V(x)$ be the Hamiltonian. Then $H(z) < 0$ implies $z \in b$, hence Theorem 3 still holds for $r = r^+ \cap r^-$, if we define

*Assumptions will be listed by A_0, A_1 ect.

$$r^{\pm} = \{z \mid z \notin b^{\pm},\ H(z) > 0\},$$

which is again S^t - invariant. To find the asymptotic behavior of an orbit in r^+ we apply the virial identity:

$$\frac{d^2}{dt^2}\ \frac{1}{2}\,x^2 = \frac{d}{dt}\,x\cdot p = p^2 + x\cdot F = 2(E - V) + x\cdot F \quad ,$$

where $E > 0$ is the constant value of H along the orbit. As we have seen, $|x_t| \to \infty$ for $t \to \infty$, therefore $\frac{d^2}{dt^2}\frac{1}{2}x^2 > \frac{E}{2}$ for sufficiently large t, so that

$$|x_t| > \varepsilon\ t$$

for some $\varepsilon > 0$ and large t. We now distinguish 2 cases:

$|F(x)| < a|x|^{-n}$ $\qquad$ A_1: $1 < n \leq 2$ (long range)

for $|x| > r$ $\qquad$ A_2: $2 < n$ (short range)

From $|\dot{p}| = |F(x_t)| < a(\varepsilon t)^{-n}$, we see that in both cases

$$q = \lim_{t\to\infty} p_t \neq 0$$

exists, in fact $|q - p_t| = 0(t^{-n+1})$.

Short range case. Then

$$y = \lim_{t\to\infty} (x_t - qt)$$

exists, and the map $(x_0,p_0) \to (y,q)$ will of course be the inverse of the wave operator ω^+ (to be constructed later). The name "scattering states" for the elements in r^+ is thereby justified.

Longe range case. To discuss the asymptotic behaviour of x_t, we assume:

A_3: $|F(x) - F(y)| \leq a|x|^{-(n+1)}|x - y|$ $\qquad$ for $|x|,|y| > r$

A_4: $|F_{\perp}(x)| \leq a|x|^{-(n+1)}$ $\qquad$ for $|x| > r$

where $F_{\perp}(x)$ is the component of $F(x)$ $\perp$ to x. A_4 is convenient but not indispensable [11] and implies that the angular momentum (whose time-derivative is $x \wedge F$) has a limit as $t \to \infty$. Let $F_{\parallel}(x)$ be the radial component of $F(x)$.

Theorem 3. ($A_{0,1,3,4}$) Let q be the asymptotic momentum of the orbit z_t in r^+, and $g_t(q)$ a solution to

$$\ddot{g} = \frac{1}{|q|} F_{\parallel}(qg); \quad \dot{g}_t \to 1 \text{ for } t \to \infty.$$

Then there exists

$$y = \lim_{t\to\infty} (x_t - qg_t(q)).$$

Proof. Let $y_t = x_t - qg_t$, then

$$\ddot{y}_t = F(y_t + qg_t) - F(qg_t) + F_{\perp}(qg_t).$$

Since $|x_t|$ and $g_t \to \infty$ for $t \to \infty$ we can apply $A_{3,4}$ for large t and find

$$|\ddot{y}_t| \leq \frac{c_1}{t^{n+1}} |y_t| + \frac{c_2}{t^{n+1}} \tag{1}$$

Suppose $|y| \leq c_3 t^m$ $(m > 0)$ for large t, which certainly holds for $n = 1$. Then (1) implies

$$|\ddot{y}_t| \leq c_4 t^{m-(n+1)} \to |y_t| < c_5 t^{m-(n-1)} \tag{2}$$

as long as $m - (n - 1) > 0$. Now begin with $m = 1$ and choose n such that $(n - 1)^{-1} \neq$ integer. After a finite number of steps we arrive at $|y_t| \leq ct^m$ with $0 < m < n-1$ and then (2) shows that $|\ddot{y}_t| = O(t^{-\alpha})$, $\alpha > 2$. By construction, $\dot{y}_t = (\dot{x}_t - q) + q(1 - \dot{g}_t) \to 0$ as $t \to \infty$, hence $|\dot{y}_t| = O(t^{-\alpha+1})$ and y_t converges.

The choice of $g_t(q)$. Existence of g_t is clear: it can in fact be computed by quadrature from

$$\dot{g}_t^2 = 1 - \frac{2}{|q|^2} V(qg) \tag{3}$$

with g_0 sufficiently large and $\dot{g}_0 > 0$. Also, it is clear that if $\bar{g}_t$ is equivalent to g_t in the sense that for all $q \neq 0$

$$\lim_{t\to\infty} (g_t(q) - \bar{g}_t(q))$$

exists, then Theorem 3 also holds with $\bar{g}_t$ instead of g_t, of course with a different limit y. Hence we can replace V in (3) by any potential $\bar{V}$ such that $\bar{V} - V$ has short range. Since A_4 states that the non-central part of F has short range, we can choose $\bar{V}$ to be spherically symmetric and as smooth as V. Prescribing $g_0(q)$ as a smooth function of $|q|$ with sufficiently large values, we then obtain an asymptotic flow of C^1 - maps of T' into T', defined for $t > 0$ by:

$$S^t_{as} : (x,p) \to (x + pg_t(|p|),\ p),$$

which satisfies the canonical equations

$$\dot{x} = p\dot{g}_t(|p|) = \frac{\partial H_{as}}{\partial p}$$

$$-\dot{p} = 0 = \frac{\partial H_{as}}{\partial x}$$

with $H_{as}(p;t) = \int^{|p|}{}_{as}\ s\ \dot{g}_t(s)$. Therefore, S^t_{as} is canonical but not a semigroup, since H_{as} is time-dependent. Of course, it is also largely non-unique.

Existence of the wave operators. We only consider the long range case. We have seen that the orbits in r^+ have well defined asymptotic data $(y,q) \in T'$, where y is defined relative to a given choice of g_t. Now we prove the converse: given $(y,q) \in T'$ relative to a certain g_t, there exists a unique solution of

$$\ddot{x}_t = F(x)$$

with

$$\lim_{t\to\infty} |x_t - y - qg_t(q)| = 0$$

Again we have a free choice among equivalent g_t, and we revert to the definition given in Theorem 3. The problem is now posed as an integral equation for $w_t = x_t - y - qg_t(q)$:

$$w_t = \int_t^\infty dt_1 \int_{t_1}^\infty dt_2\ [F(w_t + y + qg_t) - F_{\parallel}(qg_t)] \equiv \mathcal{F}(w)_t.$$

Write the integrand as $F(w_t + y + qg_t) - F(qg_t) + F_{\perp}(qg_t)$. Consider an interval $T \leq t < \infty$ and let w be in the unit ball of the space of continuous, bounded, R^n - valued functions on $T \leq t < \infty$ (sup. norm). Then since $q \neq 0$, $A_{3,4}$ apply for T sufficiently large and we find

$$\|\mathcal{F}(w)\| \leq \text{const.}\ T^{-n+1}$$

and, for $w^{(1)}, w^{(2)}$ in the unit ball;

$$\|\mathcal{F}(w^{(1)}) - \mathcal{F}(w^{(2)})\| \leq \text{const.}\ \|w^{(1)} - w^{(2)}\|\ T^{-n+1}$$

Hence, for T sufficiently large, $\mathcal{F}$ is a contraction of the unit ball and has a unique fixed point w. $x_t = w_t + y + qg_t$ is then the unique solution for $T \leq t < \infty$ and extends by A_0 to a global solution, which in turn defines the wave operator:

$$\omega^+ : (y,q) \in T' \to (x_0, p_0 = \dot{x}_0) \in T.$$

ω^+ is one-to one (by Theorem 3) and can be shown to be continuous and measure-preserving. It is clear that some additional smoothness of F(x) at ∞ will make ω^+ differentiable and therefore canoncal.

Intertwining relation. Because S^t_{as} is not a group, ω^+ cannot intertwine S^t and S^t_{as}. However, since $\ddot{g}_t \to 1$ we have $g_{t+a} - g_t - a \to 0$ as $t \to \infty$. Let $(y,q) = z' \varepsilon T'$ and $z = \omega^+ z'$. Then the orbit $S^t(S^a z)$ has asymptotic behavior

$$x_t \sim y + qg_{t+a} \sim y + qa + qg_t; \quad p_t \sim q$$

and therefore the asymptotic data $(y + qa, q) = S^a_0(y,q)$, with respect to $\ddot{g}_t$. We see that

$$S^t\omega^\pm = \omega^\pm S^t_0 \ ,$$

regardless of the fact that S^t_0 does not correctly describe the asymptotic behavior of x_t. This concludes the demonstration of Theorem 1 in the long-range case.

Notes. The short range case is treated in [9][10]. Simon [9] gives examples for non-uniqueness of $(y,q) \to (x_0,p_0)$ in the absence of a Lipschitz- condition at ∞. Herbst [11] treats the long range case without the simplifying assumption A_4 in the spirit of [12], where the quantum case is discussed.

4. The N body problem. Consider N particles 1 ... N in R^3 (the dimension is irrelevant). Let m_i, x_i, p_i be the mass, position and momentum of particle i, and $z = (x_1 p_1 \ \dots \ x_N p_N) \ \varepsilon \ R^{6N}$ the canonical coordinates of the system. The Hamiltonian is

$$H(z) = \sum_{i=1}^{N} \frac{p_i{}^2}{2m_i} + \sum_{i<k} V_{ik} \ (x_i - x_k)$$

Assume A_0 for each V_{ik} and let $T = \{z \ \varepsilon \ R^{6N} : H(z) < \infty\}$. Then the dynamics is described by a 1-parameter canonical group S^t. For convenience only, we assume the potentials to be bounded so that $T = R^{6N}$.

Hilbert space. To exhibit the close analogy to the quantum case (and also to loose automatically sets of measure zero), we take the states of the system to be elements of $\mathcal{H} = L^2(T)$. The dynamics is then described by a strongly continuous 1-parameter unitary group acting on the states $\psi \ \varepsilon \ \mathcal{H}$:

$$U^t = e^{Lt} : \psi(z) \to \psi(S^{-t}z) ,$$

where formally L is the Liouville-operator

$$L = - \sum_{i=1}^{N} \frac{p_i}{m_i} \frac{\partial}{\partial x_i} + \sum_{i<k} \text{grad } V_{ik} \left(\frac{\partial}{\partial p_i} - \frac{\partial}{\partial p_k} \right) .$$

L is well- defined on $C_0^1(T)$, and, since this domain is U^t - invariant the generator of U^t is in fact the closure of its restriction to $C_0^1(T)$.

Center - of - mass frame. Instead of $(x_1 \dots x_N)$ use (X,ξ) to describe the configurations, where $X \varepsilon R^3$ is the position of the center of mass and where $\xi \ \varepsilon \ R^{3N-3}$ fixes the relative configuration. Let (P,π) be the corresponding momenta and $Z = (X,P)$, $\zeta = (\xi,\pi)$. Representing z by the pair (Z,ζ), we have

$$T = T_{CM} \times T_{int}$$

where $T_{CM} = R^6$ is the phase space of one "particle" (the center of mass) and $T_{int} = R^{6N-6}$ the phase space in which the relative motion is described. Then, if M is the total mass,

$$H(z) = \frac{P^2}{2M} + h(\zeta)$$

and the flow S^t decomposes into parts S^t_{CM}, S^t_{int} acting independently on T_{CM} and T_{int}, the part in T_{CM} being of course the same as for a free particle of mass M in R^3. In the Hilbert-space language :

$$\mathcal{H} = \mathcal{H}_{CM} \otimes \mathcal{H}_{int} ,$$

where these spaces are the L^2-spaces over T,T_{CM} and T_{int}, respectively, and

$$U^t = U^t_{CM} \otimes U^t_{int} ,$$

where the factors are generated by Liouville-operators derived from

$P^2/2M$ and $h(\zeta)$, respectively.

Bound states. In T_{int}, we construct $b_n^{\pm}, b^{\pm}, b$ with respect to S^t_{int} as in section 2, and define the subspace of bound states (in $\mathcal{H}_{int}$) by

$$\mathcal{H}_B = L^2(b).$$

Furthermore, we call $\psi \in \mathcal{H}_B$ a *compact bound state* if

$$\psi \in L^2(b_n^+ \cap b_n^-)$$

for some n, which means that for all t,

$$(U^t_{int}\ \psi)(x) = 0 \qquad \text{if } |x| > n.$$

Since $b = \bigcup_n (b_n^+ \cap b_n^-)$, the compact bound states are *dense* in $\mathcal{H}_B$.

Scattering states. Let $\alpha = (F_1 \ldots F_n)$ be a partition of $(1 \ldots N)$ into subsets ("fragments"). Each fragment F_k can be studied as a system in its own right with the appropriate Hamiltonian H_k. If we introduce center of mass coordinates (Z_k, ζ_k) for each fragment, then $\mathcal{H}$ factors into

$$\mathcal{H} = L^2(R^{6n}) \times \mathcal{H}^1_{int} \otimes \ldots \otimes \mathcal{H}^n_{int} \quad ,$$

spanned by the products

$$\psi = \Phi(Z_1 \ldots Z_n)\phi_1(\zeta_1) \ldots \phi_n(\xi_n) \quad ,$$

where $\phi_k = 1$ if F_k is a single particle. The term "channel" is reserved for those α, for which each F_k is either a single particle or $\mathcal{H}^k_B \neq \{0\}$. For each channel α, and with respect to (4), we define

$$D_\alpha = L^2(R^{6n}) \otimes \mathcal{H}^1_B \ldots \otimes \mathcal{H}^n_B \ ,$$

$$H_\alpha = \sum_k H_k = \sum_k \frac{P_k^{\ 2}}{2M_k} + h_k(\zeta_k) \quad ,$$

where $\mathcal{H}^k_B = C$ and $h_k = 0$ if F_k is a single particle. Finally, let L_α be the Liouville-operator derived from H_α. The the group $e^{L_\alpha t}$ describes the dynamics of noninteracting fragments and leaves D_α invariant.

Theorem 4. Let each V_{ik} be of *finite range*. Then the wave operators

$$\Omega^{\pm}_\alpha = s - \lim_{t \to \pm\infty} e^{-Lt} e^{L_\alpha t}$$

exist on D_α. They are isometric from D_α into $\mathcal{H}$ and satisfy

$$e^{Lt}\Omega^{\pm}_\alpha = \Omega^{\pm}_\alpha e^{L_\alpha t}$$

i.e. the ranges $R_\alpha^\pm$ of $\Omega_\alpha^\pm$ are invariant under e^{Lt} and the parts of e^{Lt} in $R_\alpha^\pm$ are unitarily equivalent to $e^{L_\alpha t}$ on D_α

Proof. The proof is purely geometrical, working only with supports. It suffices to prove convergence on the dense subset of D spanned by products

$$\psi_\alpha = F(X_1 \dots X_n) G(P_1 \dots P_n) \phi_1(\zeta_1) \dots \phi_n(\zeta_n)$$

with F of compact support, G of compact support in the open set

$$\frac{P_i}{M_i} \neq \frac{P_k}{M_k} \qquad (i = k)$$

and $\phi_1 \dots \phi_n$ = compact bound states. The picture of $e^{L_\alpha t}\psi_\alpha$ is very simple. The relative coordinates in each F_k remain bounded. The centers-of-mass X_k start in some compact $\subset R^3$ and move like free particles with different velocities. Therefore, after a finite time T, the distance between any two particles in *different* fragments becomes larger than the range the forces. Thereafter,

$$e^{L(t-T)} e^{L_\alpha T}\psi_\alpha = e^{L_\alpha t}\psi_\alpha$$

for all t > T, i.e.

$$e^{-Lt} e^{L_\alpha t}\psi_\alpha = e^{-LT} e^{L_\alpha T}\psi_\alpha$$

is independent of t.

Theorem 5. If the wave operators exist, then

$$R_\alpha^\pm \perp R_\beta^\pm \qquad (\alpha \neq \beta) .$$

Proof. For $\psi_{\alpha,\beta}$ in dense sets $\subset D_{\alpha,\beta}$ we have to show

$$(\Omega_\alpha^\pm\psi_\alpha, \Omega_\beta^\pm\psi_\beta) = \lim_{t\to\pm\infty} (e^{L_\alpha t}\psi_\alpha, e^{L_\beta t}\psi_\beta) = 0 .$$

Take ψ_α, ψ_β as in the proof of Theorem 4. Since $\alpha \neq \beta$, there exists a pair (i,k) of particles which are in the same fragment of one channel (say α) and in different fragments of β. Then, in the support of $e^{L_\alpha t}\psi_\alpha$ we have $|x_i - x_k|$ bounded for all t, while in the support of $e^{L_\beta t}\psi_\beta$ $|x_i - x_k|$ diverges linearly for $t \to \infty$. Hence the two supports do not intersect for t sufficiently large.

The S-operator can now be constructed as usual; see [15] or [8] for an alternative definition. Their unitarity follows from

Theorem 6. (asymptotic completeness). Let each V_{ik} be of finite range. Then

$$\bigoplus_{\alpha} R_{\alpha}^{\pm} = \mathcal{H} \quad .$$

Observe that the bound states did not get lost: if $\mathcal{H}_B \neq \{0\}$ then the trival partition α_0 into one fragment also occurs, and since $H_{\alpha_0} = H$, we have $D_{\alpha_0} = \mathcal{H}_B = R_{\alpha_0}^{\pm}$. The proof of Theorem 5 is given in the next section.

Open Questions. Of course one expects that

$$(\Omega_{\alpha}^{\pm}\psi)(z) = \psi[(\omega_{\alpha}^{\pm})^{-1}z]$$

with $\omega_{\alpha}^{\pm}$ canonical from the phase space T_{α} (carrying D_{α}) into T. This is *not* obvious, since

$$T_{\alpha} = R^{6n} \times B_1 \times \ldots \times B_n \quad ,$$

and in general we only know that the B_k are measurable, but not that $(B_k - \text{int } B_k)$ has a measure zero. (Which is a stability condition for bound states!)

If the answer to this question is affirmative, we can work with *smooth* compact bound states and try to mimic the quantum mechanical convergence proof for $\Omega_{\alpha}^{\pm}$ in the case of short range forces, namely to show that

$$\int_0^{\pm\infty} dt \, \|(L - L_{\alpha})e^{L_{\alpha}t}\psi_{\alpha}\| < \infty$$

for a suitable dense set of $\psi_{\alpha} \in D_{\alpha}$. Estimating this norm, we encounter terms like

$$\left\|\frac{\partial}{\partial\pi_k} e^{L_{int}^k} \phi_k\right\| \tag{4}$$

where ϕ_k is a smooth bound state of F_k. Now $\exp(L_{int}^k)$ preserves smoothness, but the derivatives may blow up exponentially. For the proof to work as in the quantum case, one needs an upper bound of order $|t|$ for (4).

These questions do not appear or can be solved [8] for fragments consisting of less than 3 particles.

5. Proof of asymptotic completeness. We consider only the case $t \to \infty$. The proof is by induction in N : we assume that the theorem holds for all subsystems of less than N particles. The idea is simple: we show that, if the system is not bound, it breaks up into two clusters C_1, C_2 which are and *remain* separated by a distance a > range of the forces. By the induction assumption , these clusters decay into bound fragments.

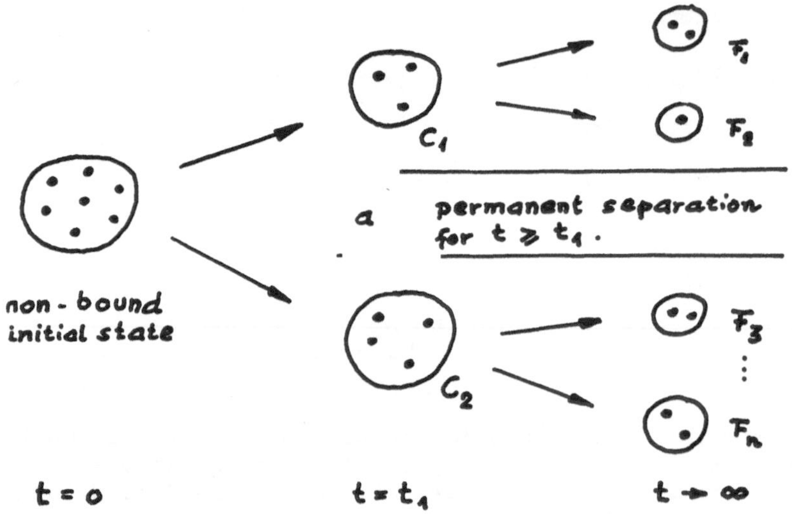

(To see that *permanent* break-up can take a long time, consider the 1-dim case of two heavy elastic spheres with a light one bouncing back and forth inbetween!)

To formalize this, let $D = (C_1, C_2)$ be any partition of (1 ... N) into 2 clusters. Define the function

$$\rho_D(z) = \min_{\substack{i \in C_1 \\ k \in C_2}} |x_i - x_k|$$

on T and let

$$A_D = \{z \mid \rho_D(S^t z) \geq a \text{ for } 0 \leq t < \infty\}.$$

Since ρ is continuous, A_D is closed, and

$$\sigma_D = \{z \mid \rho_D(S^t z) \geq a \text{ for } t \to \infty\} = \bigcup_{n=1,2,\dots} S^{-n} A_D$$

is measurable. Finally, let

$$\sigma = \bigcup_D \sigma_D$$

Lemma 2.

$$L^2(\sigma) \subset \sum_{\alpha \neq \alpha_0} R_\alpha^+ \equiv R,$$

where α_0 is the trivial channel : $R_{\alpha_0}^{\pm} = \mathcal{H}_B$.

Proof. We need only prove that $L^2(\sigma_D) \subset R$ or $L^2(A_D) \subset R$ since R is U^t - invariant. But for $\psi \in L^2(A_D)$,

$$e^{Lt}\psi = e^{L_D t}\psi \qquad (t > 0)$$

where L_D is the Liouville-operator for *noninteracting* clusters C_1, C_2. By induction hypothesis

$$e^{L_D t}\psi \to \sum_{\alpha \neq \alpha_0} e^{L_\alpha t}\psi_\alpha \qquad (\psi_\alpha \in D_\alpha)$$

for $t \to \infty$, therefore $\psi \in R$.

Lemma 3. Complement of $b^+ \subset \sigma$.

Proof. The proof extends to the end of this section. The Lemma proves asymptotic completeness because it shows that

$$\mathcal{H}_B^\perp = L^2(b^+)^\perp \subset L^2(\sigma) \subset R,$$

and since we know from Theorem 5 that $\mathcal{H}_B \subset R^\perp$, we have $\mathcal{H}_B = R^\perp$, which is another way of writing Theorem 6.

We consider a fixed $z \in b^+$ and its trajectory $S^t z = z_t$. We have to show that $z_{t_1} \in A_D$ for some t_1. Let $r_D(t)$ be the distance between the centers of mass of C_1 and C_2, and

$$r(t) = \max_D r_D(t)$$

Observe that r is also an upper bound on the distance of any one particle from the center of mass of all the others. Therefore,

$$\limsup_{t\to\infty} r(t) = \infty$$

Note that r(t) is continuous but not always differentiable. We now need some geometric estimates for N-point-configurations in R^3:

Lemma 4. There exists a constant λ, $0 < \lambda < 1$, depending only on the masses, such that $r_D > \lambda r$ implies $\rho_D > (1 - \lambda)r$. The proof is given at the end of this section.

Lemma 5. The function $r(t)$ is convex in any t-interval where $(1 - \lambda)r(t) > a$.

Proof. The proof is based on the observation that the distance between freely moving masspoints is a convex function of time. In particular, this applies to $r_D(t)$ as long as $\rho_D(t) > a$.

Let I be an open t-interval where $(1 - \lambda)r(t) > a$. For any $t_0 \in I$, choose D maximal in the sense that $r_D(t_0) = r(t_0)$. Then $r_D(t_0) - \lambda r(t_0) > a$, and by continuity $r_D(t) - \lambda r(t) > 0$ for $|t - t_0| < \varepsilon$. By Lemma t, $\rho_D(t) > a$ and $r_D(t)$ is convex for $|t - t_0| < \varepsilon$. This shows that for each $t_0 \in I$, $r(t)$ is locally bounded below by a convex function $r_D(t)$ taking the same value as $r(t)$ at t_0. Hence $r(t)$ is convex in I.

We now complete the proof of Lemma 3. Pick a t-interval in which $r(t)$ is strictly increasing and $(1 - \lambda)r(t) > a$. This exists since $r(t)$ is continuous and $\limsup r(t) = \infty$ $(t \to \infty)$. Then it follows from Lemma 5 that $r(t)$ remains convex, strictly increasing and $(1 - \lambda)r(t) > a$ forever. Shifting the time scale if necessary, we may assume these properties of r for all $t > 0$. For $t > 0$, let $\dot{r}(t)$ be the left derivative of $r(t)$. $\dot{r}(t)$ is non-decreasing and bounded by energy conservation. Therefore it exists

$$\lim_{t\to\infty} \dot{r}(t) = \dot{r}(\infty) \quad .$$

Now pick t_1 so large that $\dot{r}(t_1) > \lambda\dot{r}(\infty)$. Let D be maximal at time t_1: $r_D(t_1) = r(t_1)$. Then

$$\dot{r}(t_1) \leq \dot{r}_D(t_1) \quad .$$

We claim that, for this D, $\rho_D(t) > a$ for all $t \geq t_1$, which means that $z_{t_1} \in A_D$ and proves Lemma 3.

Assume the converse: $\rho_D(t) \geq a$ for some $t > t_1$. Then, by Lemma 4, $r_D(t) \leq \lambda r(t)$. Let t_2 be the first zero of $r_D(t) - \lambda r(t)$ to the right of t_1. Then

$$\dot{r}_D(t_2) \leq \lambda\dot{r}(t_2) \quad .$$

On the other hand, for $t_1 \leq t \leq t_2$ we have $r_D(t) - \lambda r(t) \geq 0$, hence $\rho_D(t) \geq a$ (Lemma 4) and therefore $r_D(t)$ is convex in this interval. This gives

$$\dot{r}_D(t_1) \leq \dot{r}_D(t_2) \quad .$$

Combining the 3 estimates for derivatives, we obtain

$$\dot{r}(t_1) \leq \dot{r}_D(t_1) \leq \dot{r}_D(t_2) \leq \lambda\dot{r}(t_2) \leq \lambda\dot{r}(\infty)$$

which contradicts the definition of t_1.

Proof of Lemma 4. Consider a fixed configuration $(x_1 \ldots x_N)$ and a partition $D = (C_1, C_2)$. Let X_i be the center of mass of C_i and choose a ξ - axis with origin X_1 such that its positive part goes through X_2. The ξ-coordinates of X_1, X_2 are 0, r_D respectively.

Now choose a particle of mass $m_1 \varepsilon C_1$ and obtain $D' = (C_1', C_2')$ from D by assigning this particle to the second cluster. Then the centers of mass X_1', X_2' of C_1', C_2' have the ξ-coordinates

$$\Xi_1' = -\xi_1 m_1 (M_1 - m_1)^{-1}$$

$$\Xi_2' = (r_D M_2 + \xi_1 m_2)(M_2 + m_1)^{-1} \quad ,$$

where M_i=mass of C_i. Now observe that the difference $\Xi_2' - \Xi_1'$ must not exceed r. This gives

$$\zeta_1 \leq (M_1 - m_1)(M_2 + m_2)(m_1 M)^{-1}[r - M_2(M_2 + m_1)^{-1} r_D] \ ,$$

where $M = M_1 + M_2$. Similarly, for a particle of mass m_2 in C_2:

$$\zeta_2 \geq r_D - (M_2 - m_2)(M_1 + m_2)(m_2 M)^{-1}[r - M_1(M_1 + m_2)^{-1} r_D]$$

In particular, this holds for the two particles with distance ρ_D. For these we have

$$\begin{aligned}\rho_D &\geq \xi_2 - \xi_1 \\ &\geq r_D\{1 + (m_1 m_2 M)^{-1}[m_1 M_1 (M_2 - m_2) + m_2 M_2 (M_1 - m_1)\} \\ &\quad -r(m_1 m_2 M)^{-1}[m_1(M_1 + m_2)(M_2 - m_2) + m_2(M_2 + m_1)(M_1 - m_1)\} \\ &= (1 + \alpha)r_D - \beta r.\end{aligned}$$

Observe that

$$1 + \alpha - \beta = \frac{m_1 + m_2}{M} > 0.$$

Now let $0 < \lambda < 1$, and $r_D \geq \lambda r$. Then we have

$$\rho_D \geq [\lambda(1 + \alpha) - \beta]r \geq (1 - \lambda)r$$

provided that also

$$\lambda(2 + \alpha) \geq 1 + \beta .$$

The smallest possible value of λ is therefore

$$\lambda = \max \frac{1 + \alpha}{2 + \beta} < 1$$

where the maximum is to be taken over all possible values of M_1, M_2, m_1, m_2. The case N = 2 ($\alpha=\beta=0, \rho_D=r_D=r$) shows that this construction is optimal.

References.

1. R. Clausius, Ann. der Physik 105, (1858) 239 (on the mean free path).
2. R. Clausius, Sitz. Ber. der Niederrhein. Gesellschaft, Bonn. (1870) 114.
3. J. C. Maxwell, Phil. Mag. 19, (1860) 19 and 20, (1860) 21.
4. L. Boltzmann, Wien. Ber. 66, (1872) 275.
5. See e.g. J. O. Hirschfelder, C. F. Curtiss, R. B. Bird *The Molecular Theory of Gases and Liquids* J. Wiley & Sons, New York 1954.
6. See J. R. Dorfman, E. G. D. Cohen J. Math. Phys. 8, (1967) 282.
7. J. M. Cook in 1965 *Cargèse Lectures in Theoretical Physics*, ed. F. Lurçat, Gordan & Breach, New York 1967.
8. W. Hunziker, Comm. Math. Phys. 8, (1968) 282.
9. B. Simon, Comm. Math. Phys. 23. (1971) 37.
10. R. T. Prosser, J. Math. Phys. 13, (1972) 186.
11. I. Herbst, University of Mighigan preprint 1973.
12. V. S. Buslaev, V. B. Mateev, Theor. Math. Phys. 1, (1970) 367 (English transl.).
13. C. L. Siegel, *Vorlesungen über Himmels-mechanik* ,Springer, Berlin 1956.
14. See the article by R. B. Lavine in this volume.
15. J. M. Jauch, Helv. Physica Acta 31, (1958) 661.

Ref. [1]-[3] are translated in S. G. Brush, *Kinetic Theory* Vol.1, Pergamon Press 1965.

SOME QUESTIONS IN NON-RELATIVISTIC QUANTUM SCATTERING THEORY

W.O. Amrein

University of Geneva

Geneva, Switzerland

1. Introduction.

The basic objects in scattering experiments are the incident beam, the target and the counters. The counters are assumed not to influence the scattering of the beam by the target and are usually completely omitted in the description of the scattering process. The beam as well as the target usually consists of an assembly of objects (e.g. a beam of identical particles, a solid target). One may introduce a Hilbert space $\mathcal{H}$ formed by the states of one unit of the beam and one unit of the target. The time-evolution of these states is described by a strongly continuous one parameter group $\{V_t\}$ of unitary operators. One then considers the behavior of such states at large times which should somehow be comparable to a free behavior (determined by a unitary one-parameter group $\{U_t\}$ which describes the evolution of states in $\mathcal{H}$ in the absence of interaction between beam and target). From this one hopes to prove the existence of an S-operator (i.e. roughly speaking a mapping which assigns to each possible incoming state the corresponding outgoing state). Finally, by taking into account the particular conditions of the experimental setup, one wishes to arrive at an expression for the scattering cross section. This last step usually involves averaging over various states of the beam and/or the target.

Let us point out some of the major mathematical problems which arise in this program. By Stone's theorem [2], the group $\{V_t\}$ is generated by a selfadjoint operator H, the Hamiltonian, i.e. $V_t = \exp(-iHt)$. Similarly $U_t = \exp(-iH_o t)$, where H_o is the operator representing the kinetic energy of the considered units of beam and target (if inelastic scattering or reactions are possible, there

J. A. LaVita and J.-P. Marchand (eds.), Scattering Theory in Mathematical Physics, 97–140.

are many different groups $\{U_t^\alpha\}$ of free evolutions, each corresponding to a channel labelled by α [3]). The difference between H and H_o is the operator describing the interaction between these units. The following problems then arise:

A. Selfadjointness of the Hamiltonian: A unitary time-evolution requires a *selfadjoint* Hamiltonian H. Usually one supposes given the kinetic energy operator H_o and the interaction V(H_o being selfadjoint and V symmetric). Can one then define a selfadjoint sum H of H_o and V. If this definition does not lead to a unique selfadjoint sum, is there a distinguished solution?

B. Spectrum of the Hamiltonian: Spectral properties of the Hamiltonian are very important in scattering theory, as we shall see later. Some important questions are: Is H semibounded? What are the essential spectrum, the absolutely continuous spectrum and the eigenvalues of H ? Is the singularly continuous spectrum of H empty? can there be eigenvalues embedded in the continuous spectrum? How do the eigenfunctions behave at large distances ?

C. Scattering States for the Hamiltonian: Characterize the set of states which are suitable for the description of scattering, i.e. which, under the time-evolution V_t, propagate *away* from the interaction region as $t \to -\infty$ and $t \to +\infty$. Is the set of scattering states identical with the subspace of absolute continuity of H?

D. Asymptotic Condition: In what sense do the scattering states become free as $t \to \pm\infty$? Can one define wave operators? Other mathematical problems are: Asymptotic behavior of momentum observables in the Heisenberg picture, asymptotic completeness of the wave operators, unitarity of the scattering operator S.

E. Stationary State Scattering Theory: Here one wishes to derive Lippmann-Schwinger type equations for the wave operators and and the S-operator, to study the convergence of perturbation series and the possibility of expanding the wave operators in terms of nonnormalizable eigenfunctions of H, and to discuss analyticity properties of scattering amplitudes.

F. Scattering Cross Sections: What are the formulae for a scattering cross section in terms of the scattering operator S for various experimental situations? Under what conditions is the operator S-I an integral operator with continuous kernel in the momentum representation? (In many cases the non-forward cross section is proportional to the square of the kernel of S-I.) Other questions: behavior of cross sections as a function of various parameters; time delay.

In these lectures we shall be considering mainly Problems C and D, and we shall usually be dealing with the case where H_o is the nonrelativistic kinetic energy operator. Before we begin our discussion of Problem C, we briefly make a few remarks concerning the first two problems. We should also add that the formalism of scattering theory is useful not only for the description of the type of experiment mentioned above but for many other physical problems.

There exist very general results about Problem A for Schrödinger and Dirac operators (some recent references are [4]-[9], [85][86]). Perhaps the main interest of this problem now lies in quantum field theory. For Schrödinger operators $H = -\sum_{k=1}^{N} \frac{\partial^2}{\partial x_k^2} + V(x_1,\ldots,x_N) \equiv H_o+V$ one usually splits the interaction V into its positive and negative parts (V_+ resp. V_-). Since H_o is positive, one needs less conditions on V_+ than on V_-. Essential selfadjointness of H with $D(H)=C_o^\infty(\mathbb{R}^N)$ can be proved under the following conditions [7]: $V = V_1 + V_2$ (V_1,V_2 may be different from V_+,V_-) such that $V_1 \in L^2_{\ell oc}(\mathbb{R}^N)$ and is not too negative: $V_1(x) \geq -w(|x|)$ where w(r) is monotone non-decreasing and $w(r) = o(r^2)$ as $r \to \infty$, if $N \geq 5$: $V_2 \in L^{N/2}(\mathbb{R}^N)$ if $N \leq 4$: V_2 satisfies some conditions of Stummel type. The same result holds if H_o is replaced by $-\sum_{k=1}^{N} (\frac{\partial}{\partial x_k} - ib_k(x))^2$ where the b_k are real and in C^1. (Note that essential selfadjointness means that the closure of H is selfadjoint, i.e. this method leads to a unique selfadjoint operator obtained from H_o and V.)

For N = 3, the above conditions require in particular that $V \in L^2_{\ell oc}(R^3)$. If V has a singularity of the type $V(r) = \gamma r^{-\alpha}$ this requires $\alpha < 3/2$. For a wide class of more singular potentials a distinguished selfadjoint and semibounded Hamiltonian can be defined by means of quadratic forms (cf. [4], [6]). However if $\alpha > 2$ and γ negative (i.e. for highly singular attractive potentials) $H_o + V$ is not essentially selfadjoint and there is no distinguished selfadjoint extension [10].

There have appeared numerous interesting results about Problem B in recent years (some references are [4],[5],[11]-[18], [88] and the review article by Kato [19]), but there are still many open questions, particularly for many-body systems. This problem is discussed in the lectures of Combes and Lavine, so we shall not dwell on it here.

2. Scattering States.

The basic picture of a scattering process is the following: particles or other quantum-mechanical systems which are originally independent are made to collide and then (maybe forming some new composite systems) expected to again separate and move independently of each other. States which behave in this way under the evolution group $V_t = \exp(-iHt)$ may be called *scattering states* for the Hamiltonian H. They must be distinguished from the *bound states* in which all the particles are initially close together and will stay together at all times. (One could also envisage a third possibility: states which are such that the particles are located far apart from each other at $t \to -\infty$ but close together at $t \to +\infty$, or such that the particles stay close together at $t \to -\infty$ but move apart from each other in the course of time. However there are no examples known where such a situation would occur.)

Consider for simplicity an n-body system of spinless particles in the center-of-mass frame. The Hilbert space is $\mathcal{H} = L^2(\mathbb{R}^N)$, $N = 3(n-1)$, and the vectors $x \in \mathbb{R}^N$ correspond to linearly independent relative positions between the particles. Let F_r be the projection operator in $\mathcal{H}$ onto the set of states localized in the sphere $|x| \leq r$ (F_r is the multiplication operator by the characteristic function χ_r of this sphere in $L^2(\mathbb{R}^N)$.) For a bound state f, all these relative positions will be small at all times, i.e. given any $\varepsilon > 0$, there exists $r < \infty$ such that

$$\int_{|x|>r} d^N x |(V_t f)(x)|^2 < \varepsilon \qquad \text{for all } t \in \mathbb{R}$$

This leads to the following definition of the subspace of bound states [20]:

$$M_0 = \{f \in \mathcal{H} \mid \lim_{r\to\infty} \sup_{t\in\mathbb{R}} \|(I - F_r)V_t f\|^2 = 0\} \tag{1}$$

For a scattering state g, the n particles will form at least two clusters which are expected to move away from each other. Thus at least one relative coordinate will become large as $t \to +\infty$, which means that

$$\lim_{t\to\infty} \int_{|x|\leq r} d^N x |(V_t g)(x)|^2 = 0 \qquad \text{for every } r < \infty .$$

This gives the following definition of the subspaces $M_\infty^\pm$ of scattering states (for $t \to \pm\infty$ respectively)

$$M_\infty^\pm = \{g \in \mathcal{H} \mid \lim_{t\to\pm\infty} \|F_r V_t g\|^2 = 0 \quad \text{for every } r < \infty\} \tag{2}$$

For later purposes it is convenient to introduce two larger subspaces $\overline{M}_\infty^\pm$ of scattering states by requiring that only the time-average of $\|F_r V_t g\|^2$ tends to zero:

$$\overline{M}_\infty^\pm = \{g \in \mathcal{H} \mid \lim_{T\to\infty} \frac{\pm 1}{T} \int_0^{\pm T} dt \|F_r V_t g\|^2 = 0 \quad \text{for every } r<\infty\} \tag{3}$$

In textbooks bound states are usually identified with superpositions of eigenvectors of H and scattering states with states belonging to the subspace of absolute continuity of H. In fact this latter definition of these states is assumed in practically all presentations of scattering theory, so that for instance the wave operators $\text{s-}\lim_{t\to\pm\infty} \exp(iHt)\exp(-iH_0t)$ are supposed to exist on the subspace $\mathcal{H}_{ac}(H_0)$ of absolute continuity of H_0. However, from the physicist's point of view it would be more natural to require convergence on the subspaces $M_\infty^\pm$ or on $\overline{M}_\infty^\pm$. Scattering theory from this point of view has recently been developed by Wilcox [21].

Now it turns out that in practically all cases of physical interest the above two methods of defining bound states and scattering states (i.e. either in terms of the position operators of in terms of the spectral properties of H) lead to the same subspaces. It is this point that we shall discuss in some detail here. For this, we denote by $\mathcal{H}_p(H)$ (or in short by $\mathcal{H}_p$) the closed subspace of $\mathcal{H}$ spanned by the set of all eigenvectors of H, and we define $\mathcal{H}_c(H) \equiv \mathcal{H} \ominus \mathcal{H}_p(H)$. $\mathcal{H}_c$ can be decomposed into the singularly & absolutely continuous subspaces of H: $\mathcal{H}_c(H) = \mathcal{H}_{sc} + \mathcal{H}_{ac}$ (cf. [22]). We shall denot by E_c the projection operator onto $\mathcal{H}_c(H)$ and by $E(\lambda)$ the spectral projection of H corresponding to the interval $(-\infty,\lambda]$.

We shall first state the basic theorem:

<u>Theorem.</u> [23]: Let $\mathcal{H}$ be a separable Hilbert space, H a selfadjoint operator acting in $\mathcal{H}$, and $\{F_r\}$, $r > 0$, a family of orthogonal projections such that $\text{s-}\lim_{r\to\infty} F_r = I$. Let M_0, $M_\infty^\pm$ and $\overline{M}_\infty^\pm$ be defined as in (1) - (3). Suppose that there exists a family $\{S_n\}$, $n=1,2,\ldots$, of bounded linear operators on $\mathcal{H}$ such that

(i) each S_n commutes with H

(ii) there exists $\text{s-}\lim_{n\to\infty} S_n \equiv S$

(iii) the range of S is dense in $\mathcal{H}$

(iv) $F_r S_n E_c$ is compact for all $r,n < \infty$.

Then $\mathcal{H}_p(H) = M_0$ and $\mathcal{H}_c(H) = \overline{M}_\infty^+ = \overline{M}_\infty^-$. If in addition $\mathcal{H}_{sc}(H)=\phi$ then $\mathcal{H}_c(H) \equiv \mathcal{H}_{ac}(H) = M_\infty^+ = M_\infty^-$.

<u>Remark 1.</u> Some useful operators S_n satisfying (i)-(iii) are the following:

(a) $S_n \equiv S = (H-z)^{-M}$ with M a positive integer and $z\varepsilon\rho(H)$ (the resolvent set of H)

(b) $S_n = E(n) - E(-n)$, i.e. the spectral projection of H corresponding to the finite interval $(-n,n]$.

Remark 2. If $\mathcal{H}_{sc} \neq \emptyset$, the identity of $\mathcal{H}_{ac}$ and $M_\infty^\pm$ is not necessarily true.

Proof. We shall frequently use:

$$\|f + g\|^2 \le \|f + g\|^2 + \|f - g\|^2 = 2\|f\|^2 + 2\|g\|^2 \tag{4}$$

(a) M_0, $M_\infty^\pm$, $\overline{M_\infty^\pm}$ *are closed linear subspaces of* $\mathcal{H}$

We indicate the proof for M_∞^+:

Linearity: Let f, g $\in M_\infty^+$. From (4)

$$\|F_r V_t(\alpha f+g)\|^2 \le 2|\alpha|^2 \|F_r V_t f\|^2 + 2\|F_r V_t g\|^2$$

Each term on the right-hand side converges to zero as $t \to \infty$, i.e. $\alpha f + g \in M_\infty^+$.

Closedness: Let $f_n \in M_\infty^+$ (n=1,2,...) and $\text{s-}\lim_{n\to\infty} f_n = f$. By (4):

$$\|F_r V_t f\|^2 \le 2\|F_r V_t(f-f_n)\|^2 + 2\|F_r V_t f_n\|^2 \le 2\|f-f_n\|^2 + 2\|F_r V_t f_n\|^2$$

Given $\varepsilon > 0$, chose n such that $\|f-f_n\|^2 < \varepsilon/4$ and T such that for $t > T$: $\|F_r V_t f_n\|^2 < \varepsilon/4$. It follows that $\|F_r V_t f\|^2 < \varepsilon$ and hence that $f \in M_\infty^+$.

(b) $M_0 \perp M_\infty^\pm$ *and* $M_0 \perp \overline{M_\infty^\pm}$: We prove for instance $M_0 \perp M_\infty^+$. Let $f \in M_0$, $g \in M_\infty^+$.

$$|(f,g)|^2 = |(V_t f, V_t g)|^2 = |((I-F_r)V_t f, V_t g) + (V_t f, F_r V_t g)|^2$$
$$\le 2\|g\|^2 \|(I-F_r)V_t f\|^2 + 2\|f\|^2 \|F_r V_t g\|^2$$

where we have used (4) and Schwarz's inequality. Given $\varepsilon > 0$ we may first choose r such that $\|(I-F_r)V_t f\|^2 < \varepsilon/(4\|g\|^2)$ for *all* t and then T such that for $t > T$: $\|F_r V_t g\|^2 < \varepsilon/(4\|f\|^2)$. It follows that $|(f,g)| < \varepsilon$ for arbitrary $\varepsilon > 0$, i.e. $f \perp g$.

(c) $\mathcal{H}_p \subset M_0$: Clearly each eigenvector of H belongs to M_0. Since $\mathcal{H}_p(H)$ is the closed subspace spanned by all the eigenvectors of H and M_0 is a closed linear subspace, the conclusion follows.

(d) $M_\infty^\pm \subset \mathcal{H}_c$ and $\overline{M_\infty^\pm} \subset \mathcal{H}_c$: This follows directly from (b) and (c).

(e) If $f \in \mathcal{H}_c(H)$, one has for every $e \in \mathcal{H}$

$$\lim_{T\to\infty} \frac{1}{T} \int_0^T dt\, |(e, V_t f)|^2 = 0$$

This is a standard result of ergodic theory. A proof is given in [23].

(f) $\mathcal{H}_c \subset \overline{M}_\infty^\pm$: (i) and (ii) imply that $S\mathcal{H}_c \subset \mathcal{H}_c$. It then follows from (iii) that $\mathcal{D} \equiv S\mathcal{H}_c$ is dense in $\mathcal{H}_c$. Since $\overline{M}_\infty^\pm$ are closed, it suffices to prove that $\mathcal{D} \subset \overline{M}_\infty^\pm$.

Let $r < \infty$, $g \in \mathcal{D}$ and $f \in \mathcal{H}_c$ such that $g = Sf$. Write $g = (S-S_n) f + S_n f$ and apply (4):

$$\frac{1}{T}\int_0^T dt\, \|F_r V_t g\|^2 \leq \frac{2}{T}\int_0^T dt\, \|F_r V_t (S-S_n) f\|^2 + \frac{2}{T}\int_0^T dt\, \|F_r V_t S_n f\|^2$$

$$\leq \|(S-S_n)f\| \frac{2}{T}\int_0^T dt + \frac{2}{T}\int_0^T dt\, \|F_r S_n E_c V_t f\|^2 \qquad (5)$$

where we have used (i) again. In view of (ii), the first term on the right-hand side can be made arbitrarily small by choosing n sufficiently large. For the second term we use hypothesis (iv). Since $F_r S_n E_c$ is compact, it is the limit *in operator norm* of a sequence of finite rank operators T_N ([2], Thm. VI. 13). T_N has the form

$$T_N f = \sum_{i=1}^{N} (h_i, f) g_i \qquad \text{with } g_i, h_i \in \mathcal{H} \text{ and } N < \infty .$$

Thus, using also (4)

$$\|F_r S_n E_c V_t F\|^2 \leq 2\|F_r S_n E_c - T_N\|^2 \|V_t f\|^2 + 2\|T_N V_t f\|^2$$

and hence

$$\frac{2}{T}\int_0^T dt\, \|F_r S_n E_c V_t f\|^2 \leq 4\|F_r S_n E_c - T_N\|^2 \|f\|^2 + \frac{4}{T}\int_0^T dt\, \Big\|\sum_{i=1}^{N} (h_i, V_t f) g_i\Big\|^2$$

$$\leq 4\|F_r S_n E_c - T_N\|^2 \|f\|^2 + 2^{N-1}\frac{4}{T}\sum_{i=1}^{N}\int_0^T dt\, |(h_i, V_t f)|^2 \|g_i\|^2$$

where we have used (4) repeatedly in the last step. The first term on the right-hand side of the last inequality can be made arbitrarily small by taking N sufficiently large. The second term is a finite sum of terms each of which converges to zero as $T \to \infty$ by (e), since $f \in \mathcal{H}_c$.

With this, we see that the right-hand side of (5) can be made arbitrarily small by a suitable choice of n,N and T. It follows that $g \in \overline{M}_\infty^+$.

The inclusion $g \in \overline{M}_\infty^-$ follows in the same way.

(g) (d) and (f) imply that $\mathcal{H}_c = \overline{M}_\infty^\pm$. From this and (b) one deduces that $M_0 \subset \mathcal{H}_p$. Combining this with (c) one obtains $\mathcal{H}_p = M_0$. This proves the first part of the Theorem.

(h) $\mathcal{H}_{ac} \subset M_\infty^\pm$: Let $g \in \mathcal{D}_o = S\mathcal{H}_{ac}$, i.e. $g = Sf$ with $f \in \mathcal{H}_{ac}$. Then $\{V_t f\}$ converges weakly to zero as $t \to \pm\infty$, i.e. $\lim_{t\to\pm\infty} |(e, V_t f)|^2 = 0$ for every $e \in \mathcal{H}$. By using this property in the place of (e), part (f) of the proof goes through without any time average, i.e. $\lim_{t\to\pm\infty} \|F_r V_t g\|^2 = 0$ for each $r < \infty$.

Since $\mathcal{D}_o$ is dense in $\mathcal{H}_{ac}$, the conclusion follows.

(j) If $\mathcal{H}_{sc}(H) = \emptyset$: $\mathcal{H}_c = \mathcal{H}_{ac}$. If follows from (g) that

$$M_\infty^\pm \subset \bar{M}_\infty^\pm = \mathcal{H}_c = \mathcal{H}_{ac} \quad .$$

This implies together with (h) that $\mathcal{H}_{ac} = M_\infty^\pm$ and completes the proof of the Theorem. ■

The above theorem is useful for relating bound states and scattering states to spectral properties. In order to apply it to a given Hamiltonian H one has to verify compactness of certain operators (cf. requirement (iv)) and to check whether H has any singularly continuous spectrum or not. We now look at the question of compactness.

There are various criteria for compactness of an operator. For later purposes it is useful to introduce the following four classes of bounded linear operators in the Hilbert space $\mathcal{H}$: the set $\mathcal{B}_1$ of all trace-class operators, the $\mathcal{B}_2$ of all Hilbert-Schmidt operators, the set $\mathcal{B}_o$ of all compact operators and the set $\mathcal{B} = \mathcal{B}(\mathcal{H})$ of all bounded operators. These have the following properties [2] :

(P1) $\mathcal{B}_1 \subset \mathcal{B}_2 \subset \mathcal{B}_o \subset \mathcal{B}$.

(P2) Each of the sets $\mathcal{B}_1$, $\mathcal{B}_2$, $\mathcal{B}_o$ and $\mathcal{B}$ is invariant under addition, i.e. $\mathcal{B}_1 + \mathcal{B}_1 \subset \mathcal{B}_1$ etc. .

(P3) Each of the sets $\mathcal{B}_1$, $\mathcal{B}_2$, $\mathcal{B}_o$ and $\mathcal{B}$ is invariant under the operation $A \to A^*$, i.e. $A \in \mathcal{B}_1 \Leftrightarrow A^* \in \mathcal{B}_1$ etc.

(P4) Each of the sets $\mathcal{B}_1$, $\mathcal{B}_2$, $\mathcal{B}_o$ and $\mathcal{B}$ is invariant under multiplication from the left or from the right by an operator $B \in \mathcal{B}$.

(P5) If $C \in \mathcal{B}_2$ and $D \in \mathcal{B}_2$, then $CD \in \mathcal{B}_1$.

(P6) If $C_n \in \mathcal{B}_o$ (n=1,2,...,) and $\lim_{n\to\infty} \|C - C_n\| = 0$, then $C \in \mathcal{B}_o$.

(P7) If $\{f_n\}$, n=1,2,..., is a sequence of vectors in $\mathcal{H}$ converging weakly to zero and $C \in \mathcal{B}_o$, then $\lim_{n\to\infty} \|Cf_n\| = 0$.

We denote by Q_j (j=1,...,N) the multiplication operator in $L^2(\mathbb{R}^N)$ by the variable x_j and by P_j the corresponding momentum operator, i.e. we require

$$[Q_j, Q_k] = [P_j, P_k] = 0, \quad [Q_j, P_k] = i\delta_{jk}$$

Note that the operators F_r are functions of the Q_j.

Our first application of the Theorem concerns *free Hamiltonians*. By this we mean that H_o is a selfadjoint function $\phi(P)$ of the momentum operators $P_1, \ldots, P_N$. We can show that $\mathcal{H}_p(H_0) = M_0$ and $\mathcal{H}_c(H_0) = \overline{M}_\infty^\pm$ for *any* such operator. For this we set $S_n = S = (|P|^N + 1)^{-1}$, where $|P|^2 = \sum_{i=1}^{N} P_j^2$. S is the resolvent of the selfadjoint operator $|P|^N$ and commutes with P_j. This verifies (i)-(iii). Next we evaluate the Hilbert-Schmidt norm of $F_r S$ in the momentum representation:

$$\|F_r S\|_{HS}^2 = \int d^N p \int d^N p' |\tilde{\chi}_r(p'-p) \cdot (|p|^N + 1)^{-1}|^2$$
$$= \|\tilde{\chi}_r\|_{L^2}^2 \cdot \|(|p|^N + 1)^{-1}\|_{L^2}^2$$

Here $\tilde{\chi}_r$ denotes the Fourier transform of χ_r. Both χ_r $(r < \infty)$ and $(|p|^N + 1)^{-1}$ belong to $L^2(\mathbb{R}^N)$, so that $\|F_r S\|_{HS}^2$ is finite. Thus $F_r S$ is a Hilbert-Schmidt operator. By (P1) and (P4), $F_r S E_c$ is compact and the result it proved.

Next let us look at simple perturbations of a free Hamiltonian. Suppose that $F_r(\Phi(P)+i)^{-1}$ is compact (this is true if $\lim_{|p|\to\infty} |\Phi(p)| = \infty$, cf. [23] or part (c) of the proof of Proposition 1 below) and that $H = \Phi(P) + V$ is selfadjoint with domain $D(H) = D(\Phi(P)) \cap D(V)$ (this is true if V is relatively bounded with respect to $H_o = \Phi(P)$ with relative bound less than 1). Then $M_o(H) = \mathcal{H}_p(H)$ and $\overline{M}_\infty^\pm(H) = \mathcal{H}_c(H)$. To see this, it suffices to take $S_n = S = (H+i)^{-1}$ in the Theorem and to remark that the operator $\Phi(P)(H+i)^{-1}$ is defined everywhere and hence bounded. One then has the following factorization

$$F_r S E_c = F_r(\Phi(P) + i)^{-1}(\Phi(P) + i)(H + i)^{-1} E_c$$

and it follows from (P4) that $F_r S E_c$ is compact.

We now deal in more detail with the case of non-relativistic Hamiltonians. An n-particle Hamiltonian in the center-of-mass frame then has the form ([4], p. 190ff.)

$$H = \sum_{j,k=1}^{n-1} a_{jk} \vec{P}_j \cdot \vec{P}_k + V \tag{6}$$

where the matrix a_{jk} is real, symmetric and positive definite. We shall assume that in removing the center of mass one has chosen as new variables some relative positions $\vec{x}_1, \ldots, \vec{x}_{n-1}$ between the particles in such a way that $\vec{x}_1,\ldots,\vec{x}_{n-1}$ are linear combinations of the particle positions $\vec{r}_1,\ldots,\vec{r}_n$. $\vec{Q}_1,\ldots,\vec{Q}_{n-1}$ are the multiplication operators by $\vec{x}_1,\ldots,\vec{x}_{n-1}$ in $L^2(R^{3n-3})$ and $\vec{P}_1,\ldots,\vec{P}_{n-1}$ the corresponding momentum operators. The interaction operator is usually a function of $\vec{Q}_1,\ldots\vec{Q}_{n-1}$.

Since the compactness requirement in our Theorem is only a *local* one (i.e. $r < \infty$), one expects that the behavior of V at infinity does not affect the proof of compactness. This indeed is so. We shall illustrate this for the case where V is a sum of pair potentials each of which is a function of the relative position $\vec{r}_i - \vec{r}_j$ of the corresponding pair. We label the possible pairs by an index s and thus have in the center-of-mass frame:

$$V = \sum_s V_s\Big(\sum_{j=1}^{n-1} a_j^s \vec{Q}_j \Big) \qquad \text{with some constants } a_j^s \ .$$

Proposition 1. Suppose that each V_s belongs to $L^2_{\ell oc}(\mathbb{R}^3)$ as a function of its argument. Denote by $\hat{H}$ the symmetric operator defined by (6) on $D(\hat{H}) \equiv C_o^\infty(\mathbb{R}^{3n-3})$ and by H an arbitrary selfadjoint extension of $\hat{H}$. Then $F_r(H + i)^{-1}$ is compact ($r < \infty$).

Proof [5] :

(a) The function $\sum_{j,k=1}^{n-1} a_{jk}\, \vec{p}_j\cdot\vec{p}_k$ ($p \in R^{3n-3}$) is positive and continuous on the unit sphere of $\mathbb{R}^{3n-3}$. Hence it assumes a positive minimum C on this sphere. It follows that

$$H_o \equiv \sum_{j,k=1}^{n-1} a_{jk}\, \vec{P}_j\cdot\vec{P}_k \geq C \sum_{j=1}^{n-1} |\vec{P}_j|^2 \equiv C|P|^2 \tag{7}$$

(b) Let $\Phi \in C_o^\infty(R^{3n-3})$ and $\alpha > 3/4$, and denote by $\Phi(Q)$ the multiplication operator by $\Phi(x)$ in $L^2(R^{3n-3})$. Then

$$V\Phi(Q)(H_o + C)^{-\alpha} \in \mathcal{B}(\mathcal{H}).$$

To see this, let V_s be one of the pair potentials. V_s depends only on three arguments. We may choose a new coordinate system such that $\sum_{j=1}^{n-1} a_j^s \vec{Q}_j = A\vec{Q}_1{}'$ ($\vec{Q}_j{}'$ are the position operators in the new coordinate system) with $A^2 = \sum_{j=1}^{n-1} (a_j^s)^2$. The primed coordinate system can be chosen such that the (linear) coordinate transformation is orthogonal. This means that $|P|^2 = |P'|^2$.

We shall write $L^2(\mathbb{R}^{3n-3})$ as a tensor product $L^2(\mathbb{R}^{3n-3}) = L^2(\mathbb{R}^3) \otimes L^2(\mathbb{R}^{3n-6})$ where the variable in $\mathbb{R}^3$ is $\vec{x}_1{}'$ and those in R^{3n-6} are $\vec{x}_2{}',\ldots,\vec{x}_{n-1}{}'$. V_s is of the form $V_s(A\vec{Q}_1{}') \otimes I$.

Let R denote the radius of a shpere in $\mathbb{R}^{3n-3}$ which contains the support of Φ in its interior, and choose a function $\psi \in C_0^\infty(\mathbb{R}^3)$ such that $\psi(\vec{x}) = 1$ if $|\vec{x}| \leq R$. Denote by Ψ the operator $\psi(\vec{Q}_1') \otimes I$. Clearly $\Phi(Q) = \Phi(Q)\Psi$.

The function $V_s(A\vec{x})\Psi(\vec{x})$ is square integrable by our assumption on V_s. Since the function $(1 + |p|^2)^{-\alpha} \in L^2(\mathbb{R}^3)$ for $\alpha > 3/4$, one easily verifies that the Hilbert-Schmidt norm of

$$T \equiv V_s(A\vec{Q}_1')\psi(\vec{Q}_1')(I + |\vec{P}_1'|^2)^{-\alpha}$$

is finite, i.e. the above operator is bounded in $L^2(\mathbb{R}^3)$. It follows that $T \otimes I \equiv V_s\Psi[(I + |\vec{P}_1'|^2)^{-\alpha} \otimes I]$ is bounded in $L^2(\mathbb{R}^{3n-3})$. Since $\Phi(Q) = \Phi(Q)\Psi$, the same is true for $V_s\Phi(Q)[(I + |\vec{P}_1'|^2)^{-\alpha} \otimes I]$. statement is equivalent to

$$\|V_s\Phi(Q)f\|^2 \leq \|[(I+|\vec{P}_1'|^2)^{-\alpha} \otimes I]f\|^2 \quad \text{for all } f \in D([I+|\vec{P}_1'|^2]^\alpha \otimes I)$$

Since $(I + |\vec{P}_1'|^2)^\alpha \otimes I \leq (I + |P'|^2)^\alpha = (I + |P|^2)^\alpha \leq [\frac{1}{C}(C + H_o)]^\alpha$ this implies

$$\|V_s\Phi(Q)f\|^2 \leq \|(H_o+C)^\alpha f\|^2 \cdot C^{-2\alpha} \qquad \text{for all } f \in D((H_o+C)^\alpha)$$

i.e. $V_s\Phi(Q)(H_o+C)^{-\alpha} \in \mathcal{B}(\mathcal{H})$. Since this holds for all pairs s, the assertion is proved.

(c) $F_r(H_o+C)^{-\beta}$ is compact for each $\beta > 0$. To see this, let χ_K be the characteristic function of the sphere $|p| \leq K$, $p \in \mathbb{R}^{3n-3}$, and $D_K = \chi_K(p)$ the corresponding projection operator. Since $\chi_K(p)(\sum_{jk} a_{jk}\vec{p}_j\cdot\vec{p}_k + C)^{-\beta} \in L^2(\mathbb{R}^{3n-3})$, the operator $F_r(H_o+C)^{-\beta}D_K$ is Hilbert-Schmidt. Furthermore using also (a) :

$$\|F_r(H_o+C)^{-\beta} - F_r(H_o+C)^{-\beta}D_K\| \leq \|F_r\|\,\|(H_o+C)^{-\beta}(I-D_K)\|$$

$$= \sup_{|p|\geq K}\Big(\sum_{j,k} a_{jk}\vec{p}_j\cdot\vec{p}_k + C\Big)^{-\beta} \leq \sup_{|p|\geq K}(C|p|^2+C)^{-\beta} = (CK^2+C)^{-\beta}$$

This converges to zero as $K \to \infty$. Hence (P6) implies compactness of $F_r(H_o+C)^{-\beta}$.

(d) Let $\Phi \in C_0^\infty(\mathbb{R}^{3n-3})$ and $\overline{\Phi}$ its complex-conjugate. Let $h \in \mathcal{H}$ and $g = (H+i)^{-1}h$. Then $g \in D(H) \subset D(\hat{H}^*)$, and $\hat{H}^*g = Hg$.

For $f \in \mathcal{S}$ (the Schwartz space of test functions), consider

$$(\Phi(Q)(H+i)^{-1}h, (H_o+C)f) = (g, \overline{\Phi}(Q)(H_o+C)f)$$

$$= (g, (H_o+C)\overline{\Phi}(Q)f) + \sum_{j,k} a_{jk}((\vec{\nabla}_j\cdot\vec{\nabla}_k\Phi)(Q)g, f) + 2\sum_{j,k} a_{jk}((\vec{\nabla}_j\Phi)(Q)g, \vec{\nabla}_k f)$$

Since $\Phi f \in C_0^\infty$, the first term on the right-hand side is equal to

$$(g,(H_0+C)\bar{\Phi}(Q)f)=(\Phi(Q)\hat{H}^*g,f)+C(\Phi(Q)g,f)-(g,V\bar{\Phi}(Q)f)$$

Thus, by applying the inequality of Schwarz to every scalar product:

$$|(\Phi(Q)(H+i)^{-1}h,(H_0+C)f)| \leq C_1\|f\|+C_2\|V\bar{\Phi}(Q)f\|+C_3\sum_{j=1}^{n-1}\|\vec{P}_jf\| \quad (8)$$

Let $\alpha > 3/4$. By estimating each term on the right-hand side of (8), we wish to show that

$$|(\Phi(Q)(H+i)^{-1}h,(H_0+C)f| \leq C_0 \|(H_0+C)^\alpha f\| \quad (9)$$

For the first term, this follows from the fact that $C \leq (H_0+C)^\alpha$. For the second term we use (b):

$$\|V\bar{\Phi}(Q)f\| \leq \|V\bar{\Phi}(Q)(H_0+C)^{-\alpha}\| \, \|(H_0+C)^\alpha f\|$$

Finally for the third term we can use (a) :

$$\|\vec{P}_jf\|^2 = \int d^Np|\vec{p}_j|^2|\tilde{f}(p)|^2 \leq \int d^Np|p|^2|\tilde{f}(p)|^2$$

$$\leq \frac{1}{C}\int d^Np\,(\sum_{j,k} a_{jk}\vec{p}_j\cdot\vec{p}_k + C)|\tilde{f}(p)|^2= \frac{1}{C}\|(H_0+C)^{1/2}f\|^2\leq \frac{1}{C}\|(H_0+C)^\alpha f\|$$

where $N = 3n - 3$. We now assume $3/4 < \alpha < 1$. We write $e = (H_0+C)^\alpha f$. Then (9) reads

$$|(\Phi(Q)(H+i)^{-1}h,\ (H_0+C)^{1-\alpha}e)| \leq C_0 \|e\|$$

Since $(H_0+C)^{1-\alpha}$ maps $\mathcal{S}$ onto $\mathcal{S}$, $e \to (\Phi(Q)(H+i)^{-1}h,\ (H_0+C)^{1-\alpha} e)$ defines a bounded linear functional on $\mathcal{S}$ which can then be extended to $L^2(\mathbb{R}^{3n-3})$. By the theorem of Riesz there exists $u \in L^2(\mathbb{R}^{3n-3})$ such that

$$(\Phi(Q)(H+i)^{-1}h,\ (H_0+C)^{1-\alpha}e) = (u,e) \equiv ((H_0+C)^{\alpha-1}u,\ (H_0+C)^{1-\alpha}e)$$

for all $e \in \mathcal{S}$. Thus $\Phi(Q)(H+i)^{-1}h-(H_0+C)^{\alpha-1}u$ is orthogonal to $\mathcal{S}$, hence equal to zero, which implies that $u = (H_0+C)^{1-\alpha}\Phi(Q)(H+i)^{-1}h$, i.e. $(H_0+C)^{1-\alpha}\Phi(Q)\ (H+i)^{-1} \in \mathcal{B}(\mathcal{H})$ $(3/4 < \alpha < 1)$.

(e) Given $r < \infty$, choose $\Phi \in C_0^\infty$ such that $\Phi(x) = 1$ for $|x| \leq r$. Then $F_r = F_r\Phi(Q)$. Hence

$$F_r(H+i)^{-1} = F_r(H_0+C)^{\alpha-1}(H_0+C)^{1-\alpha}\Phi(Q)(H+i)^{-1}.$$

Choose $3/4 < \alpha < 1$. By (c), $F_r(H_0+C)^{\alpha-1}$ is compact, hence it follows from (d) and (P4) that $F_r(H+i)^{-1}$ is compact. ■

Compactness of $F_r(H+i)^{-1}$ can also be proved if the interaction V contains k-body potentials (k>2) satisfying suitable local assumptions. The result also holds if H_o is replaced by

$$\sum_{j,k=1}^{n-1} a_{jk}(\vec{P}_j-\vec{A}_j(Q))(\vec{P}_k-\vec{A}_k(Q)) \text{ with } A_j^{(i)} \in C^1(R^{3n-3})(i=1,2,3;\ j=1,\dots,n-1).$$

In fact it is even possible to show that the final result of part (d) of the above proof is valid for $\alpha=0$ (cf.[24], Lemma 3 or [5] Chapter 9.2)

There remains the question of how to take into account the influence of local singularities of the interaction. We have only partial answers to this. For the sake of simplicity, we shall restrict ourselves here to the 2-body case in the center-of-mass frame. Formally tne Hamiltonian is then given by $H = \vec{P}^2 + V(\vec{Q})$. If V is a positive function, H can be defined as a Friedrichs extension. Suppose that $\mathbb{Q} \equiv D(\sqrt{H_o}) \cap D(\sqrt{V})$ is dense in $L^2(\mathbb{R}^3)$. Then the quadratic form

$$\langle f,g\rangle \equiv (\sqrt{H_o}\, f,\sqrt{H_o}\, g) + (\sqrt{V}\, f,\sqrt{V}\, g) \qquad f,g \in \mathbb{Q}$$

defines a positive selfadjoint operator with $D(\sqrt{H}) = \mathbb{Q}$ and such that $\langle f,g\rangle = (\sqrt{H}\, f,\sqrt{H}\, g)$ ([2], Ch. VIII. 6). It follows that $D(H) \subset D(\sqrt{H}) \subset D(\sqrt{H_o})$. Hence $(\sqrt{H_o} + I)(H+I)^{-1} \in \mathcal{B}(\mathcal{H})$. Since $F_r(H+I)^{-1} = F_r(\sqrt{H_o} + I)^{-1} (\sqrt{H_o} + I)(H+I)^{-1}$ and $F_r(\sqrt{H_o} + I)^{-1}$ is compact, (P4) implies compactness of $F_r(H+I)^{-1}$ for $r < \infty$. In a similar way one can treat the case where V includes hard cores ([25], [23]).

So far we have obtained compactness for arbitrary positive local singularities and for other local singularities of V which are square-integrable. If the singularity is of the type $V(r) = \gamma r^{-\alpha}$ ($r = |x|$), the L^2 - requirement means $\alpha < 3/2$. If one defines the Hamiltonian by means of quadratic forms as above, one can extend this to values $\alpha < 2$ (This method works for all interactions belonging to the Rollnik class which was discussed in much detail by Simon [4]). If $\alpha > 2$ and $\gamma < 0$, $H_o + V$ is unbounded below and symmetric, and none of its selfadjoint extensions is distinguished. The case $\alpha = 2$ lies on the borderline. If $|\gamma|$ is sufficiently large ($\gamma < 0$), the same difficulties as for $\alpha > 2$ are found. In this case (i.e. for $\alpha = 2$) Nelson [9] obtained a non-selfadjoint extension of H + V by imposing a boundary condition at the origin for the corresponding Schrödinger equation with imaginary mass. The time-evolution is absorptive in that case. This result is interesting because absorption occurs exactly for those values of the angular momentum which in the classical theory give rise to a collision of the particle with the scattering center.

If V is spherically symmetric and its singularity is purely attractive, one can verify that $\mathcal{M}_o(H) = \mathcal{H}_p(H)$ and $\overline{\mathcal{M}_\infty^\pm}(H) = \mathcal{H}_c(H)$ if

H is a *spherically symmetric* selfadjoint extension of $H_o + V$ by using methods of the theory of ordinary differential operators to establish these identities in each partial-wave subspace. We shall indicate how this result is obtained under the following assumptions on V :

There exists $\varepsilon > 0$ such that

(i) $V(r) < 0$, $\frac{dV(r)}{dr} > 0$ and $V(r) \in C^2$ for $r \in (0,\varepsilon]$

(ii) $\int_0^\varepsilon dr\, r^{-2}|V(r)|^{-1/2} < \infty$

(iii) $\int_0^\varepsilon dr|V(r)|^{-1/4}|\frac{d^2}{dr^2}|V(r)|^{-1/4}| < \infty$

(iv) $V(r) \in L^2([a,\infty)) \cap L^1([a,\infty)]$ for every $a > 0$

These conditions include all potentials of the form $V(r)=\gamma r^{-\alpha}$ with $\gamma < 0$ and $\alpha > 2$ (the case $\alpha = 2$ can be treated separately [26]). One can adapt the proof to more general situations, in particular for potentials $V(r) = V_1(r) + V_2(r)$ where V_1 satisfies (i)-(iv) and V_2 satisfies only some integrability conditions (these require in particular that V_2 be less singular than V_1 [27]). We may add that one can also prove that $\mathcal{H}_{sc}(H) = \phi$ under the assumptions (i)-(iv) [26].

Diagonalization of the angular momentum operators $\vec{L}^2$ and L_z leads to a decomposition of the Hilbert space $\mathcal{H} = L^2(R^3)$ into a direct sum $\mathcal{H} = \bigoplus_{\ell m} \mathcal{H}_{\ell m}$ ($\ell = 0,1,2,\ldots$; $m = \ell,\ell-1,\ldots,-\ell$), where each $\mathcal{H}_{\ell m}$ is isomorphic to $L^2(\mathbb{R}^+)$. If we identify $\mathcal{H}_{\ell m}$ with $L^2(\mathbb{R}^+)$, we have to investigate the differential operator

$$- \frac{d^2}{dr^2} + V(r) + \frac{\ell(\ell+ 1)}{r^2} \tag{10}$$

in $L^2(\mathbb{R}^+)$.

Proposition 2. (Pearson [27]): Let V satisfy the hypotheses (i)-(iv), denote by $\hat{H}$ the symmetric operator defined by (10) on $D(\hat{H}) = C_o^\infty(\mathbb{R}^+ - \{0\}) \subset L^2(\mathbb{R}^+)$ and by H a selfadjoint extension of $\hat{H}$. Then $F_r(H-i)^{-2}$ is a compact operator in $L^2(\mathbb{R}^+)$.

Proof.

(a) We first investigate the behavior of the solutions of

$$(- \frac{d^2}{dr^2} + V(r) + \frac{\ell(\ell + 1)}{r^2} - i) u(r) = 0 \tag{11}$$

First notice that, as a consequence of assumption (iv), (11) defines a regular differential operator in $L^2([a,b])$ for any $a > 0$, $b < \infty$, i.e. any solution of (11) is twice differentiable and in particular bounded in [a,b] ([28], §15).

Next let us look at the behavior of u(r) near r = 0. For $r \in (0,\varepsilon]$ we introduce the new variable

$$z = \int_r^\varepsilon |V(\rho)|^{1/2}\, d\rho \quad , \quad \frac{dz}{dr} = -|V(r)|^{-1/2} \tag{12}$$

Hypothesis (ii) implies that $\int_0^\varepsilon |V(\rho)|^{1/2}\, d\rho = \infty$, since one has from Schwarz's inequality

$$\infty = \int_0^\varepsilon r^{-1}\, dr = \int_0^\varepsilon dr\, r^{-1}|V(r)|^{-1/4}\,|V(r)|^{1/4}$$
$$\leq \left[\int_0^\varepsilon dr\, r^{-2}|V(r)|^{-1/2} \int_0^\varepsilon d\rho\, |V(\rho)|^{1/2}\right]^{1/2}$$

It follows that $z \to \infty$ as $r \to 0$ and hence (12) defines a bijection from $(0,\varepsilon]$ onto $[0,\infty)$.

Let us make the substitution $u(r) = |V(r)|^{-1/4}\, v(r)$ and write w(z) for the function v(r) expressed in the variable z: w(z) = v(r). A simple calculation shows that the equation (11) is transformed into

$$\frac{d^2w(z)}{dz^2} + (1 + P(z))w(z) = 0 \tag{13}$$

where

$$P(z) = |V(r)|^{-3/4} \frac{d^2}{dr^2} |V(r)|^{-1/4} + [i - \frac{\ell(\ell+1)}{r^2}]|V(r)|^{-1}$$

expressed as a function of z by change of variable. Note that $P(z) \in L^1([0,\infty))$, since

$$\int_0^\infty |P(z)|dz = \int_\varepsilon^0 |P(z)| \frac{dz}{dr}\, dr$$
$$\leq \int_0^\varepsilon dr|V(r)|^{-1/4}\left|\frac{d^2}{dr^2} |V(r)|^{-1/4}\right|$$
$$+ \int_0^\varepsilon dr|V(r)|^{-1/2} + \ell(\ell+1)\int_0^\varepsilon dr\, r^{-2}|V(r)|^{-1/2}$$

and each term on the right hand side is finite by assumptions (iii) and (ii).

Equation (13) is equivalent to the following Volterra integral equation:

$$w(z) = \alpha \cos z + \beta \sin z + \int_z^\infty dt\, \sin(z-t)P(t)w(t) \tag{14}$$

Choose $0 \leq M < \infty$ such that $\int_M^\infty |P(z)|\, dz \equiv K < 1$ and let $\delta(0<\delta\leq\varepsilon)$ be such that $M = \int_\delta^\varepsilon |V(\rho)|^{1/2}\, d\rho$. By standard estimates one deduces that the iteration solution of (14) converges uniformly on $[M,\infty)$ and that

$$|w(z)| \leq (|\alpha|+|\beta|)\sum_{n=0}^{\infty} K^n = \frac{|\alpha| + |\beta|}{1 - K} \equiv K_o \quad \text{for } z \in [M,\infty).$$

It follows that $|v(r)| \leq K_o$ for $r \in [0,\delta]$. Therefore one has for every solution of (11)

$$|u(r)| \leq K_o |V(r)|^{-1/4} \quad \text{for} \quad r \in [0,\delta] \tag{15}$$

By hypothesis (i), $|V(r)|^{-1/2}$ is an increasing function of r in $[0,\varepsilon]$. Thus for $r \leq \delta$:

$$\int_o^r d\rho |u(\rho)|^2 \leq K_o^2 \int_o^r d\rho |V(\rho)|^{-1/2} \leq K_o^2 \delta |V(r)|^{-1/2} \tag{16}$$

(b) We shall now look at the behavior of u(r) at infinity and choose two particular linearly independent solutions of (11). First we notice that one solution is given by the Volterra integral equation

$$\phi(r)=\exp(\tfrac{i-1}{\sqrt{2}}\, r)+\tfrac{i-1}{\sqrt{2}} \int_r^\infty dt\, \sin[\tfrac{1+i}{\sqrt{2}}(r-t)][V(t)+\tfrac{\ell(\ell+1)}{t^2}]\phi(t) \tag{17}$$

This can be solved by iteration with $\phi_o(r) = \exp(\frac{i-1}{\sqrt{2}} r)$.
Let $J = \|V(t)+\ell(\ell+1)t^{-2}\|_{L^1([R,\infty))}$ $(R>0)$. J is finite by hypothesis (iv). Since $|\sin[\frac{1+i}{\sqrt{2}}(r-t)]| < \exp(\frac{t-r}{\sqrt{2}})$ for $t \geq r$, it follows that for $r > R$

$$|\phi_1(r)-\phi_o(r)| \leq \int_r^\infty dt \exp(\frac{t-r}{\sqrt{2}})[V(t) - \frac{\ell(\ell+1)}{t^2}] \exp(\frac{-t}{\sqrt{2}})$$
$$\leq J \exp(\frac{-r}{\sqrt{2}})$$

and by iteration

$$|\phi_n(r)-\phi_{n-1}(r)| \leq J^n \exp(\frac{-r}{\sqrt{2}})$$

Given $\varkappa$ such that $0 < \varkappa < 1$, we may choose R so large that $J < \varkappa(1+\varkappa)^{-1}$. The iteration solution converges uniformly on $[R,\infty)$ and

$$|\phi(r)| \leq \frac{1}{1-J} \exp(\frac{-r}{\sqrt{2}}) = K_1 \exp(\frac{-r}{\sqrt{2}}) \qquad r \in [R,\infty) \tag{18}$$

It follows that

$$|\phi(r)-\phi_o(r)| \leq \exp(\frac{-r}{\sqrt{2}})\sum_{n=1}^{\infty} J^n = \exp(\frac{-r}{\sqrt{2}})\frac{J}{1-J} < \varkappa \exp(\frac{-r}{\sqrt{2}}) \tag{19}$$

Hence, if we set $K_2 = 1 - \varkappa > 0$

$$|\phi(r)| \geq |\phi_o(r)|-|\phi(r)-\phi_o(r)| \geq K_2 \exp(\frac{-r}{\sqrt{2}}) \qquad r \in [R,\infty) \tag{20}$$

(19) also implies that

$$\phi(r) = \exp(\frac{i-1}{\sqrt{2}} r)\eta(r) \quad \text{with} \quad \lim_{r\to\infty} \eta(r) = 1 \tag{21}$$

Now define

$$\psi(r) = C\phi(r) \int_R^r [\phi(t)]^{-2} \, dt \tag{22}$$

By (20), the integral exists for $r \geq R$. One easily verifies that $\psi(r)$ also satisfies the differential equation (11), so we take as our second solution the function $\psi(r)$ which on $[R,\infty)$ coincides with (22). It follows from (20) that for $r > R$

$$|\psi(r)| \leq |\phi(r)| \int_R^r K_2^{-2} \exp(\sqrt{2}\, t) \, dt$$

$$\leq \text{const} \exp(\tfrac{-r}{\sqrt{2}}) [\exp(\sqrt{2}\, r) - \exp(\sqrt{2}\, R)] \leq K_3 \exp(\tfrac{r}{\sqrt{2}}) \tag{23}$$

From (21) one deduces that there exists $K_4 > 0$ such that for $r > R_o$

$$|\psi(r)| \geq K_4 \exp\left(\frac{r}{\sqrt{2}} \right) \tag{24}$$

We shall choose the constant C in (22) such that we have for the Wronskian

$$\psi'(r)\phi(r) - \phi'(r)\psi(r) = 1 \tag{25}$$

(c) For $g \in \mathfrak{H} = L^2(\mathbb{R}^+)$, define the function

$$(Tg)(r) = \phi(r) \int_o^r \psi(t) g(t) \, dt + \psi(r) \int_r^\infty \phi(t) g(t) \, dt \tag{26}$$

Since ψ is bounded on $[0,A]$ for every $A < \infty$ and ϕ is bounded on $\mathbb{R}^+$ and belongs to $L^2(\mathbb{R}^+)$: $(Tg)(r)$ is bounded on $[0,A]$ and hence $Tg \in L^2([0,A])$ for every $A < \infty$.

For $r > R$ we have with (23) and Schwarz's inequality

$$\left|\phi(r) \int_o^r \psi(t) g(t) \, dt \right|$$

$$\leq |\phi(r)| \, (\bar{\psi}, g)_{L^2([0,R])} + |\phi(r)| \frac{K_3}{\sqrt{2}} [\exp(\sqrt{2}\, r) - \exp(\sqrt{2}\, R)]^{\frac{1}{2}} \, \|g\|$$

It follows from (18) that this is bounded for $r > R$. Similarly one sees that

$$\left|\psi(r) \int_r^\infty \phi(t) g(t) \, dt \right|$$

is bounded at infinity. If follows that $(Tg)(r)$ is bounded on R^+.

Now let $h = (H-i)^{-1} g$. Then h satisfies $(\hat{H}^* - i)h = g$, i.e.

$$\left(-\frac{d^2}{dr^2} + V(r) + \frac{\ell(\ell+1)}{r^2} - i \right) h(r) = g(r) \tag{27}$$

As a consequence of the normalization (25), a particular solution of this equation is $h(r) = (Tg)(r)$. Hence

$$(H-i)^{-1}g = Tg + B_1\phi + B_2\psi \tag{28}$$

Since $(Tg)(r)$ is bounded at infinity and $(H-i)^{-1}g$ and ϕ belong to $L^2(\mathbb{R}^+)$, it follows from (24) that $B_2 = 0$. Therefore $Tg=(H-i)^{-1}g+B_1(g)\phi\varepsilon L^2(\mathbb{R}^+)$, i.e. T defines a bounded linear operator on $\mathcal{H}$.

Let P_1 be the orthogonal projection onto the two-dimensional subspace spanned by the vectors $[T^*-(H+i)^{-1}]\phi$ and $(H+i)^{-1}[T^*-(H+i)^{-1}]\phi$. It follows that for $g \perp P_1\mathcal{H}$:

$$0 = ([T^*-(H+i)^{-1}]\phi,g) = (\phi,B_1(g)\phi) = B_1(g)(\phi,\phi) .$$

Therefore $B_1(g) = 0$ and $(H-i)^{-1}g = Tg$. Similarly one concludes that $(H-i)^{-2}g = T^2g$.

(d) Let P_2 be the orthogonal projection onto the two-dimensional subspace spanned by the vectors ϕ and $(H+i)^{-1}\bar{\phi}$ ($\bar{\phi}$ is the complex conjugate of ϕ). For $g \perp P_2\mathcal{H}$ the second integral in (26) can be rewritten such as to give

$$(Tg)(r) = \phi(r) \int_0^r \psi(t)g(t)\, dt - \psi(r)\int_0^r \phi(t)g(t)\, dt \tag{29}$$

and similarly for $[T(H-i)^{-1}g](r)$.

Let P be the projection onto the subspace spanned by $P_1\mathcal{H}$ and $P_2\mathcal{H}$ and take $g \perp P\mathcal{H}$. Let $g_1 = (H-i)^{-1}g = Tg$. By applying Schwarz's inequality in (29) and using (15) and (16) we deduce that for $r\leq\delta$

$$\begin{aligned}|g_1(r)| &\leq 2K_o|V(r)|^{-1/4}\|g\| \left[\int_0^r d\rho|V(\rho)|^{-1/2} \right]^{1/2}\\ &\leq \text{const } |V(r)|^{-1/2}\end{aligned} \tag{30}$$

Let $g_2 = (H-i)^{-2}g = Tg_1$. Using the bound (30) in (29) and the fact that $|V(r)|^{-3/4}$ is increasing in $[0,\delta]$:

$$|g_2(r)| \leq \text{const}|V(r)|^{-1/4} \int_0^r |V(\rho)|^{-3/4}\, d\rho \leq \text{const } V(r)^{-1} \tag{31}$$

Thus $|V(r)g_2(r)| <$ const for $r \in [0,\delta]$. Futhermore $V(r)g_2(r)\varepsilon L^2([\delta,\infty))$, since $V\varepsilon L^2([\delta,\infty))$ and g_2 is bounded. Thus $g_2 \in D(V)$, i.e. $V(H-i)^{-2}(I-P) \in \mathcal{B}(\mathcal{H})$. Thus, if we rewrite (27) with $h = g_2$:

$$\left(-\frac{d^2}{dr^2} + \frac{\ell(\ell+1)}{r^2}\right)g_2(r) = g_1(r) - V(r)g_2(r) + ig_2(r)$$

i.e. $g_2(r)$ belongs to the domain of $H_{o,\ell} \equiv \frac{-d^2}{dr^2} + \frac{\ell(\ell+1)}{r^2}$. If $\ell > 0$, $H_{o,\ell}$ is the restriction of $H_o = \vec{P}^2$ to the subspace $\mathcal{H}_{\ell m}$. For $\ell=0$, $-\frac{d^2}{dr^2}$ is *not* essentially selfadjoint in $L^2(\mathbb{R}^+)$, and the restriction $\overline{H}_{o,o}$

of $H_o = \vec{P}^2$ to $\mathcal{H}_{o,o}$ is characterized by the following boundary condition: $f \varepsilon \mathcal{H}_{o,o}$ belongs to $D(\overline{H}_{o,o})$ if $-\frac{d^2}{dr^2} f(r) \varepsilon L^2(\mathbb{R}^+)$ and $f(0)=0$. Since $g_2 = Tg_1$, it follows immediately from (29) and (15) that $g_2(0) = 0$ and therefore $g_2 \varepsilon D(\overline{H}_{o,o})$.

We conclude that $H_{o,\ell}(H-i)^{-2}(I-P) \varepsilon \mathcal{B}(\mathcal{H})$ for every $\ell=0,1,\ldots$

(e) $F_r(H-i)^{-2} = F_r(H_{o,\ell}+I)^{-1}(H_{o,\ell}+I)(H-i)^{-2}(I-P)+F_r(H-i)^{-2}P$
Since $F_r(H_{o,\ell}+I)^{-1}$ is compact, the first term on the righthand side is compact by (P4). The second term is different from zero only on a subspace of finite dimension, i.e. it is of trace class and hence compact by (P1). It follows that $F_r(H-i)^{-2}$ is compact, and this concludes the proof. ∎

3. Wave Operators.

The so-called *asymptotic condition* is the requirement that at very large (positive and negative) times a scattering system should behave like a free system. The physical idea behind this is that before and after the interaction takes place the target and the beam should be completely independent of each other and that it is therefore possible to consider the prepared and the detected particles as *free* particles, as is usually done in the interpretation of experiments.

In a simple scattering system with only elastic scattering (e.g. a particle interacting with a potential, or equivalently a 2-body system in the center-of-mass frame), the free evolution U_t is obtained from the kinetic energy operator $H_o = \vec{P}^2/(2m)$, i.e. $U_t = \exp(-iH_ot)$ (we usually set the mass equal to 1/2). The simplest way of expressing the asymptotic condition is to require that for each scattering state f of the total Hamiltonian there exist two other states $f_\pm$ such that V_tf approaches (in the Hilbert space norm) U_tf_- at large negative and U_tf_+ at large positive times:

$$\lim_{t\to\pm\infty} \|V_tf - U_tf_\pm\| = 0$$

i.e. we require the existence of $\text{s-}\lim_{t\to\pm\infty} U_t^*V_tf$ for all scattering states of H. One easily verifies that $f_\pm$ are scattering states of H_o [23], i.e. $f \varepsilon M_\infty^\pm(H) \Rightarrow f_\pm \varepsilon M_\infty^\pm(H_o) \equiv \mathcal{H}_{ac}(H_o)$. In view of our earlier results we shall henceforth assume $M_\infty^\pm(H) = \mathcal{H}_{ac}(H)$.

Since one would like to admit each scattering state of H_o as a possible initial state, the asymptotic condition at negative times is usually written with H and H_o interchanged, i.e. one requires the existence of

$$\Omega_-(H,H_o) = \underset{t\to -\infty}{\text{s-lim}}\ V_t{}^*U_tE_{ac}(H_o) \qquad (32)$$

and for the sake of symmetry the same condition at $t \to +\infty$. $\Omega_\pm$ are partially isometric operators (they are isometric, i.e. $\Omega_\pm^*\Omega_\pm = I$, if $E_{ac}(H_o) = I$). Only those scattering states of H which lie in the range of $\Omega_\pm$ are needed for the description of the scattering process, but in order for this description to be consistent one has to require that the range of Ω_- be contained in that of Ω_+. A theory of this type is called *asymptotically complete* if the range of both Ω_+ and Ω_- is the set of *all* scattering states of H, i.e. $\Omega_-\Omega_-{}^* = \Omega_+\Omega_+{}^* = E_{ac}(H)$ (cf. [4] for various definitions of asymptotic completeness). The proof of asymptotic completeness is often quite hard.

From the wave operators one immediately obtains the scattering operator which is the correspondence $f_- \to f_+$, i.e. $S = \Omega_+{}^*\Omega_-$. If $E_{ac}(H_o) = I$ and range $\Omega_- \subset$ range Ω_+, S is easily seen to be isometric. It is unitary if range Ω_- = range Ω_+ . (A detailed description of time-dependent scattering theory can be found in [29].)

A different formulation of the asymptotic condition is obtained if one characterizes the free system by some set of observables instead of using the free evolution. By a free observable we mean an operator $A \in \mathcal{B}(\mathcal{H})$ acting in the subspace $\mathcal{H}_{ac}(H_o)$ of scattering states of H_o which commutes with H_o (i.e. $AE_{ac}(H_o) = A$ and $A \in \{H_o\}'$). These operators are constants of the free motion, i.e. $U_t{}^*AU_t = A$ for all t. If the total evolution becomes asymptotically free, one expects such observables to be *asymptotically* constants of the total motion, i.e. $V_t{}^*AV_t$ should converge as $t \to \pm\infty$ on the scattering states of H. The most natural way to write this condition is to require the existence of

$$\lim_{t\to\pm\infty} (V_tf, AV_tf) \qquad \text{for all } f \in \mathcal{H}_{ac}(H) \qquad (33)$$

which is equivalent to the existence of $\underset{t\to\pm\infty}{\text{w-lim}}\ E_{ac}(H)V_t{}^*AV_tE_{ac}(H)$. There are various results on a scattering theory starting from (33) but the theory is not yet complete [30].

There have been several investigations starting from strong limits (cf. [31]-[36]). This has the advantage that the strong limit preserves the algebraic structure of the chosen set A_o of free observables (whereas w-lim A_nB_n need not be equal to w-lim $A_n\cdot$w-lim B_n).

In this latter approach it is still possible to deduce (under suitable additional hypotheses) the existence of generalised wave operators (i.e. partial isometries with initial set $\mathcal{H}_{ac}(H_o)$ and range $\mathcal{H}_{ac}(H)$) $\Omega_\pm$ such that

$$A_{\pm} \equiv \underset{t\to\pm\infty}{\text{s-lim}}\, V_t^{*} A V_t E_{ac}(H) = \Omega_{\pm} A \Omega_{\pm}^{*} \qquad (A \in A_o) \qquad (34)$$

and the scattering operator is then again given by $S = \Omega_+^*\Omega_-$. The correspondence $S \to S^*AS$ gives the change of the observable A as a consequence of the interaction. S is partially isometric (it is initary if $\Omega_\pm$ are isometric). As can be seen directly from (34), the generalised wave operators will be determined only up to multiplication from the right by a unitary operator in $\mathcal{H}_{ac}(H_o)$ which commutes with A_o, and similarly for S. If the ordinary wave operators (32) exist (and are asymptotically complete), they will be identical with the generalized ones apart from the above-mentioned indeterminacy, since then
$\Omega A\Omega^* = \text{s-lim} V_t^*U_tAU_t^*V_tE_{ac}(H) = \text{s-lim } V_t^*AV_tE_{ac}(H)$ if $A \in \{H_o\}'$.

The proof of the existence of the generalized wave operators from the assumption of the existence of the limits in (34) is not trival. Of course one need not require convergence of $V_t^*AV_tE_{ac}(H)$ for all operators in $\{H_o\}'$. However A_o should not be too small. What is needed is that it has cyclic vectors with respect to $\mathcal{H}_{ac}(H_o)$, i.e. vectors $e \in \mathcal{H}_{ac}(H_o)$ such that the set $\{Ae|A \in A_o\}$ is dense in $\mathcal{H}_{ac}(H_o)$. This is always true if the commutant A_o' of A_o is abelian on $\mathcal{H}_{ac}(H_o)$ ([37], I.1.4, Prop. 5 and I.2.1., Cor. of Prop. 3). (This last requirement roughly means that A_o contains at least one complete set of commuting observables.) Once the wave operators are defined on a dense set, they can be extended by continuity to $\mathcal{H}_{ac}(H_o)$. The construction of the wave operators then is as follows: suppose that the two sets $A_\pm \equiv \{A_\pm|A \in A_o\}$ also admit cyclic vectors with respect to $\mathcal{H}_{ac}(H)$ and that there is a cyclic vector e for A_o and a pair $e_\pm$ of cyclic vectors for $A_\pm$ respectively such that

$$(e,Ae) = (e_\pm,A_\pm e_\pm) \qquad \text{for all } A \in A_o \qquad (35)$$

Then one defines $\Omega_\pm$: $Ae \to A_\pm e_\pm$ $(A \in A_o)$.

The above construction defines an isometric linear operator if the correspondences $A \to A_\pm$ have the following properties:
$\lambda A + B \to \lambda A_\pm + B_\pm$, $AB \to A_\pm B_\pm$ and $A^* \to A_\pm^*$. These follow easily from the existence of $\underset{t\to\pm\infty}{\text{s-lim}}\, V_t^*AV_tE_{ac}(H)$, $A \in A_o$, cf. [32]. (The isometric property of $\Omega_\pm$ follows from (35).)

The conditions which have to be added to existence of the limits (34) in order to obtain generalized wave operators can be some what relaxed if A_o is a von Neumann algebra (i.e. closed in the weak operator topology [37]). One will still require A_o' to be abelian. One can then prove that $A_\pm$ are also von Neumann algebras,[35][38] [39]. If one assumes that $A_\pm'$ are also abelian, the requirement (35) is always satisfied and need not be added as a seperate condition. (More precisely: there exist maximal projection operators

$E_\pm \in A_o'$ such that (35) is verified for all $A \in A_o E_\pm$ and such that $\underset{t\to\pm\infty}{\text{s-lim}}\, V_t^* A V_t E_{ac}(H) = 0$ for all $A \in A_o(I-E_\pm)$ [32]. $E_\pm \mathcal{H}_{ac}(H_o)$ are then the initial domains of $\Omega_\pm$.)

In practical cases it is often difficult to verify (35) or to prove the existence of the limits (34) on a von Neumann algebra. Sometimes their existence can be proved on a C* - algebra, but this does not guarantee (35) even if A_o and A_+ have abelian commutants. The following example illustrates the difficulties that can arise [40]. Let C be the set of all complex-valued continuous functions on [0,1]. Let A_o be the C* - algebra of multiplication operators by functions $f \in C$ in $L^2([0,1])$ and A_+ the C* - algebra of multiplication operators by functions $f \in C$ in $L^2(\Delta)$, where Δ is a measurable subset of [0,1] with $0 < \mu(\Delta) < 1$ but such that its closure is equal to [0,1]. (Example: let $\{r_i\}$, i=1,2,..., be the set of rational numbers in [0,1]. Choose an open interval Δ_i of length 3^{-i} centered at the point r_i and let $\Delta = \bigcup_{i=1}^{\infty} \Delta_i \cap [0,1]$. Clearly $\bar{\Delta} = [0,1]$, $\mu(\Delta) > 0$ and $\mu(\Delta) < \sum_{i=1}^{\infty} 3^{-i} = \tfrac{1}{2}$.) The norm in A_o is $\|f\|_o = \sup_{x\in[0,1]} |f(x)|$ and that in A_+ is $\|f\|_+ = \sup_{x\in\Delta} |f(x)|$. One has $\|f\|_o = \|f\|_+$. Hence there is an isomorphism ω from A_o onto A_+ given by $\omega(f) = f$. Note that both A_o and A_+ have cyclic vectors (take $e(x) = 1$ for $x \in [0,1]$ resp. $x \in \Delta$). Suppose now that ω is implemented by a unitary operator Ω from $L^2([0,1])$ to $L^2(\Delta)$. Let P_Δ be the projection operator onto $L^2(\Delta)$ in $L^2([0,1])$. One can check that one must have $\Omega(I-P_\Delta)\Omega^* = 0$, i.e. $\Omega(I-P_\Delta) = 0$. Since $P_\Delta \neq I$, this contradicts the unitarity of Ω. (In this example ω can be extended in a natural way to certain projections in A_o''. This need not be the case in other situations.)

The same two methods of formulating the asymptotic condition can be applied in many-body systems. Since one may now have inelastic scattering and creation and annihilation of composite particles, one has to introduce a set of free evolutions $U_t^{(\alpha)}$ or a set of algebras A_α (α in some index set J) according to what fragments are observed. This occurs because if one has more than two particles, they may be grouped into two or more clusters, and the Hamiltonian describing the relative movement of the particles in a given cluster (i.e. the sum of the kinetic energy operators of each particle of the cluster and interaction *between* these particles) may have bound states. Since the individual clusters may move apart from each other, the whole system may still be in a scattering state of the total Hamiltonian.

One is thus led to introduce the channels, i.e. certain subspaces $\mathcal{H}_\alpha$ of $\mathcal{H}$, each of which is given by a decomposition of (1,...,n) into clusters $(C_1,\dots,C_m)$ and a set $(g_1,\dots,g_m)$ of eigenvectors of $C_1,\dots,C_m$ (n=number of particles, $m \leq n$, $g_k = 1$ if C_k

consists of a single particle). Since the binding energies may be degenerate, one chooses from the outset a maximal set of mutually orthogonal bound states for each possible fragment and thus obtains a countable set of channels (cf. [3] for details). The states in channel α are defined as $f_\alpha(\vec{x}_1,\ldots,\vec{x}_n) = h_\alpha(\vec{y}_1,\ldots,\vec{y}_m)\prod_{k=1}^{m} g_k(z_k)$ where $\vec{y}_k \in \mathbb{R}^3$ is the coordinate of the center-of-mass of C_k and $z_k \in R^{3(\ell-1)}$ (ℓ = number of particles in C_k) are internal coordinates of C_k. The free channel Hamiltonian is

$$H_\alpha = \sum_{k=1}^{m} \left(\frac{1}{2M_k} \vec{P}_k^2 + E_k \right) \qquad [\, U_t^{(\alpha)} = \exp(-iH_\alpha t)\,]$$

where $\vec{P}_k$ is the total momentum, M_k the total mass and E_k the bound state energy of C_k. The algebras $\mathcal{A}_\alpha$ should contain operators of the form $A_\alpha \otimes I$ where A_α acts in $\mathcal{H}_\alpha$ and commutes with H_α, and I is the identity operator with respect to the internal coordinates of $C_1,\ldots,C_m$.

The standard wave operators are then defined as

$$\Omega_\pm^{(\alpha)} = \operatorname*{s-lim}_{t\to\pm\infty} V_t^* U_t^{(\alpha)} E_\alpha \qquad (E_\alpha = \text{projection onto } \mathcal{H}_\alpha)$$

One expects that for $\alpha \neq \beta$ the range of $\Omega_+^{(\alpha)}$ will be orthogonal to that of $\Omega_+^{(\beta)}$ (and similarly with the minus sign). An intuitive motivation for this is as follows: if at large t one knows the system to be composed of certain fragments in definite bound states, far apart from each other and moving away from each other, the probability that at the same time the system will be composed of some other fragments or the same fragments in different bound states should be practically zero (if this were not so, one would not be able to distinguish uniquely between the different channels by measurements). The requirement of the orthogonality of the ranges of $\Omega_+^{(\alpha)}$ and $\Omega_+^{(\beta)}$ is easily seen to be equivalent to

$$\lim_{t\to\infty} (V_t f_\alpha, V_t f_\beta) \equiv \lim_{t\to\infty} (U_t^{(\alpha)} f_\alpha, U_t^{(\beta)} f_\beta) = 0 \text{ for all } f_\alpha \in \mathcal{H}_\alpha, f_\beta \in \mathcal{H}_\beta.$$

In non-relativistic n-body scattering this can be proved from the particular form of H_α (cf. [3]), whereas in an abstract formulation it has to be postulated [41][39].

The spaces $\mathcal{H}_\alpha$ are not all mutually orthogonal. In fact for the case where each particle forms its own fragment, i.e. $(C_1,\ldots,C_n) = (1,\ldots,n)$, the corresponding channel subspace is identical with $\mathcal{H}$. It is useful to introduce an auxiliary Hilbert space $\mathcal{H}' = \bigoplus_{\alpha\in J} \mathcal{H}_\alpha$ and to define $\Omega_\pm f \equiv \bigoplus_\alpha \Omega_\pm^{(\alpha)} f_\alpha \in \mathcal{H}$ $(f = \bigoplus_\alpha f_\alpha \in \mathcal{H}')$ and $S = \Omega_+^* \Omega_-$ [3][42]. S is then an operator in $\mathcal{H}'$. It commutes with $H_0' = \bigoplus_\alpha H_\alpha$, and it is unitary if $\bigoplus_\alpha R_+^{(\alpha)} = \bigoplus_\alpha R_-^{(\alpha)}$ ($R_\pm^{(\alpha)}$ denotes the range of

$\Omega_\pm^{(\alpha)}$. The theory is asymptotically complete if $R_+^{(\alpha)} = R_-^{(\alpha)} = E_{ac}(H)$ (in the center-of-mass system). One can avoid the introduction of $\mathcal{H}'$ by defining a different S-operator $\hat{S} = \sum_\alpha \Omega_+^{(\alpha)}\Omega_-^{(\alpha)*}$ which commutes with H and is unitary on $\mathcal{H}_{ac}(H)$ if the theory is asymptotically complete [41].

Similarly one can develop the algebraic scheme of multichannel scattering [39].

Let us now point out some methods of proving the asymptotic condition. We again start with the simple case of scattering of a particle by a potential and we assume $H_o = \vec{P}^2$. Three types of convergence criteria will be distinguished, viz. criteria based (a) on the rate of decay as $t \to \pm\infty$ of quantities like $\|V\Phi(\vec{Q})\exp(-iHt)f\|$, (b) on trace conditions and (c) on smoothness properties. Yet another method is to start with stationary state scattering theory and to prove that the wave operators obtained there coincide with the strong limits of the time-dependent theory [43].

The simplest criterion is the following (Kupsch and Sandhas [44]): one chooses a C^∞-function Φ with $|\Phi(\vec{x})| \leq 1$, $\Phi(\vec{x}) = 0$ for $|\vec{x}| \leq r$, $\Phi(\vec{x}) = 1$ for $|\vec{x}| \geq R$ (R>r). One has

$$V_t^* U_t f = V_t^*\Phi(\vec{Q})U_t f + V_t^*(I-\Phi(\vec{Q}))U_t f \tag{36}$$

The second term converges strongly to zero as $t \to \pm\infty$, since $I-\Phi(\vec{Q}) \leq F_R$ and $F_R U_t$ converges strongly to zero as was seen earlier. For the first term one writes

$$\begin{aligned} V_t^*\Phi(\vec{Q})U_t f - V_\tau^*\Phi(\vec{Q})U_\tau f &= \int_\tau^t ds\, \frac{d}{ds} V_s^*\Phi(\vec{Q})U_s f \\ &= \int_\tau^t ds\, V_s^*[H\Phi(\vec{Q}) - \Phi(\vec{Q})H_o]U_s f \\ &= \int_\tau^t ds\, V_s^* V\Phi(\vec{Q})U_s f - \int_\tau^t ds\, V_s^*(\Delta\Phi)(\vec{Q})U_s f \\ &\quad - 2i\int_\tau^t ds\, V_s^*(\vec{\nabla}\Phi)(\vec{Q})U_s\vec{P}f \end{aligned} \tag{37}$$

If one can show for a dense set $\mathcal{D}$ of vectors f that each term on the right-hand side converges strongly to zero as $t,\tau \to \pm\infty$, it follows that $\{V_t^*U_t f\}$ is a Cauchy sequence and hence that $\Omega_\pm f$ exists for $f \in \mathcal{D}$, and therefore for all $f \in \mathcal{H}$. We shall take $\mathcal{D} = \mathcal{J}$. One has

$$\left\|\int_\tau^t ds\, V_s^*(\Delta\Phi)(\vec{Q})U_s f\right\| \leq \int_\tau^t ds \|(\Delta\Phi)(\vec{Q})U_s f\| \leq C\int_\tau^t ds \|F_R U_s f\| \tag{38}$$

since $(\Delta\Phi)(\vec{x})$ is bounded and different from zero only for $r<|\vec{x}|<R$. We know already that the last integrand converges to zero for $s \to \pm\infty$, but in order to obtain convergence of the integral we have to estimate its rate of decay. For $H_o=\vec{P}^2$ this can easily be done since one knows that [29]

$$(U_s f)(\vec{x}) = (4\pi i s)^{-3/2} \int d^3y \exp[\tfrac{i}{4s}(\vec{x}-\vec{y})^2] f(\vec{y}) \tag{39}$$

It follows that

$$\|F_R U_s f\| \leq \text{const}|s|^{-3/2} \|f\|_{L^1} \|\chi_R\|_{L^2}$$

and hence (38) converges to zero as $t,\tau \to \pm\infty$. In the same way one treats the last term in (37).

The same argument applies to the first term in (37) if one assumes $V(\cdot)\Phi(\cdot) \in L^2(\mathbb{R}^3)$. One sees from this that only the behavior of V at *large* distance matters to prove existence of wave operators. By using Hölder's inequality one can improve this result by assuming only that $V\Phi \in L^p(\mathbb{R}^3)$ with $2 \leq p < 3$. This includes all potentials that converge to zero at infinity faster than $\text{const}|x|^{-1-\varepsilon}$ for some $\varepsilon > 0$. Thus:

Proposition 3. Let $V(\cdot)(1-\chi_R(\cdot)) \in L^p(\mathbb{R}^3)$ for some $R < \infty$ and some p, $2 \leq p < 3$. Then $\Omega_\pm(H,H_o)$ exist. (H may be a selfadjoint extension of $-\Delta + V$ if the latter is densely defined or a Friedrichs extension of it.)

For non-oscillating potentials this cannot be improved much, even by other means of proving convergence, since for Coulomb potentials $V(\vec{x}) = \lambda|\vec{x}|^{-1}$ the strong limits of $V_t{}^*U_t$ do not exist[29].

The above method is not so useful for proving asymptotic completeness of the wave operators. Completeness is easily seen to be equivalent to the existence of $\underset{t\to\pm\infty}{\text{s-lim}}\, U_t{}^*V_t E_{ac}(H)$. This amounts to interchanging U_t and V_t in (36). One would then need estimates on the decay rate of quantities of the form $F(\vec{Q})\exp(-iHt)f$ which are not as easy to get because one does not have an explicit expression similar to (39) for $\exp(-iHt)f$.

A different and more powerful method for proving existence and asymptotic completeness of wave operators is to use trace conditions. Here one does not use the explicit form of H_o in the existence proof (though it is sometimes needed in verifying the trace condition), so that it also follows (by interchanging H and H_o) that the wave operators are complete. Some results of this type can be found in [22], [45]-[47]. In [45] it is proved that it suffices to require that $H-H_o \in \mathcal{B}_1$ or that $(H-z)^{-M}-(H_o-z)^{-M} \in \mathcal{B}_1$ for some $z \in \rho(H) \cap \rho(H_o)$ and some positive integer M.

If $V \in L^2(\mathbb{R}^3)$, one has $D(V) \supset D(H_o)$ [22] and

$$(H-z)^{-1} - (H_o-z)^{-1} = -(H-z)^{-1}V(H_o-z)^{-1} \tag{40}$$

If $V \in L^1(\mathbb{R}^3)$, one may write $V(\vec{x}) = |V(\vec{x})|^{1/2} \cdot (\text{sign}V(\vec{x})|V(\vec{x})|^{1/2})$

and each factor belongs to $L^2(\mathbb{R}^3)$. Thus if $V \in L^1(\mathbb{R}^3) \cap L^2(\mathbb{R}^3)$ it follows that $(H_o-z)^{-1}|V|^{1/2} \in \mathcal{B}_2$ and the same for $\text{sign}V \cdot |V|^{1/2}(H-z)^{-1} = \text{sign}\, V \cdot |V|^{\frac{1}{2}} (H_o-z)^{-1}(H_o-z)(H-z)^{-1}$, since the first factor is Hilbert-Schmidt and the second one in $\mathcal{B}(\mathcal{H})$, cf. (P4). It follows from (P5) that the operator (40) is of trace class. Hence $\Omega_\pm(H,H_o)$ are complete, cf. [87].

By using the chain rule, Simon ([4], page 109) has generalized this result to more singular potentials. His requirements are that V must satisfy the Rollnik condition and decrease faster than $c|\vec{x}|^{-1-\varepsilon}$ $(\varepsilon>0)$ at infinity. If V has an attractive singularity of form $V(r) = \gamma r^{-\alpha}$, this again means $\alpha < 2$.

A more powerful trace condition is the following:

Proposition 4. (Pearson [47]): Let H_1 and H_2 be selfadjoint in $\mathcal{H}$ and $E_j(\cdot)$ $(j=1,2)$ their spectral families. Suppose $H_2E_2(\Delta)E_1(\Delta) - E_2(\Delta)H_1E_1(\Delta)$ is of trace class for every bounded measurable set Δ. Let $f \in \mathcal{H}_{ac}(H_1)$. Then the strong limit of $\exp(iH_2t)\exp(-iH_1t)f$ as $t\to\infty$ exists iff, given $\varepsilon > 0$, there exists $T > 0$ and a bounded measurable set Δ_o such that $\| E_2(\mathbb{R}-\Delta_o)\exp(-iH_1t)f \| < \varepsilon$ for all $t > T$.

This can be used (with $H_1 = H$ and $H_2 = H_o$) to prove asymptotic completeness of $\Omega_\pm(H,H_o)$ for highly singular potentials [27],[48] (their existence follows from Prop. 3). We shall just indicate (cf. [27]) how this applies to the Hamiltonians considered in Prop. 2. The proof relies on the following two facts:

(α) There exists a trace class operator P such that $H_o(H-i)^{-2}(I-P) \in \mathcal{B}(\mathcal{H})$ and $V(H-i)^{-2}(I-P) \in \mathcal{B}(\mathcal{H})$ (cf. part (d) of the proof of Prop. 2).

(β) If Δ is a bounded measurable set of $\mathbb{R}$, then $E_o(\Delta)F_R \in \mathcal{B}_1$ for every $R < \infty$ (cf. [27]).

Let Δ be a bounded measurable set of $\mathbb{R}$ and write

$$E_o(\Delta)HE(\Delta)-H_oE_o(\Delta)E(\Delta)=[E_o(\Delta)H-H_oE_o(\Delta)](H-i)^{-2}(I-P)(H-i)^2E(\Delta)$$
$$+ [E_o(\Delta)H-H_oE_o(\Delta)](H-i)^{-2}P(H-i)^2E(\Delta) \tag{41}$$

The second term is of trace class by (P4), since $P \in \mathcal{B}_1$ and the other two factors belong to $\mathcal{B}(\mathcal{H})$. Using (α), we may rewrite the first term as

$$E_o(\Delta)V(H-i)^{-2}(I-P)(H-i)^2E(\Delta) = E_o(\Delta)F_RV(H-i)^{-2}(I-P)(H-i)^2E(\Delta)$$
$$+ E_o(\Delta)[(I-F_R)V](H-i)^{-2}(I-P)(H-i)^2E(\Delta)$$

By (β) and (α) the first term is the product of $E_o(\Delta)F_R \in \mathcal{B}_1$ with an operator belonging to $\mathcal{B}(\mathcal{H})$, hence it belongs to $\mathcal{B}_1$ by (P4). In the second term we factorize $(1-\chi_R)(\vec{x})V(\vec{x}) = V_1(\vec{x})V_2(\vec{x})$ with $V_i \in L^2(\mathbb{R}^3)$ as was done above. This term then has the form

$$\underbrace{E_o(\Delta)V_1}_{\in\mathcal{B}_2}\ \underbrace{V_2(H_o+I)^{-1}}_{\in\mathcal{B}_2}\ \underbrace{(H_o+I)(H-i)^{-2}(I-P)}_{\in\mathcal{B}(\mathcal{H})\text{ by }(\alpha)}\ \underbrace{(H-i)^2E(\Delta)}_{\in\mathcal{B}(\mathcal{H})}$$

It follows from (P4) and (P5) that it is also of trace class. Combining all these results, we have shown that

$$E_o(\Delta)HE(\Delta) - H_oE_o(\Delta)E(\Delta) \in \mathcal{B}_1.$$

Next, for $\lambda \in \mathbb{R}$, let $E_o(\overline{\lambda})$ be the spectral projection of H_o corresponding to the set (λ,∞). For $f \in D(H^2) \cap \mathcal{H}_{ac}(H)$, set $g = (H-i)^2 f$. Then

$$E_o(\overline{\lambda})\exp(-iHt)f = E_o(\overline{\lambda})(H-i)^{-2}(I-P)\exp(-iHt)g + E_o(\overline{\lambda})(H-i)^{-2}P\ \exp(-iHt)g$$

The first term may be rewritten as

$$E_o(\overline{\lambda})(H_o+I)^{-1}(H_o+I)(H-i)^{-2}(I-P)\exp(-iHt)g$$

and its norm is less than $(\lambda+1)^{-1}\|(H_o+I)(H-i)^{-2}(I-P)\|\,\|g\|$. Given $\varepsilon > 0$, this is less than $\varepsilon/2$ if λ is sufficiently large. For the second term one notices that $\{\exp(-iHt)g\}$ converges weakly to zero since $g \in \mathcal{H}_{ac}(H)$. Also $E_o(\overline{\lambda})(H-i)^{-2}P \in \mathcal{B}_1 \subset \mathcal{B}_o$, hence by (P7) $E_o(\overline{\lambda})(H-i)^{-2}P\ \exp(-iHt)g$ converges strongly to zero as $t \to \pm\infty$. Hence there exists T such that this term is less than $\varepsilon/2$ if $|t|>T$. This verifies all conditions of Prop. 4, and completeness of $\Omega_\pm(H,H_o)$ is proved.

Another interesting result based on a trace condition similar to the one of Prop. 4 but involving two Hilbert spaces (cf. [46]) has recently been obtained by Matveev and Skriganov [49]. These authors have proved existence and completeness of wave operators for a class of potentials that oscillate very rapidly at infinity. We may mention the following two examples: $V(\vec{x}) = g(\theta,\phi)\sin r^m$ with $m > 4$ and $V(\vec{x}) = g(\theta,\phi)\exp(r^k)\ \sin[\exp(r^m)]$ with $m > k > 0$ and g a smooth function defined on the unit sphere.

The third time-dependent method of proving existence of wave operators is based on the notion of *smooth operators* [50] and has been developed by Lavine in [51]-[53] and [16]. Let H be selfadjoint and A closed. A is called H-smooth if

$$\|A\|_H^2 \equiv \sup_{f\neq 0} \frac{1}{2\pi\|f\|^2}\int_{-\infty}^{\infty} dt\,\|A\exp(-iHt)f\|^2 < \infty \tag{42}$$

Note that $\|A\|_H < \infty$ implies that for any fixed $f \in \mathcal{H}$, $\exp(-iHt)f \in D(A)$ for almost all t. This notion is important for scattering theory since it is related to the absolutely continuous spectrum of H and to asymptotic convergence. In fact it is not hard to prove that, if $\|A\|_H < \infty$, then $Af = 0$ for every $f \perp \mathcal{H}_{ac}(H)$ (cf. [51], Thm 2.1 or [50], Thm 5.8). If f is an eigenvector of H, the preceding assertion is immediately clear from (42). It follows that smoothness properties are also useful for proving absolute continuity of the continuous spectrum of Hamiltonians, cf. [16].

As regards asymptotic convergence, the following fact is interesting:

Proposition 5. (Lavine [52]): Let H_1 and H_2 be as in Prop. 4, $B \in \mathcal{B}(\mathcal{H})$ and Δ a bounded open interval. Suppose there exist T_j (j=1,2) such that $T_j(H_j+i)^{-1} \in \mathcal{B}(\mathcal{H})$ and $T_jE_j(\Delta)$ is H_j - smooth, and for $f_j \in D(H_j)$:

$$(Bf_1,H_2f_2) - (H_1f_1,B^*f_2) = (T_1f_1,T_2f_2) \qquad (43)$$

Then $B_\pm(\Delta) = \underset{t\to\pm\infty}{\text{s-lim}} \exp(iH_2t)B\exp(-iH_1t)E_1(\Delta)$ and $B_\pm^!(\Delta) = \underset{t\to\pm\infty}{\text{s-lim}} \exp(iH_1t)B^*\exp(-iH_2t)E_2(\Delta)$ exist, and $B_\pm^!(\Delta)=B_\pm(\Delta)^*$.

Note that if B maps $D(H_1)$ into $D(H_2)$, (43) essentially means that $H_2B - BH_1 = T_2{}^*T_1$. If B = I and $D(H_1) = D(H_2)$, this requires a factorization of V into $V_1{}^*V_2$ where one of the V_j is H-smooth and the other one H_o-smooth (e.g. on all intervals (a,b) with b > a > 0 and not containing any eigenvectors of H in order to prove completeness for Schrödinger operators).

The proof of Prop. 5 is short , in contradistinction to that of Prop. 4 which is rather lengthy. However the verification of the hypotheses of Prop. 5 requires special techniques which were developed by Kato [50] and Lavine. In [16] Lavine considered the following class of potentials $V = V_1 + V_2$:

$$V_1 \text{ satisfies } \left|\frac{\partial V_1}{\partial r}\right| \le C(1+r)^{-\gamma},\ \gamma>1, \text{ and } \lim_{|\vec{x}|\to\infty} V_1(\vec{x})=0 \qquad (44)$$

$$V_2(\vec{x})=(1+r)^{-\gamma}(V_p(\vec{x})+V_\infty(\vec{x})),\ \gamma>1, \text{ with}$$

$$V_p \in L^p(\mathbb{R}^3),\ p > 3/2, \text{ and } V_\infty \in L^\infty(\mathbb{R}^3). \qquad (45)$$

V_1 may be a long range potential without oscillation at infinity, and V_2 is a short range potential ((45) allows local singularities of the type $V(r) = \gamma r^{-\alpha}$ with $\alpha < 2$. Consequently the Hamiltonian may have to be defined by means of quadratic forms.) If $h(\vec{x})$ is real and such that $|h(\vec{x})|^2$ satisfies (45) (i.e. replace V_2 by $|h|^2$ in (45)), it is then proved that $|\text{Im}z|\ \|(H-z)^{-1}h(\vec{Q})g\|^2 \le C\|g\|^2$ for all $g \in D(h(\vec{Q}))$ and $z \in \{z \mid 0 < |\mathrm{Im} z| < 1, a < \mathrm{Re}\, z < b\}$ if $[a,b] \subset (0,\infty)$ contains no eigenvalues of H. Since

$$\| A \|_H^2 = \frac{1}{\pi} \sup_{\substack{f \in D(A^*) \\ \mathrm{Im}\, z \neq 0}} \frac{1}{\|f\|^2} |\mathrm{Im} z| \, \| (H-z)^{-1} A^* f \|^2$$

([50], Thm. 5.1), this implies that $|V_2|^{\frac{1}{2}}$ is (H_o+V_1)-smooth and $(H_o+V_1+V_2)$-smooth on (a,b) ([52], Lemma 2.1). It then follows from Prop. 5 that, if $V_1 = 0$ and V_2 satisfies (45), the wave operators $\Omega_\pm(H,H_o)$ exist and are asymptotically complete. Also, if $H=H_o+V_1+V_2$ and $H' = H_o + V_1 + V_2'$ where V_1 satisfies (44) and V_2, V_2' satisfy (45), $\Omega_\pm(H',H)$ exist and are complete. The reader may find more details in Lavine's lectures in this volume.

Let us now turn to potentials for which the ordinary wave operators (32) do not exist. These are often called long-range potentials because the interaction is still effective at large distances, even if $V(\vec{x})$ vanishes at infinity (remember that at large times the position probability density of a scattering state is practically zero in any finite region), in such a way that the *wave functions* never behave like free wave functions. (Another example where the limits (32) do not exist is a potential V such that $V(\vec{x}) \to C \neq 0$ as $|\vec{x}| \to \infty$. Then H will not have the same continuous spectrum as H_o.)

In order to prove convergence of $V_t{}^*AV_tE_{ac}(H_o)$ or to obtain generalized wave operators, the same three methods as above can be applied. The method of estimating the decay rate of certain operators has been applied successfully to many situations and smoothness properties have also been used. Trace criteria have been employed by Lavine in [31]. The convergence proof has been carried through for two different kinds of objects: (a) for $V_t{}^*AV_tE_{ac}(H)$ where A belongs to some algebra of free observables, (b) for $V_t{}^*\hat{U}_tU_t$ where $\hat{U}_t$ is a dressing transformation which may depend on time.

By using smoothness techniques Lavine [16] obtains the existence of the homorphisms $\omega_\pm(A) \equiv \operatorname{s-lim}_{t\to\pm\infty} V_t{}^*AV_tE_{ac}(H)$ for the class of potentials (44), (45) with the additional hypothesis that $|\vec{\nabla}V_1| < C(1+r)^{-\gamma}$, $\gamma > 1$, for all operators A belonging to the C*-algebra of all bounded continuous functions of P_1,P_2,P_3. By assuming in addition spherical symmetry of the long-range part V_1 of the interaction, Thomas [36] has recently shown existence of wave operators $\Omega_\pm$ satisfying (34) for all A which are bounded continuous functions of P_1,P_2,P_3 or bounded operator-valued functions of the angular momentum operators L_1,L_2,L_3 or operators belonging to the C*-algebra generated by the above two classes of operators. His contruction is as follows: Let $H_1 = H_o + V_1$ and $W_t = \exp(-iH_1t)$. (44) implies that $V_1(H_o-z)^{-1}$ is compact for $z \in \rho(H_o)$. Since

$$(H_1-z)^{-1}E_{ac}(H_1)-W_t{}^*(H_o-z)^{-1}W_tE_{ac}(H_1)=-(H_1-z)^{-1}W_t^*V_1(H_o-z)^{-1}W_tE_{ac}(H)$$

and $W_tE_{ac}(H_1)$ converges weakly to zero, it follows from (P7) that $\omega_\pm((H_o-z)^{-1}) = (H_1-z)^{-1}E_{ac}(H_1)$ for $z \in \rho(H_o) \cap \rho(H_1)$. On the other hand $\omega_\pm(\exp(i\vec{a}\cdot\vec{P}))$ are unitary operators which depend continuously on $\vec{a} \in \mathbb{R}^3$ in the strong operator topology. Hence by Stone's theorem there exist selfadjoint operators $P_\pm^j$ (j=1,2,3) on $\mathcal{H}_{ac}(H_1)$ such that $\omega_\pm(\exp(i\vec{a}\cdot\vec{P})) = \exp(i\vec{a}\cdot\vec{P}_\pm)$.

Let $g(\vec{x})$ be the Fourier transform of $(\vec{p}^2-z)^{-1}$. Then

$$\omega_\pm((H_o-z)^{-1})=\omega_\pm(\int d^3a \exp(i\vec{a}\vec{P})g(\vec{a}))=\int d^3a \exp(i\vec{a}\vec{P}_\pm)g(\vec{a})=(|\vec{P}_\pm|^2-z)^{-1}$$

where the interchange of the integral and the strong limit in $\omega_\pm$ can be justified by applying the Lebesgue dominated convergence theorem. It follows that $H_1E_{ac}(H_1) = |\vec{P}_\pm|^2$.

By using methods from the theory of ordinary differential equations one then proves that the reduction of $H_1E_{ac}(H_1)$ to any partial wave subspace $\mathcal{H}_{\ell m}$ is unitarily equivalent to multiplication by r in $L^2(\mathbb{R}^+)$. From this one deduces that there exists a unitary operator χ: $\mathcal{H} \to \mathcal{H}_{ac}(H_1)$ such that $\chi H_o\chi^{-1} = H_1E_{ac}(H_1)$ and $\chi L_j\chi^{-1} = L_j$ (j=1,2,3). If one then defines $\vec{K}_\pm = \chi^{-1}\vec{P}_\pm\chi$, one has $|\vec{K}_\pm|^2 = H_o$. By using this and the fact that the operators $\{\exp(i\vec{a}\vec{K}_\pm), \exp(i\vec{b}\vec{L})\}$ satisfy the same commutation relations as $\{\exp(i\vec{a}\vec{P}), \exp(i\vec{b}\vec{L})\}$ and all commute with H_o, one can prove that there exist unitary operators $\Gamma_\pm$ such that $[\Gamma_\pm, \exp(i\vec{b}\vec{L})] = 0$ and $\Gamma_\pm^{-1} \exp(i\vec{a}\vec{K}_\pm)\Gamma_\pm = \exp(i\vec{a}\vec{P})$. Then $\Omega_\pm^{(1)} = \chi\Gamma_\pm$ is the desired generalized wave operator for H_1. For $H = H_1 + V_2$ one may use one of Lavine's results cited above, namely the existence and completeness of $\Omega_\pm(H,H_1)$. It follows that $\omega_\pm(A) = \Omega_\pm A\Omega_\pm^*$ where $\Omega_\pm = \Omega_\pm(H,H_1)\Omega_\pm^{(1)}$.

Let us now look at the method of dressing transformations. Suppose there is a family of unitary operators $\hat{U}_t$ such that $\hat{U}_t$ commutes with all operators of some algebra A_o of free observables and such that

$$\Omega_\pm = \underset{t\to\pm\infty}{\text{s-lim}}\ V_t{}^*\hat{U}_tU_tE_{ac}(H_o) \tag{46}$$

exist. $\Omega_\pm$ are isometric. Define $T_t = \hat{U}_tU_t$ and $P_\pm = \Omega_\pm\Omega_\pm{}^*$. Then for $A \in A_o$

$$\underset{t\to\pm\infty}{\text{s-lim}}\ V_t{}^*AV_tP_\pm = \underset{t\to\pm\infty}{\text{s-lim}}\ V_t{}^*T_tAT_t{}^*V_tP_\pm = \Omega_\pm A\Omega_\pm{}^*$$

i.e. A_o satisfies the algebraic asymptotic condition and $\Omega_\pm$ are the corresponding wave operators. They are asymptotically complete if $P_\pm = E_{ac}(H)$.

It is easy to deduce the following property of the dressing operator $\hat{U}_t$ [32]:

$$\underset{t\to\pm\infty}{\text{s-lim}} (\hat{U}_{t+\tau} - \hat{U}_t) = 0 \qquad (\tau\epsilon\mathbb{R}) \tag{47}$$

This means that $\hat{U}_t$ cannot oscillate too much: The family $\hat{U}_t \equiv I$ (i.e. the (trivial) dressing transformations for short-range potentials) would satisfy (47), but $\hat{U}_t = U_t = \exp(-iH_0t)$ would not. Thus if one writes $\hat{U}_t = \exp(-iX_t)$, (47) roughly means that X_t is a small correction relative to H_0t at large times. It also follows from the above remark that T_t cannot be a group as a function of t unless $\hat{U}_t = U_0$ for all t.

It is interesting to remark that the existence of a family $\{\hat{U}_t\}$ with the above properties is almost equivalent to the algebraic asymptotic condition with A_0 a von Neummnn algebra. In fact the algebraic asymptotic condition (34) together with the requirement that $A_\pm'$ be abelian imply the existence of a family $\{\hat{U}_t\}$ such that (47) holds and such that the limits $\Omega_\pm$ in (46) coincide with the generalized wave operators obtained from the algebraic theory ([32], Thm. 2, [35][39]).

Some technical points are worth noticing here: From the algebraic asymptotic condition alone it does not follow that U_t are unitary. However there is a dense set $\mathcal{D}$ such that the limits (46) exist and are isometric on $\mathcal{D}$ and $\lim_{t\to\pm\infty} \|\hat{U}_t h\| = \|h\|$ for $h \in \mathcal{D}$. Necessary and sufficient conditions for $\hat{U}_t$ to be unitary were given by Mourre [35]. Also $\hat{U}_t$ need not satisfy (47) if there is an energy renormalization, i.e. if $V_t\Omega_\pm = \Omega_\pm W_t$ with $W_t \neq U_t$. However in this case one would write $T_t = W_t\hat{W}_t$ and then $\hat{W}_t$ verifies (47) [32]. It is also clear that the operators $\hat{U}_t$ are not uniquely determined by the algebraic asymptotic condition. Essentially, $\hat{U}_t$ is determined up to multiplication by an operator $Z_t \in A_0'$ such that $\underset{t\to\pm\infty}{\text{s-lim}}\, Z_t = Z_\pm$ where $Z_\pm$ are unitary and in A_0' .

The convergence of $V_t{}^*T_t$ is proved by showing that

$$\int_{\pm T}^{\pm\infty} \left\| \frac{d}{dt} V_t{}^*T_tf \right\| dt < \infty \tag{48}$$

for f in some dense set (cf. our earlier discussion of this method). This requires rather delicate estimates because of the extra factor $\hat{U}_t$ in T_t (cf. Buslaev and Matveev [54], Alsholm [55] and further references given in [55]). By writing $\hat{U}_t = \exp(-iX_t)$, $X_t = X_t(\vec{P})$, one has (cf. [55])

$$\left\| \frac{d}{dt} V_t{}^*T_tf \right\| = \left\| (V - \frac{\partial X_t}{\partial t}) U_t\hat{U}_tf \right\| = \left\| [V(\vec{Q} + \frac{\vec{P}t}{m} + \vec{\nabla}_pX_t) - \frac{\partial X_t}{\partial t}]f \right\|$$

since T_t commutes with $\frac{\partial X_t}{\partial t}$ and

$$\vec{Q}U_t\hat{U}_t = i\vec{\nabla}_p U_t\hat{U}_t = U_t\hat{U}_t\vec{Q} + U_t\hat{U}_t\left(\frac{\vec{p}t}{m} + \vec{\nabla}_p X_t\right)$$

If one develops for large t

$$V\left(\vec{x} + \frac{\vec{p}t}{m} + \vec{\nabla}_p X_t\right) \approx V\left(\frac{\vec{p}t}{m} + \vec{\nabla}_p X_t\right) + \vec{x}\cdot(\vec{\text{grad}}\, V)\left(\frac{\vec{p}t}{m} + \vec{\nabla}_p X_t\right)$$

this suggests to choose X_t such that

$$\frac{\partial X_t(\vec{p})}{\partial t} = V\left(\frac{\vec{p}t}{m} + \vec{\nabla}_p X_t(\vec{p})\right) \tag{49}$$

Indeed if $V(\vec{x}) \sim |\vec{x}|^{-\alpha}$ at infinity, $|\text{ grad } V \,| \sim |\vec{x}|^{-1-\alpha}$, and this should permit to obtain a finite integral in (48) for any $\alpha > 0$.

Eq. (49) can be viewed as a Hamilton-Jacobi equation $\partial S/\partial t + H(-\vec{\nabla}_p S,\vec{p},t) = 0$ for the function $S(\vec{p},t) = -\vec{p}^2 t/(2m) - X_t(\vec{p})$. Let us consider its successive iterations:

$$X_t^{(0)}(\vec{p}) = 0,\quad X_t^{(k)}(\vec{p}) = \int_0^t ds\, V\left(\frac{\vec{p}s}{m} + \vec{\nabla}_p X_s^{(k-1)}(\vec{p})\right) \tag{50}$$

The first iteration is just

$$X_t^{(1)}(\vec{p}) = \int_0^t ds\, V\left(\frac{\vec{p}s}{m}\right)$$

This gives the dressing operator first introduced by Dollard [56] for the Coulomb potential $V(r) = \gamma r^{-1}$:

$$X_t^{(D)}(p) = \sqrt{\frac{m}{2}}\,\gamma\,(\text{sign } t)\, H_o^{-\frac{1}{2}} \log(H_o|t|)$$

For a short range potential $\lim_{t\to\pm\infty} X_t^{(1)}(p)$ is finite, hence it is asymptotically constant in this approximation and may be replaced by I. For $V(\vec{x}) = |\vec{x}|^{-\alpha}$ one may check that the integrand in the second iteration behaves like $at^{-\alpha} + bt^{-2\alpha}$, and the term in $t^{-2\alpha}$ gives an essential contribution (i.e. nonconstant at $|t| \to \infty$) to the integral in (50) if $\alpha \leq \frac{1}{2}$. It has been possible to prove existence of the generalized wave operators (46) by taking $\hat{U}_t = \exp(-iX_t^{(k)}(\vec{p}))$ for a suitable k and assuming various bounds on the partial derivatives up to order N of the long-range part of V, cf. [54][55]. The choice of k and N depends on the rate of decrease of $V(\vec{x})$ at infinity. If $|V(\vec{x})| \leq \text{const } |\vec{x}|^{-\alpha}$, $\alpha > \frac{1}{2}$, it suffices to take $k = 1$.

Even more general results have been obtained with time-independent dressing transformations. This method was first used by Mulherin and Zinnes [57] (cf. also [58]) for the Coulomb interaction $V(r) = \gamma r^{-1}$. They give densely defined operators $K_\pm$ such that

$$\underset{t\to\pm\infty}{\text{s-lim}}\ V_t{}^*KU_t \equiv \Omega_\pm \tag{51}$$

are isometric and coincide with the generalized wave operators of Dollard. $K_\pm$ have the form

$$(K_\pm f)(\vec{x}) = (2\pi)^{-3/2} \int d^3p\ \exp(i\vec{p}\cdot\vec{x} \mp i\ \frac{m\gamma}{p}\ \ln(pr \pm \vec{p}\cdot\vec{x}))\tilde{f}(\vec{p})$$

The exponential is nothing else but the first term in the expansion at $|\vec{x}| \to \infty$ of the stationary Coulomb wave functions in powers of $|\vec{x}|^{-1}$, but written for all $\vec{x}$. The analogous formula for a short-range potential would be the Fourier transformation, i.e. $K_\pm = I$

For long-range potentials the operators $K_\pm$ will not commute with U_t nor with V_t, since $V_t{}^*U_t$ does not converge strongly. Of course the problem of finding $K_\pm$ such that (51) holds is highly non-unique, since even for a short-range potential one could take $K_\pm = I$ or $K_\pm = \Omega_\pm$. Also, the relation of the wave operators (51) to those of the algebraic asymptotic condition is not immediately clear. One will have to either verify that $\underset{t\to\pm\infty}{\text{s-lim}}\ [A,K_\pm U_t] = 0$ for all $A \in \mathcal{A}_o$ and $\underset{t\to\pm\infty}{\text{w-lim}}\ U_t{}^*K_\pm{}^*K_\pm U_t = I$ or exhibit a time-dependent dressing transformation $\hat{U}_t$ of the type discussed earlier such that $\underset{t\to\pm\infty}{\text{s-lim}}\ (U_t{}^*K_\pm U_t - \hat{U}_t) = 0$.

The method of Mulherin and Zinnes has been generalized by Matveev and Skriganov [89] and recently by Georgescu [59] to a very large class of spherically symmetric potentials, which in particular are not required to satisfy any differentiability condition. We summarize some of Georgescu's results. His assumptions are $V = V_1 + V_2$ with

$$V \in L^1_{\ell oc}((0,\infty)) \quad \text{and} \quad \int_0^1 r|V(r)|\ dr < \infty$$

$$\lim_{r\to\infty} V_1(r) = 0, \quad V_1(r) \ \text{continuous near infinity and}$$

$$\int_1^\infty |dV_1(r)| < \infty$$

$$V_2 \in L^1((a,\infty)) \quad \text{for some } a < \infty.$$

(V_1 is the long-range part, V_2 the short-range part of V). This defines a selfadjoint operator in each partial wave subspace $\mathcal{H}_{\ell m}$ (a boundary condition at $r = 0$ is needed for $\ell = 0$). By using methods developed by I. S. Kac, one can obtain information about eigenfunctions and the spectral representation of the differential operator $H_{\ell m}$ which are similar to known results about differential operators (cf. [28] for the latter).

For $\lambda \in \mathbb{C}$, there is a unique solution $\psi_\lambda^{(\ell)}(r)$ of the differential equation $[-\ d^2/dr^2 + \ell(\ell+1)r^{-2} + V(r)]f = \lambda f$ such that

$\lim_{r \to 0} r^{-\ell-1} f(r) = 1$. Let

$$[a_\ell(\lambda)]^2 = 2 \lim_{r\to\infty} \frac{1}{r} \int_0^r |\psi_\lambda^{(\ell)}(s)|^2 \, ds$$

and

$$\Phi_\ell(\lambda,r) = \pi^{-\frac{1}{2}} \lambda^{-\frac{1}{4}} a(\lambda)^{-1} \psi_\lambda^{(\ell)}(r)$$

It can then be proved that the spectral representation of $H_{\ell m} E_{ac}(H_{\ell m})$ is of the form [59]

$$\hat{f}(\lambda) = \int_0^\infty \Phi_\ell(\lambda,r) f(r) \, dr \qquad [f \in L^2(\mathbb{R}^+)] \tag{52}$$

with the inverse transformation

$$f(r) = \int_0^\infty \Phi_\ell(\lambda,r) \hat{f}(\lambda) \, d\lambda$$

In particular the absolutely continuous spectrum of $H_{\ell m}$ is $[0,\infty)$ and simple.

The asymptotic behavior of $\Phi_\ell(\lambda,r)$ is as follows

$$\Phi_\ell(\lambda,r) \underset{r\to\infty}{\sim} \pi^{-\frac{1}{2}} \lambda^{-\frac{1}{4}} \sin[\delta_\ell(\lambda) + \int_{r_\ell(\lambda)}^r ds \sqrt{\lambda - V_1(s)} \;]$$

The splitting up of the argument of the sine function into the sum of a "phase shift" and an integral is not unique. Essentially the only requirement is that $r_\ell(\lambda)$ be such that the square root is real (if V_1 is negative, one may choose $r_\ell(\lambda) = 0$). To define $K_\pm$ in $L^2(\mathbb{R}^3)$ one will choose $r_\ell(\lambda)$ independent of ℓ . Once this choice is made, all the phase shifts are determined. Conversely: the choice of one phase shift determines $r(\lambda)$ and thus all the other phase shifts. This indeterminacy is exactly the same as that obtained from the algebraic asymptotic condition with $A_o = \{H_o\}'$.

The dressing transformations $K_\pm$ in $L^2(\mathbb{R}^3)$ are now defined by $(m = \frac{1}{2})$

$$(K_\pm f)(\vec{x}) = (2\pi)^{-3/2} \int d^3p \, \exp[\pm i \int_{r(p^2)}^r \sqrt{p^2 - V_1(s)} \, ds \pm ipr] \exp(i\vec{p}\cdot\vec{x}) \tilde{f}(\vec{p}) \tag{53}$$

and the wave operators in $\mathcal{H}_{\ell m}$ by

$$(\Omega_\pm f)(r) = \int_0^\infty d\lambda \, \Phi_\ell(\lambda,r) \exp[\pm i \delta_\ell(\lambda)] \hat{f}_\ell^o(\lambda)$$

where $\hat{f}_\ell^o(\lambda)$ is defined by (52) but with $\Phi_\ell(\lambda,r)$ replaced by the corresponding function for the operator $H_{\ell m}^o = -d^2/dr^2 + \ell(\ell+1)r^{-2}$. One can then prove (51) by using estimates on the asymptotic behavior of $\Phi_\ell(\lambda,r)$ and a generalization of the Riemann-Lebesgue Lemma.

One notices that the quantity

$$\int_{r(p^2)}^{r} \sqrt{p^2 - V_1(s)}\; ds - pr \tag{54}$$

appearing in (53) is the action of the Maupertuis-Euler-Lagrange principle for a classical motion with potential V_1 minus the same action for $V_1 = 0$ [60]. With a suitable modification of the second term in (54) one can also use this method if a part of the potential is included in H_o.

This gives the existence of wave operators for the class of potentials mentioned above. The next step is to verify (34) for $A \in \{H_o\}'$ by finding $\hat{U}_t^{\pm}$ such that $\underset{t\to\pm\infty}{\text{s-lim}}(U_t{}^*K_{\pm}U_t - \hat{U}_t^{\pm}) = 0$. At the moment of writing, this proof is not yet complete . However, if one develops the square-root in (54) into a series

$$\sqrt{p^2 - V_1(s)} = p\left[1 + \sum_{n=1}^{\infty} c_n \left(\frac{V_1(s)}{p^2}\right)^n\right] \tag{55}$$

and if $|V_1(r)| \leq cr^{-\alpha}$ with $\alpha > \frac{1}{2}$, one may neglect all the terms with $n > 1$ in (54) (these terms would simply give a constant contribution to the integral in (53) as $r \to \infty$). Then

$$\int_{r(p^2)}^{r} ds\; p\left(1 - \frac{V_1(s)}{2p^2}\right) - pr = \text{const} - \frac{1}{2p}\int_{r(p^2)}^{r} V_1(s)\; ds$$

By setting $r = pt$ this is seen to coincide (up to an additive constant in t) with $X_t^{(1)}(p)$ given by (50), and for these cases the identity of $\underset{t\to\pm\infty}{\text{s-lim}}\, V_t{}^*K_{\pm}U_t$ and $\underset{t\to\pm\infty}{\text{s-lim}}\, V_t{}^*T_t$ can be directly verified.

The wave operators obtained by Georgescu are also asymptotically complete. Indeed the restriction of $\Omega_{\pm}$ to each $\mathfrak{H}_{\ell m}$ is asymptotically complete, as is easily seen from the following facts (cf. [26]): the absolutely continuous spectrum of $H_{\ell m}$ is $[0,\infty)$ and simple, $H_{\ell m}$ is unitarily equivalent to $H^o_{\ell m}$ (i.e. $H_{\ell m}\Omega_{\pm} = \Omega_{\pm}H^o_{\ell m}$) and $H^o_{\ell m}$ is unitarily equivalent to multiplication by r in $L^2(\mathbb{R}^+)$. One can also use the chain rule to combine Georgescu's result with the preceding results and obtain generalized wave operators for non-spherically symmetric longe-range potentials. As a last remark, we may mention that Georgescu's assumptions on V suggest that the algebraic asymptotic condition might be violated by potentials which go to zero at infinity but which are not of bounded variation at infinity. Similar conclusions were arrived at by Thomas [36] who gives as an example of such a potential $V(r) = r^{-1} \sin r$.

For many-body systems the known results about wave operators do not present such a complete picture. Many-body problems with long-range potentials have hardly been considered, although it is clear from Dollard's analysis [29][56] of the many-body problem with Coulomb potentials how the existing results about the 2-body case have to be extended to the n-body case. If the interaction consists of short-range potentials V_{ij} depending on the relative position $\vec{r}_i - \vec{r}_j$ of the corresponding pair, the existence of the channel wave operators can be proved along the lines of Prop. 3 if one assumes suitable local conditions on V_{ij} and $V_{ij}(I-F_R) \in L^2(\mathbb{R}^3) + L^p(\mathbb{R}^3)$, $2 < p < 3$ [3] [25]. Asymptotic completeness however is difficult to establish. Lavine [53] has proved completeness by using smoothness methods for the case where the V_{ij} are continuously differentiable *repulsive* short-range potentials. Similarly Iorio and O'Carrol [90] obtained completeness in the weak coupling limit. In these cases the problem is much simplified since there is only one scattering channel. Most other results have been deduced by stationary methods [61]-[64]. Here it is required that $V_{ij} \in L^2(\mathbb{R}^3)$, that $\tilde{V}_{ij}(\vec{p})$ be sufficiently smooth and decrease sufficiently rapidly at infinity (cf. [61], p. 7). There are also assumptions on the eigenvalues of H embedded in its continuous spectrum and on the absence of such eigenvalues for the Hamiltonians H_C describing any m-particle subsystem ($2 \leq m < n$). These assumptions have to be verified separately (cf. [15] and references given there for known results). Completeness results with non-local interactions were announced by Van Winter [91].

Let us conclude with a few remarks about Problems E and F (more details can be found in the lectures of Combes). Stationary state scattering theory has been developed on two levels, namely as relations between operators in Hilbert space and as integral equations in the $\vec{X}$ - or $\vec{P}$ - representation. The idea in the first approach is to start from $\Omega_\pm = \underset{t\to\pm\infty}{\text{s-lim}}\, V_t{}^*U_t$ (we assume $\mathcal{H}_{ac}(H_o)=\mathcal{H}$), to rewrite it as an integral over time, then to replace one of the evolution groups by its spectral integral and finally to interchange the spectral integral with that over time. This gives for instance

$$\Omega_+ = \underset{\varepsilon\downarrow 0}{\text{s-lim}}\; \varepsilon\int_0^\infty dt\, \exp(-\varepsilon t)V_t^*U_t = \underset{\varepsilon\downarrow 0}{\text{s-lim}}\; \varepsilon\int_0^\infty dt\, \exp(-\varepsilon t)V_t^*\int_{\mathbb{R}}\exp(-i\lambda t)dE_0(\lambda)$$

$$= \underset{\varepsilon\downarrow 0}{\text{s-lim}}\; \varepsilon\int_{\mathbb{R}}\int_0^\infty dt\, V_t^*\exp(-i\lambda t-\varepsilon t)dE_0(\lambda) = \underset{\varepsilon\downarrow 0}{\text{s-lim}}\; i\varepsilon\int_{\mathbb{R}}(H-\lambda+i\varepsilon)^{-1}dE_0(\lambda) \tag{56}$$

where the spectral integrals are defined as strong Riemann integrals. Similarly one obtains the Lippmann-Schwinger equations, e.g. (we set $R_z \equiv (H-z)^{-1}$, $R_z^o \equiv (H_o - z)^{-1}$)

$$\Omega_+ = I - \underset{\varepsilon\downarrow o}{\text{s-lim}} \int_{\mathbb{R}} R^o_{\lambda-i\varepsilon}V\Omega_+\, dE_o(\lambda) \quad [\text{on } D(H_o)] \tag{57}$$

and (on $D(H_o)$)

$$S-I = \underset{\varepsilon_1 \downarrow o}{\text{s-lim}} \ \underset{\varepsilon_2 \downarrow o}{\text{s-lim}} \int_{\mathbb{R}} (R^o_{\lambda-i\varepsilon_1} - R^o_{\lambda+i\varepsilon_1})(V - VR_{\lambda+i\varepsilon_2} V)\, dE_o(\lambda) \tag{58}$$

(cf. [3][42],[65]-[68][92][93] and further references given in [66]). The only assumption needed in addition to the asymptotic condition is that $D(V) \supset D(H_o) = D(H)$.

In the second approach one shows under more particular assumptions on V (smoothness of $V(\vec{x})$ or $\tilde{V}(\vec{k})$, or integrability conditions) that the operators appearing in the equations given above are integral operators in certain spectral representations and that the wave functions satisfy the corresponding integral equations in that representation. Some recent references are [3][4][43][68]-[72][94], and [61]-[63], [73] for the many-body case.

For long range potentials one can apply the method indicated above to $\Omega_\pm = \underset{t\to\pm\infty}{\text{s-lim}}\, V_t{}^* K_\pm U_t$ and obtain equations similar to (56), (57) but where V is replaced by $HK_+ - K_+H_o$ (one may run into domain problems here). It is also possible to show that $\Omega_\pm$ satisfy homogeneous Lippmann-Schwinger equations [30][74]. For this, one writes (on $D(H_o) = D(H)$) $\Omega_\pm = R^o_z(H_o - z)\Omega_+ = R^o_z(-V + H - z)\Omega_+$ and then

$$\Omega_+ = \int_{\mathbb{R}} \Omega_+\, dE_o(\lambda) = -\int_{\mathbb{R}} R^o_{\lambda-i\varepsilon} V\Omega_+\, dE_o(\lambda) + \int_{\mathbb{R}} R^o_{\lambda-i\varepsilon}\Omega_+ H_o\, dE_o(\lambda)$$

$$-\int_{\mathbb{R}} R^o_{\lambda-i\varepsilon}\Omega_+\lambda\, dE_o(\lambda) + i\varepsilon \int_{\mathbb{R}} R^o_{\lambda-i\varepsilon}\, dE(\lambda)\Omega_+$$

where we have used $H\Omega_+ = \Omega_+ H_o$. The second and third term on the right-hand side cancel. Thus, for any $\varepsilon > 0$

$$\Omega_+ = -\int_{\mathbb{R}} R^o_{\lambda-i\varepsilon} V\Omega_+\, dE_o(\lambda) + i\varepsilon \int_o^\infty dt\, \exp(-\varepsilon t) U_t^* V_t \Omega_+ \tag{59}$$

Now $U_t{}^* V_t \Omega_+ = U_t{}^*(V_t\Omega_+ - T_t) + U_t{}^* T_t$. The first term converges strongly to zero, cf. (46). Hence its time average in (59) converges strongly to zero as $\varepsilon \to +0$. It remains to investigate

$$\varepsilon \int_o^\infty dt\, \exp(-\varepsilon t)\hat{U}_t\, dt \tag{60}$$

as $\varepsilon \to +0$. For a short-range potential, $\hat{U}_t = I$, hence the integral in (60) equals I for all ε , and (59) leads again to (57). For various long range potentials one can check [30][74] that (60) converges *weakly* to zero by using the explicit form of $\hat{U}_t$ (this does not follow from the asymptotic condition). Hence

$$\Omega_+ = \underset{\varepsilon\downarrow 0}{\text{w-lim}} \int_{\mathbb{R}} R^o_{\lambda-i\varepsilon} V\Omega_+\, dE_o(\lambda) \qquad \text{on } D(H_o) = D(H)$$

It is also interesting to note that for a long-range potential the right hand side of (58) is equal to zero (with weak limits), i.e. $V - VR_{\lambda+io}V$ does not give the scattering amplitude. Expressions for the scattering amplitude can be found in [93].

The results about Problem F, although quite sufficient for practical purposes [75], are the least satisfactory ones from the mathematical point of view. One always has to make assumptions which permit to obtain δ - functions, and it is often not clear how these limits are defined mathematically (e.g. [76], eqs. (5.24), (5.26); [3], bottom of page 35). One may distinguish two types of methods for obtaining cross sections: that of Jauch [41] which involves averaging over an infinitely extended and very thin (no multiple scattering!) target with uncorrelated scattering centers, and that of Green and Lanford [76] using asymptotic values of momentum observables (cf. also [77]-[79]). Hunziker's method [3] uses both these ideas and the averaging is done over displacements of the beam orthogonally to its direction of propagation rather than over the target. In our view it might be of interest to study some new ways of defining cross sections.

The method of Green and Lanford can also be used for long range potentials, and one can show (at least in the two-body case) that the indeterminacy of the S-operator has no influence on the cross section or the position probability density of states at large times [33], [80].

Other mathematical problems which have received some attention recently are the definition of the time delay [81] and the behaviour of (relativistic) cross sections at high energies [82][83].

Acknowledgments. It is a pleasure to thank the Department of Theoretical Physics of the University of Geneva, in particular Professor J. M. Jauch, for maintaining an interesting research group in scattering theory, and V. Georgescu, D. B. Pearson and L. E. Thomas for assistance in the preparation of parts of the notes and for letting me use some of their unpublished results.

References.

An extensive list of references about scattering theory can be found in the London lectures of Hepp [1] and in the books of Newton [84] and Simon [4]. An excellent presentation of the various aspects of scattering theory is given in the Boulder lectures of Hunziker [3].

1. K. Hepp, Quantum Scattering Systems: A Guide to the Literature in *Mathematics of Contemporary Physics*, R. F. Streater, ed., Academic Press, London (1972).

2. M. Reed and B. Simon, *Methods of Modern Mathematical Physics*, Vol. I, Academic Press, New York (1972).

3. W. Hunziker, Mathematical Theory of Multiparticle Quantum Systems, in *Lectures in Theoretical Physics*, Vol. X A, W. E. Brittin, ed., Gordon and Breach, New York (1968).

4. B. Simon, *Quantum Mechanics for Hamiltonians Defined as Quadratic Forms*, Princeton University Press, Princeton (1971).

5. M.Schechter, *Spectra of Partial Differential Operators*, North Holland, Amsterdam (1971).

6. W. G. Faris, Helv. Phys. Acta 45, 1074-1088 (1972).

7. T. Kato, Israël J. Math 13, 135-148 (1972).

8. U. -W. Schmincke, Math.Z. 126, 71-81 and 129, 335-350 (1972).

9. E. Nelson, J. Math. Phys. 5, 332-343 (1964).

10. H. Behncke, Nuovo Cim. 55A, 780-785 (1968).

11. W. Hunziker, Helv. Phys. Acta 39, 451-462 (1966).

12. E. Balslev and J. M. Combes, Comm. Math. Phys. 22, 280-294 (1971).

13. E. Balslev, Ann. Phys. 73, 49-107 (1972).

14. K. Jörgens and J. Weidmann, *Spectral Properties of Hamiltonian Operators*, Lectures Notes in Math. Vol. 313, Springer, Berlin (1973).

15. S. Albeverio, Ann. Phys. 71, 167-276 (1972).

16. R. Lavine, J. Funct. Anal. 12, 30-54 (1973).

17. L. E. Thomas, Helv. Phys. Acta 45, 1057-1065 (1972) and Time-Dependent Approach to Scattering from Impurities in Crystals, Univ. of Geneva, preprint.

18. T. Kato, in *Proc. Internat. Conf. on Funct. Anal. and Related Topics*, p. 206-215, Tokyo Univ. Press, Tokyo (1970).

19. T. Kato, Suppl. Progr. Theor. Phys. 40, 3-19 (1967)

20. D. Ruelle, Nuovo Cim. 61A, 655-662 (1969).

21. C. H. Wilcox, J. Funct. Anal. 12, 257-274 (1973).

22. T. Kato, *Perturbation Theory for Linear Operators*, Springer, New York (1966).

23. W. O. Amrein and V. Georgescu, On the Characterization of Bound States and Scattering States in Quantum Mechanics, Helv. Phys. Acta, to appear.

24. T. Ikebe and T. Kato, Arch. Ratl. Mech. Anal. 9, 77-92 (1962).

25. W. Hunziker, Helv. Phys. Acta 40, 1052-1062 (1967).

26. W. O. Amrein and V. Georgescu, Asymptotic Completeness of Nonrelativistic Wave Operators with Highly Singular Potentials, in preparation.

27. D. B. Pearson, Time-Dependent Scattering Theory for Highly Singular Potentials in preparation.

28. M. A. Naimark, *Linear Differential Operators*, Part II, Harrap, London (1967).

29. J. D. Dollard, Rocky Mountain J. Math. 1, 5-88 (1971).

30. V. Georgescu, private communication.

31. R. Lavine, J. Funct. Anal. 5, 368-382 (1970).

32. W. O. Amrein, Ph.A. Martin and B. Misra, Helv. Phys. Acta 43, 313-344 (1970).

33. J.V. Corbett, Phys. Rev. D 1, 3331-3344 (1970).

34. E. Prugovecki, Nuovo Cim.B 4, 105-123 (1971).

35. E. Mourre, Quelques résultats dans la théorie algébrique de la diffusion, Ann. Inst. Henri Poincaré, to appear.

36. L. E. Thomas, On the Algebraic Theory of Scattering, Univ. of Geneva, preprint.

37. J. Dixmier, Les algèbres d'opérateurs dans l'espace de Hilbert, Gauthier-Villars, Paris (1969).

38. W. W. Zachary, Comment on a Paper of Amrein, Martin and Misra, Helv. Phys. Acta, to appear.

39. W. O. Amrein, V. Georgescu and Ph. A. Martin, On the Algebraic Multichannel Scattering Theory, in preparation.

40. R. Lavine and D. B. Pearson, private communications.

41. J. M. Jauch, Helv. Phys. Acta 31,127-158 and 661-684 (1958).

42. C. Chandler and A. G. Gibson, Transition from Time-Dependent to Time-Independent Multichannel Quantum Scattering Theory, J. Math. Phys., to appear.

43. T. Kato and S. T. Kuroda, in *Proc. Internat. Conf. on Funct. Anal. and Related Fields*, F. E. Browder, ed., p. 99-131, Springer, Berlin (1970).

44. J. Kupsch and W. Sandhas, Comm. Math. Phys. 2, 147-154 (1966).

45. C. R. Putnam, *Commutation Properties of Hilbert Space Operators and Related Topics*, Springer, Berlin (1967).

46. A. L. Belopol'skii and M. S. Birman, Math. USSR-Izvestija 2, 1117-1130 (1968).

47. D. B. Pearson, J. Math Phys. 13, 1490-1499 (1972).

48. D. B. Pearson,and D. H. Whould, Nuovo Cim. 14A, 765-780 (1973).

49. V. B. Matveev and M. M. Skirganov, Soviet Math. Dokl. 13, 185-188 (1972).

50. T. Kato, Math. Ann. 162, 258-279 (1966).

51. R. Lavine, Comm. Math. Phys. 20 301-323 (1971).

52. R. Lavine, Indiana Univ. Math. J. 21, 643-656 (1972).

53. R. Lavine, J. Math. Phys. 14, 376-379 (1973).

54. V. S. Buslaev and V. B. Matveev, Theor. Math. Phys. 2, 266-274 (1970).

55. P. K. Alsholm, Wave operators for Long Range Scattering, preprint.

56. J. D. Dollard, J. Math Phys. 5, 729-738 (1964).

57. D. Mulherin and I. I. Zinnes, J. Math. Phys. 11, 1402-1408 (1970).

58. C. Chandler and A. G. Gibson, Time-Dependent Multichannel Coulomb Scattering Theory, Univ. of New Mexico, Albuquerque, preprint.

59. V. Georgescu, Ph. D. Thesis, Univ of Geneva, in preparation.

60. L. D. Landau and E. M. Lifshitz, *Mechanics*, Pergamon Press, Oxford (1960).

61. L. D. Faddeev, *Mathematical Aspects of the Three-body Problem in the Quantum Theory of Scattering*, Israël Program of Scientific Translations, Jerusalem (1965).

62. K. Hepp, Helv. Phys. Acta 42, 425-458 (1969).

63. A. Schtalheim, Helv. Phys. Acta 44, 642-661 (1971).

64. J. M. Combes, Nuovo Cim. 64A, 111-144 (1969).

65. D. B. Pearson, Nuovo Cim. 2A, 853-880 (1971).

66. W. O. Amrein, V. Georgescu and J. M. Jauch, Helv. Phys. Acta 44, 407-434 (1971).

67. C. Chandler and A. G. Gibson, Helv. Phys. Acta 45, 734-737 (1972).

68. E. Prugovecki, Multichannel Stationary Scattering Theory in Two -Hilbert Space Formulation, J. Math. Phys., to appear.

69. T. Kato and S. T. Kuroda, Rocky Mountain J. Math. 1, 127-171 (1971).

70. P. Alsholm and G. Schmidt, Arch. Ratl. Mech. Anal. 40, 281-311 (1971).

71. P. A. Rejto, J. Math. Anal. Appl. 17, 453-462 and 20, 145-187 (1967).

72. J. S. Howland, J. Math. Anal. Appl. 20, 22-47 (1967).

73. R. G. Newton, J. Math. Phys. 12, 1552-1567 (1971) and Ann. Phys. 74, 324-351 (1972).

74. E. Prugoveki and J. Zorbas, Modified Lippmann-Schwinger Equations for Two-body Scattering Theory with Long-Range Potentials, Univ. of Toronto, preprint.

75. J. R. Taylor, *Scattering Theory,* Wiley, New York (1972).

76. T. A. Green and O. E. Lanford, J. Math. Phys. 1, 139-148 (1960).

77. J.D. Dollard, Comm. Math. Phys. 12, 193-203 (1969).

78. J. D. Dollard, Scattering into Cones II: n-Body Problems, J. Math. Phys., to appear.

79. J. M. Jauch, R. Lavine and R. G. Newton, Helv. Phys. Acta 45, 325-330 (1972).

80. Ph. A. Martin, Helv. Phys. Acta 45, 794-801 (1972).

81. J. M. Jauch and J. P. Marchand, Helv. Phys. Acta 40, 217-229 (1967).
J. M. Jauch, B. N. Misra and K. B. Sinha, Helv. Phys. Acta 45, 398-426 (1972). See also Ph. Martin and B. Misra, these proceedings.

82. Ph. A. Martin and B. Misra, On Trace-class Operators of Scattering Theory and the Asymptotic Behavior of Scattering Cross Sections at High Energy, J. Math. Phys., to appear.

83. Ph. A. Martin, On the High Energy Limit in Potential Scattering, Nuovo Cim., to appear.

84. R. G. Newton, *Scattering Theory of Waves and Particles,* McGraw Hill, New York (1966).

85. K.Gustafson and P. A. Rejto, Israel J. Math. 14, 63-75 (1973).

86. P.A. Rejto, On a Theorem of Titchmarsh-Weidmann Concerning Absolutely Continuous Operators, University of Minnesota, Minneapolis, preprint.

87. S. T. Kuroda, Nuovo Cim. 12, 431-454 (1959).

88. S. Agmon , J. d'Analyse Math. 23, 1-26 (1970).

89. V. B. Matveev and M. M. Skriganov, Theor. Math. Phys. 10, 156-164 (1972).

90. R. I. Iorio and M. O'Carroll, Comm. Math. Phys. 27, 137-145 (1972).

91. C. van Winter, Complex Dynamical Variables and Asymptotic Completeness in Multichannel Scattering, Lectures given at this conference.

92. E. Prugovecki, *Quantum Mechnics in Hilbert Space*, Academic Press, New York (1971).

93. C. Chandler and A. G. Gibson, Time-Independent Multichannel Scattering Theory for Charged Particles, Univ. of New Mexico Albuquerque, preprint.

94. N. Shenk and D. Thoe, Rocky Mountain J. Math. 1, 89-125 (1971).

COMMUTATORS AND LOCAL DECAY

Richard Lavine

University of Rochester

1. Introduction.

A typical problem in mathematical scattering theory is, given a pair of differential equations involving space and time which are asymptotically the same at large values of the space variable, to show that pairs of solutions become asymptotically the same in the distant future or past. For this it is necessary to prove *local decay*: the solutions eventually leave the region where the difference between the equations is large. For quantum mechanical two-body problems with potentials approaching zero without too much oscillation at infinity, local decay can be proved by a method, to be explained below, whose essential element is closely related to the generator of the group of dilations of space. The results of Balslev and Combes [1] and Van Winter [2] have already made it clear that this group is a powerful tool in scattering theory.

We shall be concerned mostly with local decay of solutions of the Schrödinger equation

$$i\,\frac{\partial\psi(t)}{\partial t} = (H_0 + V)\psi(t)\ , \quad (t)\ \varepsilon\ L^2(R^n)$$

where $H_0 = -\sum_{j=1}^{n} \partial^2/\partial xj^2$ and V is multiplication by a function V(x) satisfying

(1.1) $V(x) = V_1(x) + V_2(x)\ ,\ \left|\frac{\partial V_1(x)}{\partial r}\right| + \left|V_2(x)\right| \le (1+|x|)^{-\alpha},\ \alpha > 1.$

Indications will be given, as far as possible, as to how the method might be applied to other problems.

Our method is especially useful in treating the case where V is *long range*, i.e., it decays at infinity at least as slowly as

J. A. LaVita and J.-P. Marchand (eds.), Scattering Theory in Mathematical Physics, 141–156.

as the Coulomb potential $V(x) = C/|x|$. For short range potentials ($V(x) \leq (1+|x|)^{-\alpha}$, $\alpha > 1$) the method provides an alternative to approaches which make heavy use of the Fourier transform or explicit presentations of solutions of "unperturbed" problems. Thus it may be possible to carry it over to more general settings. This method also lends itself naturally to a study of the *time delay* and raises questions concerning the relation of this quantity to properties of the potential, such as barriers, which should be expected to cause long time delays.

In section 1 some estimates related to local decay are discussed. In section 2 a way of proving such estimates is introduced in a way not limited to the Schrodinger equation. What is needed is a quantity that increases with time. A hint on how to find such a quantity is given in section 3, and applied to some classical examples. In section 4 this idea is carried out for the Schrödinger equation. In the last section the time delay is expressed in terms of the potential in a way that shows it is bounded for each energy (for a large class of short range potentials.) Some unsolved questions will be raised along the way.

We shall be more concerned with explanation of the basic ideas than with generality or precision. More detail can be found in the references. A general reference for notions about potential scattering is the article by Amrein.

2. Time dependent and time independent estimates

The solution $\psi(x,t)$ of the Schrodinger equation gives the probability of finding a quantum mechanical particle in a set S at time t by the formula $\int_S |\psi(x,t)|^2 dx$. Thus the expected time spent in S is

$$\int_{-\infty}^{\infty} \int_S |\psi(x,t)|^2 \, dxdt = \int_{-\infty}^{\infty} ||X_S e^{-iHt}\psi||^2 \, dt$$

(where X_S is the function equal to one on S and 0 off S).

One version of local decay would require this expression to be finite for any bounded set S. This is related to the concept of *H-smoothness* introduced and developed by Kato [3]. A bounded self-adjoint operator B on a Hilbert space is smooth on the interval (a,b) with respect to a self-adjoint operator H if

$$(2.1A) \quad \int_{-\infty}^{\infty} ||B\, e^{-iHt} E((a,b))\phi||^2 dt \leq C\, ||\phi||^2$$

for some $C > 0$ and all ϕ. ($E(\cdot)$ is the family of spectral projections for H).

The estimate (2.1A) can be proved for the Schrödinger operators arising in scattering theory when B is multiplication by a function which decays faster than $|x|^{-\alpha}$, $\alpha > 1/2$, at ∞. This implies local decay in the above sense. It is also enough for obtaining wave operators, by the following theorem.

Theorem 1.1. Let H_1 and H_2 be self-adjoint operators with spectral families $E_1(\cdot)$ and $E_2(\cdot)$, and suppose that B_j is H_j-smooth on (a,b) for $j = 1, 2$. Then the integral

$$\int_{\infty}^{\infty} E_1((a,b))e^{iH_1t} \; B_1B_2e^{-iH_2t}E_2((a,b))\phi \, dt$$

exists. If $H_1 - H_2 = B_1 B_2$ then the local wave operators

$$\Omega_{\pm} \; (H_2, H_2, (a,b)) = \text{s-}\lim_{t\to\pm\infty} \; e^{iH_1t} \; e^{-iH_2t} \; E_2((a,b))$$

exist, and are complete in the sense that their ranges both equal the range of $E_1((a,b))$.

Proof. (Sketch) The convergence of the integral can be proved by taking the inner product of the integrand with a vector, integrating between two large times, and using H_j-smoothness of B_j to show the result is small. The existence of the wave operator reduces to the existence of such an integral without the factor $E_1((a,b))$. The absence of the factor introduces only mild complications [4]. The wave operator is isometric on the range of $E_2((a,b))$, and its adjoint can be obtained by interchanging the roles of H_1 and H_2 (since the assumptions are symmetric), so Ω_+^* is also isometric on the range of $E_1((a,b))$, which gives completeness. #

The estimate (2.1A) turns out to be difficult to prove, but it is implied by the following time independent estimate, which is more convenient to deal with in many cases.

$$\text{(2.1B)} \qquad |\mathrm{Im} z| \; \|B(H-z)^{-1}\|^2 \leq C^2$$

$$\text{for} \quad a < \mathrm{Re}\, z < b, \quad \sigma < |\mathrm{Im} z| < 1.$$

This estimate says that application of the operator B mollifies the singularity produced by $(H-z)^{-1}$ when z is near the spectrum of H. The proof that (2.1B) implies (2.1A) is based on the facts that

$$\int_{-\infty}^{\infty} e^{i\lambda t} \, e^{-\varepsilon|t|} \; B \, e^{-iHt} \; \phi \, dt = iB \, \{(H - \lambda + i\varepsilon)^{-1} - (H-\lambda-i\varepsilon)^{-1} \} \, \phi,$$

and smoothness of the Fourier transform of a function implies decay of that function at infinity.

Even (2.1B) can only be proved for special operators H. But for the same operators a stronger estimate can be proved without much more work.

$$\text{(2.2)} \qquad \| \, B \, (H-z)^{-1}B \, \| \leq C$$

$$\text{for} \quad a < \mathrm{Re}\, z < b, \; 0 < |\mathrm{Im} z| < 1.$$

This implies (2.1B) because

$$|\mathrm{Im} z| \; \| B\, (H-z)^{-1} \|^2 = |\mathrm{Im} z| \; \| (H-z^*)^{-1} B \|^2$$

and

$$|\mathrm{Im} z| \; \| (H-z^*)^{-1} B \phi \|^2 = |\mathrm{Im} z| \; | \langle \phi, B(H-z)^{-1} (H-z^*)^{-1} B \phi \rangle |$$
$$= \frac{1}{2} \; | \langle \phi, B\, [(H-z)^{-1} - (H - z^*)^{-1}]\, B \phi \rangle |$$
$$= | \mathrm{Im} < \phi, B(H-z)^{-1} B \phi > |$$
$$\leq \| B(H-z)^{-1} B \| \; \|\phi\|^2$$

The estimate (2.2) can be carried over to short range perturbations of H. The argument is simple if the perturbation is not only short range, but small in size. If $H(\varepsilon) = H + \varepsilon \; B T B$ where T is bounded and self-adjoint, we have the resolvent equation

$$(I + \varepsilon \; (H-z)^{-1} \; BTB) \; (H(\varepsilon) - z)^{-1} = (H-z)^{-1} \, .$$

Multiplying on the right and left by B gives

$$(I + \varepsilon \; B(H-z)^{-1} BT) \; B \; (H(\varepsilon) - z)^{-1} \; B = B(H-z)^{-1} \; B,$$

and $I + \varepsilon \; B(H-z)^{-1} \; BT$ can be inverted by a convergent geometric series if $|\varepsilon| < \| B(H-z)^{-1} \; B \; T \|$ to get an estimate like (2.2) for $B(H(\varepsilon)-z)^{-1} \; B$.

There is another time-independent property which arises in scattering theory. H is *absolutely continuous* over the interval (a,b) on the subspace M if for every subset $S \subset (a,b)$ of Lebesque measure zero the spectral projection E(B) is zero. If B is H-smooth on (a,b), then H is absolutely continuous over (a,b) on the range of B. This can be proved easily by the spectral theorem and Fourier analysis. The problem of absolute continuity usually comes up as a stumbling block in proving local decay; since our route will be to prove (2.2), which implies local decay, absolute continuity will not concern us.

3. A way to prove time integrals are finite.

The usual proofs of the estimates in the previous section for the case when $H = H_0$ use either the Fourier transform or a explicit representation of $(H_0-z)^{-1}$ as an integral operator. The useful property of the Fourier transform is that it relates decay at infinity in space to smoothness in momentum variables, in terms of which H_0 is a multiplication operator.

It is well known that Fourier analysis and the canonical commutation relations $[P,Q] = -iI$ are closely related. In fact estimates like (2.1) and (2.2) can be proved using certain other commutation relations, which can hold when Fourier analysis is not available.

Finiteness of time integrals like (2.1A) is important in many scattering problems [5], so it may be a good thing to state the idea in a rather general way. Suppose that $u(\cdot)$ is a function of time with values in a normed vector space U (finite or infinite dimensional) which satisfies

$$\frac{du}{dt} = F(u,t). \tag{3.1}$$

If U is Hilbert space and $F(u,t) = i\,Hu$, this is the Schrödinger equation. If U is phase space and $F(x,p) = (p/m, -\text{grad}\, V(q))$, it is classical mechanics. Most equations of interest in scattering theory can be put into the form (3.1).

Suppose in addition that for some differentiable function $G: U \to R$ and non-empty set P of orbits of (3.1),

(3.2) G is bounded on each orbit of P.

(3.3) $d\,G(u) \cdot F(u,t) = h(u) \geq 0.$

for every u belonging to an orbit in P.

The expression in (3.3) is the Frechet derivative of G at u acting as a linear transformation on $F(u,t)$. It arises because

$$\frac{d\,G(u(t))}{dt} = d\,G(u(t)) \cdot F(u(t),t).$$

The only point to writing it this way is to emphasize that (3.3) can be verified without first solving the differential equation (3.1). In practice it is easier to differentiate $G(u(t))$ with respect to t, using the equation (3.1), than to calculate d G.

From (3.1), (3.2) and (3.3) it follows that

$$\begin{aligned}\int_{-T}^{T} h(u(t))\,dt &= \int_{-T}^{T} \frac{d}{dt}\, G(u(t))\,dt \\ &= G(u(T)) - G(u(-T)) \\ &\leq 2 \sup_{-\infty<s<\infty} |G(u(s))|,\end{aligned}$$

which gives

$$\int_{-\infty}^{\infty} h(u(t))\,dt \leq 2 \sup_{-\infty<s<\infty} |G(u(s))|. \tag{3.4}$$

The value $G(u(t))$ will be increasing with t along the orbit in P; thus it measures to what extent a solution is "incoming" or "outgoing."

Example 3.1. In the Lax-Phillips theory of scattering [6] the evolution in time has a representation as translation on $L^2(R;N)$: $u(t,s) = u(s-t)$. If g is real valued, increasing and bounded on $(-\infty,\infty)$, the function $G(u(t,\cdot)) = \int_{-\infty}^{\infty} g(s)|u(t,s)|^2\,ds$ satisfies (3.2) and (3.3) with $h(u(t,\cdot)) = \int g'(s)\,|u(t,s)|^2 ds$.

For the Schrodinger equation

$$(3.1') \qquad \frac{du}{dt} = -iHu$$

the choice $G(u) = \langle u, Au\rangle$ for a self-adjoint operator A has a physical interpretation as the expectation value of the quantum mechanical observable that A represents. The requirement

$$(3.2') \qquad |\langle u, Au\rangle| \leq C \langle u, (H+d)u\rangle$$

implies (3.2) if $H + d > 0$. The translation of (3.3) into this situation is

$$(3.3') \qquad \langle u, i[H,a]u\rangle \geq \|Bu\|^2.$$

The conclusion is

$$(3.4') \qquad \int_{-\infty}^{\infty} \| B e^{-iHt} u \|^2 \, dt \leq 2c \langle u, (H+d)u\rangle < \infty.$$

This can be stated as follows:

Theorem 3.1. If H, A, and B are self-adjoint operators with $H+d>0$, $-(H+d) \leq A \leq H+d$, B bounded and $i[H,A] \geq B^2$, then B is H-smooth on $(-\infty,b)$ for any finite b [7].

4. Where to look for an increasing quantity.

The biggest problem with the approach indicated in the previous section is to find G such that $G(u(t))$ increases as the time t increases. Suppose that the equation of interest describes motion with constant speed along straight lines in space. If F is a convex function of the space variable and $x(t)$ the position at time t, then $F(x(t))$ should be a convex function of time, and the first time derivative of F should increase with time. And $F(x(t))$ may remain convex under certain perturbations of this simple case.

Example 4.1. Let D be a domain in R which may have infinite boundary, but has star-shaped complement, that is, the inner product of the position vector with the normal n pointing into D is positive at each point of the boundary. Consider a light ray which is reflected so that the angle of incidence equals the angle of reflection. Take

$$F(x) = \int_0^{|x|} \arctan r \, dr, \text{ so } \nabla F(x) = \frac{x}{|x|} \arctan |x|.$$

If v is the velocity,

$$\frac{dF(x(t))}{dt} = v \cdot \nabla F(x(t)) = \frac{v \cdot x}{|x|} \arctan |x|$$

is bounded, so (3.2) is satisfied. Condition (3.3) is also fulfilled.

At a time t when no reflection occurs

$$(4.1)\qquad \frac{d^2F(x(t))}{dt^2} = \frac{\arctan|x|}{|x|}\left[v^2 - \frac{(x\cdot v)^2}{x^2}\right] + \frac{(x\cdot v)^2}{x^2(1+x^2)}$$

$$\geq \frac{v^2 - (x\cdot v)^2/x^2}{1+x^2} + \frac{(x\cdot v)^2/x^2}{1+x^2} = \frac{v^2}{1+x^2}$$

At a moment when the particle is reflected there is an instantaneous change in velocity, Δv, in the direction of the inward normal n. Thus the change $\Delta v \cdot (x/|x|) \arctan |x|$ in the value of $dF(x(t))/dt$ is positive. It follows from (4.1) that

$$\int_{-\infty}^{\infty} \frac{v^2}{(1+x(t))^2}\, dt \leq 2 \max_t \arctan |x(t)| = \pi ,$$

and the amount of time spent in any bounded set can be estimated in terms of the integral on the left.

In Hunziker's lectures on classical mechanics it was noted that for $F(x) = x^2/2$ we have (taking $p^2 + V(x) = E$ for the energy) $dF/dt = 2x\cdot p$ and

$$\frac{d^2F}{dt^2} = 4 \left\{ E - \left(V + \frac{r}{2}\frac{\partial v}{\partial r}\right)\right\} .$$

The quantity on the right side of this equation will be positive for any orbit on which

$$(4.2)\qquad V(x) + \frac{r}{2}\frac{\partial v}{\partial r} < E$$

is satisfied at all points.

Suppose that for some sufficiently large value of E the potential V satisfies (4.2) throughout space. Then the effective potential $V(x) + L^2/r^2$ has no barriers to trap a particle of Energy E. For at a point x where such a particle meets a barrier one would have $V(x) + L^2/r^2 = E$ and the effective force in the radial direction $-\partial v/\partial r - 2L^2/r^3 < 0$. Multiplying the second condition by $r/2$ and adding it to the first violates (4.2).

Thus if V satisfies this condition (4.2) which rules out the possibilities of trapping barriers, the positivity condition (3.3) on the function $G(x,p) = 2\, x\cdot p$ is satisfied. But the program in section 2 can't be carried out because the boundedness condition (3.2) is not fulfilled. The factor x is unbounded along any scattered orbit. However, the argument can be saved if $x^2/2$ is replaced by a function F whose gradient is bounded.

To simplify calculations, let us work in one dimension. Instead of looking for F we start with a positive F'', define F' as its indefinite integral, and forget about F, which will automatically be convex.

Let

$$(4.3)\qquad h_R(x) = (1 + (\tfrac{x}{R})^2)^{-\beta/2}, \quad \beta > 1,$$

$$g_R(x) = \int_0^x h_R(s)\, ds.$$

Then

$$|g_R(x)| \leq \int_0^\infty h_1(\tfrac{s}{R})\, ds \leq R \int_0^\infty h_1(s) ds.$$

For suitable R, g_R will play the role of F'. Thus we need the derivative of $2g_R \cdot P$. The calculation gives

$$\frac{d}{dt}\, 2\, g_R P = 4\, h_R P^2 - 2\, g_R V'$$

Since we have analyzed the role of the quantity $E - (V + (x/2)V')$, let us isolate it.

$$(4.4)\qquad \frac{d}{dt}\, 2\, g_R P = 4h_R\{\, E - (V + \tfrac{x}{2}\, V')\} + 2V'(xh_R - g_R).$$

Lemma 4.2. Suppose that for some $\delta > 0$

$$|x\, V'(x)| \leq c\ (1 + |x|)^{-\delta}.$$

Let h_R and g_R be defined by (4.3) with $1 < \beta < 1 + \delta$. Then

$$2\ |V'(x)\ [x\, h_R(x) - g_R(x)\ ]| \leq \varepsilon(R) h_R(x),$$

where $\varepsilon(R) \to 0$ as $R \to \infty$.

Proof: Since $x\, V'(x) \to 0$ as $|x| \to \infty$, there exists for any $\varepsilon > 0$ an $r_1 > 0$ such that for $x > r_1$ the term $|x\, V'(x)\ \ h_R(x)|$ is less than $\varepsilon\, h_R(x)$.

Consider the term $V'(x)\, g_R(x)$ for $x > r$. Note that

$$\frac{g_R(x)/x}{h_R(x)} = f(\tfrac{|x|}{R}), \quad |f(x)| = \left| \frac{\frac{1}{x} \int_0^{|x|} h_1(s) ds}{h(x)} \right| \leq C'\ (1+|x|^{\beta-1})$$

so

$$x\, V'(x)\ \frac{g_R(x)/x}{h_R(x)} \leq C\ (1+|x|)^{-\delta} + \frac{CC'(1+|x|)^{-\delta}|x|^{\beta-1}}{R^{\beta-1}}.$$

This can be made small by taking $|x| > r_2$ for some positive r_2, and R large.

Finally, we take $r = \max\{r_1, r_2\}$

$$|xh_R(x)-g_R(x)| \le \int_0^{|x|} \{h_1(\tfrac{x}{R})-h(\tfrac{y}{R})\}dy \le |x| \sup_{|y|<|x|} |h_1(\tfrac{|x|}{R})-h_1(\tfrac{y}{R})|$$

So $| V'(x)\ [x\ h_R(x) - g_R(x)]\ | \le \varepsilon(R)h_1(r) \le \varepsilon(R)h_R(x)$ for $x \le r$,

where $\varepsilon(R) \to 0$ as $R \to \infty$. #

Thus if we assume that for some value of E, and $\varepsilon > 0$

$$E - \varepsilon - (V + \tfrac{x}{2} V') > \mu > 0,$$

then R can be chosen large enough so that

$$\frac{d}{dt}\ \{ 2g_R(x)\ P \} \ge 4\ h_R(x)\ \{ E - \varepsilon - (V + \tfrac{x}{2} V')\} \ge \mu\ h_R(x).$$

It follows that

$$\mu \int_{-\infty}^{\infty} h_R(x(t))dt \le 4 \sup_{-\infty<t<\infty} g_R(x(t))\ p(t)$$

$$\le 2R \int_0^{\infty} h_1(s)ds \quad \sup_t\ (P^2(t) + 1).$$

The factor $P^2(t) + 1$ is bounded by E + constant if V is bounded below. The above inequality implies that a particle of energy E spends a finite amount of time in any interval. Perhaps there are easier ways to deduce this, but this way carries over to quantum mechanics, as we shall see in the next section.

5. The Schrödinger Equation.

The two examples in section 4 deal with the motion of points. Such problems are not hard to analyze by other means. But the method used above carries over to more exact (and complicated) theories of the same phenomena. We could study the wave equation on the domain *D* of example 4.1, and prove local decay of the energy density E(x,t) by taking G(u(t)) to be the time derivative of $\int_D F(x)\ E(x,t)dx$ [8].

To treat the Schrodinger equation $i\partial\psi/\partial t = H\psi$ with $H = H_0 + V$ on $L^2(R)$ in place of classical mechanics, we replace the position observable F(x) used in the previous section by its quantum analogue multiplication by this function. An operator T representing an observable depends on time according to the rule

$$T(t) = e^{iHt}\ T\ e^{-iHt}$$

so the time derivative is given by the commutator $i[H,T]$. Since all of our operators are made up from multiplication operators and the momentum operator $P = -\ id/dx$, all calculations are based on the fact that $[P,f] = -\ if'$ for a multiplication operator f. Thus

$$(5.1) \quad \frac{d}{dt} e^{iHt} F e^{-iHt} = e^{iHt} i[H,F] e^{-iHt} = e^{iHt} i[P^2,F] e^{-iHt} = e^{iHt} (PF' + F' P) e^{-iHt}.$$

Recall that we took $F' = g_R$ in the previous section, with g_R given by (4.3). So the time derivative of (5.1) with this choice of F' is what ought to be positive. We calculate this (with a general choice of $F' = g$, for future use) and get a result similar to (4.4).

Lemma 5.1. Let $H = H_0 + V$ acting on $L^2(R)$ and let $A = Pg + gP$ and $g(x) = \int_0^x h(s)ds$ with h twice continuously differentiable. Then

$$(5.2) \quad i[H,A] = 2[hH + Hh] - 4h(V + \frac{x}{2} V') + 2V'(xh - g) + h''.$$

Proof. The rule for the commutator of P with a multiplication operator gives

$$i[H_0,A] = P^2h + 2\ PhP + hP^2,$$

and the middle term equals $P^2h + h\ P^2 + h''$, the term h'' arising from commuting P past h. The result (5.2) follows from this and $i[V,A] = -2g\ V'$. #

Now take $h = h_R$ and $g = g_R$, as in (4.3). The term h_R'' is the only one with no counterpart in (4.4). Since $h_R''(x) = R^{-2}\ h_1''(x/R)$, this term is small compared to $h_R(x) = h_1(x/R)$, so we have a situation similar to that in the previous section.

Lemma 5.2. Let $H = H_0 + V$ and suppose that $|x\ V'(x)| \le c(1 + |x|)^{-\delta}$ $\delta > 0$. Let h_R and g_R be defined by (4.3) with $1 < \beta < 1 + \delta$, and take $A = g_R\ P + P\ g_R$. Then

$$(5.3) \quad i[H,A] \ge 2\ \{\ h_R\ H + H\ h_R\ \} - 4\ \varepsilon\ (R)h_R\ -\ 4h_R(V + \frac{x}{2}\ V'),$$

where $\varepsilon\ (R) \to 0$ as $R \to \infty$.

Proof. Use the remark in the preceding paragraph and Lemma 4.2. #

However we have a problem here which did not arise in the classical case. For high energy, H may dominate $\varepsilon(R)$ and $(V + (x/2)V')$ as before, but the symmetric product of two positive operators $h_RH + Hh_R$ need not be positive because of the fact that position and momentum operators don't commute. For this reason it turns out to be convenient to prove a time independent version of Theorem 3.2.

Theorem 5.3. Suppose that H and A are self-adjoint operators satisfying $||A\phi|| \le c\ ||H\phi|| + d||\phi||$. If, for some bounded self-adjoint operator B ,

$$i[H,A] \ge B^2\ H + HB^2\ \ -\ 2\ c\ B^2$$

then B is H-smooth on (a,b) for $c < a < b < \infty$

Proof. For simplicity, we assume A is bounded. The proof for unbounded A can be found in [4]. The hypothesis says that for ψ in the domain of H

$$\mathrm{Re} < B^2\psi, (H-c)\psi > \ \le \mathrm{Im} < A\psi, H\psi >$$

Substituting $R(z)\phi = (H - z)^{-1}\phi$ for ψ gives

$$\mathrm{Re} < B^2R(z)\phi, (H-c) R(z)\phi > \ \le \mathrm{Im} < A R(z)\phi, H R(z)\phi >.$$

The relation $H R(z)\phi = \phi + z R(z)\phi$ can be used to remove H from this inequality.

$$\mathrm{Re} < B^2R(z)\phi, \phi > + \mathrm{Re}(z - c) \ ||B R(z)\phi||^2$$
$$\le \mathrm{Im} < A R(z)\phi, \phi > + |\mathrm{Im}\ z| \ ||A|| \ ||R(z)\phi||^2$$

The identity $2 \ |\mathrm{Im}\ z| \ ||R(z)\phi||^2 = | < \phi, (R(z) - R(z)^*) \phi > |$ gives, for $\mathrm{Re}\ z > a$,

$$(5.4) \quad (a - c) \ ||B R(z) \phi||^2 \le | < B^2 R(z)\phi, \phi > | + | < A R(z)\phi, \phi >|$$
$$+ \ ||A|| \ |\mathrm{Im} < \phi, R(z)\phi > |.$$

Multiplying by $|\mathrm{Im}\ z|$ gives the criterion (2.1B) for H-smoothness because $|\mathrm{Im}\ z| \ ||R(z)|| = 1$ implies that the right-hand side is bounded for $a < \mathrm{Re}\ z < b$. #

Corollary 5.4. If $H = H_0 + V$ where V satisfies the conditons of Lemma 5.2 and $V + (x/2)\ V' \le E$ then for $E < a < b < \infty$, multiplication by f is H-smooth on (a,b) if $|f(x)| \le c\ (1 + |x|)^{-\beta}$, $\beta > \tfrac{1}{2}$.

Proof. By Lemma 5.2 and the condition $V + \frac{x}{2} V' \le E$ we have

$$i[H,A] \ge 2 \ \{ h_R(H - E - \varepsilon(R)) + (H - E - \varepsilon(R))h_R\}.$$

Theorem 5.3 says that $(h_R)^{1/2}$ is H-smooth on (a,b), and any function F dominated by $(1 + |x|)^{-\beta}$ is also dominated by h_R. #

This gives H-smoothness of certain multiplication operators and therefore local decay, and existence and completeness of local wave operators (in the case of short range potentials) for energies $E \ge V + (x/2)V'$, i.e. energies so high that no classical bound orbits are possible. A potential which approaches zero at infinity but has barriers can trap classical particles of positive energy, but not quantum mechanical particles. It may cause the particle to linger between barriers for a long time, and this destroys the convexity of the expectation value of F(x) as a function of time, and thus the argument for local decay doesn't work.

Regard the barriers as a short range perturbation of a potential V_1 satisfying $E \ge V_1 + (x/2)V_1'$. If we had the stronger estimate (2.2) on $H_1 = H_0 + V_1$, this estimate could be transferred to the perturbed operator, and we could conclude that decay takes place at energies above E even with the barrier present. In fact, the estimates (2.2) can be extracted from the inequality (5.3).

Theorem 5.5. Assume all of the hypotheses of Theorem 5.3 and in addition that B is invertible and $||BAB^{-1}\phi|| \leq C\ (||H\phi|| + ||\phi||)$. Then if $\alpha > c$

$$||B(H-z)^{-1}\ B|| \leq K$$

for all z with $a < \text{Re } z < b$ and $0 < |\text{Im } z| < 1$.

Proof. As in Theorem 5.3, we assume for simplicity that A and BAB^{-1} are bounded [9]. Replacing ϕ in (5.4) by $B\phi$ gives

$$(a - c)\quad ||B\ R(z)\ B\phi||^2 \leq \quad |< B^2\ R(z)B\phi,\ B\phi >| \quad +$$

$$|< A\ R(z)B\phi,\ B\phi >| \quad + ||A||\ |\text{Im} < \phi,\ B\ R(z)B\phi>|$$

$$\leq \quad ||B\ R(z)B\phi||\ \ (||B^2|| + ||BAB^{-1}|| + ||A||) \quad ||\phi||$$

Dividing by $||B\ R(z)B\phi||$ gives the desired result. #

Corollary 5.6. Under the conditions of Corollary 5.4, $||f(H-z)^{-1}f||$ is bounded uniformly for $a < \text{Re } z < b$ and $0 < |\text{Im } z| < 1$.

Proof. The assumption on $h_R(g_R\ P + P\ g_R)h_R$ is easy to prove.# With this estimate, the series argument given in section 2 can be used to get a similar result for a perturbation of H if the perturbation (or barrier) is small. If it is large, a more intricate argument of the sort given in Streater's article must be used. It goes, very roughly, as follows.

Suppose $H_j = H_0 + V_j$ for j = 1, 2 and assume that Corollary 5.6 applies to H_1 so that $f(H_1 - z)^{-1}f$ is bounded (with B = f). Suppose also that $V_2 - V_1 = f\ T\ f$ where T is multiplication by a real bounded function $O(|x|^{-\gamma})$, $\gamma > 1$, at ∞. (We are assuming $V_2 - V_1$ decays faster than $|x|^{-2}$ at infinity. This is not necessary but makes the argument simpler [9]). As in the series approach we start with

$$(1 + f(H_1 - z)^{-1}fT)\ f(H_2 - z)^{-1}f \quad = \quad f(H_1 - z)^{-1}\ f.$$

We assume $f(H_2 - z)^{-1}f$ is not bounded in the desired way and attempt to derive a contradiction. If $f(H_2 - z)^{-1}f$ were not properly bounded, $I + f(H_1 - z)^{-1}fT$ would be not be bounded away from zero as z approaches the real axis (say from above). An argument using the compactness of $f(H_1 - z)^{-1}f$ gives the existence of ϕ and $\lambda \in [a,b]$ such that

$$(I + f(H_1 - \lambda - i\ o)^{-1}fT)\phi = 0.$$

(We assume for simplicity the existence of point-wise boundary values). Then $\psi = (H_1 - \lambda - i\ 0)^{-1}fT\ \phi$ satisfies $(H_2 - \lambda)\ \psi = 0$ in a weak sense, but we can't say that ψ belongs to the Hilbert space, only that $f\psi$ does. If ψ were a bona fide eigenvector it would be subject to theorems [10] which deny its existence. Since $f\psi$ is in Hilbert space but ψ may not be, and f is multiplication by a function decaying at infinity, the problem with ψ is its lack of decay at infinity.

In the case of $H_1 = H_0$ better decay can be deduced for ψ using the Fourier transform. The same thing can be done in general using (5.4) again. The first point to note is that

$$\text{Im} < (H_1 - \lambda - i0)^{-1} fT\phi, fT\phi > = - \text{Im} <\phi,T\phi> = 0$$

since T is self-adjoint. Now substitute $f\,T\,\phi$ for ϕ, $h_R^{1/2}$ for B, and $\lambda + i\,0$ for z in (5.4). The fact just noted allows the last term in (5.4) to be dropped, and we have

$$\text{Re}\ (a-c)\ ||h_R^{1/2}\ \psi||^2 \leq ||h_R^{1/2}\ \psi||\ ||h_R^{1/2}\ f\ T\ \phi|| + < \psi, A_R\ f\phi >$$

Recall that $h_R(x) = (1 + (x/R)^2)^{-\alpha}$, so $||\psi||$ can be estimated by letting $R \to \infty$ in $||h_R\ \psi||$. When this is done $A_R = g_R\ P + P\ g_R$ approaches $xP + Px$. The factor x is unbounded, but this is compensated by T which decays faster than $|x|^{-1}$. (The factor P is also unbounded, but we continue to ignore this as an inessential complication. A more precise treatment is given in [9]). The result, when R goes to ∞, is two factors $||\psi||$ on the left and only one on the right, multiplied by a bounded quantity. This shows $||\psi|| < \infty$ and we have a contradiction to the nonexistence of positive eigenvalues.

Although this is a nice proof, it does not seem quite satisfactory. The boundedness of $f(H-z)^{-1}f$ is proved by contradiction; we have no idea how large the bound may be. The amount of time a particle spends in an interval can be estimated in terms of the bound on $f(H_2 -z)^{-1}\ f$, so the deficiency is reflected by the lack of an estimate of how long a particle can be delayed by a barrier, in terms of the size of the barrier. A really effective proof would give such an estimate.

6. Time Delay.

The difference between the times taken to traverse the region of interaction by a scattered particle and the free particle which it asymptotically resembles is called the time delay. It can be defined more precisely as follows. Assume that $H = H_0 + V$ where V is short range so that the wave operators $\Omega_\pm$ exist.

$$(6.1)\qquad T_R(\phi) = \int_{-\infty}^{\infty} ||X_R\ e^{-iHt}\ \Omega_-\ \phi||^2\ dt - \int_{-\infty}^{\infty} ||X_R\ e^{-iH_0t}\ \phi||^2\ dt$$

is evidently the difference in the times of sojourn in the interval (-R,R) if X_R is the characteristic function of this interval. Our previous considerations have indicated that this quadratic form is finite for all ϕ, since X_R is H-smooth, and thus the form defines a bounded operator, which we call T_R. The time delay is defined to be the limit of T_R as $R \to \infty$. The existence of this limit is not at all obvious [11]. But the commutator identity of Lemma 5.1 has not yet been exhausted; it can be used to prove the existence of a very similar limit.

Theorem 6.1. Let $H = H_0 + V$, acting on $L^2(R)$, where for some $\alpha > 1$

$$|V(x)| + |x V'(x)| \leq c(1 + |x|)^{-\alpha}.$$

Let $T_R(\cdot)$ be defined by (6.1) with $X_R(x) = X(x/R)$, for a function X which is twice continuously differentiable and satisfies $X(x) = 1$ for $|x| \leq 1$ and $X(x) = 0$ for $|x| \geq 2$. Then $T_R(\cdot)$ defines an operator T_R which commutes with $H_0 \cdot E_0((a,b))\, T_R$ is bounded for $0 < a < b < \infty$. For ϕ, ψ in the domain of $H_0\, T_R$,

$$\lim_{R\to\infty} \langle\psi, H_0 T_R\, \phi\rangle = \int_{-\infty}^{\infty} \langle\Omega_-\psi, e^{iHt}(V + \tfrac{x}{2} V')\, e^{-iHt}\, \Omega_-\phi\rangle\, dt \tag{6.2}$$

$$\equiv \langle \psi, D\phi \rangle$$

The operator $E_0((a,b))D$ is also bounded for $0 < a < b < \infty$.

Proof. (Sketch) By the proof indicated in section 5 (and given in [9]), X_R is H-smooth and H_0-smooth on (a,b) if $0 < a < b < \infty$. From this the boundedness of T_R follows. Similarly, the boundedness of D will follow from H-smoothness of $|V(x) + (x/2)\, V'(x)|^{1/2}$

It is clear from the definition (6.1) that T_R commutes with H_0. Therefore

$$2\, H_0 T_R = H_0 T_R + T_R\, H_0 = \int_{-\infty}^{\infty} \Omega_-^* e^{iHt}(X_R H + H X_R) e^{-iHt}\Omega_- dt$$

$$- \int_{-\infty}^{\infty} e^{iH_0 t}\, (X_R H_0 + H_0 X_R)\, e^{-iH_0 t} dt$$

By Lemma 5.1 with $h = X_R$, $g_R(x) = \int_0^x X_R(s)ds$ and $A_R = g_R\, P + P\, g_R$ we have

$$X_R H + H X_R = \frac{i}{2}\, [H, A_R] + 2\, X_R(V + \frac{x}{2}\, V') - V'(x\, X_R - g_R) - \frac{1}{2}\, X_R''$$

The integral of the term $\Omega_-^*\exp(iHt)\; i[H, A_R]\, \exp(-iHt)\Omega_-$ over all t is just the difference between the limits of $\Omega_-^*\exp(iHt)$ A $\exp(-iHt)\; \Omega_-$ at $t = +\infty$ and $t = -\infty$. It is not hard to see that $\Omega_-^*\exp(iHt)P\, \exp(-iHt)\Omega_-$ approaches $+\sqrt{H_0}$ on one subspace and $-\sqrt{H_0}$ on its orthocomplement as $t \to \infty$, and on the same subspaces $\Omega_-^*\exp(iHt)\; g_R\, \exp(-iHt)$ approaches $g_R(\pm\infty) = \pm R\, \pi/2$ [11]. Therefore, $\Omega_-^*\exp(iHt)\; A_R\, \exp(-iHt)\; \Omega_- \to \pi R\sqrt{H_0}$ as $t \to \infty$, and the limit is $-\pi R\sqrt{H_0}$ as $t \to -\infty$, so the integral is $2\pi R\sqrt{H_0}$.

Now $X_R H_0 + H_0 X_R = (i/2)\; [H_0,\, A_R] - X_R''/2$, and by the same argument the integral of $\exp(iH_0 t)\; i[H_0,\, A_R]\, \exp(-iH_0 t)$ is also $2\pi R\sqrt{H_0}$. Thus the contributions of these terms to (6.3) cancel and we are left with

$$2H_0T_R=\int_{-\infty}^{\infty}\Omega_-{*}e^{iHt}\{2X_R(V+\frac{x}{2}V')-V'(xX_R-g_R)-\frac{1}{2}X_R''\}e^{-iHt}\Omega_-\,dt$$
$$-\int_{-\infty}^{\infty}e^{iH_0t}\,\frac{1}{2}\,X_R''\,e^{-iH_0t}\,dt.$$

It can be argued using the dominated convergence theorem that all of these terms converge weakly to zero except the one involving $X_R(V+(x/2)V')$, which converges to the integral of $V+(x/2)V$. It follows that H_0T_R converges weakly to

$$\int_{-\infty}^{\infty}\Omega_-{*}\,e^{iHt}\,(V+\frac{x}{2}V')\,e^{-iHt}\,\Omega_-\,dt. \quad \#$$

The expression (6.2) is complicated because it involves $\exp(-iHt)$, the perturbed evolution, but some information can be derived from it. If the quantity $V+(x/2)V'$ is negative so that there are no barriers to cause delays, then by (6.2) the time delay is negative (there is no delay; in fact the particle is faster than a free particle).

The converse question, whether largeness of $V+(x/2)V'$ implies long time delays, seems harder. One should also investigate the relationship of the behavior of $V+(x/2)V'$ to the location of the poles of the analytic continuation of $(H-z)^{-1}$ that has been revealed by the complex dilation group [1]. Recall in this connection that $V+(x/2)V' = H - i[H,A]$ where $A = (1/4)(xP+Px)$ is a generator of the dilation group.

References:

1. See the article by Combes in this volume.

2. C. van Winter, preprint, University of Kentucky.

3. T. Kato, Wave operators and similarity for some non-self-adjoint operators, *Math. Ann.* 162, (1966) 258-279.

4. See R. Lavine, Commutators and scattering II; a class of one body problems, *Indiana Math. Jour.*, 21, (1972) 643-656.

5. See, for example, the article in this volume by Strauss.

6. Outlined in the article by Phillips.

7. This interpretation of a theorem of Putnam asserting absolute continuity was given in T. Kato, Smooth operators and commutators, *Studia Math*, 31, (1968) 535-546.

8. Such a decay theorem has already been proved, using the "multiplier" technique developed by Morawetz (E.C. Zachmanoglou, The decay of solutions of the initial-boundary value problem for the wave equation in unbounded regions, *Arch. Rat. Mech.* 14, (1963) 312-325) This method has much in common with that outlined in the present article, but it seems never to have been applied to the Schrödinger equation.

9. R. Lavine, Absolute continuity of positive spectrum for Schrödinger operators with long-range potentials.

10. S. Agmon, Lower bounds for solutions of Schrödinger equations, *J. Anal. Math.* 23, (1970) 1-25; B. Simon, On positive eigenvalues of one body Schrödinger operators, *Comm. Pure Appl. Math.* 12, (1969) 1123-1126.

11. See J.M. Jauch, K. Sinha, B. Misra, Time delay in scattering processes, *Helv. Phys. Acta.* 45, (1972) 398-426.

12. R. Lavine, Commutators and scattering theory I: repulsive interactions, *Comm. Math. Phys.* 20, (1971) 301-323, Lemma 5.10.

CANDIDATES FOR σ_{ac} AND H_{ac}

Karl Gustafson

University of Colorado

Boulder

ABSTRACT

The purpose of this paper is to order and briefly describe, via two diagrams (Diagram 1 and Diagram 5 below), various spectra and subspaces that are closely related to scattering spectra and scattering subspaces. This will be done from the operator-theoretic viewpoint and the orderings are obtained primarily via Fredholm theory and Fourier theory, respectively. Both self-adjoint and non-self-adjoint operators will be considered, inasmuch as non-self-adjoint operators and methods are now being used in scattering theory.

1. Candidates for σ_{ac}. This could instead be entitled: mathematically, what is the "scattering" spectrum? Although historically the "continuous" spectrum evolved via other physical and mathematical considerations, it is now best understood mathematically in terms of Fredholm theory, at least if one wishes to remain in the original space. This also makes sense physically since physicists often regard the continuous spectrum as the energy continuum, which often turns out to be the essential spectrum, the latter being exactly the complement in the complex plane of a Fredholm resolvent set.

The general situation among the "candidates" for the scattering spectrum is summarized in the following diagram, which we will describe below.

$\sigma_e^5(T) = \sigma(T) - \pi_\infty^{alg.} = \sigma_\ell(T) = \sigma_e$ (Weyl [1], Browder [2])

$\cup$

$\sigma_e^4(T) = \bigcap_K \sigma(T + K) = \bigcap_F \sigma(T + F) = \sigma_e$ (Schechter [3])

$\cup$

J. A. LaVita and J.-P. Marchand (eds.), Scattering Theory in Mathematical Physics, 157–168.

$$\cup$$
$$\sigma_e^3(T) = \sigma_e^\alpha(T) \bigcup \sigma_e^\beta(T) \equiv \sigma_e(\text{Wolf}[4]) \bigcup \sigma_e(\text{Gustafson-Weidmann}[5])$$
$$= \sigma_e(\text{Schwartz}[6])$$
$$\cup$$
$$\sigma_e^2(T) = \sigma_e^\alpha(T) \bigcap \sigma_e^\beta(T) = \sigma_e(\text{Kato}[7])$$
$$\cup$$
$$\sigma_e^1(T) = \sigma_e(\text{Goldberg}[8]) = \sigma_c(\text{Akhiezer-Glazman}[9]) \supset \pi_\infty^{geom} - \pi_\infty^{alg}$$
$$\cup$$
$$\sigma_c(T) = \sigma_c(\text{Stone [10], Dunford-Schwartz [11], Sz. Nagy [12], etc})$$
$$\cup$$
$$\sigma(T_c) = \sigma_c(\text{Hilbert [13], Riesz-Sz. Nagy [14], Kato [7], etc})$$
$$\cup$$
$$\sigma(T_{ac}) = \sigma_{ac}\ (\text{Kato [7], Halmos [15], etc})$$

Diagram 1. Candidates for the scattering spectrum.

The sets and their ordering in the above diagram can be described as complements of the following Fredholm resolvent sets in the complex plane, for T a closed and densely defined linear operator in a complex Banach space. Let us note at this point that we have used $R_\lambda = R_\lambda(T) = (T + \lambda)^{-1}$; one can of course change the + to a - throughout, if desired. Also, we have used the symbol Δ in accordance with [5, 7], whereas others (e.g., the Russians) use the symbol Φ in connection with Fredholm sets or properties. Consider the following subsets of the complex plane.

$$\Delta^c(T) = \{\lambda | R(T+\lambda) \text{ is not properly dense or } T+\lambda \text{ is not } 1-1\}.$$
$$\Delta^1(T) = \{\lambda | R(T+\lambda) \text{ is closed}\} \equiv \{\lambda | T+\lambda \text{ is normally solvable}\}.$$
$$\Delta^\alpha(T) = \Delta^1(T) \bigcap \{\lambda |\ \dim N(T+\lambda) < \infty\}.$$
$$\Delta^\beta(T) = \Delta^1(T) \bigcap \{\lambda |\ \dim N(T^* + \bar{\lambda}) < \infty\}.$$
$$\Delta^2(T) = \Delta^\alpha(T) \bigcup \Delta^\beta(T) \equiv \{\lambda | T + \lambda \text{ is semi-Fredholm}\}.$$
$$\Delta^3(T) = \Delta^\alpha(T) \bigcap \Delta^\beta(T) \equiv \{\lambda | T + \lambda \text{ is Fredholm}\}.$$
$$\Delta^4(T) = \Delta^3(T) \bigcap \{\lambda |\ \text{index}(T) \equiv \dim N(T+\lambda) - \dim N(T^* + \bar{\lambda}) = 0\}.$$
$$\Delta^5(T) = \Delta^4(T) \bigcap \{\lambda |\ \text{all nearby } \lambda \text{ are in } \rho(T)\}.$$
$$\Delta(T) = \Delta^5(T) \bigcap \{\lambda | T + \lambda \text{ is 1-1}\} \equiv \rho(T), \text{ the resolvent set for T.}$$

Then $\sigma_e^i(T) = \Delta^i(T)^C$, $\sigma(T) = \Delta(T)^C$, $\sigma_c(T) = \Delta^c(T)^C$, $\sigma(T_c) = \Delta(T_c)^C$, and $\sigma(T_{ac}) = \Delta(T_{ac})^C$, where C denotes complement. Of course $T_c \equiv T|_{H_c}^{ac}$ and $T_{ac} \equiv T|_{H_{ac}}$ are defined here only for T self-adjoint and according to the usual decompositions (e.g., see Kato [7]) $H = H_c \oplus H_p$, $H_c = H_{ac} \oplus H_{sc}$. We use the notation $\pi_\infty^{geom}(T) \equiv \pi_\infty(T)$ in accordance with Berberian [16] to mean eigenvalues that are isolated points in $\sigma(T)$ and of finite geometric multiplicity; similarly $\pi_{\infty}^{alg}(T)$ denotes isolated eigenvalues of finite algebraic multiplicity. Weyl originally considered the "limit-point" spectrum $\sigma_{\ell}(T)$, including (isolated or not) eigenvalues of infinite multiplicity. Schechter proved the characterization $\sigma_e^4(T) = \bigcap_K \sigma(T+K)$ for arbitrary T, where the intersection is taken over all compact operators K ε $\mathcal{B}$ (X), and the similar characterization taken over all finite rank operators F ε $\mathcal{B}$ (X) is easily checked. The continuous spectrum $\sigma_c(T)$ is the usual convention for Banach space operators, and is sometimes also used in Hilbert space, even for self-adjoint T; in the latter case, $\sigma_c(T)$ is then the set of λ which are points of nonconstant continuity for the spectral family $\{E(\lambda)\}$ for T. The customary decomposition $T = T_{ac} \oplus T_{sc} \oplus T_p$, e.g., see Kato [7], has the advantage of uniquely defining absolutely continuous and singularly continuous spectra $\sigma(T_{ac})$ and $\sigma(T_{sc})$ but has the disadvantage of requiring that both sets be closed and possibly overlapping.

For self-adjoint T one has $\sigma_e^2(T) = \sigma_e^5(T)$, and for many self-adjoint Schrödinger operators or other particular operators of physical interest in other applications one has (or hopes for) equality throughout Diagram 1, i.e.,

$$\sigma(T) - \pi_{\infty}^{geom}(T) = \sigma_e^5(T) = \sigma(T_{ac}) = \sigma(T) - \sigma_s(T),$$

where by $\sigma_s(T)$ we mean one of several possible support sets for the singular measure; for T in a separable Hilbert space, by the spectral theorem one can always take $\sigma_s(T)$ to be of Lebesgue measure zero. The problems and programs associated with demonstrating the above are well-known.

For general non-self-adjoint T there doesn't seem to be any universal "truth" in the situation depicted in the above diagram, other than the fact mentioned above, namely, that all versions may be viewed in terms of Fredholm theory; the preferable choice of essential spectrum seems to depend on the particular type of application. All seven essential spectra are, in general, different; for examples illustrating this, see Gustafson [17]. $\sigma_e^1(T)$, the essential spectrum closest to a continuous spectrum, is not in general closed (for a counter-example, see [17]) but usually it turns out that $\sigma_e^1(T)$ coincides, at least piecewise, with $\sigma_e^2(T)$ or one of the higher versions; for some ordinary differential operators treated as spectral operators in this way, see for example Huige [18] and the earlier references therein. The six versions $\sigma_e^2(T)$ through $\sigma_e^5(T)$ are closed sets. As shown by Wolf [4] e.g., see also Kato [7, p. 233, Thm. 5.11], $\sigma_e^{\alpha}(T)$ generalizes to the general Banach space situation the characterization of points λ in the essential spectrum in terms of the existence of corresponding non-compact singular sequences $\{\Phi_n(\lambda)\}$ of vectors; $\sigma_e^{\beta}(T)$ does this for the adjoint operator T*, and was introduced in [5] primarily to obtain the ordering given above of the various essential spectra. $\sigma_e^2(T)$ is a small but closed essential spectrum, but is still not empty for most interesting operators. $\sigma_e^3(T)$ is the most natural version to use when working with the Calkin Algebra of bounded operators modulo compact operators on a Hilbert space (e.g., see [16]). $\sigma_e^4(T)$ is large enough to include all continuous spectra and residual spectra (see [17]) for T and T*; also, as mentioned above, it is the largest subset of the spectrum of T that remains invariant under arbitrary compact perturbations. $\sigma_e^5(T)$ corresponds more closely to Weyl's original concept of the "limit-point" spectrum; as shown by Browder [2, p. 110],

$$\sigma_e^5(T) = \{\lambda \in \sigma(T) \mid \lambda \notin \{\text{poles of finite rank of } (T+\lambda)^{-1}\}\};$$

a pole will always connote algebraic multiplicities. This is the essential spectrum used in the work of Combes et. al. (see Combes [19] and the references therein) and seems to be the most prevalent version used in scattering theory.

The perturbation theorems concerning preservation of the essential spectrum under relatively compact perturbations hold for all five versions $\sigma_e^2(T)$ through $\sigma_e^4(T)$, but not in general for $\sigma_e^5(T)$; for a counter-example, one may use [7, p. 210]. For further information and references related to essential spectra and perturbation theory, we refer to [17], where, however, the emphasis was put on the version $\sigma_e^4(T)$. Since in scattering theory the version $\sigma_e^5(T)$ has emerged as the preferable one, a few additional observations concerning it will now be made here.

As concerns perturbation of $\sigma_e^5(T)$, Browder [2, p. 113, see also p. 111 for another version] showed that $\sigma_e^5(T+B) = \sigma_e^5(T)$ for T closed and densely defined in a Banach space, B a relatively compact perturbation, resolvent $\rho(T)$ not empty, and the point spectra $\sigma_p(T)$ and $\sigma_p(T+B)$ nowhere dense. Schechter [3] obtained some variations and generalizations; e.g., $\sigma_e^5(T+B) = \sigma_e^5(T)$ if B is relatively compact with respect to T, $\rho(T+B)$ and $\rho(T)$ are nonempty, and $\Delta^4(T)$ is connected. Most results of this type presuppose the showing of $\sigma_e^5 = \sigma_e^4$ for the operators involved. Let us therefore state here the following necessary and sufficient description for preservation of the essential spectrum $\sigma_e^5(T)$ in the general case when perhaps $\sigma_e^5 \neq \sigma_e^4$. We recall that for a closed operator A, $\Delta^4(A)$ is an open set with a finite or countable number of components Δ_m^4, m = 1,2,3,... .

<u>Theorem 2</u>. Let T be closed and densely defined in a complex Banach space and let B be relatively compact with respect to T. Then $\sigma_e^5(T+B) = \sigma_e^5(T)$ if and only if the following condition is satisfied for every component $\Delta_m^4(T)$: $\Delta_m^4(T) \cap \rho(T)$ is nonempty $\Leftrightarrow$ $\Delta_m^4(T) \cap \rho(T+B)$ is nonempty.

We omit the details of the proof, which may be obtained for example from the Fredholm considerations of [2, 3, 7] or similar to the considerations given below; note that $\Delta_m^4(T) = \Delta_m^4(T+B)$ for a relatively compact perturbation B of T.

These perturbation conditions are satisfied, for example, (e.g., see Gohberg and Krein [20, p. 215]), for $T = H_0$ self-adjoint under any relatively compact perturbation V, whether V is self-adjoint or not, or any scalar multiple of such an operator. The advantage therefore of using the version σ_e^5, if one can, is that, after perturbation, one has remaining outside of the essential spectrum of the perturbed operator only poles.

The set $\sigma_e^5(T)$ was also characterized by Kaashoek and Lay [21] by

$$\sigma_e^5(T) = \bigcap_C \sigma(T + C)$$

where C is compact and commutes with T. The commuting requirement on C takes into account the fact (e.g., see [7, p. 209]) that the spectrum changes continuously under perturbation by commuting operators.

Another way (see [22]) of interpreting $\sigma_e^5(T)$ is by the characterization

$$\sigma_e^5(T) = \{\lambda \in \sigma(T) \mid \lambda + T \text{ is not algebraically Fredholm}\},$$

where an operator A is called algebraically Fredholm if it is Fredholm and also $\mathrm{asc}(A) < \infty$ and $\mathrm{des}(A) < \infty$, where ascent $(A) \equiv \inf\{n \geq 0 \mid N(A^n) = N(A^{n+1})\}$ and descent $(A) = \inf\{n \geq 0 \mid R(A^n) = R(A^{n+1})\}$. By the ascent-descent duality established in [22], or by direct verification, similar to the considerations of [2, 3], one may observe that all Fredholm operators of index zero fall into exactly two classes: the algebraically Fredholm ones, with all four indices $\mathrm{asc}(A) = \mathrm{asc}(A^*) = \mathrm{des}(A) = \mathrm{des}(A^*)$ equal and finite, and the others, with all four indices equal and infinite. For $\lambda \in \Delta^4(T)$, $\lambda + T$ in the first class just mentioned means that λ is a pole, so that $\lambda \notin \sigma_e^5(T)$, and $\lambda + T$ in the second class means that λ is surrounded with "continuous" spectra, so that $\lambda \in \sigma_e^5(T)$.

In a recent paper (Salinas [23]) another characterization of $\sigma_e^5(T)$ was given for the special case of a bounded operator $T \in \mathcal{B}(H)$ on a Hilbert space, namely $\sigma_e^5(T) = \bigcap_M \sigma_M(P_M T|_M)$, where M is any subspace of finite codimension, and P_M is the orthogonal projection onto M. The result was obtained independent of Fredholm theory, in the sense that the argument ([23, Lemma 2.2]) uses a criteria for the invertibility of a two by two matrix representation of a Hilbert space bounded operator. However, this characterization is incomplete, due to the lack of clarification of a subtle but important point.

In order to gain a further understanding of $\sigma_e^5(T)$, let us examine here the considerations involved in obtaining a characterization of the type just mentioned, but more generally, for an arbitrary densely defined closed operator T in a complex Hilbert or Banach space X, and via Fredholm theory. In particular, it will become clear that what is really involved here, at least from this viewpoint, is the question of "the nongeneration of continuous spectra" from a pole; and as such, there is a significant connection to both the Weinstein-Aronszajn formulae (see [7, pp. 244-250]) and to questions of the "stability" of an isolated eigenvalue (see [7, Chapter 8]) under projection.

Letting M be a subspace of finite codimension, letting M' be one of its complementing subspaces, and letting P be the projection of **X** on M along M', the desired characterization is equivalent to showing that the set of {poles (possibly nonproper)} = $\bigcup_M \rho_M(PT|_{D(T) \cap M})$. In one direction, this is immediate; given that λ_0 is a pole of $(T + \lambda)^{-1}$, by taking I - P to be the algebraic projection for λ_0 and M = R(P), T is reduced by both M and M' = R(I - P) and $\lambda_0 \in \rho_M(PT|_{D(T) \cap M})$, e.g., see [7, pp.178-180]. Moreover, I - P commutes with T, and $\lambda_0 \in \rho(T + (I - P))$; this is another way in which one may view the characterization $\sigma_e^5(T) = \bigcap_C \sigma(T + C)$, C commuting with T, mentioned above.

In the other direction, if $\lambda_0 \in \rho_M(PT|_{D(T) \cap M})$, writing $X = M' \oplus M$, $D(T) \cap M$ is dense in M, and the closed operator $P(T + \lambda_0)P$ in X has domain $M' \oplus D(T) \cap M$, range M, and null space M'. Thus $P(T + \lambda_0)P$ is Fredholm of index zero and $X = N(P(T + \lambda_0)P) \oplus R(P(T + \lambda_0)P)$; this, as is well-known (e.g., see [3, Thm. 1.1] noting that $N(P(T + \lambda_0)P) = N((P(T + \lambda_0)P)^2))$, is sufficient for 0 to be a pole of $P(T + \lambda_0)P$. Moreover, from the Fredholmness of $P(T + \lambda_0)P$ and P and the closedness of $T + \lambda_0$ one may conclude (e.g., using Schechter [24, p. 171, p. 169]) that $T + \lambda_0$ is Fredholm of index zero, i.e., $\lambda_0 \in \Delta^4(T)$.

Alternately, one could obtain the latter fact directly from the stability of σ_e^4 under relatively compact perturbations by writing $T + \lambda_0 = P(T + \lambda_0)P + B_P$ where $B_P = [(T+\lambda_0) - P(T+\lambda_0)P] = (I-P)(T + \lambda_0) + P(T + \lambda_0)(I - P)$, since then B_P is a relatively degenerate perturbation (see [7]) of $P(T + \lambda_0)P$; in decomposing B_P in this way one assumes the technicality that $M' \subset D(T)$. A modification of this approach therefore is to decompose according to $T + \lambda_0 = P(T + \lambda_0) + (I - P)(T + \lambda_0)$; then, as above, one checks that $P(T + \lambda_0)$ is closed, $R(P(T + \lambda_0)) = R((P(T + \lambda_0))^2) = M$, $N(P(T + \lambda_0)) = N((P(T + \lambda_0))^2)$, so that $X = N((P(T + \lambda_0)) \oplus M$ and 0 is a pole for $P(T + \lambda_0)$. In particular, $P(T + \lambda_0)$ is Fredholm of index zero and, by the closed graph theorem, $B = (I - P)(T + \lambda_0)$ is a relatively degenerate perturbation of $P(T + \lambda_0)$, so that $T + \lambda_0$ is Fredholm of index zero and $\lambda_0 \in \Delta^4(T)$.

It remains, in any case, to show that λ_0 is a pole of $(T + \lambda)^{-1}$; since a neighborhood of λ_0 is in $\Delta^4(T)$ and since for λ in the deleted neighborhood, i.e., $\lambda \neq \lambda_0$, the nullity dim $N(T + \lambda)$ is constant, it suffices to show that for some λ in this deleted neighborhood, $T + \lambda$ is 1 - 1. This is in fact the crux of the matter; that is, to be able to perturb a pole (e.g., of $(P(T + \lambda_0) + \lambda)^{-1}$) without generating a full (two-dimensional) neighborhood of spectra of the perturbed operator (e.g., of $T + \lambda_0$)

about the pole. We prefer to recast this question in terms of the W-A theory, which we now recall (e.g., see Weinstein-Stenger [25]; we will follow Kato [7, pp 244-250]).

For T closed in a complex Banach space, and A a T-degenerate operator, and Δ a domain of the complex plane containing only λ in $\rho(T)$ or isolated eigenvalues of finite algebraic multiplicity, one has the formula

$$\tilde{\nu}(\lambda; T + A) = \tilde{\nu}(\lambda; T) + \nu(\lambda;\omega)$$

for all $\lambda \in \Delta$, where $\omega = \omega(\lambda; t, A) = \det(1 + A(T + \lambda)^{-1})$ is defined for $\lambda \in_{\sim} \rho(T) \cap \Delta$ and is the W - A determinant of the 1st kind,and where $\tilde{\nu}$ and ν are multiplicity functions defined (see [7, pp. 246-247]) for all $\lambda \in \Delta$. In particular, $\omega(\lambda,T,A)$ is meromorphic in Δ. Thus regarding $P(T + \lambda_0)$ as the unperturbed operator and $A = (I - P)(T + \lambda_0)$ as the perturbation, and taking Δ to be a small neighborhood consisting of 0 and resolvent points λ near the pole 0 of $(P(T + \lambda_0) + \lambda)^{-1}$, we may apply the W - A theory to $T + \lambda = P(T + \lambda_0) + (I - P)(T + \lambda_0)$ to determine whether for all $\lambda \in \Delta$, $\lambda \neq 0$, $T + \lambda$ is 1 - 1. This would correspond to the first alternative in the W - A theory; the second alternative, namely, that $\omega(\lambda) \equiv 0$ in Δ (where it is agreed that one fills in any removable singularities of ω, to avoid a technicality), must therefore be ruled out. The following example shows that this cannot in general be done.

Example 3. Let T be the shift operator on $\ell_2(-\infty, \infty)$ defined by $Te_n = e_{n-1}$, $n \neq 0$, $Te_0 = 0$; then $\Delta^4_{singular}(T)$ = open unit disc, $\sigma^2_e(T) = \sigma^3_e(T) = \sigma^4_e(T)$ = unit circle, $\sigma(T) = \sigma^5_e(T)$ = closed unit disc, $N(T) = sp\langle e_0\rangle$, $R(T) = sp\langle e_{-1}\rangle^{\perp}$. Let $M = sp\langle e_{-2}, e_{-3}, \ldots\rangle \oplus sp\langle e_1, e_2, \ldots\rangle \oplus sp\langle e_0 + e_1\rangle$, a closed subspace of codimension 1, and let P be the orthogonal projection onto M; then $0 \in \rho_M(P_M T|_M)$ but 0 is not a pole of $(T + \lambda)^{-1}$.

From the above considerations one can therefore write down, for example, characterization of σ^5_e in terms of the W - A determinant, as follows.

Theorem 4. For T a densely defined closed operator in a complex Banach space, $\sigma^5_e(T) = \bigcap \sigma_M(P_M T|_M)$, where M is a subspace of finite codimension such that the Weinstein-Aronszajn determinant $\omega(\lambda; P_M(T + \lambda_0), (I - P_M)(T + \lambda_0))$ does not vanish for small $\lambda \neq 0$, $\lambda_0 \in \sigma_M(P_M T|_M)$.

We remark that in using P T rather than P T P we have used a W - A type argument "intermediate" between the W - A theories of first and second kind; also (e.g., if applied to the more special case P T P as mentioned above), in the reverse direction from that usually taken.

2. Candidates for H_{ac}. This could instead be entitled: mathematically, what are the "scattering" states? There has been considerable recent interest in this question (e.g., see Amrein and Georgescu [26, 27] for more details, references, and applications to Hamiltonians with repulsive potentials, among others), and for that reason the observations made here will be brief and somewhat incomplete.

Given a pair of operators T_1 and T_2 (e.g., $T_2 = T_1 + V$, both self-adjoint), the choice mathematically of a scattering subspace could be described as follows, if one first adopts the time-dependent viewpoint. Wave operators would be defined by $W_{\pm} = \text{s} - \lim_{t \to \pm\infty} e^{it\, T_2} e^{-it\, T_1} P$, where P is the projection onto a scattering subspace M. Since one does not expect the eigenfunctions of T_2 and T_1 to agree, M should exclude $H_p(T_1)$. Thus, for example, in one of the earliest basic papers on the subject, Jauch [28] used $M = H_c(T_1)$; this may be regarded as the "largest" choice of M, and may in fact be the more desirable choice from the physicists viewpoint. On the other hand, choosing M smaller (e.g., this is pointed out for $M = H_{ac}(T_1)$ by Kato [7, p. 532, Remark 3.6]) may provide a better likelihood for the existence of the wave operators. In general, from a mathematical point of view, M could be any reducing subspace of T_0 such that the wave operators exist; then most of the desired properties e.g., most of those of [17, pp. 529-533], of generalized wave operators will follow, including some of the sufficient conditions for their existence. Secondly, as is well-known, e.g., see Kato [29], from the time-independent or stationary viewpoint the wave operators, and therefore the scattering subspaces, depend on considerations involving boundary values of analytic functions. Thirdly, also well-known and probably closer to the physical point of view, scattering states can be defined in terms of their ergodic properties.

Let us omit here any further discussion of wave operators and other related general considerations from scattering theory, restricting outselves to the following ordering (Diagram 5 below) of certain subspaces related to $H_{ac}(H)$, where H is an arbitrary self-adjoint operator in a complex Hilbert space.

$$H_c = \{\Phi \mid \int_0^T |\langle e^{itH}\Phi, f\rangle|^2\, dt = o(T)\} \supset \overline{M_\infty^\pm} = \{\Phi \mid \int_0^T \|F_r e^{itH}\Phi\|^2\, dt = o(T)\}$$

$$\cup \qquad\qquad\qquad\qquad \cup$$

$$M_{RL} = \{\Phi \mid \langle e^{itH}\Phi, f\rangle \to 0\} \qquad M_\infty^\pm = \{\Phi \mid \|F_r e^{itH}\Phi\| \to 0\}$$

$$\cup \qquad\qquad\qquad\qquad \cup$$

$$H_{ac} = \{\Phi \mid \langle E(\lambda)\Phi, \Phi\rangle \text{ abs. cont.}\} \subset M_\infty^\pm, \text{ given some compactness}$$

$$\cup$$

$$M = \{\Phi \mid \|R_z \Phi\| = 0(y^{-1/2})\}$$

$$= \{\Phi \in H_{ac} \mid g(\lambda) \equiv d\langle E(\lambda)\Phi, \Phi\rangle / d\lambda \in L^\infty\} \subset \{\Phi \mid \langle e^{itH}\Phi, f\rangle \in L^2\}$$

$$\cup$$

$$\{\Phi \mid g \in L^\infty \cap C^0,\ g(\pm\infty) = 0\}$$

$$\cup$$

$$\cup$$
$$\{\Phi \mid \langle e^{itH}\Phi,\Phi\rangle \in L^1\}$$
$$\cup$$
$$\{\Phi \mid g \in C_0^\infty\}$$

Diagram 5. Candidates for the scattering states.

As mentioned above, this diagram is somewhat incomplete and for that reason our remarks will be limited to a brief clarification of the diagram. However, just as the ordering of spectra given in Diagram 1 depended fundamentally on Fredholm theory, it can be said that orderings such as those in Diagram 5 depend fundamentally, although in some cases less directly, on the theory of the Fourier-Stieltjes integral.

Turning to Diagram 5, the well-known ergodic characterization of H_c given there, with $T \to \infty$, is Wiener's Theorem (e.g., see Wintner [30], Kato [7, p. 515], Katznelson [31]) applied to the measure $d\langle E(\lambda)\Phi,f\rangle$, where f denotes an arbitrary vector in the Hilbert space, and where $E(\lambda)$ denotes the spectral family for H; for the use of Wiener's Theorem in this form in recent work concerning various different aspects of scattering theory see, for example, Lax and Phillips [32, p. 145], Horwitz, LaVita, and Marchand [33, p. 2541], and Sinha [34]. For the subspaces $\overline{M^{\pm}}$ and $M_{\infty}^{\pm}$ and the inclusion $H_c \supset \overline{M_{\infty}^{\pm}}$, see Amrein and Georgescu [26, 27], the inclusion $\overline{M_{\infty}^{\pm}} \supset M_{\infty}^{\pm}$ being easily checked; Wilcox [35] used the notation H^s rather than $M_{\infty}^{\pm}$. Here the F_r, $r \geq 0$ is a family of orthogonal projections converging strongly to the identity; under certain relative compactness conditions of the F_r with respect to H or parts of H, in [26, 27, 35] the inclusion $M_{\infty}^{\pm} \supset H_{ac}$ is shown. This was done previously for special "localizing" F_r by Ruelle [36], who attributes some of the motivation to an earlier paper by Haag and Swieca [37]; from the Fourier ergodic viewpoint these ideas go far back (e.g., see Schoenberg [38]) and any further history would be too complicated for these remarks here.

The notation M_{RL} connotes the Riemann-Lebesgue Lemma and as is well-known M_{RL} is in general larger than H_{ac}.

In Gustafson and Johnson [39] the following characterization of H_{ac} in terms of the growth rate of the resolvent on individual vectors was observed.

Theorem 6 [39].

$$H_{ac} = \overline{\{\Phi \mid \|R_z\Phi\| = O(y^{-1/2})\}}$$

$$= \{\Phi \mid \operatorname{Im}\langle R_z\Phi,\Phi\rangle \text{ converges in } L^1_{loc}\}$$

Here R_z denotes the resolvent operator for H, z = x + iy, and the $O(y^{-1/2})$, $y \to 0^+$, can depend on each Φ, i.e., it need not be uniform. This characterization of H_{ac} is independent of the spectral family $\{E(\lambda)\}$ and the group exp(itH) and as such could be used to define an absolutely continuous subspace for certain types of non-self-adjoint operators T. The second formulation given in Theorem 6 is perhaps of interest in characterizing H_{ac} as vectors whose matrix elements have L^1_{loc} boundary values $y \to 0^+$, and is the way in which the theorem was proved in [39]. In retrospect, an alternate proof is possible by means of verification of the observation (see Diagram 5) that the subspace (before closure) in Theorem 6 is identical to the subspace $\{\Phi \in H_{ac} \mid d\langle E(\lambda)\Phi,\Phi\rangle /d\lambda \in L^\infty \equiv L^\infty(-\infty,\infty)\}$ shown to be dense in H_{ac} by Kato [7, p. 542], and used in connection with the Kato-Rosenbloom Lemma, the latter lemma indicated by inclusion in the diagram; for generalization of this subspace in the latter formulation see Martin and Misra [40]. We also mention that the resolvent growth rate condition of Theorem 6 is implied for all vectors Φ in R(A*) for any operator A which is H-smooth (see Kato [29, (5.4)], Lavine [41]).

The next two inclusions in the diagram follow directly by use of the Fourier transform and are part of more detailed considerations to be considered elsewhere, and we omit the details.

References:

1. H. Weyl, Über beschränkte quadratische Formen, deren Differenz vollstetig ist, Rend. Circ. Mat. Palermo 27 (1909), 373-392.

2. F.E. Browder, On the spectral theory of elliptic differential operators I, Math. Ann. 142 (1961), 22-130.

3. M. Schechter, On the essential spectrum of an arbitrary operator I, J. Math. Anal. Applic. 13 (1966), 205-215.

4. F. Wolf, On the essential spectrum of partial differential boundary problems, Comm. Pure Appl. Math. 12 (1959), 211-228.

5. K. Gustafson and J. Weidmann, On the essential spectrum, J. Math. Anal. Applic. 25 (1969), 121-127.

6. J.T. Schwartz, Some results on the spectra and spectral resolutions of a class of singular integral operators, Comm. Pure Appl. Math. 15 (1962), 75-90.

7. T. Kato, *Perturbation theory for linear operators*, Springer, Berlin, (1966).

8. S. Goldberg, *Unbounded linear operators*. McGraw-Hill, New York (1966).

9. N. Akhiezer and I.M. Glazman, *Theory of linear operators in Hilbert space*, Ungar, New York, (1961).

10. M.H. Stone, Linear transformations in Hilbert space and their applications to analysis, Amer. Math. Soc. Colloq. Publ. 15, Providence, (1932).

11. N. Dunford and J.T. Schwartz, *Linear operators*, Interscience, New York, (1958).

12. B. Sz. Nagy, *Spectraldarstellung linear transformationen des Hilbertschen Raumes*, Springer, Berlin, (1942).

13. D. Hilbert, *Grundzüge einer allgemeinen Theorie der linear Integralgleichungen*, Leipzig, (1912).

14. F. Riesz and B. Sz. Nagy, *Functional analysis*, Ungar, New York, (1955).

15. P. Halmos, *Introduction to Hilbert space and the theory of spectral multiplicity*, Chelsea, New York, (1951).

16. S.K. Berberian, The Weyl spectrum of an operator, Indiana Univ. Math J. 20 (1970), 529-544.

17. K. Gustafson, Weyl's Theorems, Proc. Oberwolfach Conf. on Linear Operators and Approximation, Int. Series. Num. Math. 20, Birkäuser, Basel, (1972), 80-93.

18. G. Huige, The spectral theory of some non-self-adjoint differential operators, Comm. Pure Appl. Math. 21 (1968), 25-50.

19. J.C. Combes, N-body systems, Proc. Denver Scattering Conf. (1973).

20. I.C. Gohberg and M.G. Krein, The basic propositions in defect numbers, root numbers, and indices of linear operators, Amer. Math. Soc. Transl. (2), 13 (1960), 185-264.

21. M.A. Kaashoek and D.C. Lay, Ascent, descent, and commuting perturbations, Trans. Amer. Math. Soc. 169 (1972), 35-47.

22. K. Gustafson, Some examples for the ascent-descent spectral theory (to appear).

23. N. Salinas, A characterization of the Browder spectrum, Proc. Amer. Math. Soc. 38 (1973), 369-373.

24. M. Schechter, *Principles of functional analysis*, Academic Press, New York, (1971).

25. A. Weinstein and W. Stenger, *Methods of intermediate problems for eigenvalues*, Academic Press, New York, (1972).

26. W. Amrein and V. Georgescu, On the characterization of bound states and scattering states in quantum mechanics, (to appear).

27. W. Amrein, Some fundamental questions in the non-relativistic quantum theory of scattering, Proc. Denver Scattering Conf. (1973).

28. J.M. Jauch, Theory of the scattering operator, Helv. Phys. Acta 31 (1958), 127-158.

29. T. Kato, Wave operators and similarity for non-self-adjoint operators, Math. Ann. 162 (1966), 258-279.

30. A. Wintner, *Asymptotic distributions and infinite convolutions*, Edwards Bros. Inc., Ann Arbor, (1938).

31. Y. Katznelson, *An introduction to harmonic analysis*, Wiley, New York, (1968).

32. P. Lax and R.S. Phillips, *Scattering theory*, Academic Press, New York, (1967).

33. L. Horwitz, J. LaVita, J.P. Marchand, The inverse decay problem, J. Math. Phys. 12 (1971), 2537-2543.

34. K. Sinha, On the decay of an unstable particle, Helv. Phys. Acta. 45 (1972), 619-628.

35. C. Wilcox, Scattering states and wave operators in the abstract theory of scattering, J. Funct. Anal. 12 (1973), 257-274.

36. D. Reulle, A remark on bound states in potential scattering theory, Nuovo Cim. 61A (1969), 655-662.

37. R. Haag and J. Swieca, When does a quantum field theory describe particles, Comm. Math. Phys. 1 (1965). 308-320.

38. I. Schoenberg, Über total monotone Folgen mit stetiger Belegungsfunktion, Math. Zeit. 30 (1929), 761-767.

39. K. Gustafson and G. Johnson, On the absolutely continuous subspace of a self-adjoint operator, Helvetica Physica Acta (to appear)

40. P. Martin and B. Misra, Kato-Rosenbloom Lemma, Proc. Denver Scattering Conf. (1973).

41. R. Lavine, Commutators and scattering theory II. A class of one body problems, Indiana Univ. Math. J. 21 (1972), 643-656.

REGULAR PERTURBATIONS

James S. Howland[†]

University of Virginia

Consider a family

$$H_\kappa = H_0 + \kappa V = \int \lambda dE_\kappa(\lambda)$$

of selfadjoint operators on $\mathcal{H}$ depending on a small parameter κ. Suppose that H_0 has an eigenvalue λ_0 of finite multiplicity m, and let P be the orthogonal projection onto the null space $N(H_0 - \lambda_0)$ of $H_0 - \lambda_0$. If λ_0 is an isolated point of the spectrum of H_0, and V is mild, the resolvent

$$R(z,\kappa) = (H_\kappa - z)^{-1}$$

is analytic in κ, and hence, for small κ, there are (counting multiplicities) exactly m eigenvalues of H_κ near λ_0, which are analytic functions of κ [6, Chapter VII]. We shall then call H_κ a <u>regular perturbation of</u> λ_0. In other circumstances, however, H_κ may have only a continuous spectrum near λ_0 for non-zero κ. Essentially two cases have been studied: 1) λ_0 isolated [1,4,6,10] and 2) λ_0 embedded in an absolutely continuous spectrum [2,5,11]. In the first case, the perturbation V must be exceedingly strong, as it is for Stark effect Hamiltonians [6,9,10], and $R(z,\kappa)$ will not be analytic at $\kappa = 0$. In the second case, however, which is related to the Auger effect [11], the perturbation may be very mild, and $R(z,\kappa)$ analytic.

Most discussions of these problems center around <u>spectral concentration at</u> λ_0, which is said to occur if there are sets J_κ, of asymptotically small measure, such that

---

[†]Supported by DA-ARO-31-124-71-G182

J. A. LaVita and J.-P. Marchand (eds.), Scattering Theory in Mathematical Physics, 169–172.

$$s - \lim_{\kappa \to 0} E_\kappa(J_\kappa) = P.$$

In the regular case, one can simply take J_κ to consist of the m eigenvalues of H_κ near λ_0; in general, it is of interest to know how small the sets J_κ may be taken. It must be stressed that spectral concentration is an <u>asymptotic property of a family of operators</u> - it has no unitarily invariant significance for a single operator. The notion originated in the investigations of Titchmarsh on ordinary differential operators [8], where the Green's function was shown to have a second sheet pole which approaches λ_0 as $\kappa \to 0$, the measure of J_κ being essentially proportional to the imaginary part of the pole. The question therefore arises whether a given instance of concentration is due in a similar manner to a resonance, that is a second sheet pole of some matrix element $(R(z,\kappa)\phi,\phi)$. Of course, literally any number can be so obtained by a judicious choice of ϕ, but if H_κ is an analytic family and $\lambda(\kappa)$ is a pole of $(R(z,\kappa)\phi,\phi)$ which is analytic in κ, with $\lambda(0) = \lambda_0$, then for κ in a suitable sector of the complex plane, $\lambda(\kappa)$ falls on the <u>first</u> sheet, and is therefore a point eigenvalue of the non-self-adjoint operator H_κ. The family $\lambda(\kappa)$ therefore has a unitarily invariant significance for the family H_κ.

The main point of this lecture is that there is a theory of regular perturbations of embedded eigenvalues which is analogous to that of isolated eigenvalues, in which eigenvalues are replaced by resonances and analyticity of $R(z,\kappa)$ in κ is replaced by analyticity of a certain matrix. The usefulness of this matrix, which was apparently first considered by M.S. Livsic [7], is due to an easily proved formula occurring frequently in the physics literature. Let H be selfadjoint and $\mathcal{K}$ a subspace of $\mathcal{H}$, which for simplicity we take to be a finite dimensional subspace of the domain of H. Let P be the orthogonal projection on $\mathcal{K}$. With respect to the decomposition $\mathcal{H} = \mathcal{K}^\perp \oplus \mathcal{K}$, write

$$H = \begin{pmatrix} T & \Gamma \\ \Gamma^* & A \end{pmatrix}.$$

If $R(z) = (H - z)^{-1}$ and $G(z) = (T - z)^{-1}$, then as operators *on* $\mathcal{K}$

$$PR(z)P = (B(z) - z)^{-1}$$

where

$$B(z) = A - \Gamma^* G(z) \Gamma. \tag{1}$$

We shall refer to $B(z)$ as the <u>Livsic matrix of</u> H <u>on</u> $\mathcal{K}$. For $\operatorname{Im} z > 0$, $B(z)$ is a dissipative operator on $\mathcal{K}$. If $B(z) = B$ is identically constant, one obtains the Lax-Phillips Theory (see [3]), so that the spectral points of B are the poles of the S-matrix.

Generalizing, if $B(z)$ has an analytic continuation from the upper half-plane into the second sheet, we shall call a zero of

$$\det(B(z) - z)$$

a <u>resonance</u> <u>of</u> H <u>on</u> $\mathcal{K}$.

Let $B(z,\kappa)$ be the Livsic matrix of H_κ on $\mathcal{K} = N(H_0 - \lambda_0)$. We shall now call H_κ a <u>regular</u> <u>perturbation</u> <u>of</u> λ_0 if $B(z,\kappa)$ has a second sheet continuation which is holomorphic in (z,κ) in a neighborhood of $(\lambda_0,0)$. The equation

$$\det(B(z,\kappa) - z) = 0$$

for resonances reduces to $(z - \lambda_0)^m$ for $\kappa=0$ because $B(z,0) = \lambda_0 I_m$. Standard arguments therefore show that for small κ, there are (counting multiplicities) exactly m resonances near λ_0, and that they have Puiseux expansions in κ. These expansions have a special form because they cannot have positive imaginary part when κ is real [5,11]. Examples can be given in which non-analytic Puiseux series do occur [5], although this is a degenerate case.

In an application, the proof that $B(z,\kappa)$ is holomorphic may not be trivial because (1) involves $G(z)$ which is probably not known explicitly. However, the cases which have so far been treated seem to fit into this framework. For example [5], if $H_\kappa = H_0 + \kappa CD$ and

$$Q(z) = D(T_0 - z)^{-1} C$$

is bounded and continues to the second sheet, then

$$B(z,\kappa) = \lambda_0 I_m + \kappa PC[I + \kappa Q(z)]^{-1} DP$$

is clearly holomorphic in (z,κ). It is hardly more difficult to treat the Balslev-Combes Hamiltonians discussed by Simon [11].

This method provides a unified approach to perturbation series, since whenever the perturbation is regular, formulas follow for the resonances which are independent of the particular machinery (Balslev-Combes, factorization, Green's function, etc.) used to obtain the continuation.

Spectral concentration can be discussed in a framework which includes isolated eigenvalues. Let H_n be a sequence of self-adjoint operators for which $R_n(z)$ converges strongly to $R_0(z)$ for non-real z [6, pp. 427-437], and assume that the Livsic matrix $B_n(z)$ of H_n on $\mathcal{K} = N(H_0 - \lambda_0)$ continues from the upper half-plane to a neighborhood of λ_0 independent of n. If $B_n(z) \to B_0(z) = \lambda_0 I_m$ uniformly, there are (counting multiplicities) exactly m resonances $z_1(n),\ldots, z_m(n)$ of H_n near λ_0.

Concentration Theorem. For $j=1,\ldots,m$, let $z_j(n) = \lambda_j(n) - i\Gamma_j(n)$, and choose $\eta_j(n)$ so that $\Gamma_j(n) = o(\eta_j(n))$. If

$$J_n = \bigcup_{j=1}^{m} (\lambda_j(n) - \eta_j(n), \lambda_j(n) + \eta_j(n))$$

then $E_n(J_n)$ converges to P strongly.

The proof follows [4].

Many additional references can be found in the articles listed below.

References:

1. H. Baumgärtel, Nichtanalytische Störungen Kompakter Operatoren, Mber. Dt. Akad. Wiss. Berlin., (to appear).
2. H. Baumgärtel, Kompakte Störungen nichtisolierter Eigenwerte endlicher Vielfachheit. Mber. Dt. Akad. Wiss. Berlin., (to appear).
3. L.P. Horwitz, J. LaVita, and J.-P. Marchand, The inverse decay problem., J. Math. Phys. 12 (1971), 2537-2543.
4. J.S. Howland, Spectral concentration and virtual poles, II. Trans. Amer. Math. Soc. 162 (1971) 141-156.
5. J.S. Howland, Puiseux series for resonances near an embedded eigenvalue., Pacific J. Math. (to appear).
6. T. Kato, Perturbation theory for linear operators, Springer-Verlag, New York, 1966.
7. M.S. Livsic, The application of non-self-adjoint operators to scattering theory., Soviet Physics JETP (English translation) 4 (1957) 91-98.
8. J.B. McLeod, "Spectral concentration, I: The one-dimensional Schroedinger operator" in "Perturbation Theory and its Applications in Quantum Mechanics," ed. C. Wilcox - Wiley, New York (1966).
9. P.A. Rejto, "Spectral concentration for the Helium Schroedinger operator." Helv. Phys. Acta 43 (1970) 652-667.
10. R.C. Riddell, Spectral concentration for self-adjoint operators, Pacific J. Math (1967) 377-401.
11. B. Simon, Resonances in n-body quantum systems with dilatation analytic potentials and the foundations of time-dependent perturbation theory, Ann. Math 97 (1973) 247-274.

A REFINEMENT OF THE KATO-ROSENBLUM LEMMA AND ITS APPLICATIONS IN SCATTERING THEORY: RELATIVISTIC POTENTIAL SCATTERING, HIGH ENERGY BEHAVIOUR OF SCATTERING CROSS SECTIONS AND THE TIME-DELAY OPERATOR

Ph. Martin* and B. Misra

University of Colorado, Boulder

1. Introduction.

In the mathematical theory of scattering processes the condition that certain pertinent operators are of trace-class plays a central role. For instance, it is shown by Kato [1], Birman and Krein [2] and others that a sufficient condition for the existence, completeness and the "invariance property" of the Möller wave operators is that the difference $R_z - R_z^0$ of the resolvents of the total and the unperturbed Hamiltonians is of trace-class (or more generally $R_z^n - R_z^{0n}$ is of trace-class for some integer n).

It is known that the proof of these fundamental results leans heavily on some variants of the following simple lemma which we shall call here:

The Kato-Rosenblum lemma [3]. Let $H = \int \lambda dE_\lambda$ be a self-adjoint operator and suppose that $||E_\lambda \Psi||^2$ and $||E_\lambda \phi||^2$ are absolutely continuous with

$$\text{ess. sup.} \frac{d||E_\lambda \Psi||^2}{d\lambda} \equiv m_\Psi^2 < \infty; \quad \text{ess. sup.} \frac{d||E_\lambda \phi||^2}{d\lambda} \equiv m_\phi^2 < \infty$$

Then for any trace-class operator T

$$\int_{-\infty}^{\infty} (\Psi, e^{iHt} Te^{-iHt} \phi)dt \leq 2\pi m_\phi m_\Psi ||T||_1$$

where $||T||_1$ denotes the *trace-norm* of T.

* Partially supported by the Swiss National Science Foundation.

J. A. LaVita and J.-P. Marchand (eds.), Scattering Theory in Mathematical Physics, 173–191.

Thus, it is not unexpected, but nevertheless pleasing, that a refinement of this lemma should serve again as the unifying thread of various new developments in scattering theory. And this will be the main theme of this paper.

The refined version of the Kato-Rosenblum lemma will be formulated in the next section as theorem 1. This theorem is a refinement of the lemma in two respects: First, the one-parameter unitary group exp(iHt) occurring in the Kato-Rosenblum lemma is replaced by a unitary representation $g \to U_g$ of an arbitrary locally compact abelian group. And second, more specific and detailed properties of the operator valued integrals $\int U_g^* T U_g \, dg$ are obtained than just the inequality of the lemma.

The remaining sections are devoted to some recent applications of theorem 1 in scattering theory. The first such application we consider (in §3) is concerned with obtaining a necessary and sufficient condition for the difference of the resolvents $R_z - R_z^0$ to be nuclear in potential scattering. In this connection we consider not only the extensively studied non-relativistic scattering problem (Schrödinger Hamiltonian) but the relatively unexplored relativistic potential scattering also. Since the term Relativistic Potential Scattering connotes various things to various people (and to some it is even a contradiction in terms!) we denote a short appendix to an explanation of this model.

The two other applications that we consider here relate to the properties of scattering cross sections (§ 4) and the time-delay operator of scattering processes (§ 5).

Finally we should mention that in the following discussion we have omitted most of the proofs and technical details and much of the motivating discussions in the interest of brevity. But the interested reader may fill in the missing details by going to the appropriate sources mentioned in the text.

2. A Refinement of the Kato-Rosenblum Lemma.

Before formulating theorem 1 we need to recall a few facts about the representations of locally compact abelian groups G in order to establish our terminology and notations.

Let G be a locally compact abelian group. Its character group will then be denoted by $\hat{G}$. If x is a character of G (i.e. if x is a continuous function on G which satisfies: $|x(g)| = 1$; $x(g_1)\, x(g_2) = x(g_1 g_2)$ for all $g_1 g_2 \in G$) then we shall write $\langle x,g \rangle$ to denote the value of the function x at the point g of G. It is clear that if g is held fixed and x runs over $\hat{G}$ then $\langle x,g \rangle$ defines a character of the dual group $\hat{G}$.

Now let $g \to U_g$ be a strongly continuous unitary representation of G which acts on the Hilbert space $\mathcal{H}$. We recall that with every such representation of G there is associated a unique projection-valued measure E(•) defined on the Borel sets of $\hat{G}$ such that

$$(\phi, U_g \Psi) = \int_{\hat{G}} \langle x, g\rangle \, d(\phi, E(\cdot)\Psi) dx$$

for every $g \in G$ and all Ψ and ϕ in $\mathcal{H}$. [SNAG Theorem].

If for every Ψ in $\mathcal{H}$ the numerical valued measure $\|E(\cdot)\Psi\|^2$ is absolutely continuous with respect to the Haar measure on $\hat{G}$ then we shall say that the representation $g \to U_g$ of G is *absolutely continuous*. In the following we shall consider such absolutely continuous representations only.

Let $g \to U_g$ be an absolutely continuous representation of G in $\mathcal{H}$. Then is is well-known [4] that there exists a direct-integral representation

$$\mathcal{H} = \int_{\oplus \hat{G}} H_x dx$$

of $\mathcal{H}$ over the measure space $(\hat{G}, dx)$ (with dx denoting the Haar measure on $\hat{G}$) such that in this representation the operators U_g (with $g \in G$) are represented by the multiplication operators with the function $\langle x, g\rangle$. Somewhat more explicitly, there exists a family H_x of Hilbert spaces labelled by characters x of G and a correspondence $\Psi \to \Psi_x$ between vectors Ψ in $\mathcal{H}$ and vector valued functions Ψ_x (with $\Psi_x \in H_x$) on $\hat{G}$ such that:

$$(\Psi, \phi) = \int_{\hat{G}} (\Psi_x, \phi_x)_x \, dx$$

for all Ψ, ϕ in $\mathcal{H}$; and

$$U_g \, \Psi \to \langle x, g\rangle \, \Psi_x$$

for every $g \in G$ and all Ψ in $\mathcal{H}$. Here $(\Psi_x, \phi_x)_x$ denotes the scalar product in the Hilbert space H_x. The direct integral representation of $\mathcal{H}$ just described is said to "diagonalize" the given representation $g \to U_g$ of the group G.

With the above preliminary remarks out of the way we can now formulate

<u>Theorem 1</u>. Let $g \to U_g$ denote an absolutely continuous representation (in the Hilbert space $\mathcal{H}$) of the locally compact Abelian group G and let the correspondence $\Psi \to \Psi_x$ yield the direct integral representation of

$$\mathcal{H} = \int_{\oplus \hat{G}} H_x \, dx$$

which diagonalizes the representation $g \to U_g$. Let D denote the set of all Ψ in $\mathcal{H}$ for which

$$M_\Psi = \underset{x \in \hat{G}}{\text{ess. sup.}} \; || \Psi_x ||_x < \infty$$

Then

(a) For every nuclear operator T in $\mathcal{H}$

(i) $B_T(\Psi,\phi) = \int_G (\Psi, U_g^* T U_g \phi)\, dg < \infty$ for all Ψ, $\phi \in D$.

Moreover, for almost all $x \in \hat{G}$, there exists nuclear operators $Q_T(x)$ acting in H_x such that

(ii) $B_T(\Psi,\phi) = \int_{\hat{G}} (\Psi_x, Q_T(x)\, \phi_x) dx$ for all Ψ, $\phi \in D$.

The family $Q_T(x)$ is essentially unique in the sense that if $Q_T'(x)$ is another family of operators satisfying (ii), then $Q_T(x) = Q_T'(x)$ for a.a. $x \in \hat{G}$.

(b) Furthermore, the following relations hold

(1) $\mathrm{Tr}\, T = \int_{\hat{G}} \mathrm{Tr}\, Q_T(x)\, dx$

(2) $\int_{\hat{G}} || Q_T(x) ||_1 \, dx \leq ||T||_1$

Here dg and dx represent (respectively) the Haar measures on G and $\hat{G}$ that are appropriately normalized relative to each other [5].

It is not the place to go into a proof of this theorem which may be found in [6,7]. Nor need we discuss its various technical corollaries. But we should draw the attention of the reader to the following remarks:

1. Although the inequality of the Kato-Rosenblum lemma does not, at first glance, seem to be included in the theorem 1 it is not hard to verify that it is indeed subsumed under this theorem.

2. The part (a) of the theorem remains true (with the set D replaced by an appropriately modified set) for a much larger class of operators T than the trace class operators and even for certain unbounded operators. For instance, if U_g is the one-parameter group exp(iHt) and T is a, possibly unbounded, operator for which $R_z^n\, T\, R_z^n$ is nuclear ($R_z \equiv (H - zI)^{-1}$) then part (a) of the theorem holds for T. This remark is of use in the applications considered in §4 and §5.

3. The part (a) of theorem (1) shows that an operator integral of the form $\int U_g^*\, T\, U_g dg$ defines a densely defined sesquilinear form

B_T on the Hilbert space. This naturally raises the question of existence and uniqueness of linear operators associated with such forms. Although this question is relevant in the application considered in §5 and other applications concerned with the study of the so called Friedrichs-Kato Γ-Operation we shall not discuss it here due to the limitations of space. We only mention that there are two natural ways to associate linear operators with such forms. One is to consider the direct integral $\int_{\oplus \hat{G}} Q_T(x)\, dx$ of operators $Q_T(x)$ whose existence is given by the theorem. And the other is to study the "closability" property of the forms B_T and then to invoke standard theorems concerning the relation between closed forms and linear operators.

4. Finally, it will be of some mathematical interest as well as of use in physical applications to know if the theorem 1 admits of an appropriate converse. That is to say, can the class of nuclear operators be characterized as the class of all operators T for which part (a) of the theorem holds for some (or every) absolutely continuous representation of a (or every) locally compact abelian group G and for which

$$\int_{\hat{G}} || Q_T(x) ||_1 \, dx < \infty$$

or is the latter class strictly larger than the class of nuclear operators? To our knowledge this question is not yet settled.

3. Necessary and Sufficient Conditions for the Difference of Resolvents and Related Expressions to be Nuclear.

As mentioned in the introduction the condition that the difference $R_z - R_z^0$ of the resolvents of the total and unperturbed Hamiltonians (or some related expression) is nuclear plays an important role in scattering theory. And yet the circumstances in which this abstract condition is actually fulfilled has not been investigated to any great extent in the literature. Such an investigation is, of course, necessary for determining the range of application of the theory based on the abstract condition that $R_z - R_z^0$ is nuclear.

The only readily available result of this kind is a sufficient condition for $R_z - R_z^0$ to be nuclear in non-relativistic potential scattering [8]. In this section we shall show that theorem 1 (especially part a) enables us to obtain a necessary and sufficient condition for $R_z - R_z^0$ to be nuclear in potential scattering.

With the term "potential scattering" we have in mind not only the non-relativistic potential scattering system whose Hamiltonian in the center of mass system is the usual Schrödinger Hamiltonian:

$$H = \frac{|\vec{p}|^2}{2m} + V(\vec{x}); \quad |\vec{p}| = \sqrt{p_1^2 + p_2^2 + p_3^2} \tag{1}$$

but also a two-body relativistic scattering system whose "center of mass" Hamiltonian is of the form

$$H = \sqrt{|\vec{p}|^2 + m_1^2} + \sqrt{|\vec{p}|^2 + m_2^2} + V(\vec{x}) \tag{2}$$

(For more explanation of the model of relativistic potential scattering see the appendix). In both these models the operators $\vec{p}$ and $\vec{x}$ satisfy the usual Weyl form of the commutation relation and they form an irreducible system.

Since we wish our results to be applicable to both the non-relativistic Hamiltonian (1) as well as the Hamiltonian (2) of the relativistic potential scattering we shall formulate our results in the following general setting:

For the unperturbed Hamiltonian H_0 we shall take the multiplication operator (in $L^2(d^3\vec{p})$) by a positive function $F(\vec{p})$. The function $F(\vec{p})$ will be assumed to be infinitely differentiable everywhere (with the exception of isolated points) and to possess the following growth property:

$$\lim_{|\vec{p}|\to\infty} \frac{F(\vec{p})}{|\vec{p}|^\alpha} = c \neq 0 \text{ for some } \alpha > 0.$$

It may be noted here that for the unperturbed non-relativistic Hamiltonian $\alpha = 2$ whereas in the relativistic case $\alpha = 1$. The interaction will be taken to be a function $V(\vec{x})$ of the operators $\vec{x}$; the operators $\vec{x}$ being the self-adjoint generators of the three-parameter group $U_{\vec{\lambda}}$:

$$(U_{\vec{\lambda}} \Psi)(\vec{p}) \equiv \Psi(\vec{p}-\vec{\lambda}) \tag{3}$$

for all Ψ in $L^2(d^3\vec{p})$. The operator $V(\vec{x})$ is supposed to be H_0-bounded with H_0-bound less than one. In addition, the potential function $V(\vec{x})$ will be assumed to satisfy certain regularity conditions which we need not specify here (see §3) of [6] for more precision) and the following growth condition:

$$\lim_{|\vec{x}|\to\infty} |\vec{x}|^\gamma V(\vec{x}) = c \quad \text{for some } \gamma > 0 \text{ and } c \neq 0 \text{ if } \gamma \leq 3.$$

With this setting we now provide a complete characterization of the class of free Hamiltonians $H_0 \equiv F(\vec{p})$ and the potentials $V(\vec{x})$ for which the difference of the resolvents of H_0 and of $H \equiv H_0 + V(\vec{x})$ is nuclear.

<u>Proposition (1)</u>. Let $H_0 \equiv F(\vec{p})$ and $V(\vec{x})$ be as described above. Then $R_z - R_z^0$ is nuclear if and only if $\alpha > 3/2$ and $\gamma > 3$.

Without going into the details of the proof of the proposition (which may be found in [7]) we shall now briefly indicate the role of the theorem 1 in this proof.

We first remark that $R_z - R^0_z$ is nuclear if and only if $R^0_z\ V\ R^0_z$ is nuclear. The proof of the necessity of the conditions of the proposition now follows from a direct application of the theorem 1 (part a).

To this end we consider the unitary operators $U_{\vec{\lambda}}$ defined by relation (3). It is clear that they form an absolutely continuous representation of the additive group R^3. Moreover it can be easily verified that in the case of the unitary group $U_{\vec{\lambda}}$ the domain D of theorem 1 contains all functions $\phi(\vec{p})\varepsilon L^2(d^3\vec{p})$ that are in the Schwartz class $\mathcal{S}$ of fast decreasing test functions. Thus, according to theorem 1, to show the necessity of the condition $\alpha > 3/2$ it is only necessary to verify that the function $(\phi, U_{\vec{\lambda}} R^0_z V R^0_z U^*_{\vec{\lambda}}\phi)$ with $\phi\ \varepsilon\ \mathcal{S}$ is not integrable unless $\alpha > 3/2$. To verify this last statement one needs to study the asymptotic behavior of the matrix element $(\phi, U_{\vec{\lambda}}\ R^0_z\ V\ R^0_z\ U^*_{\vec{\lambda}}\ \phi)$. Since $U_{\vec{\lambda}}\ R^0_z\ U^*_{\vec{\lambda}}$ is the multiplication operator by the function

$$\frac{1}{F(\vec{p}-\vec{\lambda})-z} \quad \text{and } U_{\vec{\lambda}} \equiv e^{i\vec{\lambda}\cdot\vec{x}}$$

commutes with $V(\vec{x})$ both being functions of $\vec{x}$, it is not hard to see that the asymptotic behaviour of this matrix element will be determined by α, the index which characterized the growth property of $H_0 \equiv F(\vec{p})$. In fact we show that for $\phi\ \varepsilon\ \mathcal{S}$

$$(\phi,\ U_{\vec{\lambda}}\ R^0_z\ V\ R^0_z\ U^*_{\vec{\lambda}}\phi) \rightarrow \frac{(\phi, V\phi)}{|\vec{\lambda}|^{2\alpha}}$$

as $|\vec{\lambda}| \rightarrow \infty$. This verifies that $(\phi,\ U_{\vec{\lambda}}\ R^0_z\ V\ R^0_z\ U^*_{\vec{\lambda}}\ \phi\ )$ is not integrable unless $\alpha > 3/2$ and with this establishes the necessity of the condition $\alpha > 3/2$ for $R_z - R^0_z$ to be nuclear. We make similar use of the group $V_{\vec{\lambda}} \equiv e^{i\vec{\lambda}\cdot\vec{p}}$ to show the necessity of the condition $\gamma > 3$.

The theorem 1 plays no part in the sufficiency proof of the proposition. This part of the proof follows from an extension of an argument found in [8].

We conclude this section with the following remarks stemming from the proposition (1).

1. In the theory of non-relativistic potential scattering it was previously known that the condition $V(x) = 0\ \left(\frac{1}{|\vec{x}|^{\gamma}}\right)$ with $\gamma > 3$ is sufficient for $R_z - R^0_z$ to be nuclear [8]. According to proposition 1 this condition is now found also to be necessary.

2. For the relativistic potential scattering, on the other hand, it is found from proposition (1) that $R_z - R_z^0$ is never nuclear, not even for the "nicest" possible potentials!

In this respect the mathematical theory of relativistic potential scattering differs sharply from the non-relativistic case. The Kato theory of trace-class perturbation seems to be without any application in relativistic potential scattering. (It is, of course, conceivable that there might exist potentials for which $R_z^n - R_z^{0n}$ is nuclear for suitable n so that Kato's theory would be applicable to such potentials. But if so, such potentials are yet to be unearthed).

We shall see in the next section that this contrast between the relativistic and non-relativistic potential scattering leads also to differing behaviour of the scattering cross sections at high energy.

3. In the following section we shall find it useful to have a necessary and sufficient condition for $R_z^{0\beta} V R_z^{0\beta}$ to be nuclear. By slightly extending the method of proof of proposition (1) one can show that $R_z^{0\beta} V R_z^{0\beta}$ is nuclear if and only if

$$\alpha\beta > 3/2 \text{ and } \gamma > 3.$$

4. Finiteness of Scattering Cross Section and its Behaviour at High Energy.

The problem of finding the class of potentials for which the total scattering cross section at fixed values of incident energy is finite has received a good deal of attention in the recent literature. Mathematically, this problem reduces to the problem of finding conditions on the potentials such that the "energy shell" operator $(S_\varepsilon - I) + (S_\varepsilon - I)^*$ is of trace-class or equivalently (owing to the unitarity of S_ε) $S_\varepsilon - I$ is in the Hilbert-Schmidt class. Here S_ε denotes the "energy shell" scattering operator at energy ε. For non-relativistic potential scattering this problem has been studied in [9] and [10] and for the special case of the spherically symmetric potentials in [11]. All of these results can be subsumed under the following:

Theorem (Birman and Krein) [12]. Let the resolvents of the self-adjoint operators H and H_0 satisfy the condition that $R_z^n - R_z^{0n}$ is nuclear for some n. Then the "energy shell operators" $S_\varepsilon - I$ of the scattering system defined by the free Hamiltonian H_0 and the total Hamiltonian H is nuclear for almost all ε in the spectrum of H_0.

There are available, of course, certain other results of a more specialized nature concerning the finiteness of scattering cross section which do not seem to be included under the Birman-Krein theorem. For instance, it is shown for the Schrödinger Hamiltonian [13] and for the Klein-Gordon equation with a potential

[14] that the total scattering cross section is finite for potentials that behave like $\frac{1}{r^{2+\varepsilon}}$ at infinity.

Such results as well as their proofs are of very specialized nature and the Birman-Krein theorem and certain related results given below are still the best possible abstract results concerning the finiteness of scattering cross section.

At any rate, the aim of this section is to show how the Birman-Krein theorem and the related results ensue from a direct application of theorem 1. Instead of considering the Birman-Krein theorem we shall first prove a related but slightly different result formulated as proposition (2) below. The reason for concentrating on this proposition rather than on the Birman-Krein theorem is that while we cannot specify any class of potentials for which the assumptions of the Krein-Birman theorem holds in the relativistic scattering problem we can do so for the assumption of the proposition (2). Moreover, the proposition (2) yields a "best possible" estimate of the asymptotic behaviour of scattering cross sections in the setting of the present analysis. After discussing the proposition (2) we shall, then, briefly indicate a proof of the Birman-Krein theorem based on theorem 1. For more details on the material of this section the reader is referred to [15] and [7].

Proposition (2). Let the self-adjoint operators H and H_0 satisfy the usual asymptotic and completeness condition of single channel scattering theory. Assume further that H_0 has absolutely continuous spectrum, $D(H) = D(H_0)$ and that for a negative real point c in the common resolvent set of H and H_0, $R_c^\beta \; V \; R_c^{0\beta}$ is nuclear for some $\beta > 0$ $(V \equiv H-H_0)$. Then

(i) The energy shell operator $(S_\varepsilon - I_\varepsilon)$ is nuclear for a.a. ε in the spectrum of H_0 and hence the total scattering cross section $\sigma_{tot}(\varepsilon)$ is finite for almost all values of energy ε. Moreover,

(ii) $$\int \frac{\sigma_{tot}(\varepsilon)\;|\vec{p}|^2}{(\varepsilon-c)^{2\beta}}\, d\varepsilon < \infty .$$

Now we indicate how theorem 1 is used in the proof of this proposition. We start from a well-known representation of the S operator [10]

$$(\varphi,\ (S-1)\Psi) = -i\int_{-\infty}^{\infty} (\varphi,\ e^{iHot}\ \Omega_+^*\ Ve^{-iHot}\ \Psi)\ dt \tag{4}$$

for φ, $\Psi \in D(H) = D(H_0)$. Here $\Omega\pm = \text{s-lim}_{t\to\pm\infty}\ e^{iHt}\ e^{-iHot}$ are the wave operators. Relation (4) in conjunction with the intertwining property of Ω_+^* yields

$$(\varphi,\ R_c^{0\beta}(S-1)R_c^{0\beta}\ \Psi) = -i\int_{-\infty}^{\infty}(\varphi,\ e^{iHot}\Omega_+^*\ R_c^{\ \beta}\ V\ R_c^{0\beta}\ e^{-iHot}\Psi)dt \tag{5}$$

for a suitable dense set of vectors φ and Ψ. We notice that the operator in the left-hand side of (5) commute with H_0. Therefore in the direct integral representation of the Hilbert space $\mathcal{H} = \int \mathcal{H}_\varepsilon \, d\varepsilon$ which diagonalizes H_0, it takes the form

$$(\varphi, R_c^{0\beta} (S-1) R_c^{0\beta} \Psi) = \int \frac{(\varphi_\varepsilon, (S_\varepsilon-1)\Psi_\varepsilon)}{(\varepsilon-c)^{2\beta}} d\varepsilon \tag{6}$$

where S_ε are the energy shell components of the S operator. On the other hand, since $R_c{}^\beta V R_c^{0\beta}$ is assumed to be nuclear and e^{-iHot} constitute an absolutely continuous representation of R, it follows from theorem 1 (aii) that the right-hand side of (5) is of the form $\int(\varphi_\varepsilon, \tau_\varepsilon \Psi_\varepsilon)d\varepsilon$ where τ_ε is nuclear on $\mathcal{H}_\varepsilon$. Hence (6) implies (owing to the essential uniqueness of the family τ_ε) that $S_\varepsilon-1 = (\varepsilon-c)^{2\beta} \tau_\varepsilon$ which established the trace-class property of $S_\varepsilon-1$ for a.e ε. To prove (ii) we make use of part (b) of theorem 1 and conclude that

$$\int Tr\tau_\varepsilon d\varepsilon = Tr \, \Omega_+^* R_c^\beta V R_c^{0\beta} < \infty$$

and thus

$$2 \, Re \int Tr\tau_\varepsilon d\varepsilon = \int \frac{Tr[S_\varepsilon-1 + (S_\varepsilon-1)^*]}{(\varepsilon-c)^{2\beta}} d\varepsilon < \infty.$$

But it is known that

$$Tr \, (S_\varepsilon-1 + S_\varepsilon^*-1) = \frac{1}{\pi} |\vec{p}|^2 \sigma_{tot}(\varepsilon)$$

and relation (ii) is thereby established.

We may now indicate how a similar use of theorem 1 yields also a simple of proof of the Birman-Krein theorem. The representation (4) of S and the identity

$$R_z^n - R_z^{0n} = - [R_z^n V R_z^0 + R_z^{n-1} V R_z^{02} + \ldots + R_z V R_z^{0n}]$$

allow us to write again

$$(\phi, R_z^{0n+1} (S-I)\Psi) = \frac{1}{n} i \int (\phi, e^{iHot}[\Omega_+^* (R_z^n - R_z^{0^n})]e^{-iHot}\Psi)dt.$$

Since $\Omega_+^* (R_z^n - R_z^{0^n})$ is nuclear by virtue of the assumption of the Krein-Birman theorem we may now apply part (aii) of the theorem 1 and establish the trace-class property of the energy shell operators $S_\varepsilon-1$ as before.

We now comment briefly on the applications of the proposition (2) in the potential scattering problem.

1. In non-relativistic potential scattering it is known [see proposition (1)] that $R_z^\beta V R_z^{0\beta}$ is nuclear with β=1 for all sufficiently

short ranged potentials. Thus for such potentials $S_\varepsilon - I$ is nuclear and hence the total scattering cross section $\sigma_{tot}(\varepsilon)$ is finite for almost all values of the energy ε. This is essentially the results of [9] and [10] and we see here that they are subsumed under proposition (2).

Moreover, the finiteness of the integral (ii) with $\beta = 1$ and $\varepsilon = \frac{|\vec{p}|^2}{2m}$ implies that $\sigma_{tot}(\varepsilon) \to 0$ as $\varepsilon \to \infty$ (if $\sigma_{tot}(\varepsilon)$ is monotone and continuous for sufficiently large ε). This is, of course, a well-known result which has been obtained previously from a direct investigation of the Schrödinger equation. (Actually, the slightly sharper result that

$$\sigma_{tot}(\varepsilon) = 0\,(\tfrac{1}{\varepsilon}) \text{ as } \varepsilon \to \infty$$

is proved in [16].)

2. More interesting is the application of proposition (2) in relativistic potential scattering; if only because this problem has not been treated previously.

The first question to be decided is if there exist any class of potentials for which $R_z^\beta\, V\, R_z^{0\beta}$ is nuclear in the relativistic case. We recall that this cannot hold with $\beta = 1$. We recall also that $R_z^{0\beta}\, V\, R_z^{0\beta}$ is nuclear for all sufficiently short ranged potentials for $\beta > 3/2$. Unfortunately, for $\beta \neq 1$ the trace-class property of $R_z^\beta\, V\, R_z^{0\beta}$ does not imply the same property for $R_z^\beta\, V\, R_z^{0\beta}$ or for $R_z^\beta\, V\, R_z^\beta$. For this implication to hold we need the technical assumption that the domains of $(H_0 - zI)^\beta$ and $(H - zI)^\beta$ are identical.

When $\beta \neq 1$ this assumption is, in general, difficult to verify. But the following results, which are admittedly not the best possible results, imply that in the relativistic scattering problem there exist a large class of potentials (including all those belonging to the Schwartz class of fast decreasing functions) for which $R_z^\beta\, V\, R_z^{0\beta}$ (and also $R_z^\beta\, V\, R_z^\beta$) is nuclear for sufficiently large values of β.

<u>Lemma</u>. Let V be an H_0-bounded potential with relative bound less than one and assume that $V\, D(H_0^k) \subseteq D(H_0^k)$ for $k = 1, 2 \ldots n-1$, then $D(H^k) = D(H_0^k)$ for $k = 1, 2 \ldots n$.

For the relativistic problem, we need to have n at least equal to 2, which will be the case if V leaves the domain of H_0 invariant. It is not hard to check that the class of square integrable potentials $V(\vec{x})$ whose Fourier transform $V(\vec{p})$ satisfy the integrability condition

$$\int |\vec{p}|\, |V(\vec{p})|\, d^3p < \infty$$

have this property. In view of the foregoing remark we are assured of the application of the proposition (2) and hence of the finiteness of $\sigma_{tot}(\varepsilon)$ in relativistic potential scattering for a wide class of potentials.

As in the non-relativistic case, the finiteness of the integral (ii) again provides information about the asymptotic behaviour of $\sigma_{tot}(\varepsilon)$. In order to find the best asymptotic estimate of $\sigma_{tot}(\varepsilon)$ we have to substitute in the integral (ii) the minimum value of β for which $R_z^{\beta}\ V\ R_z^{0\beta}$ is nuclear i.e. $\beta = 3/2 + \delta$ (δ being an arbitrary positive number). (See remark 3 of §3). Convergence of the integral (ii) then requires (with

$$\varepsilon = \sqrt{|\vec{p}|^2 + m_1{}^2} + \sqrt{|\vec{p}|^2 + m_2{}^2}\ ,\ \text{that}$$

$$\lim_{\varepsilon \to \infty} \frac{\sigma_{tot}(\varepsilon)}{\varepsilon^{\delta}} = 0 \quad \text{for any} \quad \delta > 0. \tag{7}$$

We conclude this section with the following final remarks.

1. In contrast to the situation in non-relativistic scattering where $\sigma_{tot}(\varepsilon)$ is found to vanish necessarily as $\varepsilon \to \infty$, we find in (7) that $\sigma_{tot}(\varepsilon)$ need not vanish and may even increase as $\varepsilon \to \infty$. In our analysis this difference between the asymptotic behaviour in relativistic and non-relativistic cases is solely the result of the difference in the kinematics of the free particles, and is not due to any essential difference of the interaction which produces the scattering.

2. The asymptotic behaviour (7) may be compared with the well-known Froissart bound $\sigma_{tot}(\varepsilon) < c(\lg\varepsilon)^2$ as $\varepsilon \to \infty$ which has been derived from the postulates of field theory [17]. (7) is of course weaker than the Froissart result in two respects. Since our analysis is limited to single channel system (7) refers to the elastic cross section only whereas the Froissart bound applies to the total (elastic + inelastic) cross section. Secondly a growth of the order $(\lg\varepsilon)^n$ is consistent with (7) for any positive integer n whereas Froissart bound specifies n to be 2.

3. (7) may also be compared with the predictions of the Dirac and Klein-Gordon equations with a potential term. For these models it is shown ([14], [16]) that $\sigma_{tot}(\varepsilon)$ approaches a non-zero constant as $\varepsilon \to \infty$. This result is, of course, consistent with (7). But it should be emphasized that scattering theory defined by a Hamiltonian of the form (2) of §3 and the theory defined by Dirac or Klein-Gordon equation with an external potential are not equivalent mathematically. It would, thus, be of interest to establish an exact asymptotic expansion of $\sigma_{tot}(\varepsilon)$ for the Hamiltonian (2) and compare it with that found in [14] and [16].

Such a comparison would be also of interest in view of the

present experimental indication that $\sigma_{tot}(\varepsilon)$ does not seem to approach a non-zero constant but seems to rise with the energy. If this is confirmed then the scattering models based on the relativistic equation with an external potential will be found to be totally inadequate for reproducing even the gross features of high energy scattering. But to determine the adequacy or inadequacy of models based on the Hamiltonian of the form (2) one has, of course, to obtain better asymptotic estimate of $\sigma_{tot}(\varepsilon)$ than that given by (7).

5. Time-Delay of Scattering Processes.

In this last application we shall discuss the notion of time delay in scattering theory and show how theorem 1 can be used to give a precise mathematical formulation of this concept. We begin with a physically motivated (but mathematically heuristic) definition of the time delay. We want to evaluate the time spent by the two colliding particles within the region of mutual interaction and compare it with the time which they would have spent in the same region had they moved freely.

Consider simple scattering systems, with free Hamiltonian H_0 and the total Hamiltonian H. H and H_0 may have either the non-relativistic form (1) or the relativistic form (2) of §3. We consider a scattering state $\Psi_t = \exp(-iHt)\,\Omega_-\varphi$ which behaves as the freely evolving state $\varphi_t = \exp(-iH_0t)\varphi$ as $t \to -\infty$.

Let us now consider a sphere of radius r in the configuration space R^3 and the projection operator P_r on $\mathcal{L}^2(R^3)$ defined by

$$(P_r\varphi)(\vec{x}) = \begin{cases} \varphi(\vec{x}) & |\vec{x}| < r \\ 0 & |\vec{x}| \geq r \end{cases}$$

The probability of finding the two colliding particles at a distance less than r from each other at the time t is $(\Psi_t, P_r\Psi_t)$ and therefore the mean time spent by the particles in the sphere is

$$\int_{-\infty}^{\infty} (\Psi_t, P_r\Psi_t)dt.$$

The similar quantity corresponding to the free motion is

$$\int_{-\infty}^{\infty} (\varphi_t, P_r\varphi_t)dt.$$

We define the time delay of the scattering state $\Psi = \Omega_-\varphi$ in the sphere of radius r to be the difference of these two quantities:

$$T_r(\varphi,\varphi) \equiv \int_{-\infty}^{\infty} (\Psi_t, P_r\Psi_t)dt - \int_{-\infty}^{\infty} (\varphi_t, P_r\varphi_t)dt \qquad (8)$$

When it exists, expression (8) can be rewritten in the following form with the help of the intertwining relation

$$T_r(\varphi,\varphi) = \int_{-\infty}^{\infty} (\varphi, e^{iHot} (\Omega_-^* P_r \Omega_- - P_r) e^{-iHot} \varphi) dt \tag{9}$$

We shall also consider the following sesquilinear form which arises naturally from (9)

$$T_r(\Psi,\varphi) = \int_{-\infty}^{\infty} (\Psi, e^{iHot} \Gamma_r e^{-iHot} \varphi) dt \tag{10}$$

$$\text{with} \quad \Gamma_r = \Omega_-^* P_r \Omega_- - P_r \; . \tag{11}$$

The mathematical questions to be investigated at this point are

(i) Under what conditions does the formal expression (10) yield a densely defined sesquilinear form, and to what extent does this form define a time delay operator for finite spheres?

(ii) In what sense can we take the limit $r \to \infty$ in (10) and define the time delay for infinite space region?

The first part can be fully answered with the help of theorem 1. This is the content of proposition (3).

Proposition (3). Let H and H_0 be of the form (1) and (2) of §3 and satisfy the usual requirements of a simple scattering system. (They are required in the case (2) to verify certain additional technical conditions that are mentioned in the course of the proof.) Then

(i) There exists a dense set D_0 in $\mathcal{H}$ such that for φ and $\Psi \in D_0$ the integral (10) is finite and defines a symmetric sesquilinear form.

(ii) There exists an essentially unique family of trace-class operators τ_ε^r acting on the components $\mathcal{H}_\varepsilon$ of the direct integral of $\mathcal{H}$ with respect to H_0 such that

$$T_r(\Psi,\varphi) = \int (\Psi_\varepsilon, \tau_\varepsilon^r \varphi_\varepsilon) d\varepsilon \text{ for all } \varphi, \Psi \in D_0.$$

The operators τ_ε^r are self-adjoint and may be called the energy shell time delay operators for the sphere of radius r. And the trace of τ_ε^r has the physical interpretation of being the total time delay (in the sphere of radius r) when the incident kinetic energy is ε.

Proof. The operator Γ_r (11) is usually not nuclear but it follows from the remark 3 of §3 that such an operator is obtainable by multiplying Γ_r from the left and the right with suitable powers of R_z^0. In fact $R_{z^*}^{0\beta} P_r R_z^{0\beta}$ is nuclear for $\beta = 1$ in the non-relativistic scattering problem and for $\beta = 2$ in the relativistic case.

In the non-relativistic scattering problem the nuclearity of $R^{0\beta}_{z*}P_r R^{0\beta}_z$ (with $\beta = 1$) implies under the mild condition $D_H = D_{H_0}$ that $R^{\beta}_{z*}P_r R^{\beta}_z$ is also nuclear and hence the same holds for

$$R^{0\beta}_{z*}\,\Gamma_r R^{0\beta}_z = \Omega^*_-\, R^{\beta}_{z*}\, P_r\, R^{\beta}_z\, \Omega_- - R^{0\beta}_{z*}\, P_r\, R^{0\beta}_z\ .$$

In the relativistic scattering problem, on the other hand, the trace-class property of $R^{0\beta}_{z*}\, P_r\, R^{0\beta}_z$ (with $\beta = 2$) does not generally imply the same property for $R^{\beta}_{z*}\, P_r\, R^{\beta}_z$. But as remarked in § 4 this implication holds for at least a large class of potentials.

Thus, at any event, we may conclude that $R^{0\beta}_{z*}\,\Gamma_r\, R^{0\beta}_z$ is nuclear for suitable β; this being true rather generally for non-relativistic scattering and for at least a very wide class of potentials, if not generally, in the relativistic case. Hence we may apply theorem 1 to the nuclear operator $R^{0\beta}_{z*}\,\Gamma_r\, R^{0\beta}_z$ and the unitary group exp(-iHot) in both the relativistic and non-relativistic scattering problem.

Let Q^r_ε be the family of trace class operators given by the theorem 1 (part aii and b) in this special case and D the dense set of vectors of this theorem. Now we define $D_0 \equiv (R^{0\beta}_z\,\phi\,|\,\phi\varepsilon D)$ and $\tau^r_\varepsilon \equiv (\varepsilon - z)^\beta Q^r_\varepsilon$. Then D_0 is still dense and one verifies easily all the claims of the proposition.

The time-delay operator τ^r for finite regions of space is then obtained by taking the "direct integral" of the energy shell operator τ^r_ε . This, then, completes the discussion of the question (i).

Question (ii), that is, the analysis of the limit $r\to\infty$ is, however, more delicate and theorem 1 alone does not suffice for this purpose. For details of this analysis (in the non-relativistic case) the reader is referred to [6]. Here we mention only that it relies strongly on the following result due to Krein and Birman.

Theorem (Krein and Birman) [18, 2]. Let H and H_0 be self-adjoint operators for which the difference of the resolvents $R_z - R^0_z$ is of trace class. Then there exists an essentially unique measurable function $\xi(\varepsilon)$ such that

$$\mathrm{Tr}(R_z - R^0_z) = -\int \frac{\xi(\varepsilon)}{(\varepsilon - z)^2}\, d\varepsilon$$

and

$$\xi(\varepsilon)(1 + \varepsilon^2)^{-1} \in L_1(-\infty, \infty)\ .$$

Moreover, the function $\xi(\varepsilon)$, called the *spectral displacement function*, is related to the "energy shell" scattering operator S_ε of the scattering system defined by the free Hamiltonian H_0 and the total Hamiltonian H as follows:

$$\xi(\varepsilon) = \frac{1}{\pi} \operatorname{Tr} \delta_\varepsilon$$

where δ_ε is the *phase-shift operator* related to S_ε through:

$$S_\varepsilon = e^{-2i\delta_\varepsilon}.$$

Using this result and after a lengthy series of manipulations one proves that:

(i) $\lim_{r\to\infty} \operatorname{Tr} \tau_\varepsilon^r \equiv \tau_\varepsilon$ exists in the sense of distribution limit.

(ii) $\tau_\varepsilon = 2 \frac{d}{d\varepsilon} (\operatorname{Tr} \delta_\varepsilon) + \sum_i n_i \, \delta(\varepsilon - \varepsilon_i).$

where n_i and ε_i are (respectively) the multiplicities and eigenvalues of H in the positive interval of the real line.

This last formula is, essentially, the Eisenbud-Wigner formula for the time delay which had been derived earlier with plausible but mathematically heuristic reasoning [19].

As a final remark we note that the proof of the existence and of the trace-class property of the energy shell time-delay operators τ_ε^r for finite regions of space is entirely independent of any trace-class condition on $R_z - R_z^0$. In fact, as we have seen, this proof applies also in the relativistic potential when $R_z - R_z^0$ is never of trace-class. In contrast, the existing analysis [6] of the limit $r\to\infty$ leans heavily on the condition that $R_z - R_z^0$ is nuclear and thus it has no direct application in relativistic potential scattering. It would thus be of interest to extend this analysis so that it applies to the relativistic scattering problem as well. It is, of course, known that the Krein-Birman theorem cited above admits of a generalization in which the trace-class condition on $R_z - R_z^0$ is relaxed to a condition that $R_z^n - R_z^{0n}$ is nuclear for sufficiently large n [18]. And it is conceivable (although we have not verified the details) that the analysis of [6] can be carried through under such a less restrictive assumption. But even then we have no guarantee for the applicability of this analysis in relativistic potential scattering since we do not at present know of any potentials which would satisfy, at least, this relaxed condition.

Appendix: Relativistic Potential Scattering.

We consider two relativistic particles of mass m_1 and m_2 and spin zero. They are described in Hilbert spaces $\mathcal{H}_i$ $i = 1, 2$ carrying irreducible representations $U_i(\Lambda)$ of the Poincaré group $\mathcal{P}$ corresponding to mass m_1 and m_2. The basic observables are the momenta $\vec{p}_i$ and their canonically conjugate operators $\vec{q}_i$ which are the well-known Newton-Wigner position operators [20].

$\mathcal{H}_i$ can be realized as the spaces $\mathcal{L}^2(\vec{p}_i, d^3p_i)$, in which the

momenta and free Hamiltonians act as multiplication operators by $\vec{p}_i$ and $\sqrt{|\vec{p}_i|^2 + m_i^2}$ respectively, and $\vec{q}_i$ are the differential operators $i\,\vec{\nabla}_{\vec{p}_i}$. The two-particle system is described in the tensor product space $\mathcal{H} = \mathcal{H}_1 \otimes \mathcal{H}_2$, which carries the tensor product representation
$U_0(\Lambda) = U_1(\Lambda) \otimes U_2(\Lambda)$ with generators $H_0 = \sqrt{|\vec{p}_1|^2 + m_1^2} + \sqrt{|\vec{p}_2|^2 + m_2^2}$ $\vec{P} = \vec{p}_1 + \vec{p}_2$, $\vec{J}$ and $\vec{K}_0$ of the time translations, rotations and pure Lorentz transformations respectively.

We will say that we have a *relativistically invariant two-body scattering theory* if

(i) There exists on $\mathcal{H}$ another representation $U(\Lambda)$ of the Poincaré group with generators $(H, \vec{P}, \vec{J}, \vec{K})$ such that

(ii) $\lim_{t\to\pm\infty} \exp(iHt)\exp(-iH_0t) = \Omega\pm$ where $\Omega\pm$ have the usual properties of wave operators,

(iii) $U(\Lambda)\Omega\pm = \Omega\pm U_0(\Lambda)$ for all $\Lambda \in \mathcal{P}$.

The group $\exp(-iHt)$ describes the interacting motion of the two particles and

(i) expresses mathematically the restricted principle of relativity in quantum mechanics by requiring that H together with suitable operators $\vec{P}$, $\vec{J}$, $\vec{K}$ generate a unitary representation of the Poincaré group.

(ii) is the usual requirement of asymptotically free motion which characterizes scattering processes.

And (iii) which is equivalent with $[S, U_0(\Lambda)] = 0$, insures the covariance of the scattering amplitude and cross section.

Clearly (i) - (iii) impose restrictions on the possible choice of the total Hamiltonian H. A general class of Hamiltonians satisfying these conditions can be constructed as follows.

It is convenient first to introduce the center of mass momentum $\vec{P} = \vec{p}_1 + \vec{p}_2$ and the relative momentum $\vec{p}$.† Now the two particle Hilbert space $\mathcal{H}$ can be identified with

$$\mathcal{L}^2(\vec{P}, d^3\vec{P}) \otimes \mathcal{L}^2(\vec{p}, d^3\vec{p}).$$

† $\vec{p}$ is implicitly defined by the equation

$$\vec{p} = \frac{\varepsilon_2\vec{p}_1 - \varepsilon_1\vec{p}_2}{\varepsilon_1 + \varepsilon_2} \quad \text{with } \varepsilon_i = \tfrac{1}{2}\left(\sqrt{|\vec{p}_i|^2 + m_i^2} + \sqrt{|\vec{p}|^2 + m_i^2}\right),\ i = 1,2.$$

With this identification the free Hamiltonian H_0 acts as the multiplication operator by the function $(|\vec{P}|^2 + h_0^2)^{1/2}$ with $h_0 = (|\vec{p}|^2 + m_1^2)^{1/2} + (|\vec{p}|^2 + m_2^2)^{1/2}$, and the relative Newton-Wigner position operator $\vec{q}_1 - \vec{q}_2$ acts as $I \otimes \vec{x}$ where $\vec{x}$ is the operator $i\nabla_{\vec{p}}$ on $\mathcal{L}^2(\vec{p}, d^3p)$. The total Hamiltonian H is now chosen to be of the form $H = (|\vec{P}|^2 + h^2)^{1/2}$ where $h = h_0 + V$ and V, like h_0, does not act on the variable $\vec{P}$, but acts only in $\mathcal{L}^2(\vec{p}, d^3p)$. If h and h_0 satisfy the asymptotic condition and a weaker form of the so called "invariance principle" of wave operators† , then H and H_0 satisfy too the asymptotic condition and the resulting S-operator satisfies the covariance condition $[S, U^0(\Lambda)] = 0$. [21, 22] Thus the problem of relativistic scattering is reduced to a study of the system (h, h_0) in the Hilbert space $\mathcal{L}^2(\vec{p}, d^3p)$ of relative coordinates as in the non-relativistic situation.

The class of Hamiltonians $H = (|\vec{P}|^2 + h^2)^{1/2}$ described above is known in the literature as the Bakamjian-Thomas class. This class is the most general class of two-body relativistic Hamiltonian in the sense that for every scattering theory satisfying (i) - (iii) there always exists a Hamiltonian in the Bakamjian-Thomas class which yields the same S-operator as that of the given theory. For more details the reader may refer to [21].

We speak of *relativistic potential scattering* if the operator V is a function of the relative position operator $\vec{x}$ alone.

Finally it should be noted that relativistic potential scattering as described here is mathematically not equivalent to the scattering theory based on relativistic equation of particles (such as Dirac or Klein-Gordon equations) with an added potential term.

† This means that together with h and h_0, $\sqrt{h^2 + \lambda^2}$ and $\sqrt{h_0^2 + \lambda^2}$ (λ, a scalar) also satisfy the asymptotic condition and the resulting wave operators are the same as those defined by h and h_0.

References:

1. T. Kato, Perturbation Theory of Linear Operators (Springer, New York 1966), Chap. X, § 4

2. M.S. Birman and M.G. Krein, Soviet Math Doklady 3 740 (1962)

3. T. Kato, loc. cit., Chap. X, §6, lemma 4.5; C.R. Putnam Commutation Properties of Hilbert Space Operators (Springer 1967) lemma 5.3.2

4. M.A. Naimark and V. Fomin, Amer. Math Soc. Transl. Series 2, Vol. 5, (1957), p. 35

5. L.H. Loomis, Introduction to Abstract Harmonic Analysis (Van Nostrand 1953)

6. J.M. Jauch, K. Sinha and B. Misra, Helv. Phys. Acta. 45 3, (1972) 398

7. Ph. Martin and B. Misra, On Trace Class Operators of Scattering Theory and the Asymptotic Behaviour of Scattering Cross Section of High Energy, to appear in Journ. Math Phys.

8. T. Kato, loc. cit. Chap. X, §4, Th. 4.9 and Ex. 4.10

9. V.S. Buslaev, Dokl. Akad. Nauk. SSSR 143 (1962) 1067

10. T. Ikebe, Pac. Journal of Math, 15 No. 2 (1965) 511

11. T.A. Green and O.E. Lanford III, J. Math Phys. 1 (1966) 139

12. M.S. Birman and M.G. Klein, loc. cit. Th. 3, p. 742

13. B. Misra, D. Speiser and G. Targonski, Helv. Phys. Acta 36 (1963) 963

14. Ph. Martin, On the High Energy Limit in Potential Scattering, to appear in the Nuovo Cimento

15. J.M. Jauch and K. Sinha, Helv. Phys. Acta 45 4 (1972) 580

16. W. Hunziker, Helv. Phys. Acta. 36 (1963) 838

17. A. Martin and F. Cheung "Analyticity Properties and Bounds of the Scattering Amplitude", Gordon Breach (1970) Chap. 6

18. M.G. Klein, Soviet Math Doklady 3 (1962) 707

19. See for instance Goldberger and Watson, Collision Theory (Wiley, New York 1964) p. 490

20. T.D. Newton and E.P. Wigner, Rev. Mod. Phys. 21 (1969) 400

21. R. Fong and S. Sucher, J. Math Phys. 5 (1964) 456

22. F. Coester, Helv. Phys. Acta 38 (1965) 7

THE GEL'FAND-LEVITAN METHOD IN THE INVERSE SCATTERING PROBLEM

Roger G. Newton*

Indiana University, Bloomington

1. Introduction.

A wave packet of quantum mechanical particles develops in time according to the Schrödinger equation

$$i\partial_t \Psi_t = H\Psi_t.$$

If we want to describe a physical situation in which the packet is prepared in the distant past and there consists of free particles far removed from the scattering center, then we demand as an initial condition that

$$\Psi_t \xrightarrow[t \to -\infty]{s} \Psi_t^{(in)}$$

where

$$i\partial_t \Psi_t^{(in)} = H_0 \Psi_t^{(in)}$$

and H_0 is the "free" hamiltonian that describes the kinetic energy of the particles. It differs from the full hamiltonian H by the interaction H' that describes the force exerted by the scattering center on the particles,

$$H = H_0 + H'.$$

In the distant future the state vector approaches again a state of free particles,

$$\Psi_t \xrightarrow[t \to +\infty]{s} \Psi_t^{(out)}$$

* Partially supported by the National Science Foundation and the U.S. Army Research Office, Durham, North Carolina.

J. A. LaVita and J.-P. Marchand (eds.), Scattering Theory in Mathematical Physics, 193–235.

where

$$i\partial_t \Psi_t^{(out)} = H_0 \Psi_t^{(out)} .$$

(We shall be concerned with elastic scattering only.) The connection between the in and out states is given by the unitary scattering operator,

$$\Psi_t^{(out)} = S\Psi_t^{(in)} . \qquad (1.1)$$

Since practically all scattering experiments are performed with beams of (almost) mono-energetic particles, and since calculations are facilitated thereby, it is convenient to perform a Fourier transformation

$$\Psi(E) = \int_{-\infty}^{\infty} dt e^{iEt} \Psi_t$$

so that $\Psi(E)$ satisfies the time-independent Schrödinger equation

$$H\Psi(E) = E\Psi(E)$$

and similarly

$$H_0 \Psi_{in,out}(E) = E\Psi_{in,out}(E) .$$

$\Psi_{in}(E)$ describes a beam of particles in an idealized sense if the momentum of the particles is prescribed, that is, if we choose (in the coordinate representation)

$$\Psi_{in} = \Psi_0 = e^{i\underline{k}\cdot\underline{r}} .$$

In that case $\Psi(E)$ is labeled by the momentum $\underline{k}$ of the incoming particles and $\Psi(\underline{k},\underline{r})$ satisfies the Lippmann-Schwinger equation

$$\Psi = \Psi_0 + G_0^+ H' \Psi$$

where G_0^+ is the boundary value of the resolvent

$$G_0(E) = (E - H_0)^{-1}$$

as E approaches the positive real axis (i.e., the continuous spectrum of H_0) from above.

A stationary-phase argument shows that as $t \to +\infty$, the Fourier-integral expression for Ψ_t, and hence of $\Psi_t^{(out)}$, has its dominant contributions from the region of large distances. That is to say, we get Ψ_{out} from the large distance behavior of the L.S. equation, which is

$$\Psi(\underline{k},\underline{r}) = e^{i\underline{k}\cdot\underline{r}} + e^{ikr} r^{-1} A(k;\hat{r},\hat{k}) + o(r^{-1}) \qquad (1.2)$$

(We write k for $|\underline{k}|$, r for $|\underline{r}|$, $\hat{k}$ for $\underline{k}/k$, and the relationship between E and k is that $E = k^2$.) The scattering amplitude A is also expressed as $A(\underline{k}',\underline{k})$, where $\underline{k}' = k\hat{r}$, so that $\underline{k}$ expresses the momentum of the incoming particles and $\underline{k}'$ that of the outgoing ones. It is directly related to the experimentally measured differential scattering cross section

$$\frac{d\sigma}{d\Omega} = |A(\underline{k}', \underline{k})|^2 .$$

On the other hand, the kernel of the scattering operator in the momentum representation (the S-matrix) is given by

$$S(\underline{k}', \underline{k}) = \delta(\underline{k} - \underline{k}') - \frac{1}{\pi i} \delta(k^2 - k'^2) A(\underline{k}', \underline{k}) \quad (1.3)$$

which connects A with the S-matrix.

The inverse scattering problem consists of the attempt to reconstruct the interaction H' from the scattering amplitude A. From the point of view of the Schrödinger equation as a differential equation it is the problem of reconstructing the differential (or integro-differential) equation from asymptotic information on the solutions along the continuous spectrum. As always, there are three aspects to the problem: existence, uniqueness, and construction procedure.

Suppose we pose the problem in all generality: Find an operator H' such that the outgoing-wave solutions of

$$(H_0 + H')\Psi = E\Psi$$

have the asymptotic behavior

$$\psi \sim e^{i\underline{k}\cdot\underline{r}} + e^{ikr}r^{-1}A(k;\hat{r},\hat{k}) + o(r^{-1})$$

where A is given. This problem obviously has a vast amount of non-uniqueness. I need only take any reasonable function $\psi(\underline{k},\underline{r})$, twice differentiable with respect to $\underline{r}$, that has the given asymptotic form, and then let H' be <u>defined</u> by the Schrödinger equation, i.e.,

$$H' = [(E - H_0)\psi]/\psi.$$

This is an almost trival and uninteresting problem, both from the physical and from the mathematical point of view. The interaction constructed in that way will, of course, generally depend on $\underline{k}$.

The inverse problem becomes physically interesting and mathematically non-trivial if we demand that H' be <u>local</u> and independent of E or $\underline{k}$. We are now dealing with a differential equation of the Sturm-Liouville type, to be reconstructed from spectral information of a specific kind.

Let us first think of the situation in one dimension. Here the asymptotic data are given by the transmission and reflection coefficients, both as functions of the energy, and for each side of incidence. Thus we are given two (or four) functions (restricted by unitarity) of a single parameter, and we are looking for a single function of one parameter. This is not too unreasonable and from somewhat restricted data. might be expected to yield a unique result in a reasonable class of potential functions. In two dimensions, the scattering amplitude is a function of three variables: energy and two angles, one for the incident and one for the outgoing direction. The potential is a function of only two variables, distance and angle. Hence one must expect that only a very restricted family of functions can arise as the scattering amplitude in two dimensions of a Schrödinger equation with a local potential.

In three dimensions there are even more restrictions. The scattering amplitude is a function of five parameters (energy and four angles) whereas the potential is a function of only three parameters. Thus there is a very serious existence problem: What functions $A(k; \hat{k}', \hat{k})$ are such that they can possibly be scattering amplitudes belonging to a Schrödinger equation with a local potential? This existence problem arises even in the largest class of potentials for which an S-matrix exists. If we restrict the class of potentials which for practical and technical reasons we always have to do, the existence problem becomes even more severe.

A specific restriction of great physical interest is to assume that the potential is spherically symmetric, i.e., a function of the distance r only. In that case A can be a function of only one angle, the scattering angle θ between the incoming and outgoing particles, and the energy. If we expand it in a Legendre series,

$$A(k, \cos\theta) = \sum_{\ell=0}^{\infty} (2\ell+1) a_\ell(k) P_\ell(\cos\theta)$$

then the unitarity of the S-matrix implies that $a_\ell(k)$ be of the form

$$a_\ell = \frac{1}{2ik} (e^{2i\delta_\ell} - 1)$$

where the phase shift δ_ℓ is real. The inverse scattering problem can now be posed in this form: given $\delta_\ell(k)$, find $V(r)$. Since V is a one-parameter function, we might expect to be able to pose two independent problems: (a) Given $\delta_\ell(k)$ for all $0 \leq k < \infty$ and one ℓ; or, (b) Given $\delta_\ell(k)$ for one k and $\ell = 0, 1, \ldots$, what is the potential? The first is the inverse problem at fixed angular momentum; the second, at fixed energy.

Another point that arises in connection with the existence problem is that of the bound states, i.e., the point spectrum. The question is whether the point spectrum can be arbitrarily assumed in addition to the scattering information or not. If the S-matrix contains all the possible information on the physical system then the bound states should be obtainable from it, and the scattering and bound state parts of the spectrum are not independent. If the point spectrum can be obtained from the scattering, that is of considerable physical interest. If they are independent, on the other hand, then that automatically means that in the absence of any information on the bound states there is always an infinite class of scattering-equivalent potentials, quite apart from any other kind of ambiguity that might arise.

Let us go into the bound-state question right-away in a bit more detail. If the potential is spherically symmetric and its first and second absolute moments are finite,

$$\int_0^\infty dr|V|\ r^\alpha < \infty, \quad \alpha = 1, 2, \tag{1.4}$$

then each phase shift obeys the Levinson theorem [1]: If $\delta_\ell(k)$ is defined to be continuous then

$$\delta_\ell(0) - \delta_\ell(\infty) = \pi n_\ell, \tag{1.5}$$

where n_ℓ is the number of bound states of angular momentum ℓ, i.e., the number of points in the discrete spectrum of H in the subspace of the Hilbert space belonging to the ℓth irreducible representation of the rotation group O(3). [There is a subtle exception to the above statement of Levinson's theorem in the case of a "zero-energy resonance" which we won't worry about now.] This theorem constitutes a certain relation between the continuous and the point spectrum. It allows, in principle, a determination of the number of bound states of angular momentum ℓ from the ℓth phase shift. In the stated class of potentials, it is the only relation. That is, the location of the point spectrum may be assigned independently of the phase shift, as we shall see.

On the other hand, suppose that we are given the full scattering amplitude as a function of energy and angle. If the potential is such that

$$\int (d\underline{r})|V|\ r^\alpha < \infty, \quad \alpha = 0, -1,$$

(but not necessarily spherically symmetric) then the forward scattering amplitude $A(k) \equiv A(k;\hat{k},\hat{k})$ is the boundary value of an analytic function regular in the upper half plane, except for poles on the positive imaginary axis at $k = iK_n$, where $-K_n^2$ are the eigenvalues of H. [2] The order of each pole equals the degeneracy of the corresponding eigenvalue, and as $|k| \to \infty$ with $\text{Im}\, k \geq 0$ we have

$$A'(k) \equiv A(k) - B = o(1)$$

where

$$B = -(1/4\pi) \int (d\underline{r}) V(\underline{r})$$

is the Born approximation. This implies that we may obtain the positions of the poles from A(k) on the real axis and hence we may obtain the point spectrum from the forward scattering amplitude. This may be done, for example, by Fourier transformation. For $t > 0$

$$\frac{1}{2\pi i} \int_{-\infty}^{\infty} dk A'(k) e^{ikt} = \sum_n R_n e^{-K_n t} \tag{1.6}$$

where R_n is the residue of the nth pole of A(k).

Thus, the full scattering amplitude is not independent of the point spectrum and we cannot expect to be able to prescribe them separately.

2. Fixed angular momentum.

We will now discuss in some detail the inverse scattering problem for spherically symmetric potentials in three dimensions at fixed angular momentum, and for simplicity we will take $\ell = 0$. [3]

If V is a function of r only, then the Schrödinger equation is separable. Assuming the solution $\psi(\underline{r})$ is of the form

$$\psi(\underline{r}) = \psi_\ell(r) r^{-1} Y_\ell^m(\theta, \phi)$$

where Y_ℓ^m is a spherical harmonic, the radial function $\psi_\ell(r)$ must satisfy the radial Schrödinger equation

$$-\psi_\ell'' + \ell(\ell+1) r^{-2} \psi_\ell + V\psi_\ell = k^2 \psi_\ell,$$

and it must be regular at $r = 0$. It follows from a comparison of the asymptotic form of ψ_ℓ

$$\psi_\ell = \text{const.} \sin(kr + \delta_\ell - (1/2)\pi\ell) + o(1)$$

with the Legendre expansion of the scattering amplitude that δ_ℓ here is the phase shift. We take $\ell = 0$ and then drop the subscript.

Let $\phi(k, r)$ satisfy the second order differential equation

$$\phi'' + k^2\phi - V\phi = 0 \tag{2.1}$$

and the boundary condition

$$\phi(k, 0) = 0, \quad \phi'(k, 0) = 1. \tag{2.2}$$

(Primes indicate differentiation with respect to r.) We shall always assume conditions (1.4). Then it is not difficult to show that for each fixed r, $\phi(k, r)$ is an entire analytic function of k, and its asymptotic behavior is such that for $|k| \to \infty$,

$$\phi(k, r) = k^{-1}\sin kr + o(k^{-1}\sin kr). \tag{2.3}$$

In addition, of course, $\phi(k, r)$ is an <u>even</u> function of k. It follows from these facts that the support of the Fourier sine-transform of $k\phi(k, r) - \sin kr$, considered as a function of k, is the interval from 0 to r; that is, there exists a function $K(r, r')$ such that

$$\phi(k, r) = k^{-1} \sin kr - \int_0^r dr' K(r, r')k^{-1} \sin kr'.$$

This is the crucial observation, and it follows directly from the analyticity and asymptotic properties of ϕ. Note that

$$\phi_0(k, r) = k^{-1}\sin kr \tag{2.4}$$

is the solution of the Schrodinger equation with $V = 0$, with the same boundary conditions. We can write

$$\phi(k, r) = \phi_0(k, r) - \int_0^r dr' \, K(r, r') \, \phi_0(k, r') \tag{2.5}$$

as a representation for ϕ. What is more,

$$K(r, o) = 0. \tag{2.5'}$$

If we now use the Schrödinger equation for ϕ and ϕ_0 we find, by integration by parts, that $K(r, r')$ must satisfy the hyperbolic partial differential equation

$$[\partial_y{}^2 - \partial_x{}^2 + V(x)] \, K(x, y) = 0 \tag{2.6}$$

and the boundary condition on a characteristic

$$-2\partial_x \, K(x,x) = V(x). \tag{2.7}$$

The integral representation (2.5) for ϕ may be regarded as a Volterra integral equation for ϕ_0. Since the functions $\phi_0(k, r)$ span the space, (1-K) maps the whole space onto a subspace,[0] the orthogonal complement of the space spanned by the point spectrum.

The Volterra equation then is an equation for the inverse mapping, defining K' such that

$$\phi_0(k, r) = \phi(k, r) - \int_0^r dr' K'(r, r')\phi(k, r'). \tag{2.8}$$

That K' is again a Volterra operator follows from the fact that the Volterra equation for it can be solved by iteration.

Now let us look at the asymptotic form of ϕ for large r. That is easily obtained from the integral equation

$$\phi(k, r) = k^{-1}\sin kr + k^{-1}\int_0^r dr' \sin k(r-r')V(r')\phi(k, r'). \tag{2.9}$$

Note that this too is a Volterra equation and thus can be solved by iteration under very general conditions on $V(r)$. For large r we obtain

$$2ik\phi(r) = e^{ikr}\,\Delta(-k) - e^{-ikr}\,\Delta(k) + o(1) \tag{2.10}$$

where

$$\Delta(k) = 1 + \int_0^\infty dr e^{ikr}\,V(r)\phi(k, r). \tag{2.11}$$

The definition of the phase shift implies that for large r

$$\phi(r) = \text{const.}\,(e^{-ikr} - Se^{ikr}) + o(1) \tag{2.12}$$

where

$$S = e^{2i\delta}\,. \tag{2.13}$$

We therefore find that

$$S(k) = \Delta(-k)/\Delta(k). \tag{2.14}$$

The function $\Delta(k)$ is known as the Jost function. Because of the analyticity of $\phi(k, r)$ and the fact that $e^{ikr}\phi(k, r)$ and its k-derivative, for k in the upper half of the complex plane, are bounded as functions of r (which follows from the integral equation for ϕ), $\Delta(k)$ is an analytic function of k regular in the upper half plane. Furthermore, in $\text{Im } k \geq 0$

$$\lim_{|k| \to \infty} \Delta(k) = 1. \tag{2.15}$$

Since ϕ and V are real, (2.11) shows that

$$\Delta(-k) = \overline{\Delta}(k)$$

and therefore (2.14) and (2.13) can be written

$$\Delta(k) = |\Delta(k)|e^{-\delta(k)}. \tag{2.16}$$

We must now establish a completeness relation for $\phi(k, r)$. The simplest thing to do is to define a regular solution $\psi(k, r)$ of the Schrödinger equation by means of the integral equation

$$\psi(k,r) = \phi_0(k,r) + \int_0^\infty dr' g^+(k;r,r')V(r')\psi(k,r') \tag{2.17}$$

where $g^+(k;r,r')$ is the kernel of the boundary value of the resolvent

$$g^+ = (k^2 - H_0)^{-1} .$$

Here H_0 is the self-adjoint extension of $-d^2/dr^2$ on $L^2(0,\infty)$. The kernel of g^+ is given by

$$g^+(k;r,r') = -k^{-1} e^{ikr_>} \sin kr_<$$

where

$$r_< = \min(r,r'), \quad r_> = \max(r,r').$$

It is easy to show by evaluating $\psi'(k, o)$ that

$$\phi(k,r) = \Delta(k)\psi(k,r). \tag{2.18}$$

One can also show, but less easily, that $\Delta(k)$ is equal to the <u>Fredholm determinant</u> of the integral equation (2.17) for ψ. From this fact it follows that the zeros of $\Delta(k)$ in the upper half plane occur exactly at those values $k = iK_n$ for which $-K_n^2$ is an eigenvalue of the operator $H = H_0 + V$, and each zero has a multiplicity equal to the degeneracy of the corresponding eigenvalue.

One now proves the completeness of the functions $\psi(k,r)$, together with the eigenfunctions of H. (Alternatively one proves the completeness of the ϕ's directly.) I am putting it this way, because the completeness relation for ψ has the simple form

$$\frac{2}{\pi}\int_0^\infty dk k^2 \psi(k,r)\psi^*(k,r') + \sum_n \psi^{(n)}(r)\psi^{(n)}(r') = \delta(r - r') \tag{2.19}$$

where $\psi^{(n)}(r)$ is the nth normalized eigenfunction of H,

$$\int_0^\infty dr|\psi^{(n)}(r)|^2 = 1.$$

Hence we may write the completeness in terms of ϕ as a Stieltjes integral

$$\int d\rho(E)\phi(k,r)\phi(k,r') = \delta(r - r') \tag{2.20}$$

where the <u>spectral function</u> $\rho(E)$ is defined by

$$\frac{d\rho}{dE} = \begin{cases} \dfrac{k}{\pi|\Delta(k)|^2}, & E \geq 0, \\[2ex] \sum_n \dfrac{\delta(E - E_n)}{N_n^2}, & E < 0, \end{cases} \tag{2.21}$$

with

$$N_n^2 = \int_0^\infty dr\, \phi^2\, (iK_n, r). \tag{2.22}$$

There is, of course, a similar, simpler completeness of the functions ϕ_0,

$$\int d\rho_0\, (E)\phi_0(k,r)\phi_0(k,r') = \delta(r - r') \tag{2.23}$$

with

$$\frac{d\rho_0(E)}{dE} = \begin{cases} \dfrac{k}{\pi}, & E \geq 0 \\[1ex] 0, & E < 0. \end{cases} \tag{2.24}$$

Let us now use the two representations (2.5) and (2.8) together with the two completeness relations (2.20) and (2.23). Multiplying (2.5) by ϕ_0 and integrating with $d\rho_0$ yields, by (2.23)

$$\int d\rho_0\, \phi(k,r)\phi_0(k,r') = \delta(r - r') - K(r,r')$$

whereas multiplying (2.8) by ϕ and integrating with $d\rho$ gives, by (2.20),

$$\int d\rho\phi(k,r)\phi_0(k,r') = \delta(r - r') - K'(r',r)$$

Subtracting these two results, we get

$$\int d\,(\rho - \rho_0)\phi(k,r)\phi_0(k,r') = K\,(r\,,r') - K'(r',\,r)$$

and hence for $r' < r$,

$$K(r,r') = \int d(\rho - \rho_0)\phi(k,r)\phi_0(k,r'). \tag{2.25}$$

On the other hand, multiplying (2.5) by ϕ_0 and integrating with $d(\rho - \rho_0)$ gives, according to (2.25), for $r' \leq r$

$$K(r,r') = g(r,r') - \int_0^r dr'' K(r,r'')g(r'',r') \tag{2.26}$$

where

$$g(r,r') = \int d(\rho - \rho_0)\phi_0(k,r)\phi_0(k,r'). \tag{2.27}$$

Equation (2.26) is the Gel'fand-Levitan equation. [4] It is a linear integral equation, generally of the Fredholm type, whose kernel is given by (2.27). If the spectral function ρ is given then $g(r,r')$ is known. The solution of (2.26) then gives the potential, by (2.7), and the solution of the Schrödinger equation, by (2.5).

We can prove directly that the homogeneous version of (2.26) has only the trivial solutions. That involves again the completeness. Suppose that

$$\chi_t(r) = - \int_0^t dr'' \chi_t(r'')g(r'',r).$$

We multiply this by $\chi_t(r)$ and integrate from o to t, using (2.27):

$$\int_0^t dr[\chi_t(r)]^2 + \int d\rho \left[\int_0^t dr\phi_0(k,r)\chi_t(r)\right]^2$$

$$= \int d\rho_0 \left[\int_0^t dr\phi_0(k,r)\chi_t(r)\right]^2 .$$

But the completeness relation (2.23) implies that

$$\int_0^t dr[\chi_t(r)]^2 = \int d\rho_0 \left[\int_0^t dr\phi_0(k,r)\chi_t(r)\right]^2$$

and hence

$$\int d\rho \left[\int_0^t dr\,\phi_0(k,r)\chi_t(r)\right]^2 = 0 .$$

The monotonicity of ρ and the completeness of ϕ_0 then allow us to conclude that $\chi_t(r) \equiv 0$. Hence the Fredholm alternative implies that (2.26) has a unique solution.

So we now have the method to infer V from the spectral function ρ. How do we find ρ from δ? According to (2.21) we must find $|\Delta(k)|$. Eq. (2.16) shows that this amounts to finding the magnitude of an analytic function, regular in the upper half plane and subject to (2.15), if its phase on the real axis is given. If there are no zeros, this is accomplished by a Hilbert transform of the logarithm. If there are zeros at $K = iK_n$ one readily finds that

$$|\Delta(k)| = \prod_n \left(1 - \frac{E_n}{E}\right) \exp\left[\frac{1}{\pi}\int_0^\infty \frac{dE'\ \delta(E')}{E - E'}\right] \tag{2.28}$$

the integral being meant in the sense of a Cauchy principal value. Thus, if we know $\delta(E)$ and the eigenvalues E_n then (2.28) allows us to construct the positive-energy part of the spectral function via (2.21). The normalization parameters N_n are still free, however, and if there are n bound states then there is an n-parameter family of phase-equivalent spectral functions, and hence an n-parameter family of phase-equivalent underlying potentials.

This is the inverse scattering problem that was solved first and is most thoroughly investigated. In order to generalize the procedure it is important to understand its structure. There are several different ways of approaching the same equations and of deriving them. Some of them are useful for generalizations in one direction, others for carrying similar methods into other directions. Let me give you another approach to the same equations that will be useful in applying an analogous procedure in the inverse problem at fixed energy.

We start out by defining the integral kernel $g(r,r')$ as in (2.27) and show that it satisfies the partial differential equation

$$(\partial_x^2 - \partial_y^2)g(x,y) = 0 \tag{2.29}$$

analogous to (2.6). Then we set up the Gel'fand-Levitan equation (2.26) and prove that it has a unique solution. In this case that proof involves the completeness, but in other cases it may be carried out by other means. Next, one shows by integrations by parts and (2.29) that if $V(r)$ is defined by (2.7) then the function defined by the left-hand side of (2.6) satisfies the homogeneous version of (2.26). Hence it must vanish. The boundary conditions (2.5′) follows from (2.26) and (2.27). One then defines ϕ by (2.5) and shows, by means of (2.6) and (2.5′), that it is a regular solution of the radial Schrödinger equation (2.1) with the potential (2.7). The expansion (2.25) follows from inserting (2.27) in (2.26). Finally one must connect the input ρ in (2.27) with the resulting ϕ. That is done either by using the completeness of ϕ and ϕ_0 together with (2.5), or by other means.

The main disadvantage of this way of proceeding is that unless the existence of the representation (2.5) for a given potential can be proved (which in the method presented here in detail is done first), there is no guarantee that the inversion procedure leads back to the potential one started with. So the uniqueness is in doubt.

The method given works for each angular momentum ℓ, in a similar way. This means, in other words, that the potential $V(r)$ may be constructed from any chosen phase shift δ_ℓ. If there are no bound states of that angular momentum then the inversion is unique. Now the Bargmann inequality [5]

$$n_\ell \leq \int_0^\infty drr|V|/(2\ell + 1)$$

implies that for potentials in the class (1.3) there always is an angular momentum ℓ above which there are no bound states. Hence there always exists an ℓ value such that V may be uniquely reconstructed without knowledge of bound states. In principle, that ℓ-value can even be recognized from the phase shifts by means of Levinson's theorem.

3. Fixed Energy

The inverse problem we want to discuss now is one in which we assume we are given all phase shifts δ_ℓ, $\ell = 0, 1, \ldots$ at one given energy. [6] So we will set $k = E = 1$, which means using distance units equal to k^{-1}. The radial Schrödinger equation may be written in the form

$$D(x)\phi_\ell(x) = \ell(\ell + 1)\phi_\ell(x) \tag{3.1}$$

where the differential operator D is given by

$$D(x) = x^2 \frac{d^2}{dx^2} + 1 - V(x) \tag{3.2}$$

and it will be convenient to use the boundary conditions

$$\lim_{x\to 0} \quad \phi_\ell(x)x^{-\ell-1}(2\ell+1)!! = 1. \tag{3.3}$$

The inversion procedure will follow more or less the previous model, except for two things.

The procedure of Sec. 2 consisted of two separate steps: (1) From the phase shift to the spectral function; (2) from the spectral function to the potential. The main difficulty in the present problem is that we have no completeness relation for the solutions of (3.1) and the solutions for different ℓ-values are not orthogonal in any useful sense. As a result, while we are still going to require two steps analogous to (1) and (2) above, the intermediate point will not be a spectral function. Furthermore, we cannot use a completeness relation as a tool. Analyticity, which was an important argument to get to the integral representation (2.5), can be used here only after an analytic interpolation of the discrete set of phase shifts, due to Regge. [6] It has been used to prove the existence of an analog of (2.5), namely (3.13) below, by Loeffel. [6] However, we shall proceed differently.

The logic of our procedure will follow the outline given near the end of Sec. 2. Let us define $\phi_\ell^{(o)}$ to be the solutions of (3.1) and (3.3) with $V = 0$. These are the spherical Riccati-Bessel functions

$$\phi_\ell^{(o)}(x) = u_\ell(x) = ((1/2)\pi x)^{1/2} \, J_{\ell+1/2}(x).$$

We define the function

$$g(x,y) = \sum_{\ell=0}^{\infty} c_\ell u_\ell(x) u_\ell(y) \tag{3.4}$$

with the real coefficients c_ℓ that decrease sufficiently rapidly so that $g(x,y)$ has all the necessary properties. If (3.4) can be twice differentiated term by term, then $g(x,y)$ satisfies the hyperbolic equation

$$[D_o(x) - D_o(y)]g(x,y) = 0 \tag{3.5}$$

where

$$D_o(x) = x^2 \left(\frac{d^2}{dx^2} + 1\right), \tag{3.6}$$

and the boundary conditions

$$g(x,o) = g(o,x) = 0. \tag{3.7}$$

Next we set up the analog of the Gel'fand-Levitan equation

$$K(x,y) = g(x,y) - \int_0^x dzz^{-2}K(x,z)g(z,y). \tag{3.8}$$

Let us assume for the moment that this equation has a unique solution. This will have to be proved later. If follows from (3.8) that its solution $K(x,y)$ is such that

$$K(x,o) = 0. \tag{3.9}$$

Next we define the function

$$\xi(x,y) \equiv D(x)K(x,y) - D_o(y)K(x,y) \tag{3.10}$$

in which $K(x,y)$ is the unique solution of (3.8) and the function $V(x)$ in $D(x)$ is given by

$$V(x) = -2x^{-1}\frac{d}{dx}[x^{-1}K(x,x)]. \tag{3.11}$$

Then it is a matter of straight-forward but somewhat tedious differentiations and integrations by parts, as well as the use of (3.7) and (3.9), to show that $\xi(x,y)$ satisfies the homogeneous version of (3.8) for each x as a function of y. Since (3.8) has a unique solution, it follows that $\xi \equiv 0$ and hence

$$[D(x) - D_0(y)]\,K(x,y) = 0. \tag{3.12}$$

We now define

$$\phi_\ell(x) = u_\ell(x) - \int_0^x ddy^{-2}K(x,y)u_\ell(y) \tag{3.13}$$

and show directly by means of (3.12) that ϕ_ℓ satisfies (3.1), with V given by (3.11). Furthermore, $\phi_\ell(o) = 0$.

We are now in the position of starting with a set $\{c_\ell\}$ of coefficients in (3.4) and ending up, after solving (3.8), with a potential (3.11) and a regular solution (3.13) of the Schrödinger equation with that potential. We must next connect the c_ℓ with the scattering data. If we insert (3.4) in (3.8) we obtain the expansion

$$K(x,y) = \sum_{\ell=0}^{\infty} c_\ell\phi_\ell(x)u_\ell(y) \tag{3.14}$$

where the ϕ_ℓ are defined in (3.13). If we resubstitute this expansion in (3.13) we obtain an infinite set of algebraic equations,

$$\phi_\ell(x) = u_\ell(x) - \sum_{\ell'=0}^{\infty} L_{\ell\ell'}(x) c_{\ell'} \phi_{\ell'}(x) \tag{3.15}$$

where

$$\begin{aligned} L_{\ell\ell'}(x) &= \int_0^x dy y^{-2} u_\ell(y) u_{\ell'}(y) \\ &= \frac{u_\ell(x) u'_{\ell'}(x) - u_{\ell'}(x) u'_\ell(x)}{(\ell' - \ell)(\ell' + \ell + 1)} \ . \end{aligned} \tag{3.16}$$

Let us now allow x to become large. Then

$$\lim_{x \to \infty} L_{\ell\ell'}(x) = \begin{cases} i^{\ell' - \ell - 1} M_{\ell\ell'} & , \ell \neq \ell' \\ \frac{1}{2}\pi/(2\ell + 1) & , \ell = \ell' \end{cases} \tag{3.17}$$

where

$$M_{\ell\ell'} = \begin{cases} 1/(\ell' - \ell)(\ell' + \ell + 1) & , \ell' - \ell \text{ odd,} \\ 0 & , \ell' - \ell \text{ even.} \end{cases} \tag{3.18}$$

In the limit as $x \to \infty$ the asymptotic form of $\phi_\ell(x)$ is

$$\phi_\ell(x) = A_\ell \sin(x - (1/2)\pi\ell + \delta_\ell) + o(1). \tag{3.19}$$

Substitution of this and of (3.17) in (3.15) gives the infinite set of equations

$$A_\ell e^{i\delta_\ell} = 1 - (1/2)\pi \frac{c_\ell A_\ell e^{i\delta_\ell}}{2\ell + 1} + i\Sigma_{\ell'} M_{\ell\ell'} c_{\ell'} A_{\ell'} e^{i\delta_{\ell'}}$$

and their complex conjugates. In order to solve them for c_ℓ, with the δ_ℓ given, we set

$$b_\ell = c_\ell A_\ell \tag{3.20}$$

and multiply by $e^{-i\delta_\ell}$, getting

$$A_\ell = e^{-i\delta_\ell} - (1/2)\pi \frac{b_\ell}{2\ell+1} + i\Sigma_{\ell'} M_{\ell\ell'} b_{\ell'} e^{i(\delta_{\ell'} - \delta_\ell)} \ .$$

The real and imaginary parts of these equations are

$$\sin \delta_\ell = \sum_{\ell'} M_{\ell\ell'} b_{\ell'} \cos(\delta_{\ell'} - \delta_\ell) \tag{3.21}$$

and

$$A_\ell = \cos \delta_\ell - (1/2)\pi \frac{b_\ell}{2\ell+1} - \sum_{\ell'} M_{\ell\ell'} b_{\ell'} \sin(\delta_{\ell'} - \delta_\ell) \tag{3.22}$$

since the A_ℓ are real.

Now the idea is to solve (3.21) for the b_ℓ. Then (3.22) gives the A_ℓ, and (3.20) give the c_ℓ. That route then leads from the phase shifts δ_ℓ in (3.21) to the c_ℓ and hence to a $V(r)$.

Let's look at the two uniqueness questions that arise in this, namely, that of the solution of (3.8) and that of the solution of (3.21).

Suppose that $\chi(x,y)$, as a function of y, is a solution of the homogeneous form of (3.8),

$$\chi(x,y) = -\int_0^x dz \chi(x,z) g(z,y) z^{-2}.$$

Substitution of the expansion (3.4) then leads to the expansion

$$\chi(x,y) = \sum_\ell c_\ell \chi_\ell(x) u_\ell(y)$$

where χ_ℓ is given by

$$\chi_\ell(x) = -\int_0^x dy y^{-2} \chi(x,y) u_\ell(y) = -\sum_{\ell'} L_{\ell\ell'}(x) c_{\ell'} \chi_{\ell'}(x).$$

Thus either the infinite matrix $L_{\ell\ell'}(x)c_{\ell'}$ has the eigenvalue 1, or $\chi_\ell(x) \equiv 0$. But it is easy to see that the positive definite matrix $L(x)$ is such that $\mathrm{tr}(Lc)^2 = \mathrm{tr}(L^{1/2}cL^{1/2})^2 \to 0$ as $x \to 0$. Hence there must be an interval such that in it $L(x)$ cannot have the eigenvalue 1 and hence $\chi_\ell(x) \equiv 0$ there. Thus the solution $K(x,y)$ of (3.8), for $x \leq y$, starts out uniquely at $x = 0$ and the solution that is <u>analytic</u> on the real x-axis remains unique.

As for existence of a solution, a sufficient condition for square integrability of the kernel $g(x,y)$ is that there exist a number p such that

$$|c_\ell| < c\ell^p. \tag{3.23}$$

In that case the Fredholm alternative assures the existence of a solution.

Then there is the infinite set of equations (3.21). If we let M be the matrix whose elements are equal to $M_{\ell\ell'}$; a, the column matrix whose elements are $b_\ell \cos \delta_\ell$; and e, the column matrix whose elements are all equal to unity, then (3.21) may be written

$$(1 + R)a = M^{-1}(\tan \delta)e, \tag{3.24}$$

where

$$R = M^{-1}(\tan \delta)M(\tan \delta)$$

and $(\tan \delta)$ is the diagonal matrix whose diagonal elements are equal to $\tan \delta_\ell$. Thus once the inverse M^{-1} is constructed, there remains only the inversion of $1 + R$, and the properties of R depend on the behavior of the assumed phase shifts. For example, if one assumes (as in a practical case one surely always will) that there exists an integer p such that for $\ell \geq p$ we have $\delta_\ell = 0$, then (3.24) constitutes a set of p equations and the b_ℓ for $\ell \geq p$ are explicitly given in terms of the b_ℓ, $\ell < p$. In general $1+ R$ may be inverted by the Fredholm method if the δ_ℓ decrease sufficiently rapidly. So the solution of (3.21) depends essentially on the construction of M^{-1}.

It can be proved that an inverse of M exists, and in fact it can be explicitly constructed. But it can also be proved that this inverse is not unique. There exists a vector that is annihilated by it. Hence for sufficiently rapidly decreasing phase shifts, in general (except for the sets for which $\det(1 + R) = 0$) a solution of (3.21) exists, but it is not unique. Even if we set all $\delta_\ell = 0$, so that there is no scattering at all, there is a non-trival solution, and the c_ℓ behave sufficiently well for the inversion procedure to work. In fact, at every energy there exists a one-parameter family of <u>transparent</u> potentials. They decrease at infinity like $r^{-3/2}$ and oscillate. Among the one-parameter family of potentials constructed by the present method there is always one and only one that decreases faster than $r^{-2+\varepsilon}$, provided that the phase shifts decrease asymptotically faster than

$\ell^{-3-\eta}$ (for arbitrarily small positive ε and η).

There is another source of nonuniqueness in this inversion procedure. The beginning ansatz (3.4) was arbitrary. We could just as well have summed ℓ over a set larger than the non-negative integers. Indeed, the sum may be replaced by an integral. Since the input is restricted to phase shifts of integral values of ℓ, this introduces a large source of non-uniqueness.

Let me now just outline what has been proved rigorously in this program, primarily by Loeffel and by Sabatier. The existence of the representation (3.13) has been proved for all potentials for which

$$\int_{x_0}^{\infty} dx|V| + \int_0^{x_0} dx x^{1-\varepsilon} |V| < \infty \tag{3.25}$$

for some $x > o$ and $\varepsilon > o$, by Loeffel. The transformation kernel $K(x,y)$ has been thoroughly studied from the point of view of the partial differential equation (3.12) together with the boundary conditions (3.9) and (3.11), by Sabatier. He proves that if the phase shifts are bounded by $c\ell^{-1-\varepsilon}$ for some $c > o$, $\varepsilon > o$, then "almost always" there exists a potential in a class F of potentials that is dense in the class (3.25) with respect to the metric induced by this norm. The "almost always" refers to the possibility that R of (3.24) has the eigenvalue 1 and (3.24) has no solution for any choice of M^{-1}. The inversion is not unique in that class, but Sabatier shows how to construct all phase-equivalent potentials in F.

4. The Three-Dimensional Inverse Problem.

The one-dimensional inverse problem is sufficiently similar to the "fixed angular momentum" case discussed in Sec. 2 that we shall not pursue it here.[7] Before we go to the three-dimensional case it will be useful to understand the structure of the Gel'fand-Levitan method in the abstract.

If we consider $\phi(k,x)$ and $\phi_0(k,x)$ as vectors ϕ_k and ϕ_k^0 in a linear space then the representation (2.5) may be written

$$\phi_k = U\phi_k^0 \quad . \tag{4.1}$$

Since the ϕ_k^0 form a complete set (in the Fourier-integral sense) in L^2, (4.1) is equivalent to U being an intertwining operator

$$HU = UH_0 \tag{4.2}$$

for H and H_0. If we write the completeness relations (2.20) and (2.23) symbolically

$$\int d\rho\, \phi_k \bar{\phi}_k = 1, \quad \int d\rho_0\, \phi^o_k \bar{\phi}^o_k = 1$$

and insert (4.1) we get

$$UgU^* = 1 - UU^*$$

where

$$g = \int d(\rho - \rho_0)\, \phi^o_k \bar{\phi}^o_k .$$

This could be done with <u>any</u> intertwining operator (4.2). However, the operator U of (2.5) has the special property of being of the form

$$U = 1 - K \tag{4.3}$$

where K is <u>triangular</u>, that is,

$$K(x,y) = 0 \quad \text{for} \quad y > x. \tag{4.4}$$

Because of this we can construct an inverse

$$U^{-1} = 1 - K'$$

from the range of U to all of $\mathcal{H}$ and K′ of (2.8) is also triangular. Hence we have

$$Ug = K - K'^*$$

which for $y < x$ reads

$$K(x,y) = g(x,y) - \int dz K(x,z) g(z,y).$$

This is the Gel'fand-Levitan equation. Thus the essence of the method is the existence of an intertwining operator (4.2) of the form (4.3), where K is <u>triangular</u>. The difficulty that has been in the path of generalizing the method to more than one dimension has been to find a generalization of this triangularity property. [8] It has been overcome by a very clever new Green's function introduced by Faddeev in 1966. [9]

To give you a bit of the history of this: In the winter of 1972, I used Faddeev's new Green's function to do the 3-dimensional inverse problem, and when I was essentially finished, I received a preprint from Faddeev, [10] dated summer, 1971, in which he had done almost everything I had done and more. What I will present, then, is a mixture of Faddeev's and my own work, in which Faddeev has clear priority. None of this has been published yet, primarily because of an open question in it that has not been solved. I shall point it out when I get to it.

We are now working with the full three-dimensional Schrödinger equation

$$(\Delta + k^2 - V)\ \psi = 0 \tag{4.5}$$

and we shall not assume that V is spherically symmetric. A Green's function of the Helmholtz equation is any solution of the inhomogeneous equation

$$(\Delta + k^2)G(\underline{r},\underline{r}') = \delta(\underline{r} - \underline{r}'). \tag{4.6}$$

The new Green's function of Faddeev is given by

$$G_{\underline{q}}(K^2;\underline{r} - \underline{r}') = \frac{1}{(2\pi)^3} \int (d\underline{p})\ \frac{e^{i(\underline{p} + i\underline{q}) \cdot (\underline{r} - \underline{r}')}}{K^2 - (\underline{p} + i\underline{q})^2} \tag{4.7}$$

where $\underline{q}$ is a fixed real vector. It is easy to see that it satisfies (4.6). Let me mention parenthetically that it is the resolvent of $-\Delta$ on the Hilbert space $L^2(R^3)$ with the weight function $e^{2\underline{q}\cdot\underline{r}}$ (where $-\Delta$ is not self-adjoint).

We now define the unit vector $\hat{\gamma}$ by $\underline{q} = q\hat{\gamma}$, and we write $a_0 = \underline{a} \cdot \hat{\gamma}$, $\underline{a}_\perp = \underline{a} - a_0\hat{\gamma}$ for any vector $\underline{a}$. After a shift in the variables of integration we can then write

$$G_{\underline{q}}[(\underline{k} + i\underline{q})^2, \underline{r}] \equiv G_{\hat{\gamma}}(\mu^2, s;\underline{r})$$

$$= \frac{1}{(2\pi)^3} \int (d\underline{p})\ \frac{e^{i\underline{p} \cdot \underline{r} + isr_0}}{\mu^2 - p^2 - 2sp_0} \tag{4.8}$$

where $\mu^2 = k_\perp^2$, $s = k_0 + iq$. It is understood that $q \geq 0$ and hence $\text{Im}\ s \geq 0$. One can easily see from this integral representation that $G_{\hat{\gamma}}(\mu^2, s;\underline{r})$, for fixed μ, $\hat{\gamma}$, and $\underline{r}$, is an analytic function of s regular in the upper half plane, but there is a cut from $-\infty$ to $+\infty$ and the values in the two half planes are not one another's continuations. We are interested in $G_{\hat{\gamma}}$ only in the upper half plane and particularly, its boundary value for real s.

It is a straight-forward exercise to give the following representations:

$$G_{\hat{\gamma}}(\mu^2,s;\underline{r}) = -(4\pi r)^{-1} e^{irs} + \tilde{G}_{\hat{\gamma}}(\mu^2,s;\underline{r}) \tag{4.9}$$

$$\tilde{G}_{\hat{\gamma}}(\mu^2,s;\underline{r}) = \frac{\mu}{4\pi} \int_{r_0/r}^{1} du \frac{e^{irsu}}{(1-u^2)^{1/2}} J_1 [\mu r(1-u^2)^{1/2}] \tag{4.10}$$

and

$$G_{\hat{\gamma}}(\mu^2,s;\underline{r}) = G^+(k,r) + B^0_{\hat{\gamma}}(\underline{k},\underline{r}) \tag{4.11}$$

$$B^0_{\hat{\gamma}}(\underline{k},\underline{r}) = i(2\pi)^{-2} \int (d\underline{p})\delta(k^2 - p^2)\theta[\hat{\gamma}\cdot(\underline{p}-\underline{k})]e^{i\underline{p}\cdot\underline{r}} \tag{4.12}$$

where G^+ is the usual outgoing-wave Green's function

$$G^+(k,r) = -\frac{e^{ikr}}{4\pi r}$$

and θ is the Heaviside function

$$\theta(x) = \begin{cases} 1, & x \geq 0 \\ 0, & x < 0. \end{cases}$$

A number of useful estimates can be proved for $G_{\hat{\gamma}}$ which we shall not write down here. Note that (4.9) and (4.10) show that

$$G_{\hat{\gamma}}(\mu^2, -\bar{s};\underline{r}) = \overline{G}_{\hat{\gamma}}(\mu^2, s;\underline{r}) \tag{4.13}$$

and (4.11) and (4.12) show that

$$G_{\hat{\gamma}}(0,s;\underline{r}) = G^+(s,r). \tag{4.14}$$

Let us now look at the asymptotic values of $G_{\hat{\gamma}}$ for large r and real s. They are obtained by the method of stationary phase from (4.11). Let $C^+_{\hat{\gamma},\hat{k}}$ be the cone

$$C^{+}_{\hat{\gamma},\hat{k}} = \{\underline{r} | r_0 > 0,\ r_{\perp}/r_0 < \mu/s\}, \tag{4.15}$$

and $C^{-}_{\hat{\gamma},\hat{k}}$ the cone

$$C^{-}_{\hat{\gamma},\hat{k}} = \{r | r_0 < 0,\quad |r_{\perp}/r_0| < \mu/s\}. \tag{4.15´}$$

Then we find that for $r \to \infty$

$$G_{\hat{\gamma}}(\mu^2,s;\underline{r}) = \begin{cases} o(r^{-1}) & ,\ \underline{r} \in C^{+}_{\hat{\gamma},\hat{k}} \\ -(2\pi r)^{-1}\cos kr + o(r^{-1}), & \underline{r} \in C^{-}_{\hat{\gamma},\hat{k}} \\ -(4\pi r)^{-1}e^{ikr} + o(r^{-1}) & ,\ \underline{r} \notin C^{+}, C^{-},\ s > o \\ -(4\pi r)^{-1}e^{-ikr} + o(r^{-1}) & ,\ \underline{r} \notin C^{+}, C^{-},\ s < 0 \end{cases}$$

This may be written in the compact form

$$G_{\hat{\gamma}}(\mu^2,s;\underline{r}) = -(4\pi r)^{-1}[e^{ikr}\theta(sr - kr_0) + e^{-ikr}\theta(-sr - kr_0)] + o(r^{-1}) \tag{4.16}$$

We now use this Green's function to set up an integral equation for a solution of the Schrodinger equation (4.5):

$$\phi_{\hat{\gamma}}(\underline{k},\underline{r}) = \phi_{\hat{\gamma}}(\underline{k}_{\perp},s;\underline{r})$$

$$= e^{i\underline{k}_{\perp}\cdot\underline{r}_{\perp} + isr_0} + \int (d\underline{r}')G_{\hat{\gamma}}(\mu^2,s;\underline{r} - \underline{r}')V(\underline{r}')\phi_{\hat{\gamma}}(\underline{k},\underline{r}'). \tag{4.17}$$

Multiplying by $|V|^{1/2}e^{-isr_0}$ and defining $\phi'_{\hat{\gamma}} = |V|^{1/2}e^{-isr_0}\phi_{\hat{\gamma}}$,

$$\Gamma_{\hat{\gamma}}(\mu^2,s;\underline{r},\underline{r}') = |V(\underline{r})|^{1/2}G_{\hat{\gamma}}(\mu^2,s;\underline{r} - \underline{r}')e^{is(r_0' - r_0)}|V(\underline{r}')|^{1/2}v(\underline{r}')$$

with

$$v(\underline{r}) = V(\underline{r})/|V(\underline{r})|$$

we obtain the equation

$$\phi'_{\hat{\gamma}}(\underline{k}_{\perp},s;\underline{r}) = |V(r)|^{1/2}e^{i\underline{k}_{\perp}\cdot\underline{r}_{\perp}} + \int (d\underline{r}')\Gamma_{\hat{\gamma}}(\mu^2,s;\underline{r},\underline{r}')\phi'_{\hat{\gamma}}(\underline{k}_{\perp},s;\underline{r}'). \tag{4.18}$$

The previously mentioned estimates on the Green's function can then be used to show easily that if

$$\int (d\underline{r}) |V| < \infty \tag{4.19}$$

and for all $\underline{r}$

$$\int (d\underline{r}') \frac{|V|}{|\underline{r} - \underline{r}'|} \leq M < \infty \tag{4.19}$$

then for all $\hat{\gamma}$ and μ and for all s with $\text{Im } s \geq 0$, the kernel $\Gamma_{\hat{\gamma}}$ is in L^2:

$$\int (d\underline{r})(d\underline{r}') |\Gamma_{\hat{\gamma}}(\mu^2, s; \underline{r}, \underline{r}')|^2 < \infty . \tag{4.20}$$

Hence the integral equation (4.18) has a unique solution in $L^2(R^3)$ for all s with Im s $\geq$ 0, except at those values of s where the homogeneous form of (4.18) has a solution. These are the "exceptional points where the modified Fredholm determinant of (4.18) vanishes:

$$D_{\hat{\gamma}}(\underline{k}) = D_{\hat{\gamma}}(\mu, s) = \det_2(1 - G_{\hat{\gamma}}(\underline{k})V) = 0. \tag{4.21}$$

Now $G_{\hat{\gamma}}$ was shown to be an analytic function of s regular in the upper half plane. Hence so is $\Gamma_{\hat{\gamma}}$. It then follows easily that if V satisfies (4.19), (4.19′), and

$$\int (dr) r |V| < \infty \tag{4.19}$$

then $\Phi_{\hat{\gamma}}'$ is an analytic function of s, meromorphic in the upper half plane with poles of finite order at the exceptional points $s = s_n$ defined by (4.21).

As for the behavior of $\Phi_{\hat{\gamma}}'$ for large $|s|$, one can show that if the potential satisfies (4.19) and (4.19′′) then for $\text{Im } s \geq 0$

$$\lim_{|s| \to \infty} \|\Gamma_{\hat{\gamma}}\| = 0, \tag{4.22}$$

the Fredholm determinant $D_{\hat{\gamma}}$ approaches unity,

$$\lim_{|s|\to\infty} D_{\hat{\gamma}}(\mu,s) = 1, \tag{4.23}$$

and

$$\phi^{\sim}_{\hat{\gamma}}(\underline{k}_{\perp},s;\underline{r}) = |V(\underline{r})|^{1/2} e^{i\underline{k}_{\perp}\cdot\underline{r}_{\perp}} + o(1). \tag{4.24}$$

It follows that there exists a semi-circle in the upper half-plane around the origin outside of which there are no exceptional points. If there are no exceptional points on the real axis then the total number of exceptional points must be <u>finite</u>.

Now a few words about those exceptional points. Their number or physical significance is not known. It follows from (4.13) that for $\mu \neq 0$ they occur in pairs symmetric with respect to the imaginary s-axis. Because of (4.14), as $\mu \to 0$, they approach the zeros of the modified Fredholm determinant

$$D(s) = \det_2[1 - G(s)V] \tag{4.25}$$

of the Lippmann-Schwinger equation, and hence in that limit they approach the values $s_n = iK_n$, where $-K_n^2 = E_n$ are the eigenvalues of $H = -\Delta^2 + V$. It is not known whether it is possible for exceptional points to occur on the real axis.

We now form the function $\phi_{\hat{\gamma}}$ from (4.17) and $\phi^{\sim}_{\hat{\gamma}}$, namely

$$\phi_{\hat{\gamma}}(\underline{k}_{\perp},s;\underline{r}) = e^{i\underline{k}_{\perp}\cdot\underline{r}_{\perp} + isr_0} + e^{isr_0}\int (d\underline{r}')G_{\hat{\gamma}}(\mu^2,s;\underline{r}-\underline{r}')e^{is(r_0'-r_0)} \cdot |V(\underline{r}')|^{1/2} v(\underline{r}')\phi^{\sim}_{\hat{\gamma}}(\underline{k}_{\perp},s;\underline{r}').$$

The estimates on $G_{\hat{\gamma}}$ then show that this gives $\phi_{\hat{\gamma}}$ as a continuous, twice differentiable function of $\underline{r}$ for all s, save for the <u>exceptional points</u>. We may consequently conclude the following important properties of $\phi_{\hat{\gamma}}(\underline{k}_{\perp},s;\underline{r})$.

The function $\phi_{\hat{\gamma}}$ defined by the integral equation (4.17) is a solution of the Schrödinger equation (4.5). For each fixed μ^2, $\hat{\gamma}$, and $\underline{r}$, it is an analytic function of s meromorphic in the upper half plane and continuous on the real axis, save for the exceptional points there, if any. Moreover the function

$$\hat{\phi}_{\hat{\gamma}}(\underline{k}_{\perp},s;\underline{r}) = D_{\hat{\gamma}}(\mu,s)\phi_{\hat{\gamma}}(\underline{k}_{\perp},s;\underline{r}) \tag{4.26}$$

is continuous on the real axis and holomorphic in the upper half-plane. What is more, it follows from (4.23) and (4.24) that

$$\hat{\phi}_{\hat{\gamma}}(\underline{k}_{\perp},s;\underline{r}) = e^{i\underline{k}_{\perp}\cdot\underline{r}_{\perp} + isr_0} + o(e^{isr_0}) \tag{4.27}$$

as $|s| \to \infty$. Consequently there exists a representation of the form

$$\hat{\phi}_{\hat{\gamma}}(\underline{k},\underline{r}) = e^{i\underline{k}\cdot\underline{r}} - \int_{r_0}^{\infty} dr_0{}' \tilde{K}_{\hat{\gamma}}(\underline{r};\underline{k}_{\perp},r_0')e^{isr'_0} \, . \tag{4.28}$$

This is the analog of (2.5) and it follows here by similar arguments. Note again here the important fact that the transformation kernel $\tilde{K}_{\hat{\gamma}}$ is <u>triangular</u> with respect to the direction $\hat{\gamma}$ in R^3. That is the consequence of the definition of $\phi_{\hat{\gamma}}$ by means of Faddeev's Green's function.

In order to make things more symmetric we may Fourier transform the dependence of $\hat{\phi}_{\hat{\gamma}}$ on $\underline{k}_{\perp}$ at the same time and write

$$\begin{aligned}\hat{\phi}_{\hat{\gamma}}(\underline{k},\underline{r}) &= e^{i\underline{k}\cdot\underline{r}} - \int(d\underline{r}')K_{\hat{\gamma}}(\underline{r},\underline{r}')\theta[\hat{\gamma}\cdot(\underline{r}'-\underline{r})]e^{i\underline{k}\cdot\underline{r}'} \\ &= \int(d\underline{r}')U_{\hat{\gamma}}(\underline{r},\underline{r}')e^{i\underline{k}\cdot\underline{r}'}\end{aligned} \tag{4.29}$$

where

$$U_{\hat{\gamma}}(\underline{r},\underline{r}') = \delta(\underline{r}-\underline{r}') - K_{\hat{\gamma}}(\underline{r},\underline{r}')\theta[\hat{\gamma}\cdot(\underline{r}'-\underline{r})].$$

Similarly, (4.28) may be written

$$\hat{\phi}_{\hat{\gamma}}(\underline{k},\underline{r}) = \int_{-\infty}^{\infty} dt\,\tilde{U}_{\hat{\gamma}}(\underline{r};\underline{k}_{\perp},t)e^{ist}$$

with

$$\tilde{U}_{\hat{\gamma}}(\underline{r},\underline{k}_{\perp},t) = \delta(t-r_0)e^{i\underline{k}_{\perp}\cdot\underline{r}_{\perp}} - \tilde{K}_{\hat{\gamma}}(\underline{r};\underline{k}_{\perp},t)\theta(t-r_0). \tag{4.30}$$

Equations (4.29) and (4.30) may be inverted to read

$$U_{\hat{\gamma}}(\underline{r}.\underline{r}') = (2\pi)^{-3}\int(d\underline{k})\hat{\phi}_{\hat{\gamma}}(\underline{k},\underline{r})e^{-i\underline{k}\cdot r'} \tag{4.31}$$

$$\tilde{U}_{\hat{\gamma}}(\underline{r},\underline{k}_\perp,t) = (2\pi)^{-1}\int_{-\infty}^{\infty} dk_0\hat{\phi}_{\hat{\gamma}}(\underline{k},\underline{r})e^{-ik_0t} . \tag{4.32}$$

Thus $\tilde{U}_{\hat{\gamma}}$ obeys the wave equation

$$[\Delta - \partial_t^2 + k_\perp^2 - V(\underline{r})]\tilde{U}_{\hat{\gamma}}(\underline{r};\underline{k}_\perp ,t) = 0. \tag{4.33}$$

It represents a wave, made up according to (4.32) out of solutions of the time-dependent Schrödinger equation, that has a sharp plane wave front normal to the vector $\hat{\gamma}$ and advancing at uniform speed. Of course, it does not satisfy the time-dependent Schrödinger equation.

Let us now return to $\phi_{\hat{\gamma}}$ and look at its asymptotic behavior for large r. If we define

$$\chi_{\hat{\gamma}}(\underline{k},\underline{r}) = \phi_{\hat{\gamma}}(\underline{k},\underline{r}) - e^{i\underline{k}\cdot\underline{r}} \tag{4.34}$$

then for real s we obtain from (4.16) and (4.17),

$$\chi_{\hat{\gamma}}(\underline{k},\underline{r}) = r^{-1}e^{ikr}\theta(sr - kr_0)h_{\hat{\gamma}}(\hat{r},\underline{k})$$

$$+ r^{-1}e^{-ikr}\theta(-sr - kr_0)h_{\hat{\gamma}}(-\hat{r},\underline{k}) + o(r^{-1}) \tag{4.35}$$

where

$$h_{\hat{\gamma}}(\hat{k}',\underline{k}) = -(4\pi)^{-1}\int(d\underline{r})e^{-i\underline{k}'\cdot\underline{r}}V(\underline{r})\phi_{\hat{\gamma}}(\underline{k},\underline{r}) \tag{4.36}$$

and $\underline{k}' = \hat{k}'k$. It follows from (4.13) that

$$h_{\hat{\gamma}}(-\hat{k}', -\underline{k}) = h_{\hat{\gamma}}(\hat{k}' ,\underline{k}). \tag{4.37}$$

Note that if there are exceptional points on the real axis then at the exceptional values of s there are non-trivial solutions of the homogeneous version of (4.17) and their asymptotic behavior for large r is given by (4.35). For such a solution there would then exist a double cone outside of which there would be only outgoing (or incoming, depending on whether $s \gtrless 0$) spherical waves at infinity, and in half of which the function would be $o(r^{-1})$.

We must now relate $\phi_{\hat{\gamma}}$ to the usual outgoing wave solution of the Schrödinger equation, for two reasons: One is to derive a completeness relation, and the other, to connect $\phi_{\hat{\gamma}}$ with the scattering amplitude.

The functions $\phi^{\pm}(\underline{k},\underline{r})$ are the well-known solutions of the Lippmann-Schwinger equation

$$\phi^{\pm}(\underline{k},\underline{r}) = e^{i\underline{k}\cdot\underline{r}} + \int(d\underline{r}')G^{\pm}(k;\underline{r}-\underline{r}')V(\underline{r}')\phi^{\pm}(\underline{k},\underline{r}') \quad (4.38)$$

where

$$G^{\pm}(k,\underline{r}-\underline{r}') = -\frac{e^{\pm ik|\underline{r}-\underline{r}'|}}{4\pi|\underline{r}-\underline{r}'|} .$$

The asymptotic form of ϕ^{+} is given by (1.2), where the scattering amplitude is given by

$$A_k(\hat{k}',\hat{k}) = A(\hat{k}',\underline{k}) = A(\underline{k}',\underline{k})$$

$$= -(4\pi)^{-1}\int(d\underline{r})e^{-ik\hat{k}'\cdot\underline{r}}\,V(\underline{r})\,\phi^{+}(\underline{k},\underline{r}). \quad (4.39)$$

It obeys the symmetry

$$A(-\underline{k}, -\underline{k}') = A(\underline{k}',\underline{k}) \quad (4.40)$$

on account of the reality of the potential.

As Eq. (1.3) shows, the S-matrix may be defined as an integral operator on the unit sphere, whose kernel is given by

$$S_k(\hat{k}',\hat{k}) = \delta(\hat{k}',\hat{k}) - (k/2\pi i)\,A_k(\hat{k}',\hat{k}) \quad (4.41)$$

in terms of the scattering amplitude and the solid-angle δ-function $\delta(\hat{k}',\hat{k})$. Eq. (1.1) implies that S connects ϕ^{-} with ϕ^{+} in the sense that

$$\phi^{+}(k\hat{k},r) = \int(d\hat{k}')S_k(\hat{k}',\hat{k})\phi^{-}(k\hat{k}',r). \quad (4.42)$$

The functions ϕ^{+} and ϕ^{-} separately span the space of solutions of (4.5). Therefore $\phi_{\hat{\gamma}}$ must be representable in terms of either

ϕ^+ or ϕ^- with the same value of k^2:

$$\phi_{\hat{\gamma}}(\underline{k},\underline{r}) = \int (d\underline{\hat{k}}')\Delta^{\pm}_{\hat{\gamma}}(k;\hat{k}',\hat{k})\phi^{\pm}(k\hat{k}',\underline{r}). \tag{4.43}$$

The fact that

$$\chi^{\pm}(\underline{k},\underline{r}) = \phi(\underline{k},\underline{r}) - e^{i\underline{k}\cdot\underline{r}} \tag{4.44}$$

and $\chi_{\hat{\gamma}}$ of (4.34) is $O(r^{-1})$ as $r \to \infty$ (so that both $\phi_{\hat{\gamma}}$ and $\phi^{\pm}$ have the same plane wave as their leading terms) implies that $\Delta^{\pm}_{\hat{\gamma}}$ must be of the form

$$\Delta^{\pm}_{\hat{\gamma}}(k;\hat{k}',\hat{k}) = \delta(\hat{k}',\hat{k}) \pm (k/2\pi i)q^{\pm}_{\hat{\gamma}}(k;\hat{k}',\hat{k}) \tag{4.45}$$

where the thus defined kernels $q^{\pm}_{\hat{\gamma}}$ can be expected to be smooth in some appropriate sense. Since (4.13) and (4.17) imply that

$$\phi_{\hat{\gamma}}(-\underline{k},\underline{r}) = \overline{\phi}_{\hat{\gamma}}(\underline{k},\underline{r}) \tag{4.46}$$

and (4.38) implies that

$$\phi^{+}(-\underline{k},\underline{r}) = \overline{\phi}^{-}(\underline{k},\underline{r}) \tag{4.47}$$

it follows from (4.43) that

$$\overline{\Delta}_{\hat{\gamma}}^{+}(k;-\hat{k}',-\hat{k}) = \Delta_{\hat{\gamma}}^{-}(k;\hat{k}',\hat{k}), \tag{4.48}$$

and therefore

$$q_{\hat{\gamma}}^{-}(k;\hat{k}',\hat{k}) = \overline{q}_{\hat{\gamma}}^{+}(k;-\hat{k}',-\hat{k}). \tag{4.49}$$

We shall henceforth write simply $\Delta_{\hat{\gamma}}$ and $q_{\hat{\gamma}}$ for $\Delta_{\hat{\gamma}}^{+}$ and $q_{\hat{\gamma}}^{+}$ when there is no ambiguity.

Insertion of (4.45) in (4.43) and the use of (4.34) and (4.44) gives

$$\begin{aligned}\chi_{\hat{\gamma}}(\underline{k},\underline{r}) = \chi^{\pm}(\underline{k},\underline{r}) &\pm (k/2\pi i)\int (d\underline{\hat{k}}')e^{ikk'\cdot\underline{r}}q_{\hat{\gamma}}^{\pm}(k;\hat{k}',\hat{k}) \\ &\pm (k/2\pi i)\int (d\underline{\hat{k}}')\chi^{\pm}(k\hat{k}',\underline{r})q_{\hat{\gamma}}^{\pm}(k;\hat{k}',\hat{k}).\end{aligned} \tag{4.50}$$

If we let $r \to \infty$ and use the asymptotic forms (1.1) and (4.35) we obtain

$$q_{\hat{\gamma}}(k;\hat{k}',k) = \theta[\hat{\gamma}\cdot(\hat{k}'-\hat{k})]h_{\hat{\gamma}}(k;\hat{k}',\hat{k}) \quad (4.51)$$

and

$$h_{\hat{\gamma}}(k;\hat{k}',\hat{k}) = A_k(\hat{k}',\hat{k}) + (k/2\pi i)\int (d\hat{k}'')A_k(\hat{k}',\hat{k}'')q_{\hat{\gamma}}(k;\hat{k}'',\hat{k}). \quad (4.52)$$

Substitution of (4.51) in (4.52) gives

$$h_{\hat{\gamma}}(k;\hat{k}',\hat{k})=A_k(\hat{k}',\hat{k})+(k/2\pi i)\int(d\hat{k}'')A_k(\hat{k}',\hat{k}'')h_{\hat{\gamma}}(k;\hat{k}'',\hat{k})\theta[\hat{\gamma}\cdot(\hat{k}''-\hat{k})]. \quad (4.52')$$

This may be regarded as an integral equation for $h_{\hat{\gamma}}$ or for $q_{\hat{\gamma}}$, if the scattering amplitude A is given.

Let us separate the components of $\underline{k}$ and $\underline{k}'$ along $\hat{\gamma}$ and perpendicular to $\hat{\gamma}$, and write

$$q_{\hat{\gamma}}(k;\hat{k}',\hat{k}) = q_{\hat{\gamma}}(\underline{k}_\perp',s';\underline{k}_\perp,s)$$

with the understanding that $k_\perp'^2 + s'^2 = k_\perp^2 + s^2$. Then (4.51) shows that the kernel $q_{\hat{\gamma}}$ is triangular in s. This implies that the inverse of $\Delta_{\hat{\gamma}} = \Delta_{\hat{\gamma}}^{+\gamma}$ obeys a Volterra equation and hence exists. The Fredholm determinant of $\Delta_{\hat{\gamma}}$ can be explicitly evaluated as

$$\det \Delta_{\hat{\gamma}}(k) = \exp[(k/2\pi i)\int(d\hat{k})q_{\hat{\gamma}}(k;\hat{k},\hat{k})] \quad (4.53)$$

and this cannot vanish. We may therefore invert (4.43) and obtain

$$\phi^{\pm}(\underline{k},\underline{r}) = \int(d\hat{k}')\Delta_{\hat{\gamma}}^{\pm\,-1}(k;\hat{k}',\hat{k})\phi_{\hat{\gamma}}(k\hat{k}',\underline{r}). \quad (4.54)$$

Substitution in (4.42) gives the decomposition

$$S_k(\hat{k}',\hat{k}) = \int(d\hat{k}'')\Delta_{\hat{\gamma}}^{-}(k;\hat{k}',\hat{k}'')\Delta_{\hat{\gamma}}^{+\,-1}(k;\hat{k}'',\hat{k}). \quad (4.55)$$

This representation of the S-matrix, first given by Faddeev, is the three-dimensional analog of (2.14). The significance of this particular representation, as compared to many others one may write down, is contained in (4.48) and the triangular nature of $q_{\hat{\gamma}}$. The third significant property of $\Delta_{\hat{\gamma}}(k;\hat{k}',\hat{k}) = \Delta_{\hat{\gamma}}(\underline{k}',\underline{k})$ is that if we set $(\underline{k}'-\underline{k})\cdot\hat{\gamma} = s'-s = 0$, then it follows from the analyticity properties of $\phi_{\hat{\gamma}}$ and (4.36) that $D_{\hat{\gamma}}h_{\hat{\gamma}}$ and hence $D_{\hat{\gamma}}q_{\hat{\gamma}}$, is the boundary value of an anlytic function of s regular in the upper half plane.

The decomposition (4.55) may be simplified by using the

orthogonality

$$(2\pi)^{-3}\int(d\underline{r})\overline{\phi}^{\pm}(\underline{k},\underline{r})\,\phi^{\pm}(\underline{k}',\underline{r}) = \delta(\underline{k}-\underline{k}') \qquad (4.56)$$

of the functions $\phi^{\pm}$. Use of this and of (4.43) shows that

$$(2\pi)^{-3}\int(d\underline{r})\overline{\phi}_{-\hat{\gamma}}(\underline{k},\underline{r})\phi_{\hat{\gamma}}(\underline{k}',\underline{r})$$

$$= k^{-2}\delta(k-k')\int(d\hat{k}'')\overline{\Delta}^{\pm}_{-\hat{\gamma}}(k;\hat{k}'',\hat{k})\Delta^{\pm}_{\hat{\gamma}}(k;\hat{k}'',\hat{k}')$$

$$= k^{-2}\delta(k-k')[\delta(\hat{k},\hat{k}') + \xi^{\pm}].$$

But (4.51) and (4.48) show that $\xi^{\pm} = 0$ for $(\hat{k}'-\hat{k})\cdot\hat{\gamma} > 0$ or $(\hat{k}'-\hat{k})\cdot\hat{\gamma} < 0$, respectively, while $\xi^{\pm}$ is "smooth". Consequently we must have

$$(2\pi)^{-3}\int(d\underline{r})\overline{\phi}_{-\hat{\gamma}}(\underline{k},\underline{r})\phi_{\hat{\gamma}}(\underline{k}',r) = \delta(\underline{k}-\underline{k}') \qquad (4.56)$$

and

$$\int(d\hat{k}'')\overline{\Delta}_{-\hat{\gamma}}^{\pm}(k;\hat{k}'',\hat{k})\Delta_{\hat{\gamma}}^{\pm}(k;\hat{k}'',\hat{k}') = \delta(\hat{k}',\hat{k}); \qquad (4.57)$$

in other words,

$$\Delta_{\hat{\gamma}}^{\pm\,-1} = \Delta_{-\hat{\gamma}}^{\pm *}. \qquad (4.58)$$

As a result, the decomposition (4.55) may also be written

$$S_k(\hat{k}',\hat{k}) = \int(d\hat{k}'')\overline{\Delta}_{\hat{\gamma}}(k;-\hat{k}',-\hat{k}'')\overline{\Delta}_{-\hat{\gamma}}(k;\hat{k},\hat{k}''). \qquad (4.59)$$

In a similar manner we find

$$(2\pi)^{-3}\int(d\underline{r})\overline{\phi}_{\hat{\gamma}}(\underline{k},\underline{r})\phi_{\hat{\gamma}}(\underline{k}',r) = k^{-2}\delta(k-k')\int(d\hat{\underline{k}}'')\overline{\Delta}_{\hat{\gamma}}^{\pm}(k;\hat{k}'',\hat{k})\cdot$$

$$\cdot\,\Delta^{\pm}_{\hat{\gamma}}(k;\hat{k}'',\hat{k}'). \qquad (4.60)$$

Now let us look at the completeness relation. We have from (4.54)

$$\int(d\underline{k})\phi^{+}(\underline{k},\underline{r})\overline{\phi}^{+}(\underline{k},\underline{r}') = \int(d\underline{k})(d\hat{k}')(d\hat{k}'')\Delta^{-1}_{-\hat{\gamma}}(k;\hat{k}',\hat{k})\overline{\Delta}_{\hat{\gamma}}^{-1}(k;\hat{k}'',\hat{k})\,\cdot$$

$$\cdot\,\phi_{-\hat{\gamma}}(k\hat{k}',\underline{r})\overline{\phi}_{\hat{\gamma}}(k\hat{k}'',\underline{r}') = \int(d\underline{k})\phi_{-\hat{\gamma}}(\underline{k},\underline{r})\overline{\phi}_{\hat{\gamma}}(\underline{k},\underline{r}')$$

because of (4.58). But (4.43) shows that

$$\phi_{-\hat{\gamma}}(\underline{k},\underline{r}) = \int(d\hat{k}')\phi_{\hat{\gamma}}(k\hat{k}',\underline{r})M_{\hat{\gamma}}'(k;\hat{k}',\hat{k})$$

where

$$M_{\hat{\gamma}}'(k;\hat{k}',\hat{k}) = \int (d\hat{k}'')\overline{\Delta}_{-\hat{\gamma}}(k;\hat{k}'',\hat{k}')\Delta_{-\hat{\gamma}}(k;\hat{k}'',\hat{k})$$

because of (4.58). Therefore

$$(2\pi)^{-3}\int (d\underline{k})\,\phi^{+}(\underline{k},\underline{r})\overline{\phi}^{+}(\underline{k},\underline{r}')$$

$$= (2\pi)^{-3}\int (d\underline{k})(d\hat{k}')\phi_{\hat{\gamma}}(k\hat{k}',\underline{r})M_{\hat{\gamma}}'(k;\hat{k}',\hat{k})\overline{\phi}_{\hat{\gamma}}(\underline{k},\underline{r}').$$

But the well-known completeness of the functions ϕ^{+} implies that the left-hand side is equal to the kernel of the operator $1 - \Lambda$, where Λ is the projection onto the subspace spanned by the eigenvectors of H,

$$\Lambda(\underline{r},\underline{r}') = \sum_n \phi_n(\underline{r})\phi_n(\underline{r}') \tag{4.61}$$

where ϕ_n is a normalized eigenfunction of H. We therefore have the completeness relation

$$(2\pi)^{-3}\int (d\underline{k})(d\hat{k}')\phi_{\hat{\gamma}}(k\hat{k}',\underline{r})M_{\hat{\gamma}}'(k;\hat{k}',\hat{k})\overline{\phi}_{\hat{\gamma}}(\underline{k},\underline{r}') + \sum_n \phi_n(\underline{r})\,\phi_n(\underline{r}') = \delta(\underline{r}-\underline{r}'). \tag{4.62}$$

We now want to express the ϕ_n in terms of the functions $\hat{\phi}_{\hat{\gamma}}$. Since for $\underline{k} = k\hat{\gamma}$ the function $\phi_{\hat{\gamma}}$ goes over into ϕ^{+},

$$\phi_{\hat{\gamma}}(k\hat{\gamma},\underline{r}) = \phi^{+}(\underline{k},\underline{r}) \tag{4.63}$$

we must have for $k = k_n = iK_n$

$$\hat{\phi}_{\hat{\gamma}}(i\hat{\gamma}K_n,\underline{r}) = C_n(\hat{\gamma})\phi_n(\underline{r}). \tag{4.64}$$

The representation (4.29) therefore implies

$$\phi_n(\underline{r}) = \int (d\underline{r}')U_{\hat{\gamma}}(\underline{r},\underline{r}')\alpha_{\hat{\gamma}}^{(n)}(\underline{r}') \tag{4.65}$$

where

$$\alpha_{\hat{\gamma}}^{(n)}(\underline{r}) = e^{-K_n\hat{\gamma}\cdot\underline{r}}/C_n(\hat{\gamma}). \tag{4.66}$$

Note that the integral converges because of the triangularity of $K_{\hat{\gamma}}$, even though $\alpha_{\hat{\gamma}}^{(n)}$ increases exponentially as $\hat{\gamma}\cdot\underline{r}\to\infty$. Thus we may write (4.62)

$$(2\pi)^{-3}\int(d\underline{k})(dk')\hat{\phi}_{\hat{\gamma}}(k\hat{k}',\underline{r})M_{\hat{\gamma}}(k;\hat{k}',\hat{k})\overline{\hat{\phi}}_{\hat{\gamma}}(\underline{k},\underline{r}')$$
$$+\ \sum_n\hat{\phi}_{\hat{\gamma}}(iK_n\hat{\gamma},\underline{r})\overline{\hat{\phi}}_{\hat{\gamma}}(iK_n\hat{\gamma},\underline{r}')/\ |C_n(\hat{\gamma})|^2 = \delta(\underline{r}-\underline{r}') \tag{4.67}$$

where

$$M_{\hat{\gamma}} = \frac{1}{D_{\hat{\gamma}}}\ \Delta_{-\hat{\gamma}}^{\ *}\ \Delta_{-\hat{\gamma}}\frac{1}{\overline{D}_{\hat{\gamma}}} \equiv 1 + m_{\hat{\gamma}}\ , \tag{4.68}$$

in the sense of operators on the unit sphere. Insertion of (4.29) and (4.65) in (4.67) leads to

$$\int(d\underline{x})(d\underline{y})U_{\hat{\gamma}}(\underline{r},\underline{x})W_{\hat{\gamma}}(\underline{x},\underline{y})\overline{U}_{\hat{\gamma}}(\underline{r}',\underline{y}) = \delta(\underline{r}-\underline{r}') \tag{4.69}$$

in which the kernel $W_{\hat{\gamma}}$ is given by

$$W_{\hat{\gamma}}(\underline{r},\underline{r}') = (2\pi)^{-3}\int(d\underline{k})(d\underline{k}')\delta(k^2-k'^2)M_{\hat{\gamma}}(\underline{k},\underline{k}')e^{i(\underline{k}\cdot\underline{r}-\underline{k}'\cdot\underline{r}')}$$
$$+\ \sum_n\alpha_{\hat{\gamma}}^{(n)}(\underline{r})\overline{\alpha}_{\hat{\gamma}}^{(n)}(\underline{r}') \tag{4.70}$$

and we have written

$$M_{\hat{\gamma}}(\underline{k},\underline{k}') = 2k^{-1}M_{\hat{\gamma}}(k;\hat{k},\hat{k}')$$

with the understanding that $k = k'$. With the definition (4.68) this becomes

$$W_{\hat{\gamma}}(\underline{r},\underline{r}') = \delta(\underline{r}-\underline{r}') + g_{\hat{\gamma}}(\underline{r},\underline{r}')$$

where

$$g_{\hat{\gamma}}(\underline{r},\underline{r}') = (2\pi)^{-3}\int(d\underline{k})(d\underline{k}')\delta(k^2-k'^2)m_{\hat{\gamma}}(\underline{k},\underline{k}')e^{i(\underline{k}\cdot\underline{r}-\underline{k}'\cdot\underline{r}')}$$
$$+\ \sum_n\alpha_{\hat{\gamma}}^{(n)}(\underline{r})\overline{\alpha}_{\hat{\gamma}}^{(n)}(\underline{r}') \tag{4.71}$$

and again $m_{\hat{\gamma}}(\underline{k},\underline{k}') = 2k^{-1}m_{\hat{\gamma}}(k;\hat{k},\hat{k}')$.

We are now in the position outlined at the beginning of this Section, and we obtain the analog of the Gel'fand-Levitan equation from (4.69) and (4.71) by the triangularity of the kernel of $U - 1$ in (4.29): For $\underline{r}\cdot\hat{\gamma} < \underline{r}'\cdot\hat{\gamma}$ we have

$$K_{\hat{\gamma}}(\underline{r},\underline{r}') = g_{\hat{\gamma}}(\underline{r},\underline{r}') + \int(d\underline{x})\theta[\hat{\gamma}\cdot(\underline{x}-\underline{r})]K_{\hat{\gamma}}(\underline{r},\underline{x})g_{\hat{\gamma}}(\underline{x},\underline{r}'). \tag{4.72}$$

Use of (4.71) in (4.72) leads to the expansion

$$K_{\hat{\gamma}}(\underline{r},\underline{r}') = (2\pi)^{-3}\int(d\underline{k})(d\underline{k}')\delta(k^2 - k'^2)m_{\hat{\gamma}}(\underline{k},\underline{k}')\hat{\Phi}_{\hat{\gamma}}(\underline{k},\underline{r})e^{-i\underline{k}'\cdot\underline{r}'}$$
$$+ \sum_n \phi_n(\underline{r})\,\overline{\alpha}_{\hat{\gamma}}^{(n)}(\underline{r}') \tag{4.73}$$

which is the analog of (2.25). Hence $K_{\hat{\gamma}}$ satisfies the hyperbolic differential equation

$$[\Delta - \Delta' - V(\underline{r})]\, K_{\hat{\gamma}}(\underline{r},\underline{r}') = 0 \tag{4.74}$$

whereas $g_{\hat{\gamma}}$ obeys the equation

$$(\Delta - \Delta')g_{\hat{\gamma}}(\underline{r},\underline{r}') = 0. \tag{4.75}$$

The usual method, use of (4.74) and (4.75) together with integration by parts, leads to

$$2\frac{\partial}{\partial r_0}\, K_{\hat{\gamma}}(\underline{r};\underline{r}'_\perp,r_0) = V(\underline{r})\delta(\underline{r}_\perp - \underline{r}'_\perp) \tag{4.76}$$

which is obviously the analog of (2.7). It may also be written

$$2\delta[(\underline{r} - \underline{r}')\cdot\hat{\gamma}]\,\hat{\gamma}\cdot(\underline{\nabla} + \underline{\nabla}')\, K_{\hat{\gamma}}(\underline{r},\underline{r}') = V(\underline{r})\delta(\underline{r} - \underline{r}'). \tag{4.76'}$$

Note the remarkable fact that the left-hand side of (4.76') has to come out independent of $\hat{\gamma}$. Otherwise the whole procedure would break down.

One may also, perhaps more conveniently if less elegantly, work with the kernel $\tilde{K}_{\hat{\gamma}}$ of (4.28). If we write

$$b_{\hat{\gamma}}(\underline{k}'_\perp,t';\underline{k}_\perp,t) = (2\pi)^{-1}\int dk_0 dk_0'\;\delta(k^2 - k'^2)m_{\hat{\gamma}}(\underline{k}',\underline{k})e^{i(k_0't' - k_0t)}$$
$$+ \sum_n (2\pi)^2\delta(\underline{k}'_\perp)\delta(\underline{k}_\perp)\alpha_{\hat{\gamma}}^{(n)}(t')\overline{\alpha}_{\hat{\gamma}}^{(n)}(t), \tag{4.77}$$

$$a_{\hat{\gamma}}(\underline{r};\underline{k}_\perp,t) = \int(d\underline{k}'_\perp)e^{i\underline{k}'_\perp\cdot\underline{r}_\perp}\, b_{\hat{\gamma}}(\underline{k}'_\perp,r_0;\underline{k}_\perp,t) \tag{4.78}$$

then the analog of the Gel'fand-Levitan equation reads

$$\tilde{K}_{\hat{\gamma}}(\underline{r},\underline{t}) = a_{\hat{\gamma}}(\underline{r},\underline{t}) - \int(d\underline{t}')\theta[(\underline{t}' - \underline{r})\cdot\hat{\gamma}]\tilde{K}_{\hat{\gamma}}(\underline{r},\underline{t}')b_{\hat{\gamma}}(\underline{t}',\underline{t}) \tag{4.79}$$

and the potential is given by

$$V(\underline{r}) = 2\frac{\partial}{\partial r_0}\,\tilde{K}_{\hat{\gamma}}(\underline{r};\underline{k}_\perp,r_0)e^{-i\underline{k}_\perp\cdot\underline{r}_\perp}. \tag{4.80}$$

A further step necessary in the inversion procedures is to construct the Fredholm determinant $D_{\hat{\gamma}}$ from the data. It is needed in (4.68). We start out with the use of (4.11),

$$1 - G_{\hat{\gamma}}V = 1 - G^{+}V - B^{\circ}_{\hat{\gamma}}V = (1 - G^{+}V)[1 - (1 - G^{+}V)^{-1}B^{\circ}_{\hat{\gamma}}V]$$
$$= (1 - G^{+}V)(1 - B_{\hat{\gamma}}{}^{+}V)$$

where according to (4.12)

$$B_{\hat{\gamma}}{}^{+}(\underline{k};\underline{r},\underline{r}) = i(2\pi)^{-2}\int(d\underline{p})\delta(k^2 - p^2)\theta[\hat{\gamma}\cdot(\underline{p} - \underline{k})]\phi^{+}(\underline{p},\underline{r})e^{-i\underline{p}\cdot\underline{r}'}. \tag{4.81}$$

We take modified Fredholm determinants, for which one easily proves

$$\det_2[(1 - A)(1 - B)] = \det_2(1 - A)\det_2(1 - B)e^{-\mathrm{tr}AB}$$

and therefore

$$D_{\hat{\gamma}}(\underline{k}) = \det_2(1 - G_{\hat{\gamma}}V) = \det_2(1 - G^{+}V)\det_2(1 - B_{\hat{\gamma}}{}^{+}V)e^{-\mathrm{tr}G^{+}VB^{+}_{\hat{\gamma}}V}$$

$$= D^{+}(k)\det(1 - B^{+}_{\hat{\gamma}}V)e^{\mathrm{tr}(VB^{\circ}_{\hat{\gamma}})}$$

where $D^{+}(k)$ is the boundary value of (4.25), and we have used the fact that

$$\det_2(1 - A) = \det(1 - A)e^{\mathrm{tr}A}$$

if trA exists. The trace in the exponential is given by

$$\beta_{\hat{\gamma}}(\underline{k}) = \mathrm{tr}(VB^{\circ}_{\hat{\gamma}}) = (i/4\pi)(k - \underline{k}\cdot\hat{\gamma})\int(d\underline{r})V(\underline{r}),$$

and it is easy to see that

$$\mathrm{tr}(VB_{\hat{\gamma}}{}^{+})^{n} = \mathrm{tr}(F_{\hat{\gamma}})^{n}$$

where

$$F_{\hat{\gamma}}(\underline{k};\hat{p},\hat{p}') = (k/2\pi i)\theta[\hat{\gamma}\cdot(\hat{p}' - \hat{k})]A_{k}(\hat{p},\hat{p}')$$

is the kernel of the integral equation (4.52'). Consequently

$$\det(1 - B_{\hat{\gamma}}{}^{+}V) = \det(1 - F_{\hat{\gamma}}) = d_{\hat{\gamma}}(\underline{k})$$

if $d_{\hat{\gamma}}(\underline{k})$ is the Fredholm determinant of (4.52'). Thus we have found

$$D_{\hat{\gamma}}(\underline{k}) = D^{+}(k)d_{\hat{\gamma}}(\underline{k})e^{\beta_{\hat{\gamma}}(\underline{k})}\ . \tag{4.82}$$

On the other hand one can show that

$$\frac{D^-(k)}{D^+(k)} = \frac{\overline{D}^+(k)}{D^+(k)} = \det S\; e^{i(k/2\pi)\int(d\underline{r})V} \tag{4.83}$$

which means that

$$\arg D^+(k) = i/2 \ln \det S - (k/4\pi)\int(d\underline{r})V(\underline{r}). \tag{4.84}$$

Note that as $k \to \pm\infty$,

$$A(\underline{k}',\underline{k}) \simeq -(1/4\pi)\int(d\underline{r})V(\underline{r})e^{i\underline{r}\cdot(\underline{k}-\underline{k}')}$$

and hence

$$\ln \det S(k) \simeq (k/2\pi i)\int(d\underline{r})V \tag{4.85}$$

so that the last term on the right of (4.84) just cancels the leading contribution of the first term as $k \to \pm\infty$.

Now $D^+(k)$ is the boundary value of an analytic function $D(k)$ regular in the upper half plane. Furthermore for $\operatorname{Im} k \geq 0$,

$$\lim_{|k|\to\infty} D(k) = 1. \tag{4.86}$$

Its zeros are confined to the imaginary axis, located at $k = iK_n$, where $E_n = -K_n^2$ are the eigenvalues of H. Therefore we find the analog of (2.28), namely

$$D^+(k) = \lim_{\varepsilon\to 0+} \prod_n \left(1 - \frac{E_n}{E}\right)\exp\left[\frac{1}{\pi}\int_0^\infty dE' \frac{\arg D^+(k')}{E' - E - i\varepsilon}\right]. \tag{4.87}$$

In the product each eigenvalue has to be included as many times as its degeneracy. Eq. (4.84) substituted in (4.87) explicitly gives $D^+(k)$. The integral of V in (4.84) is obtained from the high-energy behavior of $\det S$ according to (4.85). Eq. (4.82) then gives us $D_{\hat{\gamma}}(\underline{k})$, since $d_{\hat{\gamma}}(\underline{k})$ can be constructed from the scattering amplitude.

Finally we need the constants $C_n(\hat{\gamma})$ of (4.66) as defined by (4.64). Since ϕ^+ can be written

$$\phi^+ = \phi_0 + G^+V\phi_0,$$

where G^+ is the boundary value of the resolvent of H, and the principal part of $G^+(E)$ near $E = E_n$ is

$$G^+(E) = \frac{\phi_n\overline{\phi}_n}{E - E_n} + \ldots$$

the principal part of ϕ^+ is given by

$$\phi^{+}(\underline{k},\underline{r}) = c_n(\hat{k}) \frac{\phi_n(r)}{k - k_n} + \quad \ldots \tag{4.88}$$

where

$$c_n(\hat{k}) = \int (d\underline{r}) \phi_n(\underline{r}) V(\underline{r}) \phi_0(\underline{k}_n, \underline{r}) / 2k_n \tag{4.89}$$

with $\underline{k}_n = \hat{k} k_n$, $k_n = iK_n$.

Now, as was mentioned in the Introduction, the forward scattering amplitude

$$A(\underline{k},\underline{k}) = -(1/4\pi) \int (d\underline{r}) e^{i\underline{k} \cdot \underline{r}} V(\underline{r}) \phi^{+}(\underline{k},\underline{r})$$

is the boundary value of an anlytic function of k regular in the upper half plane with simple poles at the eigenvalues. Its principal part at $k = k_n$ is

$$A(\underline{k},\underline{k}) = -c_n(\hat{k}) \frac{\int (d\underline{r}) e^{-i\underline{k}_n \cdot \underline{r}} V\phi_n}{4\pi(k - k_n)} + \ldots = \frac{2k_n |c_n(\hat{k})|^2}{4\pi(k - k_n)} + \ldots$$

according to (4.88) and (4.89), and because k_n is purely imaginary. In other words,

$$|c_n(\hat{k})|^2 = \lim_{k \to k_n} (k - k_n)(2\pi/k_n) A(\underline{k},\underline{k}) . \tag{4.90}$$

Using (1.6) and (4.40) we get for $t > 0$.

$$2 \int_0^\infty dk \; A(\underline{k},\underline{k}) \cos kt = -\sum_n K_n^2 |c_n(\hat{k})|^2 e^{-K_n t} , \tag{4.91}$$

which allows us to obtain both the K_n and the $|c_n(\hat{k})|$. But $c_n(\hat{k})$ is related to the $C_n(\hat{\gamma})$ defined by (4.64), according to (4.26) and (4.88) as

$$C_n(\hat{\gamma}) = c_n(\hat{\gamma}) D'(iK_n) \tag{4.92}$$

where D' is the derivative of the Fredholm determinant whose boundary value is D^+. That can be explicitly obtained from (4.87).

We now collect the bits and pieces we have assembled to solve the inverse problem. The starting points are the explicit calculation of the determinant D^+ by (4.87), the bound states and normalization constant by (4.91) and (4.92), the determinant $d_{\hat{\nu}}$, and the determinant $D_{\hat{\nu}}$ by (4.82). We then solve the linear integral equation (4.52′) for $h_{\hat{\nu}}$. We then have $\Delta_{\hat{\nu}}$ and hence all the input to construct $a_{\hat{\nu}}$ and $b_{\hat{\nu}}$ of (4.77) and (4.78) with (4.68). Then we solve the Gel'fand-Levitan equation (4.79) and get the potential from (4.80).

It is therefore incumbent upon us to examine the integral equations (4.52′) and (4.79) or (4.72) for uniqueness of their solutions. If we know that the given scattering amplitude is associated with a local potential then we know that solutions exist. However, if these solutions are not unique then we do not know if the machinery will work.

First examine (4.52′). If at a given energy the scattering amplitude $A_k(\hat{k}',\hat{k})$ as an integral kernel on the unit sphere is in L^2, then (4.52′) is a Fredholm equation. That will be the case if

$$\int (d\hat{k})\,\sigma_{\text{total}}(\underline{k}) < \infty$$

since

$$\sigma_{\text{total}}(\underline{k}) = \int (d\underline{k}')\,|A_k(\hat{k}',\hat{k})|^2 .$$

If furthermore

$$\int (d\hat{k})\,A_k(\hat{k},\hat{k})$$

exists then $d_{\hat{\nu}}(\underline{k})$ exists. There is, however, still the possibility that $d_{\hat{\nu}}(\underline{k}) = 0$. If that happens, then, according to (4.82), we have $D_{\hat{\nu}}(\underline{k}) = 0$ also, and there is an exceptional point for (4.17) on the real axis. Eq. (4.68) shows that then the kernels of (4.79) and (4.72) are not locally square integrable. So exceptional points have to be ruled out by assumption. If

$$(k/2\pi)^2 \int (d\hat{k})\,\sigma_{\text{total}}(\underline{k}) < 1$$

then it follows that $d_{\hat{\nu}}(\underline{k})$ cannot vanish. But that is a very strong sufficient condition.

Let us then look at (4.72), or more directly, at (4.79). Assume that there is a solution of the homogeneous equation, $\eta(\underline{x}) = \eta(\underline{x}_\perp, x_0)$

$$\eta(\underline{x}) = -\int (d\underline{y})\,\theta(y_0 - z_0)\,\eta(\underline{y})\,b_{\hat{\nu}}(\underline{y},\underline{x})$$

for $x_0 > r_0$. Define

$$\tilde{\eta}(\underline{k}) = \int_{r_0}^{\infty} dt\eta(\underline{k}_{\perp},t)e^{ik_0t}$$

$$\eta_n = \int_{r_0}^{\infty} dt\eta(o,t)\alpha_{\hat{\gamma}}^{(n)}(t) = (1/C_n)\int_{r_0}^{\infty} dt\eta(o,t)e^{-K_nt} .$$

Then we get

$$\int(d\underline{k}_{\perp})\int_{r_0}^{\infty} dt|\eta(\underline{k}_{\perp},t)|^2 = (1/2\pi)\int(d\underline{k})|\tilde{\eta}(\underline{k})|^2$$

$$= -\int(d\underline{k})(d\underline{k}')\delta(k^2 - k'^2)m_{\hat{\gamma}}(\underline{k}',\underline{k})\overline{\tilde{\eta}}(\underline{k})\tilde{\eta}(\underline{k}')-(2\pi)^2\Sigma|\eta_n|^2.$$

Use of (4.68) and the positive definiteness of M then implies that $\tilde{\eta} \equiv 0$, and $\eta_n = 0$, and hence $\eta \equiv 0$. Therefore the solution of (4.79) is unique.

We are thus assured that if we start with a (reasonable) scattering amplitude as a function of angle and energy, and we know that there is an underlying local potential, that we will reconstruct it by the given method. However, there is a large existence problem, as was already pointed out in the Introduction. There is, first of all, the difficulty that the present version of the Gel'fand-Levitan equation seems to be a singular integral equation. At least its kernel has not been shown to be compact. So the existence of a solution is not necessarily assured by the absence of a non-trivial solution of the homogeneous equation.

As for the general existence problem, Faddeev has given a necessary and sufficient condition for the existence of an underlying local potential, but he has given only a rough outline of a proof. The condition that the function $h_{\hat{\gamma}}(k;\hat{k}',\hat{k}) = h_{\hat{\gamma}}(\underline{k}',\underline{k})$ be such that $h_{\hat{\gamma}}(\underline{k}_{\perp}',s;\underline{k}_{\perp},s)D_{\hat{\gamma}}(k_{\perp},s)$ is the boundary value of an analytic function of s regular in the upper half plane was already noticed to be necessary for the existence of a local potential. It turns out to be sufficient as well.

5. The Case of Axial Symmetry.

As a special case of the three-dimensional inverse problem let us now assume that the potential has axial symmetry. We should then obviously choose $\hat{\gamma}$ to be in the direction of the axis of symmetry, and we shall do so. We shall take it as the z-axis of the coordinate system.

The scattering amplitude can now be expanded in a Fourier series relative to its dependence on the azimuthal angles, ϕ and ϕ':

$$A_k(\hat{k}',\hat{k}) = (1/2\pi)\sum_{-\infty}^{\infty} e^{im(\phi'-\phi)} A_m(k;x',x)$$

where $x = \cos\theta$, $x' = \cos\theta'$. Similarly the function $h_{\hat{\gamma}}$ can be expanded

$$h_{\hat{\gamma}}(k;\hat{k}',\hat{k}) = (1/2\pi)\sum_{-\infty}^{\infty} e^{im(\phi'-\phi)} h_m(k;x',x).$$

Eq. (4.52′) now becomes

$$h_m(k;x,y) = A_m(k;x,y) + (k/2\pi i)\int_y^1 dz A_m(k;x,z)h_m(k;z,y) \quad (5.1)$$

and q_m is given by

$$q_m(k;x',x) = \theta(x'-x)h_m(k;x',x). \quad (5.2)$$

We then form

$$\Delta_m(k;x',x) = \delta(x-x') + (k/2\pi i)q_m(k;x',x) \quad (5.3)$$

in terms of which

$$\Delta_{\hat{\gamma}}(k;\hat{k}',\hat{k}) = (1/2\pi)\sum_{-\infty}^{\infty} e^{im(\phi'-\phi)} \Delta_m(k;x',x).$$

If we assume in addition that the potential has <u>reflection symmetry</u> with respect to the (x,y)-plane, so that the symmetry axis has no preferred sense, then $\hat{\gamma}$ could be chosen equally well along the negative z-axis and we must have

$$\Delta_{-\hat{\gamma}}(k;\hat{k}',\hat{k}) = (1/2\pi)\sum_{-\infty}^{\infty} e^{im(\phi'-\phi)} \Delta_{-m}(k;-x',-x). \quad (5.4)$$

We also expand the spectral kernel (4.68)

$$M_{\hat{\gamma}}(k;\hat{k}',\hat{k}) = (1/2\pi)\sum_{-\infty}^{\infty} e^{im(\phi'-\phi)} M_m(k;x',x)$$

If we write

$$D_{\hat{\gamma}}(\underline{k}) = D_{\hat{\gamma}}(k_\perp,k_0) = D(k,x)$$

because it does not depend on the direction of $\underline{k}_\perp$, we get from (4.68)

$$M_m(k;x,y) = \delta(x-y) + \lambda_m(k;x,y)$$

$$= \frac{\int_{-1}^{1} dz \bar{\Delta}_{-m}(k;-z,-x)\Delta_{-m}(k;-z,-y)}{D(k,x)D(k,y)} . \tag{5.5}$$

In order to expand the kernel $g_{\hat{\gamma}}$ we expand

$$e^{i\underline{k}_\perp \cdot \underline{r}_\perp} = \exp[ik_\perp r_\perp \cos(\phi_k - \phi_r)] = \sum_{-\infty}^{\infty} i^m e^{im(\phi_k - \phi_r)} J_m(k_\perp r_\perp)$$

and obtain from (4.71)

$$g_{\hat{\gamma}}(\underline{r},\underline{r}') = (1/2\pi)^2 \sum_{-\infty}^{\infty} e^{im(\phi' - \phi)} \int_0^\infty dk k^2 \int_{-1}^{1} dx \int_{-1}^{1} dy\ \lambda_m(k;x,y)$$

$$J_m[kr_\perp(1-x^2)^{1/2}] J_m[kr'_\perp(1-y^2)^{1/2}] e^{ik(xr_0 - yr'_0)}$$

$$+ \sum_n e^{-K_n(r_0 + r'_0)} / |C_n|^2 .$$

Here, of course, $r_0 = z$. So if we expand

$$g_{\hat{\gamma}}(\underline{r},\underline{r}') = (1/2\pi) \sum_{-\infty}^{\infty} e^{im(\phi' - \phi)} g_m(r_\perp, z; r'_\perp, z')$$

then

$$g_m(r,z;r',z') = (1/2\pi) \int_0^\infty dk k^2 \int_{-1}^{1} dx \int_{-1}^{1} dy\ \lambda_m(k;x,y)$$

$$J_m[kr(1-x^2)^{1/2}] J_m[kr'(1-y^2)^{1/2}] e^{ik(xz - yz')}$$

$$+ \delta_{mo} \sum_n e^{-K_n(z + z')} / |C_n|^2 . \tag{5.6}$$

Note that only g_0 contains explicit reference to the bound states.

Finally we expand

$$K_{\hat{\gamma}}(\underline{r},\underline{r}') = (1/2\pi) \sum_{-\infty}^{\infty} e^{im(\phi' - \phi)} K_m(r_\perp, r_o; r'_\perp, r'_o)$$

and (4.72) becomes

$$K_m(r,z;r',z') = g_m(r,z;r',z')$$
$$- \int_0^\infty dr''r'' \int_z^\infty dz''K_m(r,z;r'',z'')g_m(r'',z'';r',z'). \qquad (5.7)$$

The potential is obtained from (4.76) as

$$2 \frac{\partial}{\partial z} K_m(r,z;r',z) = V(r,z)r^{-1}\delta(r - r'). \qquad (5.8)$$

Notice that we can reconstruct the same potential from any m-value, but the point spectrum is needed only if $m = 0$ is used.

The next step in symmetry assumptions is to take V to be spherically symmetric. However, in the present method there appears to be no further simplification when that step is taken. At the present time there exists no known bridge from this three-dimensional procedure to the fixed-angular momentum method of Section 2.

REFERENCES

L. N. Levinson, Kgl. Danske Videnskab, Selskab, Mat.-fys. Medd. 25, 9 (1949).

2. See R.G. Newton, Scattering Theory of Waves and Particles, McGraw-Hill, N.Y. (1966), pp. 286ff and references given on p. 296.

3. R. Jost and W. Kohn, Kgl. Danske Videnskab. Selskab, Mat.-fys. Medd. 27, 9 (1953) For generalizations to $\ell \neq 0$, see R.G. Newton, op. cit., pp. 613ff; references will be found there on p. 632.

4. I.M. Gel'fand and B.M. Levitan, Doklady Akad. Nauk SSSR 77, (1951) 557; and Isvest. Akad, Nauk SSSR 15, (1951) 309 [Am. Math. Soc. Transl. 1, 253].

5. V. Bargmann, Proc. Natl. Acad. Sci. U.S. 38, 961 (1952); J. Schwinger, Proc. Natl. Acad. Sci. U.S. 47, 122 (1961).

6. T. Regge, Nuovo Cimento 14, 951 (1959); R.G. Newton, J. Math. Phys. 3, 75 (1962), and 8, 1566 (1967); P.C. Sabatier, J. Math. Phys. 7, 1515 and 2079 (1966); 8, 905 (1967); 9, 1241 (1968); 12, 1393 (1971); 13 676 (1972); P.J. Redmond, J. Math. Phys. 5, 1547 (1964); J.J. Loeffel, Ann. Inst. Henri Poincare 8, 339 (1968)

J.R. Cox and K.W. Thompson, J. Math. Phys. 11, 804 and 815 (1970); C.P. Sabatier and Q. Van Phu, Phys. Rev. D4, 127 (1971); for generalizations, see C. Coudrey and M. Coz, Ann. Phys. (N.Y.) 61, 488 (1970) and J. Math. Phys. 12, 1166 (1971); M. Hooshyar, J. Math. Phys. 12, 2243 (1971) and 13, 1931 (1972).

7. I. Kay and H.E. Moses, Nuovo Cimento 3, 276 (1956); L.D. Faddeev, Trudy Mat. Inst. Steklov 73, 314 (1964) [Am. Math. Soc. Transl. Ser. 2, 65, 139].

8. The method given by I. Kay and H.E. Moses, Nuovo Cimento 22, 683 (1961) and Comm. Pure and Appl. Math. 14, 435 (1961), if it works, does not necessarily lead to a local potential, even if it exists.

9. L.D. Faddeev, Doklady Akad. Nauk SSSR, 165, 514 (1965) and 167, 69 (1966) [Soviet Phys. Doklady, 10, 1033 and 11, 209 (1966)].

10. L.D. Faddeev, Three-dimensional inverse problem in the quantum theory of scattering, preprint ITP-71-106E, Kiev, 1971.

11. We shall always denote the unit vector in the direction of the vector $\underline{r}$ by $\hat{r}$, and integration over the directions of $\underline{r}$ by $(d\hat{r})$.

PERTURBATIONS OF THE LAPLACIAN BY COULOMB LIKE POTENTIALS

J.C. Guillot and K. Zizi

Universités de Dijon et Paris-Nord

France

We are concerned with some perturbations of the Coulomb potential. The starting point is the following self-adjoint operator in $L^2(R^n)$, $(n \geq 3)$

$$H_0 = -\Delta - \frac{\alpha}{|x|}$$

defined on $H^2(R^n)$, the second Sobolev space, where $x \in R^n$ and $\alpha \in R$.

<u>Notations</u>. Let $E_0(\cdot)$ be the spectral measure associated with H_0, $R_0(z)$ the resolvent of H_0, namely, $R_0(z) = (H_0 - z)^{-1}$, $z \in \rho(H_0)$, where $\rho(H_0)$ is the resolvent set of H_0. Let $\sigma(H_0)$ (resp. $\sigma_p(H_0)$, $\sigma_c(H_0)$, $\sigma_{ac}(H_0)$) be the spectrum of H_0 (resp. the point, continuous and absolutely continuous spectra) and $H_{0,ac}$ the absolutely continuous part of H_0.

<u>Spectral properties of H_0</u>. Let us recall the main facts.

(i) <u>The Spectrum</u>. We have $\sigma_c(H_o) = \sigma_{ac}(H_o) = [o,\infty)$ for every $\alpha \in R$. Furthermore

if $\alpha \leq 0$, then $H_0 \geq 0$ and $\sigma(H_0) = \sigma_{ac}(H_0)$
if $\alpha > 0$, there exists also a sequence of eigenvalues:

$$\sigma_p(H_o) = \left\{ \lambda_p = - \frac{\alpha^2}{4\left(\frac{n-1}{2} + p\right)^2};\ p = 0, 1, 2, \ldots.. \right\}$$

(ii) <u>Eigenfunction Expansions</u>. Let us write them only for $H_{0,ac}$. Let us define the following function

J. A. LaVita and J.-P. Marchand (eds.), Scattering Theory in Mathematical Physics, 237–242.

$$\varphi_c^-(x;\xi) = \left(\frac{1}{2\pi}\right)^{\frac{n}{2}} \exp\left(\frac{\pi\alpha}{4|\xi|}\right)\frac{\Gamma\left(\frac{n-1}{2} - i\frac{\alpha}{2}|\xi|\right)}{\Gamma\left(\frac{n-1}{2}\right)} e^{ix\xi} \times$$

$$\times\ {}_1F_1\left(\frac{\iota\alpha}{2|\xi|}\right);\ \frac{n-1}{2}\ ;\ i(|\xi|\cdot|x| - \xi.x))$$

where x and $\xi \in R^n$. ${}_1F_1\ (\cdot;\cdot;\cdot)$ is the confluent hypergeometric function. Define

$$\varphi_c^+(x;\xi) = \overline{\varphi_c^-(x;-\xi)}\,.$$

Then, for every $f \in L^2(R\ )$, the two mappings $F_c^{\pm}$:

$$f(x) \xrightarrow{\ \mathcal{F}_c^{\pm}\ } \hat{f}_c^{\pm}(\xi) = \text{L.i.m.} \int_{R^n} \varphi_c^{\pm}(x;\xi)\, f(x)dx$$

are partially isometric from $L^2(R^n)$ onto $L^2(R^n)$ and unitary from $E_0([o,\infty))\ L^2(R^n)$ onto $L^2(R^n)$. Each of them defines a spectral representation of $H_{0,ac}$. Let $F_c^{\pm *}$ denote the adjoint mappings.

(iii) The Wave Operators. The generalized Dollard's wave operators, namely,

$$W_{\pm}^D(H_0,\ -\Delta) = S - \lim_{t\to\pm\infty} e^{it\,H_0}\, e^{-it\Delta} \exp\left(\frac{i\epsilon(t)\alpha}{2(-\Delta)^{1/2}} \text{Log}(-4|t|\Delta)\right)$$

exist and are complete. (Here $\varepsilon(t) = +1$ for $t > 0$ and $\varepsilon(t) = -1$ for $t < 0$). Furthermore, we have the following fundamental connection between the time dependent wave operators and the stationary ones:

$$W_{\pm}^D = F_c^{\pm *}\, F_0$$

where F_0 is the usual Fourier transform. Dollard's proof of this equality is not so clear and is limited to the case $n = 3$. But we have a new rigorous proof of it which is valid for every n, in which we use the spherical symmetry of H_0 explicitly.

(iv) The Resolvent Kernel. The resolvent kernel is a function $G(x,y;k)$ such that $(R_o(z)f)(x) = \int_{R^n} G(x,y;k)f(y)dy$ for every $z \in \rho(H_0)$ with $k \in C$ such that $k^2 = z$, $\text{Im}k > o$ and $k^2 \notin \sigma_p\ (H_0)$. An expression of the function G has been calculated by L. Hostler [1] and J. Schwinger [2] independently. In fact J. Schwinger has calculated the Fourier transform of the function G. But Hostler [3], [4], [5] has given an explicit expression of G. His calculation is purely formal but all the steps used by Hostler can be justified from the mathematical point of view. In the general case, Hostler obtained an integral representation of G which holds only under the restriction $\text{Im}k > o$. We have been able to obtain an analytic continuation of this integral representation which holds for $\text{Im}k = o$, too. Let us give the expression of the function G only in the physical case $n = 3$. For that we need solutions of the following differential equation:

$$\left(\frac{d^2}{dz^2} - \frac{1}{4} + \frac{i\nu}{z}\right) u(z) = 0 ; \qquad i\nu = \frac{i\alpha}{2k} .$$

In the theory of special functions, it is known that this differential equation has two particular solutions, the so called Whittaker functions usually denoted $M_{i\nu;\,1/2}(z)$ and $W_{i\nu;1/2}(z)$. (For the of the Whittaker functions see H. Buchholz [6]. Let us consider the new variables:

$$\begin{aligned} u &= |x| + |y| + |x-y| \\ \nu &= |x| + |y| - |x-y| \end{aligned} \quad ; \quad x, y \in R^n$$

Then we have, for $n = 3$,

$$G(x,y;k) = \frac{\Gamma(1-i\nu)}{4\pi|x-y|} \det \begin{vmatrix} W_{i\nu;1/2}(-iku) ; & M_{i\nu;1/2}(-ik\nu) \\ W'_{i\nu;1/2}(-iku) ; & M'_{i\nu;1/2}(-ik\nu) \end{vmatrix}$$

where $\nu = \frac{\alpha}{2k}$ and $M'_{i\nu;1/2}(z) = \frac{d}{dz} M_{i\nu;1/2}(z)$ (resp. $W'_{i\nu;1/2}(z)$)

For every n odd, a similar, but more complicated expression of G can be given.

Applications to Scattering Theory.

(1) The first application concerns the scattering theory for the following differential operator:

$$-\Delta - \frac{\alpha}{|x|} + q(|x|) \text{ in } L^2(R^n)$$

where $q(\cdot)$ is a real valued spherical symmetric potential satisfying the following conditions

$$\int_0^1 r|q(r)|dr < \infty \quad ; \quad \int_1^\infty |q(r)|dr < \infty .$$

(2) The second one concerns the scattering theory for the $L^2(R^n)$ operator

$$H = \sum_{j=1}^n \left(\frac{1}{i}\frac{\partial}{\partial x_j} + b_j(x)\right)^2 - \frac{\alpha}{|x|} + q(x) ,$$

where $b_j(x)$ and $q(x)$ are two real valued functions satisfying certain conditions. Define

$$Q(x) = \sum_{j=1}^n \left(\frac{1}{i}\frac{\partial b_j}{\partial x_j}(x) + b_j^2(x)\right) + q(x) ; \qquad b_j(\cdot) \in C(R^n).$$

Then we suppose that

$$(1 + |x|)^{a} \quad Q(x) \in L^{p_1}(R^n)$$

$$(1 + |x|)^{a} \quad b_j(x) \in L^{p_{2j}}(R^n) \quad ; \quad 1 \leq j \leq n$$

with $a > \frac{n}{2}$; Max $(2, \frac{n}{2}) < p_1 < 2n$; $n < p_{2_j}$; $1 \leq j \leq n$.

The potentials $b_j(\cdot)$ and $q(\cdot)$ are short range perturbations of H_o. In both cases, the statements are quite similar. Let us give a theorem in the case (2) and in the physical case $n = 3$.

Theorem

(I)

(i) H is a self-adjoint operator in $L^2(R^3)$ defined on $H^2(R^3)$.

(ii) $\sigma_{ac}(H) = [o, \infty)$.

(iii) The usual wave operators $W_\pm(H, H_0)$ exist and are complete, (such that the generalized Dollard's wave operators $W^D_\pm(H_0 - \Delta)$ exist and are complete too).

(II) (Eigenfunction expansions for H_{ac}).

(i) There exists a closed null set Γ_0 of $[o, +\infty)$ ($o \in \Gamma_0$) such that, for every $\xi \in R^3(\Gamma_0) = \{ \xi : |\xi|^2 \notin \Gamma_0\}$, the Lippmann-Schwinger integral equation

$$\varphi_\pm(x;\xi) = \varphi_c(x;\xi) - \int_{R^3} \varphi_\pm(y;\xi) \left\{ \sum_{j=1}^{3} \left(2i\, b_j(y) \frac{\partial}{\partial y_j} + i \frac{\partial b_j(y)}{\partial y_j} + b_j(y)^2 \right) + q(y) \right\} G(x,y; \mp |\xi|)\, dy$$

has a unique solution in the Hilbert space

$$\mathcal{H} = \{ f : R^3 \to C \mid (1 + |x|)^{-a} f \in L^2(R^3)\} .$$

Furthermore $\varphi_\pm(\cdot;\xi) \in H^2_{loc}(R^3) \cap L^\infty(R^3)$ are continuous functions of x. $\varphi_\pm(\cdot;\xi)$ are generalized eigenfunctions of H, i.e. they satisfy

$$H \varphi_\pm(\cdot;\xi) = |\xi|^2 \varphi_\pm(\cdot;\xi)$$

in the sense of distributions.

(ii) The families $\varphi_+(\cdot;\xi)$ and $\varphi_-(\cdot;\xi)$ can be used to obtain two alternative spectral representations of H_{ac}. One defines

$$(F_\pm f)(\xi) = \text{L.i.m.} \int_{R^3} \overline{\varphi_\pm(x;\xi)}\, f(x) dx .$$

The operators F_+ and F_- are partial isometries from $L^2(R^3)$

onto $L^2(R^3)$; when restricted to the absolutly continuous subspace of H, they are unitary. The adjoint maps are

$$(F_{\pm}^{*}g)(x) = \underset{N\to\infty}{\text{L.i.m.}} \int_{\Omega_N} \varphi_{\pm}(x;\xi)g(\xi)d\xi \ ,$$

where $\Omega_N = \{ \xi \in R^n : |\xi|^2 \in K_N \}$, $\{K_N\}_{N=1}^{\infty}$ being an increasing sequence of compact sets with $\bigcup_{N=1}^{\infty} K_N = R^+ - \Gamma_0$
Moreover

$$H_{ac} = F_{\pm}^{*} M_{|\xi|^2} F_{\pm} \ ,$$

where $M_{|\xi|^2}$ is the multiplication operator by the function $|\xi|^2$ and

$$W_{\pm} = F_{\pm}^{*} \mathcal{F}_c \ .$$

This theorm is proved by using the Kato-Kuroda technique of factorization of the perturbation [7].

Remarks.

(1) In the case $b_j(x) \equiv 0$ for every j, part (I) of the theorem has been proved by R. Lavine [8]. Lavine has proved also that the positive continuous singular spectrum of H is absent. Furthermore if we suppose that q(x) is a locally Hölder continuous function except at a finite number of singularities, then it follows from S. Agmon's work [9] that there are no positive eigenvalues imbedded in the continuous spectrum. In this case, we can get a complete eigenfunction expansion for H if we add the contribution of the negative eigenvalues.

(2) As to the spectral property of H, T. Ikebe and Y. Saito [10] have proved a very general result using the limiting absorption principle.

(3) The last application deals with the exterior problem. Let Ω be an unbounded domain in R^n with boundary Γ consisting of two disjoint, compact, C^2 hypersurfaces Γ_1 and Γ_2 and suppose that the origin O belongs to the interior of the compact body determined by Γ_2. Let ν denote the unit exterior normal to Γ and let $\sigma(x)$ be an s - Hölder continuous function on Γ_1 with $1 < s \leq 1$. Let

$$\varrho(x) = (1 + |x|)^{(n+1+\varepsilon)|2}$$

with a fixed but arbitrary $\varepsilon > o$. Suppose that q(x) is a real valued function on $\Omega \cup \Gamma$ such that $\varrho(x)q(x)$ is uniformly α-Hölder continuous on $\Omega \cup \Gamma$ $(o < \alpha \leq s)$ and tends to zero as $|x| \to +\infty$. Let H be the self-adjoint operator in $L^2(\Omega)$ defined by $Hg = -\Delta g - \frac{\alpha}{|x|} g + qg$ and with domain consisting of all functions g in $L^2(\Omega)$

which have functions in $L^2(\Omega)$ for their first and second order distribution derivatives and which satisfy the boundary conditions

$$\left(\frac{\partial}{\partial\nu} - \sigma(x)\right) g(x) = 0 \quad \text{on } \Gamma_1 \quad ,$$
$$g(x) = 0 \quad \text{on } \Gamma_2 \quad .$$

The study of this operator H in the case $\alpha = o$ has been done by N. Shenk and D. Thoe [11], [12].

In our case, we derive eigenfunction expansions for H using, instead of the Lippmann-Schwinger equation, a matrix equation derived from potential theory in the way proposed by Skenk and Thoe. Then we use these eigenfunction expansions to study the wave and scattering operators associated with H.

All the details and proofs will be published later.

References:

1. L. Hostler, Bull. Am. Phys. Soc. 7 (1962) 609
2. J. Schwinger, J. Math. Phys. 5 (1964) 1606
3. L. Hostler and R.H. Pratt, Coulomb Green's function in closed form, Phys. Rev. Lett. 10 (1963) 469
4. L. Hostler, Coulomb Green's function and the Furry approximation, J. Math, Phys 5 (1964) 591
5. L. Hostler, Coulomb Green's function in f-dimensional space, J. Math. Phys. 11 (1970) 2966
6. H. Buchholz, The Confluent hypergeometric function, Springer-Tracts in Natural Philosophy, V. 15 (1969)
7. T. Kato and S.T. Kuroda, Theory of simple scattering and eigenfunction expansions, Functional Analysis and Related Topics, Springer Verlag (1970)
8. R. Lavine, Absolute continuity of positive spectrum for Schrödinger operators with long-range potentials, J. of Funct. Anal. 12 (1973) 30
9. S. Agmon, Lower bounds for solutions of Schrödinger equations, J. d'Anal. Math. 23 (1970) 1
10. T. Ikebe and Y. Saito, Limiting absorption method and absolute continuity for the Schrödinger operator, J. Math. Kyoto Univ. 12 (1972) 513
11. N. Shenk and D. Thoe, Outgoing solutions of $(-\Delta + q - k^2)u = f$ in an exterior domain. J. of Math. Anal. and Appl. 31 (1970) 81
12. N. Shenk and D. Thoe, Eigenfunction expansions and scattering theory for perturbations of $-\Delta$, J. of Math. Anal. and Appl. 36 (1971) 313

ANALYTIC PERTURBATION APPROACH TO N-PARTICLE QUANTUM SYSTEMS

J. M. Combes

Centre Universitaire de Toulon et du Var

France

1. Scattering Theory for N-Particle Quantum Systems.

a) Asymptotics of the Schrödinger equation. The dynamics of an assembly of N non-relativistic interacting particles are governed by a hamiltonian

$$H = \sum_{i=1}^{N} \frac{p_i^{\,2}}{2m_i} + \sum_{j<i} V_{ij} = H_o + V$$

where p_i and m_i are the momentum and mass of particle $i (p_i = |\underline{p}_i|)$, H_o is the total kinetic energy of the system and V_{ij} is the interaction between particles i and j. Since there are no external forces we will describe the system in its center of mass reference frame. The quantum mechanical equation of motion is the Schrödinger equation

$$i \frac{\partial \Psi_t}{\partial t} = H\Psi_t$$

where Ψ_t describes the state of the system at time t; it is an element of some representation space and H is now the linear operator associated in this representation to the total energy observable. In the so called *momentum representation* states of the system are square integrable functions $\Psi(\underline{P})$ where $\underline{P} = (\underline{p}_1, \ldots \underline{p}_N)$, $\sum_{i=1}^{N} \underline{p}_i = 0$; then H_o is just the multiplication operator by $\sum_{i=1}^{N} \frac{p_i^{\,2}}{2m_i}$.
If the interaction is local then V is the convolution operator by the Fourier transform (in distribution sense) of the potential function.

J. A. LaVita and J.-P. Marchand (eds.), Scattering Theory in Mathematical Physics, 243–272.

In the configuration representation the states are square integrable functions $\Psi(\underline{X})$ where $\underline{X} = (\underline{x}_1,\ldots\underline{x}_N)$, $\sum m_i\,\underline{x}_i = 0$, is the position vector. In this representation H_o turns out to be the $3(N-1)$ dimensional Laplacian $\sum_{i=1}^{N} \Delta_{x_i}$ and, if the interaction is local, V is simply represented as the multiplication operator by the potential function. If V is non local the corresponding operator is generally an integro-differential operator obtained from the quantization rules.

To discuss the asymptotics of N particle systems it is very convenient to notice that for any cluster decomposition D = $D = \{C_1,C_2,\ldots C_k\}$ of the N particle system and any representation one has a tensor product decomposition

$$H_N = H^{C_1} \otimes \ldots \otimes H^{C_k} \otimes H^D \tag{1}$$

where H^C is the state space for particles in cluster C (in their own center of mass reference frame) and H^D is the state space for the centers of mass of clusters in D. This decomposition comes from a splitting of the phase space into a direct sum of subspaces associated to the dynamical independent variables describing the above entities. This splitting of the phase space is conveniently (and abstractly) described in terms of the two forms $Q(\underline{X})$ and $K(\underline{P})$ defined as follows. Let $Q(\underline{X}) = \sum_{i=1}^{N} m_i\, x_i^2$ denote the quadratic form on the configuration space (isomorphic to $R^{3(N-1)}$ associated to the moment of inertia about the center of mass and let $\langle \underline{X},\underline{X}' \rangle$ be the associated scalar product; it is related to the usual euclidean scalar product on the configuration space by

$$\langle \underline{X},\underline{X}' \rangle = \underline{X}\cdot\overline{M}\underline{X}' \tag{2}$$

where $\overline{M}$ is the mass matrix. Let $D = \{C_1,C_2,\ldots C_k\}$ be a cluster decomposition of the system; according to Huygens' formula one has a splitting

$$Q(\underline{X}) = Q_D(\underline{X}) + \sum_{C\varepsilon D} Q_C(\underline{X}) \tag{3}$$

where $Q_C(\underline{X})$ is the moment of inertia of particles in cluster C about their center of mass and $Q_D(\underline{X})$ is the moment of inertia for centers of mass of clusters in D. Let then P_C and P_D denote the projection operator on the orthogonal complement (for the scalar product (2)) of the null-spaces for Q_C and Q_D. Then each position vector $\underline{X}$ splits into a sum

$$\underline{X} = \underline{X}_D + \sum_{C\varepsilon D} \underline{X}_C \tag{4}$$

with $\underline{X}_D = P_D\underline{X}$ and $\underline{X}_C = P_C\underline{X}$. Notice that if $D' \subset D$ i.e. clusters in D' are obtained by further partitioning clusters of D, then $Q_D(\underline{X}) < Q_{D'}(\underline{X})$ and accordingly $P_D < P_{D'}$.

All these facts have duel counterparts in momentum space. Let $K(\underline{P}) = \sum_{i=1}^{N} \frac{P_i^2}{2m_i}$ denote the quadratic form associated to the total kinetic energy in the center of mass reference frame. It is related to the euclidean scalar product on the configuration space by

$$K(\underline{P}) = \frac{1}{2}\,\underline{P}\cdot \overline{M}^{-1}\,\underline{P} \tag{5}$$

To a cluster decomposition D corresponds a splitting

$$K(\underline{P}) = K_D(\underline{P}) + \sum_{C\varepsilon D} K_C(\underline{P}) \tag{6}$$

where $K_D(\underline{P})$ is the kinetic energy for centers of mass of clusters in D and $\overline{K}_C(\underline{P})$ is the internal kinetic energy for particles in C. Each momentum vector splits into a sum

$$\underline{P} = \underline{P}_D + \sum_{C\varepsilon D} \underline{P}_C \tag{7}$$

where $\underline{P}_D$ and $\underline{P}_C$ are conjugate variables of $\underline{X}_D$ and $\underline{X}_C$.

To (6) corresponds a splitting of the kinetic energy operator as

$$\begin{aligned} H_0 = H_0^{C_1} \otimes I^{C_2} \otimes \ldots \otimes I^D + I^{C_1} \otimes H_0^{C_1} \otimes \ldots \otimes I^D + \ldots + \\ + I^{C_1} \otimes I^{C_2} \otimes \ldots \otimes H_0^D \end{aligned} \tag{8}$$

where H_0^D, H_0^C are the kinetic energy operators on $\mathcal{H}^D$ and $\mathcal{H}^C$ associated to K_D and K_C. When possible we will avoid this cumbersome notation and write simply $H_0 = H_0^D + \sum_{C\varepsilon D} H_0^C$, but for some later mathematical purposes it will be necessary to retain the structure (8). Also for further purposes notice that if $D' \subset D$ so that $D' = \{D(C_1), \ldots D(C_k)\}$ where D(C) is a cluster decomposition of C, one has

$$H_0^{D'} = H_0^D + \sum_{C\varepsilon D} H_0^{D(C)} \tag{9}$$

Let us now return to the Schrödinger equation; under some general conditions (see e.g. Amrein's Lectures or [1]) H is a self-adjoint operator so that the initial value problem has a well-defined solution $\Psi_t = \exp(-iHt)\,\Psi_0$. The object of time-dependent scattering theory is to study the behaviour for large times of these solutions. As shown by Amrein this problem is directly related to the study of the spectral properties of H. For one particle problem existence of a decomposition $\mathcal{H} = M_{ac}(H) \otimes M_d(H)$ implies that each state splits into an almost stationary one describing bound particles and into a decaying state in the sense that $\int_{X \leq L} |\Psi_t(\underline{X})|^2 \, d\underline{X}$ tends to zero for large times; this means that the particle should be asymptotically free. Various definitions of "asymptoticaly free" exist, depending on the strength of the potentials and can be found in Amrein's lectures. For N-particle system the same decaying property holds for states in $M_{ac}(H)$; one has to replace X by some total inter-particle distance e.g. the radius of gyration R about the total center of mass defined by $(\sum_{i=1}^{N} m_i) R^2 = Q(\underline{X})$. This quantity can tend to infinity even when some interparticle distance remains finite. This is exactly what happens in typical multichannel experiments: for large times the system is split into a set of clusters $D = \{C_1, C_2, \ldots C_k\}$ not interacting any more between them-selves. The generator of the evolution of this set α of non-inter-acting clusters is

$$H^D = H_0{}^D + \sum_{C \in D} H_0{}^C \text{ where } H^C = H_0{}^C + V^C \text{ and } V^C = \sum_{i,j \in C} V_{ij}$$

Assume in addition that particles into clusters form bound-states with wave-functions $\phi^{b(C)}$ such that

$$H^C \phi^{b(C)} = E^{b(C)} \phi^{b(C)}$$

Then the wave-functions of the system have the form

$$\Psi_\alpha = \Psi^D \otimes (\bigotimes_{C \in D} \phi^{b(C)}) \tag{10}$$

So that H finaly reduces on such a state of non-interacting bound-states to

$$H_0{}^\alpha = H_0{}^{D(\alpha)} + E_\alpha$$

where $D(\alpha) = \{C_1, C_2, \ldots C_k\}$ for $\alpha = (b(C_1), b(C_2), \ldots b(C_k))$ and $E_\alpha = \sum_{C \in D(\alpha)} E^{b(C)}$. Let P_α denote the projection operator on states like (10).

In analogy with the one and two particle case one expects that for short-range forces any state $\Psi \in M_{ac}(H)$ will have an asymptotic behaviour

$$\Psi_t \underset{t \to \infty}{\simeq} \sum_\alpha e^{-iH_0^\alpha t} \Psi_\alpha{}^\pm , \Psi_\alpha{}^\pm \in P_\alpha H \tag{11}$$

where the sum is over all possible channels, in particular the one where all particles are free. The decomposition (11) if it exists will be unique. In fact $\forall \alpha \neq \beta$ one has

$$\lim_{t\to\pm\infty} | (e^{-iH_0^\alpha t}\Psi_\alpha^\pm, e^{-iH_0^\beta t}\Psi_\alpha^\pm)| = \lim_{t\to\pm\infty} |(e^{-i(H_0^{D(\alpha)} - H_0^{D(\beta)})t} \Psi_\alpha^\pm, \Psi_\beta^\pm)| \qquad (12)$$
$$= 0$$

To show this it is sufficient to remark that in the case $D(\alpha) = D(\beta)$ the condition $\alpha \neq \beta$ implies $P_\alpha P_\beta = 0$. Otherwise for $D(\alpha) \neq D(\beta)$ the operator $H_0^{D(\alpha)} - H_0^{D(\beta)}$ is absolutely continuous so that convergence of the right member to zero is just a consequence of the Rieman-Lebesgue lemma. Relation (12) is known as "asymptotic orthogonality of channels." From (11) one is lead naturally to require as a multichannel asymptotic conditon the existence of

$$\Omega^\pm_\alpha = \text{s.lim}\, e^{iHt} e^{-iH_0^\alpha t} P_\alpha$$

For short-range forces the following general results have been proved:

Theorem (Hack [2]). Assume $V_{ij} \in L^2(R^3) + L^P(R^3)$, $2 \leq P < 3$, $\forall(i,j)$ The the wave-operators $\Omega^\pm_\alpha$ exist. They are partially isometric operators from $P_\alpha \mathcal{H}$ to $\mathcal{M}^\alpha_{ac}(H)$ satisfying

1) Intertwining property
$$\Omega^{\pm *}_\alpha H \Omega^\pm_\alpha = H_0^\alpha \cdot P_\alpha$$

2) Orthogonality
$$\Omega^{\pm *}_\alpha \Omega^\beta_\pm = \delta_{\alpha\beta} P_\alpha$$

Property (1) shows that the restriction of H to $\Omega^\pm_\alpha \mathcal{H}$ is unitarily equivalent to H_0^α so that $\Omega^\pm_\alpha \mathcal{H} \subset \mathcal{M}^\alpha_{ac}(H)$. It shows that the spectral type of H will be in general highly more complicated than in the one channel situation since one has

$$H = (\bigoplus_\alpha \Omega^\pm_\alpha H_0^\alpha \Omega^{\alpha *}_\pm) \oplus \bar{H}$$

where $\bar{H}$ is expected to be simply the discrete part of H. This direct sum decomposition comes from (2) which simply expresses asymptotic orthogonality of channels.

<u>b) Asymptotic completeness: some fundamental problems</u>. A general proof of the asymptotic completeness relation

$$\mathcal{M}_{ac}(H) = \bigoplus_\alpha \Omega^\pm_\alpha \Omega^{\pm *}_\alpha P_\alpha \mathcal{H} \qquad (13)$$

has been given by Faddeev [3] and Hepp [4] for few particles systems, $N \leq 4$. For general N particle systems it is possible to prove that it holds locally [3] [4] namely when in (13) one considers only states having an energy smaller than some threshold energy Λ. If Λ is taken to be the lowest three-body channel energy this is possible by a suitable utilisation of Kato's theorems on completeness of wave-operators ([1]). We will not give the details of these proofs but want to mention two important new facts, which together with the existence of rearrangement collisions make asymptotic completeness more difficult to reach in multichannel systems in comparison to the two-particle case.

(i) Asymptotic stability of channels. This expresses the fact that the part of Ψ_t in a given channel tends strongly to a limit for large times. Mathematicaly this can be defined as follows: A channel α is asymptoticaly stable if $\forall\ \Psi \in M_{ac}(H), \exists\ \Psi^{\pm}_{\alpha} \in P_{\alpha} H$ and $\bar{\Psi}^{\pm}_{\alpha}(t)$ such that

$$\lim_{t\to\pm\infty} ||\Psi_t - (e^{-iH_0^{\alpha}t}\Psi^{\pm}_{\alpha} + \bar{\Psi}^{\pm}_{\alpha}(t))|| = 0 \qquad (14\text{-a})$$

$$\forall\ \phi_{\alpha} \in P_{\alpha} H, \lim_{t\to\pm\infty} (e^{-iH_0^{\alpha}t}\phi_{\alpha}, \bar{\Psi}^{\pm}_{\alpha}(t)) = 0 \qquad (14\text{-b})$$

The last condition (14-b) is needed for a unique definition of $\Psi^{\pm}_{\alpha}$. Of course asymptotic completeness requires asymptotic stability of all channels. We want to stress the fact that an essential condition for stability of channels is a fast decrease property for bound-state wave-functions in configuration space. This can be understood intuitively from the fact that the strength of the interaction between the bound-states of channel α at large intercluster distances will depend strongly on the "tails" of the bound-states and should not be strong enough to produce long-range rearrangement processes under suitable decrease properties for their wave-functions.

Let us show sufficiency of such conditions in the case of two-body channels. Let us write

$$H = P_{\alpha} H P_{\alpha} + Q_{\alpha} H Q_{\alpha} + W_{\alpha}$$

where $Q_{\alpha} = I - P_{\alpha}$ and $W_{\alpha} = P_{\alpha} V_{D(\alpha)} Q_{\alpha} + Q_{\alpha}V_{D(\alpha)}P$ describes the coupling between channel α and the others.

If the wave-functions of bound-states in channel α are in the domain of any polynomial of order 2 in the relative position variables and if two-particle interactions are short-range, Kato's theorems apply with the result that

$$\Omega^{\pm *}_{\alpha,\,inel} = \text{s-}\lim_{t\to\pm\infty} e^{+i(H - W_{\alpha})t}\, e^{iHt}\, P_{ac}(H)$$

exist and are complete (here $P_{ac}(H)$ denotes the projection operator on $M_{ac}(H)$). So that $\forall\ \Psi \in M_{ac}(H)$,

$$\text{s-}\lim_{t\to\pm\infty} ||\Psi_t - [e^{-iP_{\alpha}HP_{\alpha}t} P_{\alpha}\, \Omega^{\pm *}_{\alpha,inel}\Psi + e^{-iQ_{\alpha}HQ_{\alpha}t} Q_{\alpha}\Omega^{\pm *}_{\alpha,inel}\Psi]|| = 0$$

Now $P_{\alpha}H P_{\alpha} = (H_0^{\alpha} + V_{\alpha})P_{\alpha}$ where V_{α} is the average over bound-state wave-functions of $V_{D(\alpha)}$ and is short-range if the two-particle interactions are. So that there exists $\Psi^{\pm}_{\alpha}$ such that

$$\lim_{t\to\pm\infty} || e^{-iP_{\alpha}H P_{\alpha}t} P_{\alpha} \Omega^{\pm *}_{\alpha,inel} \Psi - e^{-iH_0{}^{\alpha}t} \Psi^{\pm}_{\alpha} || = 0$$

Choosing

$$\bar{\Psi}^{\pm}_{\alpha}(t) = e^{-iQ_{\alpha}H Q_{\alpha}t} Q_{\alpha} \Omega^{\pm *}_{\alpha,inel} \Psi$$

one gets the proof of conditions (14-a) and (14-b). As far as I know there is no general time-dependent proof of asymptotic stability of channels.

(ii) Absence of continuous singular spectrum. If all channels are stable one gets $\forall\ \Psi \in M_{ac}(H)$:

$$\Psi_t \underset{t\to\pm\infty}{\simeq} \sum_{\alpha} e^{-iH_0{}^{\alpha}t} \Psi^{\pm}_{\alpha} + \bar{\Psi}^{\pm}(t)$$

where

$$\forall\alpha\ ,\ \forall\ \phi \in P^{\alpha} H,\ \lim_{t\to\pm\infty} (e^{-iH_0{}^{\alpha}t} \phi_{\alpha}, \bar{\Psi}^{\pm}(t)) = 0$$

Asymptotic completeness holds if $\text{s-lim}_{t\to\pm\infty} \bar{\Psi}^{\pm}(t) = 0$. A necessary condition for this is permanent break-up into asymptoticaly free subsystems plus the condition $M_{cs}(H^C) = \emptyset$, $\forall C$. To show this assume permanent break-up for a given state $\Psi \in M_{ac}(H)$ e.g. a behaviour

$$\Psi_t \xrightarrow[t\to\pm\infty]{} e^{-iH^D t} (\Psi^D \otimes (\bigotimes_{C\in D} \Psi^C))$$

where Ψ^D describes the asymptotic state of the centers of mass of $D = \{C_1, C_2, \dots C_k\}$ and $\Psi^C \in H^C$. According to Hack [1] the wave-operators $\Omega^{\pm}_D = \text{s-lim} \exp(iHt) \exp(-iH^D t)$ exist for interactions satisfying the conditions of Theorem 1 (notice that $M_{ac}(H^D) = H$). So that the set of vectors Ψ^C obtained from such asymptotic break-up span H^C. Then it is clear that in order for asymptotic completeness to hold i.e. that an asymptotic behaviour like (11) exists one needs each Ψ^C to be either a bound-state or a scattering state that is $H^C = M_d(H^C) \oplus M_{ac}(H^C)$ plus asymptotic completeness for the subsystem C. So $M_{cs}(H^C) = \emptyset$ $\forall C$, is necessary for asymptotic completeness. It is possible to show using explicit counter-examples [5] that $M_{cs}(H^C) \neq \emptyset$ for some C does not only break asymptotic completeness but also unitarity of the S matrix in the sense

$$\bigoplus_{\alpha} \Omega^{+}_{\alpha} \Omega^{+ *}_{\alpha} \neq \bigoplus_{\alpha} \Omega^{-}_{\alpha} \Omega^{- *}_{\alpha}$$

To conclude this review of the connection between the spectral properties of Schrödinger hamiltonians and scattering theory let us mention another requirement which appears in the time-dependent Faddeev-Hepp analysis [3], [4] and which is the absence of bound-states embedded in the continuum. The necessity of this condition in the above time dependent description of the scattering process is not clear, except for bound-states at thresholds in which case spatial decrease properties of bound-state wave-functions does not generaly hold and leads e.g. to Efimov effect. In fact it is not clear whether this assumption is of purely technical nature or a basic physical requirement.

2. Time Independent Theory.

a) Green's functions and expectation values. The traditional approach to time independent scattering theory is through the study of the Green's function $(p \,|(H - z)^{-1}q)$, p, $q \in R^{3(N-1)}$. For many particle systems this cannot be done along as general lines as the time dependent theory since the Lippman-Schwinger type equations connecting the kernels of the S-matrix elements $S_{\alpha\beta} = \Omega_+^{\alpha\,*}\,\Omega_-^{\beta}$ are not of the Fredholm type. Substitutes for these equations exist such as the celebrated Faddeev equations [3] or the Weinberg-Van Winter equations [6]. As for the two particle case the mathematical difficulties appear when Imz tends to zero since the kernel of these equations is no longer Hilbert-Schmidt or compact in $\mathcal{H}$ for physical values of the energy. To handle this situation one has to introduce auxiliary Banach spaces in which these equations have solutions in this limit as is done e.g. by Ikebe [7] in the two particle case or Faddeev [3] in the three particle case. Another technique used e.g. by Nuttall [8] and Lovelace [9] consists of investigating suitable analytic continuation of $G(\underline{P},\underline{Q},z) = (\underline{P},\ (H - z)^{-1}Q)$ outside $R^{3(N-1)} \times R^{3(N-1)} \times C_{\sigma(H)}$ (where $C_{\sigma(H)}$ is the complex plane minus the spectrum of H).

This is accomplished by allowing complex contours in multi-particle equations in order to avoid singularities of the integrands appearing in these equations. The approach presented here is more closely related to the second method. It is based on two observations.

First from a mathematical point of view one should be able to extract all the informations one needs for $\{G(\underline{P},\underline{Q},z),\ \underline{P},\ \underline{Q} \in R^{3(N-1)}\}$ from expectation values $\{(\phi,\ (H - z)^{-1}\Psi),\ \phi,\ \Psi \in \mathcal{H}\}$ since each set determines the other.

Second the contour deformation methods mentioned above suggest that useful analytic continuations of $G(\underline{P},\underline{Q},z)$ outside $R^{3(N-1)} \times R^{3(N-1)} \times C_{\sigma(H)}$ are obtained as follows:

Let G be a transformation group on $R^{3(N-1)}$ and define

$$G\gamma\ (\underline{P},\underline{Q};z) = G(\gamma(\underline{P}),\gamma(Q);z) \quad \gamma \in G$$

Then $G_\gamma\ (\underline{P},\underline{Q};z)$ is nothing else but the kernel of the resolvent $(H(\gamma) - z)^{-1}$ where $H(\gamma) = u(\gamma)Hu(\gamma)^{-1}$ and $u(\gamma)$ is a unitary representation of G on $\mathcal{H}$. By allowing now the transformations γ to map $R^{3(N-1)}$ into $C^{3(N-1)}$ and studying the spectral properties of the family $H(\gamma)$ (and accordingly the analyticity properties of $(H(\gamma) - z)^{-1}$ as an operator function of γ and z) one should be able to perform the analytic continuation program. In this way the powerful methods of analytic perturbation theory can be used as very useful substitutes for the integral equation methods. The idea that non self-adjoint operator theory could be useful for mathematical problems of scattering theory was in fact suggested by C. Dolph before 1960 in a series of papers [10]. We describe below the general group of transformations which we will use in applications.

b) The linear group. The linear group $\mathcal{L}_N$ is the set $\{(\underline{\tau}, \lambda); \underline{\tau} \in R^{3(N-1)}; \lambda \in R^+\}$ with the group law

$$(\underline{\tau}, \lambda) * (\underline{\tau}', \lambda') = (\lambda'^{-1}\, \underline{\tau} + \underline{\tau}'; \lambda\lambda') \qquad (*)$$

It as a unitary representation on $\mathcal{H} \simeq L^2(R^{3(N-1)}$ given by

$$(u(\underline{\tau}, \lambda)\phi)(\underline{X}) = \lambda^{3(N-1)/2} \exp(-i < \underline{X}, \lambda\underline{\tau} >)\phi(\lambda\underline{X}),\ \phi \in \mathcal{H}$$

This representation can be written in many ways as a tensor product of representations of subgroups. Namely let $D = \{C_1, C_2, .. C_k\}$ be a cluster decomposition; then

$$u(\underline{\tau}, \lambda) = u^D(\underline{\tau}, \lambda) \otimes (\bigotimes_{i=1}^{k} u^{C_i}(\underline{\tau}, \lambda))$$

where $u^D(\underline{\tau}, \lambda)$ and $u^C(\underline{\tau}, \lambda)$ are defined as in (15) with $\phi \in \mathcal{H}^D$ or $\phi \in \mathcal{H}^C$ respectively and where the position vectors $\underline{X}_D$ or $\underline{X}_C$ for centers of mass of clusters in D or particles in cluster C are substituted for $\underline{X}$; then only the orthogonal projection of $\underline{\tau}$ in the corresponding subspace of configuration space is involved. In fact the decomposition of $u(\underline{\tau}, \lambda)$ comes from the direct sum decomposition (4) of configuration space associated to D.

To study the action of $u(\underline{\tau}, \lambda)$ on the N particle kinetic energy operator H_0 we notice in momentum representation:

$$\widetilde{(u(\underline{\tau},\lambda)\phi)}\ (\underline{P}) = \lambda^{-\frac{3(N-1)}{2}}\ \phi(\lambda^{-1}\underline{P} + \bar{M}\underline{\tau}),\ \phi \in \mathcal{H},$$

where $\underline{P}$ is the 3(N-1) dimensional momentum vector and $\bar{M}$ is the mass operator defined by formula (2). In this representation H_0 is multiplication by the quadratic form $K(\underline{P})$ so that simple computation

(*) The two subgroups corresponding respectively to $\underline{\tau} = 0$ or $\lambda = 1$ are the usual dilation and boost group.

gives:

$$u(\underline{\tau},\lambda)H_0\, u^{-1}(\underline{\tau},\lambda) = \lambda^{-2}\; H_0(\lambda\underline{\tau}) \tag{15}$$

where $H_0(\lambda\underline{\tau})$ is the multiplication operator by $K(\underline{P} + \bar{M}\lambda\underline{\tau})$.

If we now take $(\underline{\tau},\lambda) \in C^{3(N-1)} \times (C - \{0\})$, the right member (15) defines an operator analytic function

$$H_0(\underline{\tau},\lambda) = \lambda^{-2}\, H_0(\lambda\underline{\tau})$$

Since the perturbation $K(\underline{P} + \bar{M}\,\lambda\underline{\tau})$ is $(H_0 - \varepsilon)$- bounded, elementary smallness arguments imply that $H_0(\underline{\tau},\lambda)$ is closed on $\mathcal{D}(H_0)$; so $H_0(\underline{\tau},\lambda)$ is an analytic family of type (A)[1]. The operator $H_0(\lambda\underline{\tau})$ is unitarily equivalent to $H_0(i\mathrm{Im}(\lambda\underline{\tau}))$ through $u(\mathrm{Re}(\lambda\underline{\tau}),\, 1)$. So the spectrum of $H_0(\underline{\tau},\lambda)$ is the image $P(\underline{\tau},\lambda)$ under a dilation scale λ^{-2} of the paraboloid

$$\{\zeta \in C;\ \mathrm{Re}\,\zeta \geq -\frac{Q(\sigma)}{2} \quad \text{and} \quad |\mathrm{Im}\,\zeta|^2 \leq 2Q(\underline{\sigma})\,\mathrm{Re}(\zeta) + Q^2(\sigma)\}$$

where $\sigma = \mathrm{Im}(\lambda\underline{\tau})$ and $Q(\underline{\sigma})$ is the quadratic form on configuration space defined in the appendix. To prove this statement we notice that

$$K(\underline{P} + i\bar{M}\underline{\sigma}) = K(\underline{P}) - Q(\underline{\sigma}) + 2i\,\bar{M}^{-\frac{1}{2}}\,\underline{P}\cdot\bar{M}^{\frac{1}{2}}\underline{\sigma}$$

For fixed value of $\mathrm{Re}\,\zeta = K(\underline{P}) - Q(\underline{\sigma})$, the maximum value of $\mathrm{Im}\zeta = 2\bar{M}^{-\frac{1}{2}}\,P\cdot\bar{M}^{\frac{1}{2}}\,\sigma$ when P varies on the corresponding energy sphere is $K(\underline{P})^{\frac{1}{2}}\cdot Q(\underline{\sigma})^{\frac{1}{2}}$.

One can define in the same way analytic families $H^C_0(\underline{\tau},\lambda)$ and $H^D_0(\underline{\tau},\lambda)$ for any subsystem C or cluster decomposition D, substituting in (15) H^C_0 and H^D_0 respectively for H_0, where H^C_0 and H^D_0 are respectively the internal kinetic energy operators for particles in cluster C or centers of mass of clusters in D.

Using the decomposition [3] one can see that their spectra are as above with $Q_C(\underline{\sigma})$ or $Q_D(\underline{\sigma})$ instead of $Q(\underline{\sigma})$.

We now want to fulfill our program of studying the action of $u(\underline{\tau},\lambda)$ on the resolvent $(H - z)^{-1}$. For this we make the following assumptions on the potentials V_{ij}.

Definition 1: Let G be a Lie subgroup of the linear group. Denote by $\hat{G}$ the complexified of G. Let $u(\gamma)$, $\gamma \in G$, be the representation of G as defined above. A two-body interaction V_{ij} is *G-analytic* in $V \subset \hat{G}$ if the family

$$V_{ij}(\gamma) = u(\gamma)\, V_{ij}\, u(\gamma)^{-1},\ \gamma \in G$$

has an analytic extension from $V \cap G$ to V as a Δ-compact operator, (i.e. as a compact operator from $\mathcal{D}(\Delta)$ to $L^2(R^3)$).

This assumption can in fact be weakened without altering results of this paper, provided quadratic forms are used in place of unbounded operators. The "weak G-analyticity" defined as follows would be sufficient: $V_{ij}(\gamma)$, $\gamma \in V$, is an analytic family of compact operators from the form domain for Δ to its dual space (for the topology induced by $L^2(R^3)$).

Simon has shown [11] that G-analyticity implies weak G - analyticity. We refer to his paper for the extension of our methods to such perturbations.

One can give a very simple characterization of local dilation analytic potentials as follows [21]: A sphericaly symetric potential $V(r)$, $r \in R^+$, is dilation analytic in $\mathcal{V} = C_a = \{z \mid z \neq 0$ and $|\text{Arg } z| < a\}$ if it satisfies the following conditions

1) $V(r)$ is analytic in C_a
2) $V(r)$ tends to zero uniformly along rays $e^{i\phi}R^+, |\phi| < c < a$
3) $\int_{0 < r < 1} |V(e^{i\phi}r)|^2 \, dr < \infty \qquad \forall\phi, \ |\phi| < a$

For weak G-analyticity (3) can be replaced by

3´) $\int_{0 < r < 1} |V(e^{i\phi}r)|^{3/2} \, dr < \infty \qquad \forall\phi, \ |\phi| < a$

Homogeneous potentials $gr^{-\beta}$, $0 < \beta < \frac{3}{2}$ (or 2 in the weak case), Yukawa potentials, are dilation analytic. On the other hand all multiplicative relatively compact interactions obviously are boost-analytic since they are invariant under boost-transformations; a similar property holds e.g. for electromagnetic forces.

$$-\nabla^2 \to -(\nabla - eA)^2$$

or spin-orbit forces under suitable conditions [12]; this comes from the very simple action of the boost group on momentum operators. So that all multiplicative interactions and those depending linearly in the momentum which are dilation analytic also are analytic with respect to the linear group.

We can now define a family $H(\gamma)$, $\gamma \in \mathcal{V}$, by:

$$H(\gamma) = H_0(\gamma) + \sum_{i \neq j} V_{ij}(\gamma)$$

The fundamental property is that if $\gamma \in G$, γ_0 and $\gamma_0 * \gamma^{-1}$ are in $\mathcal{V}$ then

$$H(\gamma * \gamma_0) = u(\gamma) \, H(\gamma_0) \, u(\gamma)^{-1} \tag{16}$$

In fact our assumptions imply that $H(\gamma)$ is obtained from $H_0(\gamma)$ through the addition of an $(H_0 - \varepsilon)$- bounded perturbation; so that for all $\gamma \in \mathcal{V}$ the operator $H(\gamma)$ is closed on $\mathcal{D}(H_0)$ and analytic on this domain (type [A]); identity (16) simply expresses unicity of analytic continuation in $\mathcal{V}$.

We now want to extract all possible information about the family $H(\gamma)$ from analytic perturbation theory and its application to our particular situation through the use of the W-V W equations. Let us describe rapidly these essential tools.

c) Analytic perturbation theory. The facts stated here are proved in Kato's book [1] so that we just quote it without proof.

i) Let $H(\gamma)$, $\gamma \in V$ be an analytic family of closed operators. If $\gamma_0 \in V$ and $z \in \rho(H(\gamma_0))$ (the resolvent set of $H(\gamma_0)$) it also belongs to $\rho(H(\gamma))$ for γ in a neighborhood of γ_0 and $(H(\gamma) - z)^{-1}$ is analytic in this neighborhood.

ii) Let $\pi(\gamma_0)$ be a bounded isolated part of $\sigma(H(\gamma_0))$ (the spectrum of $H(\gamma_0)$). Let Γ be some closed contour such that $\pi(\gamma_0)$ is entirely contained in the interior of Γ. Then for γ in a neighborhood of γ_0, $\sigma(H(\gamma))$ also consists of two parts separated by Γ. Furthermore the projection operators

$$P(\pi,\gamma) = - \frac{1}{2\pi} \int_\Gamma \ (H(\gamma) - z)^{-1}\, dz$$

are analytic in this neighborhood of γ_0.

iii) Assume that the part $\pi(\gamma)$ of $\sigma(H(\gamma))$ contained in the interior of Γ consists of isolated eignevalues of finite multiplicity. Then these eigenvalues are branches of one or several analytic functions which have at most algebraic singularities (they are in fact solutions of algebraic equations with analytic coefficients).

iv) Let $A(\gamma)$ be an analytic family of compact operators in V. Then either $[1 + A(\gamma)]^{-1}$ exists nowhere in V or it is a meromorphic family in V.

d) The Weinberg-Van Winter equations. The relative compactness assumption on the potentials V_{ij} imply the so called Kato inequality

$$||V\phi|| \leq \varepsilon\, ||H_0\phi|| + b(\varepsilon)\, ||\phi||,\ \forall\phi \in \mathcal{D}(H_0) \tag{17}$$

Here ε and $b(\varepsilon)$ are positive constants and ε can be choosen arbitrarily small [14]. This inequality implies in turn that $||V\, R_0(z)|| = 0\ (\text{dist}\ (z, \sigma(H_0))^{-1})$ as $z \to \infty$, $\text{Im}z \neq 0$, where $R_0(z) = (H_0 - z)^{-1}$. From this follows the convergence of the Neuman series

$$R(z) = R_0(z) \sum_{n=0}^{\infty} [-V\, R_0(z)]^n,\ \text{Re}z \ll 0,$$

where $R(z) = (H - z)^{-1}$. Using $V = \sum_{i\, j=1}^{N} V_{ij}$ we see that for large negative Rez the resolvent G is a norm convergent series of terms like

$$g = R_0\, V_{ij}\, R_0\, V_{kl} \cdots V_{rp}\, R_0 \tag{18}$$

The W-V W equation is obtained by giving prescriptions for the summation of the terms. These summation rules are based on the observation that each term like (18) defines a cluster decomposition of [1, 2, ... N] through the links (ij)(kl)...(rp): two particles are linked if the corresponding pair appears in (18) or if there exists a sequence of particles linked in this way connecting them.

This defines a cluster decomposition D(g) and gives a first classification of terms having same D's. A further classification consists of suppressing (pair interactions) x G_0 from the right to the left. Then one defines a sequence of cluster decompositions called the connectivity of g:

$$S_k(g) = (D_N, D_{N-1}, \ldots, D_K) = D(g))$$

where $D_l \subset D_{l+1}$ and $D_N = \{1, 2, \ldots N\}$

Then terms (18) are classified according to their connectivity. One can then prove that terms having a connectivity S_k sum to

$$R_{S_k} = R_{D_N} V_{D_N D_{N-1}} R_{D_{N-1}} \ldots V_{D_{k+1} D_k} R_{D_k}$$

where

$$R_D(z) = (H^D - z)^{-1} \text{ and } V_{D_{l+1}D_l} = V_{D_l} - V_{D_{l+1}},\ V_D = \sum_{C\varepsilon D} V^C.$$

Summing over sequences S gives

$$G = D + C;\ D = \sum_{S_k,\ k\geq 2} R_{S_k};\ C = \sum_{S_1} R_{S_1}$$

Since $R_{D_1} = R$ one gets finally, writing C = IG, the Weinberg-Van Winter equation

$$G(z) = D(z) + I(z)\, G(z) \tag{19}$$

Let us make some remarks:

1°) From the definitions of H^D we see that the knowledge of $(H^C-z)^{-1}$ for each $C \neq \{1,2,\ldots N\}$ allows us to derive $(H-z)^{-1}$ ($(H^D-a)^{-1}$ is obtained from $(H^C - z)^{-1}$ by convolution formulas).

2°) Weinberg equation can be derived in the same way for $G(\gamma, z) = (H(\gamma) - z)^{-1}$ $\gamma \in \mathcal{L}_N$. It reads then

$$R(\gamma, z) = D(\gamma, z) + I(\gamma, z)\, R(\gamma, z)$$

where $D(\gamma, z)$ and $I(\gamma, z)$ are obtained by substituting $H^C(\gamma)$ and $H^D(\gamma)$ for H^C and H^D in the definition of D and I.

3°) Under our assumptions on two-particle interactions and for $\gamma \in N$ one can show [13] [14] that $I(\gamma, z)$ is a compact operator for $\text{Im} z \neq 0$.

4°) One has

$$I = \sum_{S_1} R_{D_N} V_{D_N D_{N-1}} \ldots R_{D_2} V_{D_2 D_1}$$

from which one can deduce that I(z) is analytic in $\bigcup_D \rho(H^D)$. Furthermore from (17) one can deduce that $||I(z)|| \to 0$ for Re $z \to -\omega$. So that according to d) $(1 - I(z))^{-1}$ is meromorphic in $\bigcup_D \sigma(H^D)$.

Since D(z) also is analytic in this domain one obtains that $(H - z)^{-1}$ is meromorphic there. Inductively one can then show that the essential spectrum of H is $[\Lambda_0, \infty[$ where $\Lambda_0 = \inf_\alpha E_\alpha$

5°) Assume E is an eigenvalue of $H(\gamma)$ in $\bigcup_D (H^D(\gamma))$. Then the projection operator $P(E,\gamma)$ for $H(\gamma)$ on the eigenspace associated to E satisfies according to the Weinberg equation

$$P(E,\gamma) = I(E,\gamma)\ P(E,\gamma)$$

From this and compactness of $I(E,\gamma)$ one can deduce that $P(E,\gamma)$ is compact hence finite dimensional under the above conditions on E.

e) Analytic and spectral properties of Hamiltonian and Green's functions.

Definitions 2:

$\sigma(\gamma)$: spectrum of $H(\gamma)$

$\sigma_e(\gamma)$: essential spectrum of $H(\gamma)$ (complement of the set of poles for $(H(\gamma) - z)^{-1}$)

$\sigma_d(\gamma)$: discrete spectrum of $H(\gamma)$ (isolated eigenvalues with finite multiplicity)

For symmetric interactions one can show that $\sigma(\gamma) = \sigma_e(\gamma) \cup \sigma_d(\gamma)$.

For any subsystem C we denote by $H^C(\gamma)$ the family defined as $H(\gamma)$ for the system C. For any cluster decomposition D the operator

$$H^D(\gamma) = H^D{}_0(\gamma) + \sum_{C\varepsilon D} H^C(\gamma)$$

is closed on $\mathcal{D}(H_0)$. Let $\sigma^C(\gamma)$, $\sigma^D(\gamma)$ denote the spectra of $H^C(\gamma)$, $H^D(\gamma)$ respectively etc.... Then we have the following basic properties:

Theorem I. Assume V_{ij} satisfy the conditions of Def. 1$\forall i, j$ and are symmetric. Then:

1) If $E \epsilon \sigma_d(\gamma_0), \gamma_0 \ \varepsilon\ \mathcal{V}$ then E also is an eigenvalue of $H(\gamma)$ for any γ belonging to the maximal open connected subset $\mathcal{D}(\gamma_0, E)$ containing γ_0 and contained in $\{\gamma \ \varepsilon\ \mathcal{V};\ E \notin \sigma_e(H(\gamma))\}$

2) Let $P(E,\gamma)$ denote the finite dimensional projection operator for $H(\gamma)$ associated to the eigenvalue $E \ \varepsilon\ \sigma_d(\gamma)$. Then $P(E,\gamma)$ is analytic in $\mathcal{D}(\gamma_0,E)$

3.a) $\sigma_e(\gamma) = \bigcup_D (\sigma_0{}^D(\gamma) + \sum_{C\varepsilon D} \sigma_d{}^C(\gamma))$

where the union is over all many-cluster decompositions and $\sigma_0{}^D$ is the spectrum of $H_0{}^D(\gamma)$

3.b) $\sigma(\gamma)$ is contained in a sector with arbitrarily small opening angle and direction $\beta(\gamma)$ (*)

Details of the proof can be found in [15], [16] in the case where G is the dilation group. We just give an outline here.

Sketch of the proofs. 1) and 2) are consequences of unitary implementability through $u(\gamma)$ of the mapping $H(\gamma_0) \to H(\gamma + \gamma_0)$, if γ is real, and analyticity of $H(\gamma)$. These two facts and basic results from analytic perturbation theory imply local invariance of discrete spectrum and from this our statements.

The proof of 3.a) and 3.b) is inductive and uses the Weinberg equation.

$$(H(\gamma) - z)^{-1} = D(\gamma,z) + I(\gamma,z)(H(\gamma) - z)^{-1} \tag{20}$$

For N = 2 the statements are true since according to the relative compactness of the interaction, $\sigma_e(\gamma)$ coincides with the spectrum of $H_0(\gamma)$ i.e. $P(\gamma)$ which satisfies the sectoriality assumption.

We assume now that they hold for any subsystem. Both $I(\gamma,z)$ and $D(\gamma,z)$ are finite sums of products like $(H^D(\gamma) - z)^{-1} V_{ij}(\gamma)$ where $D = \{C_1, C_2, \dots C_k\}$, $k \geq 2$, is a cluster decomposition of the total system. Such factors are bounded and z-analytic in the complementary set of $\sigma^D(\gamma)$. So $I(\gamma,z)$ and $D(\gamma,z)$ are analytic in the complement of $\bigcup_D \sigma^D(\gamma)$ where the union is over all many-cluster decompositions. Now any $H^D(\gamma)$ has the following tensorial structure:

$$H^D(\gamma) = H^{C_1}(\gamma) \otimes I_{\mathcal{H}^{C_2}} \cdots \otimes I_{\mathcal{H}^{C_k}} \otimes I_{\mathcal{H}^D} + I_{\mathcal{H}^{C_1}} \otimes H^{C_2}(\gamma) \otimes \cdots$$

$$\cdots \otimes I_{\mathcal{H}^D} + \cdots + I_{\mathcal{H}^{C_1}} \otimes I_{\mathcal{H}^{C_2}} \cdots \otimes H^D_0(\gamma)$$

All factors $H^C(\gamma)$ satisfy by assumption sectoriality condition 3.b). The same property holds for $H^D(\gamma)$ whose spectrum is a paraboloid. Then theorems on tensor products of closed operators (17) (18) imply

$$\sigma^D(\gamma) = \sigma_0{}^D(\gamma) + \sum_{C \in D} \sigma^C(\gamma)$$

---

* A sector is a set $\{z \in C;\ |\arg(e^{-i\beta}(z - z_0))| < \frac{\phi}{2}\}$; ϕ is the opening angle and β the direction of the sector. The paraboloid $P(\underline{\tau},\lambda)$ is contained in sectors with arbitrarily small opening angles and direction

$$\beta(\underline{\tau},\lambda) = e^{-2i \arg \lambda}$$

From the induction assumption for $\sigma^{C}(\gamma)$ and sectoriality of $H_0^{D}(\gamma)$ for each cluster decomposition D, the obvious extension of (9) for $\gamma \neq 0$ gives finally:

$$\sigma^{D}(\gamma) = \bigcup_{D' \supset D} (\sigma_0^{D'}(\gamma) + \sum_{C \in D'} \sigma_d^{C}(\gamma))$$

It remains to use compactness of $I(\gamma,z)$; the relative compactness of all $V_{ij}(\gamma)$ imply compactness of $I(\gamma,z)$ far from $\bigcup_D \sigma^D(\gamma)$ and therefore by analytic continuation everywhere in the complement of this open connected set. Then analytic Fredholm theory allows us to conclude that $(H(\gamma) - z)^{-1}$ is meromorphic according to (19) in the complement of $\bigcup_D \sigma^D(\gamma)$; the fact that this meromorphic domain is maximal comes from the cluster properties of the system and we refer to (16) for the proof of this fact.

Concerning the sectoriality of $H(\gamma)$ we notice that $\sigma_d(\gamma)$ consists of poles of $[I - I(\gamma,z)]^{-1}$; since $||I(\gamma,z)||$ goes to zero as an inverse power of dist. $(z, \bigcup_D \sigma^D(\gamma))$ it is sufficient to prove that $\sigma_e(\gamma)$ is contained in a sector with arbitrary small aperture and direction $\beta(\gamma)$. But this holds from the induction assumption for any $\sigma^D(\gamma)$ and then from simple geometrical arguments for the finite union of these sets.

Corollary 1:

1) $\sigma_d^{C}(\underline{0},\lambda)$ and $\Sigma(\lambda) = \bigcup_D (\sum_{C \in D} \sigma_d^{C}(\underline{0},\lambda))$ are contained in $C^- \cup R$ for $0 < \mathrm{Arg}\lambda < a$

2) $\sigma_d^{C}(\underline{0},\lambda) \cap R$ and $\Sigma^r = \Sigma(\lambda) \cap R$ are independent of λ

3) The real accumulation points of $\Sigma(\lambda)$ on R belong to Σ^r

Proof:

1) It is sufficient to prove that $\sigma_d^{C}(\underline{0},\lambda_0)$ is contained in $C^- \cup R$ for $\mathrm{Arg}\lambda_0 > 0$. Assume the converse and let $E \in \sigma_d^{C}(\underline{0},\lambda_0)$, $\mathrm{Im}E > 0$; then E also is an eigenvalue of $H^C(\underline{0},\lambda)$ as long as $-a(E) < \mathrm{Arg}\lambda < a$ according to theorem (I; 1 and 3 a) where $a(E)$ is positive. In particular E is an eigenvalue of H. This is impossible since H is self-adjoint. This shows 1) by induction.

2) It is sufficient to show the statement for $\sigma_d^{C}(\underline{0},\lambda) \cap R$; this again is an elementary consequence of Th(I; 1 and 3a) and $H^{*}(\lambda)=H$

3) Since $\sigma(\underline{0},\lambda)$ is closed and $\Sigma(\lambda) \subset \sigma(\underline{0},\lambda)$ the only possible real accumulation points for $\Sigma(\lambda)$ are at $\sigma(\underline{0},\bar{\lambda}) \cap R = \Sigma^r$ by Th(I;3.a)

Remarks:

a) We will see below (proof of theorem 3) that Σ^r is nothing else but the set of threshold energies E_α.

b) This corollary is valid to some extent for any $H(\underline{\tau},\lambda)$, $\mathrm{Arg}\lambda < 0$. One can prove 1) with some slight modification of the argument.

Concerning 2) one can only show that $\sigma_d^C(\underline{\tau},\lambda) \cap R$ and $\Sigma(\underline{\tau},\lambda) \cap R$ are contained in Σ^r. As to 3) it remains true.

By suitable choices of groups G we will now try to solve the problems mentioned in Sect. I.

In the following we will denote by $\mathcal{D}$ the dense set of analytic vectors with respect to L_N (in the sense of Nelson) $\mathcal{D}_N = \{\phi \varepsilon \mathcal{H} | \phi(\gamma) = u(\gamma)\phi$ is an analytic function$\}$

Lemma: Let the interactions V_{ij} be dilation analytic in a sector C_a. Then for all $\Psi \varepsilon \mathcal{D}$ the function $(\Psi, (H - z)^{-1}\Psi)$ has a many sheeted continuation from C^+ through $\sigma(H)$ with branch points at

$$\Sigma = \bigcup_{0 \leq \mathrm{Arg}\lambda < a} \bigcup_{D} \Big(\sum_{C \varepsilon D} \sigma_d^C(\underline{0},\lambda) \Big)$$

and poles at

$$\bigcup_{0 \leq \mathrm{Arg}\,\lambda < a} \sigma_d(\underline{0},\lambda)$$

Proof. To simplify the notations we write λ for $(\underline{0},\lambda)$. Let $\Psi \varepsilon \mathcal{D}$ and

$$F(\lambda,z) = (\Psi(\bar{\lambda})), (H(\lambda) - z)^{-1} \Psi(\lambda)), \ \lambda \varepsilon R^+, \ \mathrm{Im}z > 0 \qquad (21)$$

By theorem (1; 3a) $\sigma_e(\lambda) = \bigcup_{D} (\lambda^{-2} R^+ + \sum_{C \varepsilon D} \sigma_d^C(\lambda))$

Accordingly the function $F(\lambda,z)$ is analytic in $\lambda_0(z) < \mathrm{Arg}\lambda < a$ for some $\lambda_0(z) < 0$. For λ real it is obviously constant and equal to $(\Psi,(H - z)^{-1}\Psi)$; so it is constant in this domain. Fixing now λ such that $0 < \mathrm{Arg}\lambda < a$ we get a meromorphic continuation in z from C^+ to $\complement\sigma(\lambda)$. From the constancy of $F(\lambda,z)$ in λ it is clear that these various continuations coincide in overlapping regions (shaded in Fig. 1). In fact according to corollary (1.1) $\sum_{C \varepsilon D} \sigma_d^C(\lambda) \cap R$ is independent of λ for $0 < \mathrm{Arg}\lambda < a$ and consists of the set of real thresholds. According to theorem (I.1) it is clear that all points in Σ (and not only the threshold energies in Σ^r) will be branch points of the analytically continued $(\phi,(H - z)^{-1}\phi)$. In addition there will be poles at points $E \varepsilon \sigma_d(\lambda)$ for some λ; these points are however independent of λ in the following sense (see Fig. 1): if E is an isolated eigenvalue of $H(\lambda_0)$ it will remain an eigenvalue of $H(\lambda)$ in the set $D(\lambda_0,E)$ of Th. I, that is, as long as it is not absorbed by some "scattering cut" $E_\alpha + \lambda^{-2} R^+$, $E_\alpha \in \Sigma$

Theorem 2:

1) $\mathcal{H} = \mathcal{M}_{ac}(H) \oplus \mathcal{M}_d(H)$

2) $\sigma(H|_{\mathcal{M}_{ac}(H)}) = [\,\Lambda_0, \infty[$ where $\Lambda_0 = \inf_\alpha E_\alpha$

Proof: It is sufficient to show that for a dense set of vectors in $\mathcal{H}$ the support of the discrete part of $(\phi, P(E)\phi)$, where $\{P(E), E \in R\}$ is the spectral family of H, is countable.

In fact this support is contained in the set of points where the Radon-Nikodym derivative of (,P(E)) does not exist. According to the relation

$$\frac{d}{dE}(\phi, P(E)\phi) = \lim_{\varepsilon \downarrow 0} \frac{1}{2i\pi}(\phi, (H - E - i\varepsilon)^{-1} - (H - E + i\varepsilon)^{-1}\phi)$$

such a set can consist only of the real poles of $(\phi, (H-z)^{-1}\phi)$ and of Σ^r. For $\phi \in \mathcal{D}$ the above lemma implies that this set consist only of the real poles of $F(\lambda,z)$ for some λ with $\mathrm{Arg}\lambda > 0$ plus possibly a subset of Σ^r. Since this last set is obviously countable 1) follows. 2)simply comes from the above mentioned $\sigma_e(H) = [\Lambda_0, \infty[$ and 1).

f. Bound states and resonances.

Theorem 3. Let the interaction V_{ij} be L_N-analytic in a domain $C^{3(N-1)} \times C_a$ with $a \leq \frac{\pi}{2}$. Then

a) The point spectrum of H outside the set Σ^r consists of finite multiplicity eigenvalues which can accumulate at most at Σ^r.

b) Bound-state wave-functions corresponding to non-threshold eigenvalues are L_N-analytic in a domain

$$\mathcal{V}(E) = \{(\underline{\tau},z) \mid \underline{\tau} \in T(E),\ \lambda \in C_a\}$$

where $T(E) = \{\underline{\tau};\ Q(\mathrm{Im}(\underline{\tau} e^{\frac{i\pi}{4}})) < \inf_\alpha (|E_\alpha - E| + \Gamma_\alpha)$

Here $\Gamma_\alpha = |\mathrm{Im}\, E_\alpha|$ and the infimum is taken over $E_\alpha \in \Sigma$

Proof. We use $P\{E\} = \mathrm{s\text{-}lim}_{z \to E}(E - z)(H - z)^{-1}$ where $P\{E\}$ is zero if E is not in the point spectrum of H and is the projection operator on the eigenspace associated to E otherwise. Assume $\phi, \Psi \in \mathcal{D}$; then using the same arguments as in the lemma one gets since $E \notin \Sigma^r$:

$$(\phi, P\{E\}\Psi) = \lim_{\substack{z \to E \\ z \in C^{\pm}}} (E - z)(\phi(\bar{\lambda}), (H(\lambda) - z)^{-1}\Psi(\lambda)),\ \pm\,\mathrm{Arg}\lambda \geq 0 \tag{22}$$

This limit is non zero $\forall\ \phi,\ \Psi \in \mathcal{D}$ if and only if $(H(\lambda) - z)^{-1}$ has a pole at $z = E$ for all $\lambda \in C_a$. Using $H^*(\lambda) = H(\bar{\lambda})$ (time reversal invariance) one gets that this holds if and only if E is in the point spectrum of $H(\lambda)$. For $\text{Arg}\lambda \neq 0$, E is then in $\sigma_d(\lambda)$ so that according to remark 5) following W-V W equations the residue $P\{E,\lambda\}$ of $(H(\lambda) - z)^{-1}$ is a finite rank projection operator (we assume now $E \notin \Sigma^r$); furthermore according to theorem (1 ; 2) $P\{E,\lambda\}$ is analytic in $0 < |\text{Arg}\lambda| < a$. To show that it is in fact analytic in $|\text{Arg}\lambda| < a$ and coincides for λ real with $u(\lambda)\ P\{E\}\ u^{-1}(\lambda)$ we rewrite (22) as

$$(\phi,\ P\{E,\lambda\}\Psi) = (\phi(-\bar{\lambda}),\ P\{E\}\Psi(\lambda)) \quad , \quad \phi\ \ \Psi \in \mathcal{D}$$

This function is analytic in $|\text{Arg}\,\lambda| < a$ and bounded. So that by Phragmen-Lindelöf theorem

$$(\phi,\ P\{E,\lambda\}\ \Psi) \leq \|\phi\|\ \|\Psi\| \sup_{|\text{Arg}\lambda| = \alpha < a} \|P\{E,\lambda\}\| \tag{23}$$

Now $\|P\{E,\lambda\}\|$ is constant along the rays $|\text{Arg}\lambda| = \alpha$ since these projection operators are unitarily equivalent to each other. Then (23) implies that $P\{E,\lambda\}$ has analytic expectation values for vectors in a dense set; since they are uniformly bounded as quadratic forms on $\mathcal{H}$, $(\phi,\ P\{E,\lambda\}\ \Psi)$ extends to an analytic quadratic form on $\mathcal{H} \times \mathcal{H}$ which provides us with the suitable analytic continuation of $P\{E,\lambda\}$. Now the dimension of $P\{E,\lambda\}$ is finite, hence constant; so we get

$$\dim P\{E\} = \dim(\text{norm}\lim_{\lambda \to 1} P\{E,\lambda\}) < \infty$$

This proves a). To show b) we first notice that the above proof implies dilation analyticity of any bound-state wave function ϕ_E associated to E in $|\text{Arg}\lambda| < a$. For simplicity we assume $a > \pi/4$. Let us consider $\phi_E(\underline{0},\rho\exp(\pm i\pi/4))$ (we now return to the usual notations); It is an eigenstate of $H(\underline{0},\rho\exp(\pm i\pi/4))$ corresponding to the isolated eigenvalue E. According to Th.(I; 2 and 3.a) $\phi_E(\underline{\tau},\rho\exp(i\pi/4)) = u(0,\rho)\phi_E(\underline{\tau},\rho\exp(i\pi/4))$ is analytic in $\underline{\tau}$ as long as

$$E \notin \bigcup_D \left(\sigma_0{}^D(\underline{\tau}\exp(i\pi/4)) + \sum_{C\in D} \sigma_d{}^C(\underline{\tau},\ \exp(i\pi/4))\right)$$

This condition is equivalent (using cartesian equations for the paraboloids) to

$$-\ 2\ Q_D(\underline{\sigma})\ \Gamma_\alpha + Q^2{}_D(\underline{\sigma}) - \text{Re}(E - E_\alpha)^2 < 0 \tag{24}$$

$\forall D$ and $\forall\ E_\alpha \in \sum_{C\in D} \sigma_d{}^C(\tau\ \exp(i\pi/4))$ (here $\underline{\sigma} = \text{Im}(\exp(i\pi/4\underline{\tau}))$ and $\Gamma_\alpha = \text{Im}E_\alpha$; notice that $\Gamma_\alpha \geq 0$). (24) is satisfied if:

$$Q_D(\underline{\sigma}) < |E_\alpha - E| + \Gamma_\alpha \tag{25}$$

A similar result holds for $\phi(\underline{\tau},\exp(i\pi/4))$. To conclude the proof we use again a Phragmen-Lindelof argument. Let $\phi \in \mathcal{D}$ and define:

$$F_\phi(\underline{\tau},\lambda) = (\phi, \phi_E(\underline{\tau},\lambda)), \ (\underline{\tau},\lambda) \in R^{3(N-1)} \times R^+$$

$$= (\phi(-\underline{\tau}\lambda, 1), \phi_E(\underline{0},\lambda))$$

by the usual unitarity arguments. So that $F_\phi(\tau,\lambda)$ is analytic in $C^{3(N-1)} \times C_a$; we can then apply Phragmen-Lindelöf to $F_\phi(\underline{\tau},\lambda)$ as a function of λ for fixed $\underline{\tau}$;

$$|F_\phi(\underline{\tau},\lambda)| \leq \sup_{|\mathrm{Arg}\lambda| = \pi/4} |F_\phi(\underline{\tau},\lambda)|$$

Now $F_\phi(\underline{\tau},\rho\exp(i\pi/4)) = (\phi, \Psi(\underline{\tau},\rho\exp(i\pi/4))$

Accordingly for any $\underline{\tau}$ satisfying (24) one obtains

$$|F_\phi(\underline{\tau},\rho\exp(\pm i\pi/4))| \leq ||\phi|| \ ||\Psi(\underline{\tau}, \exp(\pm i\pi/4))||$$

Since $\mathcal{D}$ is dense and $Q_D(\underline{\sigma}) \leq Q(\underline{\sigma}) \ \forall D$ this implies that the form on H given by $(\phi, \phi_E(\underline{\tau},\lambda))$ is analytic and bounded in the domain $T(E)$. This concludes the proof.

Corollary 1 [19]. Under the assumption of Theorem 3, $H\phi_E = E\phi_E$, $E \notin \Sigma$, implies that for any θ, $0 \leq \theta < 1$, ϕ_E is in the domain of $e^{\theta\sqrt{2M(|E - E_\alpha| + \Gamma_\alpha)}\,R}$ where E_α is the closest bound-state or resonance threshold and R is the radius of gyration. If $E \in [-\infty,\Lambda_0]$ the dilation analyticity assumption is not needed.

The proof is elementary and we refer to [19] and [23] for details.

Corollary 2 [20]. In theorem 3 assume that $a = \pi/2$ and $\lim_{|\mathrm{Arg}\lambda| \to \pi/2} H(\underline{\tau},\lambda)$ exist in norm resolvent sense and are boost-analytic. Then $\sigma_d(H) \cap R^+$ is empty.

Proof. The proof is inductive; we assume no positive energy thresholds for H. This is true for N = 2 since the only threshold is at zero. Let $E \in \sigma_d(H)$; then E remains an eigenvalue of $H(\underline{\tau}, \lambda)$ as $\mathrm{Arg}\lambda$ tends to $\pi/2$ since E remains isolated in this limit by the induction hypothesis. Let $\mathcal{K}$ denote the set of vectors with support in a cone outside zero i.e. $\phi \in \mathcal{K}$ implies $\mathrm{supp}\,\phi \subset \{\underline{X}, \langle\underline{X},\underline{u}\rangle > \alpha|X|$ and $Q(\underline{X}) > L$ for some $L > 0$, $0 < \alpha < 1$ and unit vector $\underline{u}\}$

Consider

$$F_\phi(\lambda) = (\phi, \phi_E(\underline{0},\lambda)) \quad , \quad |\mathrm{Arg}\lambda| < \pi/2$$

This function is analytic and has continuous boundary values as $|\mathrm{Arg}\lambda| \to \pi/2$ from our assumptions. In this limit one can write for $\zeta \in R^+$:

$$F_\phi(\zeta e^{\pm i\pi/2}) = (\phi, e^{-\langle \underline{X},\underline{u}\rangle\sigma\zeta}\phi_E(\pm\sigma\underline{u},\zeta e^{\pm i\pi/2}))$$
$$\leq e^{-\alpha L\sigma\zeta} ||\phi|| \, ||\phi_E(\pm\sigma\underline{u},\zeta e^{\pm i\pi/2})||$$

Here σ is some positive real number chosen in such a way that $\pm\sigma\underline{u}$ is in the analyticity domain for $\phi_E(\cdot, e^{\pm i\pi/2})$. In addition one has $||\phi_E(\cdot, \zeta e^{\pm i\pi/2})|| = || \phi_E(\cdot, e^{\pm i\pi/2})|| \; \forall \zeta \in R^+$ and $||\phi_E(\underline{0},\lambda)||^2 \leq ||\phi_E(\underline{0}, e^{+ i\pi/2})||^2 + ||\phi_E(\underline{0},e^{-i\pi/2})||^2$, $|\text{Arg}\lambda| \leq \pi/2$

as shown in [15]. From these inequalities it follows that F_ϕ is bounded in the right half-plane and goes to zero exponentially on the imaginary axis. By a theorem of Carlson $F_\phi = 0$ and since the linear span of $\mathcal{K}$ is dense this implies $\phi_E = 0$. This corollary covers Coulomb and Yukawa potentials but not e.g. exponentials $e^{-\kappa|X|}$. However, this last case can be handled by a more general formulation of the Corollary where norm resolvent convergence is replaced by strong resolvent convergence (called generalized convergence in [1]) plus some specific assumptions on the spectra of the limits such as $\sigma_e(\lim_{|\text{Arg}\lambda| = \pi/2} H(\underline{\tau}, \lambda)) = R^-$.

To conclude this section I would like to discuss briefly properties of the "resonance set"

$$\sigma_{res}(H) = (\bigcup_{0< \text{Arg}\lambda < a} \sigma_d(\underline{0}, \lambda)) \cap C \tag{26}$$

we will see in section III that this set contains in fact the poles on unphysical sheets of the scattering amplitudes. Resonances have a less honourable mathematical status than bound-states since they cannot be described as eigenvalues of a self-adjoint operator related to multiparticle equations of motion. They recover some in our formalism since they now are isolated eigenvalues for some $H(\gamma)$'s; this fact has an important application in the possibility of applying perturbation theory to resonances. This can be done as follows. Let

$$H_g = H + g\, W \tag{27}$$

where W is a sum of symmetric dilation analytic two-body interactions, and define $H_g(\lambda) = H(\underline{0}, \lambda) + g\, W(\underline{0}, \lambda)$.

Let $E_0 \in \sigma_{res}(H)$ and fix some λ such that $2\,\text{Arg}\lambda > -\text{Arg}\, E_0$ so that $E_0 \in \sigma_d(\underline{0}, \lambda)$. Then analytic perturbation theory tells us in a neighbourhood of $g = 0$ there will exist a family $\{E_1(g), E_2(g), \ldots E_n(g)\}$ such that $E_i(g) \in \sigma_d(H_g(\lambda))$ is a branch of some analytic function; here n is the multiplicity of E.

Puisieux expansions for these functions are available and given by ordinary perturbation theory. A case of important physical interest is when $E_0 \in \sigma_d(H) \cap \sigma_e(H)$, $E_0 \notin \Sigma^r$ and is simple. Then $E(g)$ is analytic and the two first terms of the perturbation series $E(g) = E + gE^{(1)} + g^2E^{(2)} + O(g^3)$ are just [21]

$$E^{(1)} = (\phi_{E_0}(\bar{\lambda}), W(\underline{0}, \lambda)\, \phi_{E_0}(\lambda))$$
$$= (\phi_{E_0}, W\, \phi_{E_0})$$

since ϕ_{E_0} is dilation analytic. The second term is

$$E^{(2)} = -(\, W(\bar{\lambda})\ \phi_{E_0}(\bar{\lambda}), \hat{R}(\lambda, E_0)\, W(\lambda)\, \phi_{E_0}(\lambda)\,)$$

where $\hat{R}(\lambda, z) = (H(\lambda) - z)^{-1} - (E_0 - z)^{-1} P\{E_0, \lambda\}$ is the reduced resolvent. Since $W\phi E_0$ is dilation analytic one can let $\mathrm{Arg}\lambda$ go to zero in this expression. In particular one has

$$\mathrm{Im}E^{(2)} = -\frac{1}{2} \lim_{\varepsilon \downarrow 0} (W\phi_{E_0}, [\hat{R}(1, E_0 + i\varepsilon) - \hat{R}(1, E_0 - i\varepsilon)\, W\phi_{E_0})$$
$$= -\frac{1}{\pi} \frac{d}{dE} (W\, \phi_{E_0}, P_{ac}(E)\ \phi_{E_0}) \Big|_{E = E_0}$$

where P_{ac} is the restriction of the spectral family of H to $M_{ac}(H)$. This result is known as the Fermi Golden rule. It extends to non-simple eigenvalues as shown in (21), (22).

3. Analytic Properties of Scattering Amplitudes.

a) A general approach to analyticity properties. Results of the preceding section have interesting applications to the study of analytic properties of scattering amplitudes. We will describe below a tentative approach to this problem in the case of two-body elastic scattering amplitudes. For a two-body channel $\alpha = (b(C_1), b(C_2))$ the amplitude is given by

$$T_\alpha(\underline{k}, \underline{q}) = (\phi_{\alpha,\underline{k}}, V_{D(\alpha)}\, \phi_{\alpha,\underline{q}}) + \lim_{\varepsilon \downarrow 0} (\phi_{\alpha,\underline{k}}, (H - E_\alpha(k) - i\varepsilon)^{-1}\, \phi_{\alpha,\underline{q}})$$

where $k^2 = q^2$ and

$$E_\alpha(k) = E_\alpha + \frac{k^2}{2\mu_{D(\alpha)}} \tag{28-a}$$

where $\mu_{D(\alpha)}$ is the reduced mass $\left(\frac{1}{m_{C_1}} + \frac{1}{m_{C_2}}\right)^{-1}$

$$\phi_{\alpha,\underline{k}}(\underline{X}) = (V_{D(\alpha)} \cdot \phi^{b(C_1)} \otimes \phi^{b(C_2)})(\underline{X})\, e^{-i\underline{k}\cdot\underline{X}_{D(\alpha)}} \tag{28-b}$$

Here $\underline{k}$ is the relative momentum of the two bound-states.

Let us define the non relativistic Mandelstam variables P and Δ:

$$\underline{P} = \frac{1}{2}(\underline{k} + \underline{q}) = P\underline{e}$$
$$\underline{\Delta} = \frac{1}{2}(\underline{q} - \underline{k}) = \Delta\underline{e}' \qquad (29)$$

where $\underline{e}$ and $\underline{e}'$ are orthogonal unit vectors; one has:

$$k^2 = P^2 + \Delta^2; \cos\theta = P^2 - \Delta^2 / P^2 + \Delta^2 \qquad (30)$$

where θ is the scattering angle. Analyticity of the amplitude is most often described as a function of the energy $\omega_\alpha(k) = \frac{k^2}{2\mu_{D(\alpha)}}$, scattering angle θ and momentum transfer Δ. For specific purposes, such as the derivation of Mandelstam representation, simultaneous analyticity in $\omega_\alpha(k)$ and Δ (or cosθ) is needed. We will investigate such properties in the case of two-particle amplitudes following very closely Simon's exposition of classical results in this field [24]. We will then briefly sketch the extension of these techniques to two-body elastic amplitudes. The methods exposed below are just extensions of Hunziker's approach [25] to non-forward scattering amplitudes and can be sketched as follows: consider the non-Born part of the scattering amplitude

$$t(\underline{k}, \underline{q}) = \lim_{\varepsilon \downarrow 0} (\phi_{\alpha,\underline{k}}, (H - E_\alpha(k) - i\varepsilon)^{-1}\phi_{\alpha,\underline{q}})$$

for some two-body channel α, where we have assumed that the potential is square-integrable in order to give a meaning to the expectation value.

For $\gamma \in L_N$ define

$$t_\gamma(\underline{k}, \underline{q}) = \lim_{\varepsilon \downarrow 0} (\phi_{\alpha,\underline{k}}(\gamma), (H(\gamma) - k^2 - i\varepsilon)^{-1}\phi_{\alpha,\underline{q}}(\gamma)) \qquad (31)$$

where $\phi_{\alpha,\underline{k}}(\gamma) = u(\gamma)\, \phi_{\alpha,\underline{k}}$. Obviously for k real one has $t_\gamma(\underline{k},\underline{q}) = t(\underline{k},\underline{q})$. Now let us choose linear transformations of the following form

$$\gamma(v) = (\underline{\mu}(v), \lambda(v))$$

where v stands for some of the variables k, θ or Δ.

For fixed function $\gamma = (\underline{\mu}, \lambda)$ one can try to determine an analyticity domain $\mathcal{D}$ in v for the function t_γ ; this can be done by investigating the analyticity domains for the resolvent and the two vectors $\phi_{\alpha,\underline{k}}(\gamma)$ and $\phi_{\alpha,\underline{q}}(\gamma)$ in the left member of (31).

The final aim is to show that $\mathcal{D} = \bigcup_\gamma \mathcal{D}(\gamma)$ is an analyticity domain in v for the original amplitude t, the analytic continuation in v being provided by the t_γ ' s. This is true provided

1) $\mathcal{D}$ is connected

2) The various analytic functions obtained in this way coincide in overlapping domains:

$t_{\gamma_1} = t_{\gamma_2}$ in $\mathcal{D}(\gamma_1) \cap \mathcal{D}(\gamma_2)$

This can be done with the usual unitarity argument provided one can find an essential set of points, in general a path $\Gamma \subset \mathcal{D}_{\gamma_1} \cap \mathcal{D}_{\gamma_2}$ such that $\gamma_1(v) * \gamma_2^{-1}(v) \in L_N$ for all $v \in \Gamma$.

3) The physical scattering amplitude (real values of v) can be obtained as the boundary value of some t_γ.

Let us see how this method works on some examples for two-particle scattering amplitudes.

b) Forward scattering amplitudes.

Theorem 4. Assume V = UW where U and W are spherically symmetric dilation analytic functions in L^2 and U is bounded at the origin. Then there exists a function f(k) meromorphic in the upper k-plane such that

1) $\lim_{\mathrm{Im}k\to 0} f(k) = t(\underline{k}, \underline{k})$ exists $\forall k \in R^+ - \{0\}$

2) All poles of f are on the positive imaginary axis. They are simple and can occur only at points k_0 such that $k_0^2 \in \sigma_d(H)$

3) f(k) has a meromorphic continuation through R^+. Its poles can occur only at points k_0 such that $k_0^2 \in \sigma_{res}(H)$.

Proof. We first choose $\gamma_1(k) = (-k\underline{n}, 1)$ where $\underline{n} = \underline{k}/k$ is held fixed and define the $\underline{n}$ independent function

$$\begin{aligned} f_{\gamma_1,\varepsilon}(k) &= (V, (H(\gamma_1(k)) - (k^2 + i\varepsilon))^{-1}V), \ \varepsilon > 0 \\ &= (U, M_1(k, \varepsilon)\ W) \end{aligned} \tag{32}$$

where

$$M_1(k, \varepsilon) = W(H(\gamma_1(k) - (k^2 + i\varepsilon))^{-1}U \tag{33}$$

Obviously $t(\underline{k}, \underline{k}) = \lim_{\varepsilon\downarrow 0} f_{\gamma_1,\varepsilon}(k)$ for $k \in R^+$. From the second resolvent equation one gets

$$M_1(k,\varepsilon) = M^0{}_1(k, \varepsilon) + M^0{}_1(k, \varepsilon)\ M(k, \varepsilon) \tag{34}$$

where $M^0{}_1\ (k, \varepsilon)$ is obtained by substituting H_0 for H in (33). Under our assumptions $M^0{}_1(k, \varepsilon)$ has a uniformly bounded Hilbert-

Schmidt norm for $\varepsilon > 0$ and $k \in C^{++}$. Furthermore since $(k^2 + i\varepsilon) \in \rho(H(\gamma_1(k)))$ it is analytic in k for each ε. Then $M^0{}_1(k) = \lim_{\varepsilon\to 0} M^0{}_1(k,\varepsilon)$ exists and is analytic in C^{++} with continuous boundary values on R^+. Klein-Zemach theorem [24] tells us that $\lim_{\substack{k\to\infty \\ k\in R^+}} ||M^0{}_1(k)||=0$ so that one has a meromorphic solution to (34) in the limit $\varepsilon \downarrow 0$ given by

$$M_1(k) = \lim_{\varepsilon\downarrow 0} M_1(k, \varepsilon) = [1 - M^0{}_1(k)]^{-1} M^0{}_1(k)$$

Accordingly $f_{\gamma_1}(k) = \lim_{\varepsilon\downarrow 0} f_{\gamma_1, \varepsilon}(k)$ is meromorphic in C^{++} and its boundary values on R^+ when they exist coincide with $t(k, k)$. Minus the fact that this limit exists $\forall k \in R^+$ this shows (1); the proof will be completed below by using the dilation analyticity assumption.

Let us now choose $\gamma_2(k) = (0, k^{-1})$ and define

$$f_{\gamma_2,\varepsilon}(k) = [\hat{\phi}_{\underline{k}}(\gamma_2) , (H(\gamma_2(k)) - (k^2 + i\varepsilon))^{-1} \phi_{\underline{k}}(\gamma_2))$$

This function has a meromorphic continuation in $|\mathrm{Arg} k| < a$ from above since

$$\phi_k(\gamma_2)(\underline{X}) = V_{k^{-1}}(\underline{X})\, e^{i\underline{X}\cdot\underline{n}}$$

and $(H(\gamma_2(k)) - (k^2 + i\varepsilon))^{-1}$ is meromorphic according to Theorem 1. To control the limit as $\varepsilon \to 0$ one needs as before to prove meromorphy in the Hilbert-Schmidt class of

$M_2(k) = \lim_{\varepsilon\downarrow 0} M_2(k, \varepsilon)$ where $M_2(k, \varepsilon)$ is given by $W_{\gamma_2(k)}(H(\gamma_2(k)) - (k^2 + i\varepsilon))^{-1}U_{\gamma_2(k)}$. This can be done under our dilation analyticity assumptions along the same lines as for $M_1(k)$ with the result that

$$f_{\gamma_2}(k) = \lim_{\varepsilon\downarrow 0} f_{\gamma_2,\varepsilon}(k) = \lim_{\varepsilon\downarrow 0} (U_{\gamma_2,k} , M_2(k, \varepsilon)\; W_{\gamma_2, k}) \tag{35}$$

is meromorphic in $|\mathrm{Arg} k| < a$. Since one has obviously $f_{\gamma_2}(k) = t(k,k)$ for $k \in R^+$ the two functions f_{γ_1} and f_{γ_2} are analytic continuations of each other and f simply is the unique meromorphic function which they determine in $C^{++} \cup \{\lambda; |\mathrm{Arg}\lambda| < a\}$.

Using Schwartz symmetry $\bar{f}(k) = f(-\bar{k})$ coming from the fact that f_{γ_1} has real boundary values on the imaginary axis one gets (1). To prove 2) it is sufficient to show that the poles of $M_1(k)$ coincide with $(\sigma_d(H))^{\frac{1}{2}}$. This can be done as follows: Let ϕ and Ψ be in $\mathcal{D}$;

then

$$\lim_{\varepsilon\downarrow 0} (\phi, M_1(k, \varepsilon)\Psi) = \lim_{\varepsilon\downarrow 0} (\overline{\phi(\gamma_1^{-1}(k))}, W(H - (k^2 + i\varepsilon))^{-1}U\Psi (\gamma_1^{-1}(k)))$$

and this last term has a pole at $k = k_0$ if and only if $k_0^2 \varepsilon \sigma_d(H)$. Since $\mathcal{D}$ is dense 2) follows. Finally to prove the last part of 3) one uses the same observation that f_{γ_2} has a pole in $|\mathrm{Arg}k| < a$ only if $M_2(k)$ has a pole; using again expectation values between elements of $\mathcal{D}$ one sees that such a pole at $k = k_0$ can happen only if the expectation value of the resolvent of H between elements of $\mathcal{D}$ has a pole at k_0^2 that is according to the Lemma if k_0^2 is in the discrete spectrum of some $H(0, \lambda)$ for $\mathrm{Arg}\lambda > -\mathrm{Arg}\, k_0$.

c) Fixed momentum transfer analyticity. Here we require exponential fall-off for V namely

$$V(X) = e^{-\alpha |X|} W(X) \text{ where } W \in L^2 \text{ is spherically symmetric}$$

Then we have

Theorem 5. Assume $|\Delta| < \alpha$. Then there exists a function $f(k, \Delta)$ meromorphic in the upper half-plane such that

1) for k_0 and Δ_0 reals $\lim_{\substack{k\to k\\ \Delta\to\Delta_0}} f(k, \Delta) = t(\underline{k}_0, \underline{q}_0)$ exist everywhere; ((k_0, Δ_0) and $(\underline{k}_0, \underline{q}_0)$ are related by (30))

2) All poles of $f(k, \Delta)$ for fixed $\Delta, |\Delta| < \alpha$, are on the positive imaginary axis. They are simple and can occur only at points k_0 such that $k_0^2 \varepsilon \sigma_d(H)$.

Proof. For k and q reals one has

$$t(k, q) = \lim_{\varepsilon\downarrow 0} (\phi_{\bar{k},\bar{\Delta}}, (H - (k^2 + i\varepsilon))^{-1} \phi_{k,-\Delta}) \tag{36}$$

where from (30)

$$\phi_{k,\Delta}(\underline{X}) = V(X) \exp i\, (\sqrt{k^2 - \Delta^2}\ \underline{e} + \Delta\underline{e}')\underline{X} \tag{37}$$

From the exponential fall-off property of V it is easy to see (use e.g. Rollnick method) that this limit exists and coincides with $t(k, q)$. Now let k be fixed and real; then $\phi_{k,\pm\Delta}$ are analytic in Δ as long as $(\mathrm{Im}\sqrt{k^2 - \Delta^2})^2 + (\mathrm{Im}\Delta)^2 < \alpha^2$.

It is easily shown (see [24]) that this condition holds for all $k \varepsilon R^+$ provided $|\Delta| < \alpha$. We then denote by $f(k, \Delta)$ the function defined by this analytic continuation in $R^+ \times B_\alpha$ where B_α is the open ball of radius α in C. For fixed $\Delta \varepsilon B_\alpha$ we now investigate analyticity properties in k. For this we define

$$\gamma_\mu(k) = (-\mu\, k\underline{e}, 1)$$

$$f_\gamma(k, \Delta) = \lim_{\varepsilon\downarrow 0} (\phi_{\bar{k},\bar{\Delta}}(\bar{\gamma}), (H(\gamma) - (k^2 + i\varepsilon))^{-1} \phi_{k, -\Delta}(\gamma)) \tag{38}$$

where $\phi_{k,\Delta}(\gamma) = u(\gamma) \phi_{k,\Delta}$ for real k. For fixed μ all elements in (38) are analytic in k provided

1) $k^2 \in \rho(H(\gamma_\mu(k)))$ which requires $|\mu| < 1$

2) $|\mathrm{Im}(P - \mu k)|^2 + |\mathrm{Im}\Delta|^2 < \alpha^2$

This defines for $|\mu| < 1$ an analyticity domain $\mathcal{D}(\mu)$ given by 2). We now refer to Simon [24] for the proof that under the condition $\Delta \in B_\alpha$ one as $\bigcup_{|\mu|<1} \mathcal{D}(\mu) \supset \{k| \text{ Im } k \geq 0\}$. Since $\mathcal{D}(\mu)$ contains R^+ one obtains in this way the analytic continuation for $f(k, \Delta)$. The investigation of the poles of $f(k, \Delta)$ in the upper half-plane can be done as for forward amplitudes. Remark: by assuming W is dilation analytic in L^2 one can of course prove the existence of an analytic continuation through R^+ as in the forward case.

d) The great Lehman ellipse. We now investigate analyticity in both k and $z = \cos\theta$. The assumption on V are the same as in 2 plus dilation analyticity of V in the cone $C\pi/2$. We will see that the amplitude is analytic for fixed k in the so called great Lehman ellipse (the small one being given by the analyticity of the Born part of the amplitude). Such a result is the basis of dispersion relations for partial amplitudes, Froissard bounds on cross-sections and Regge theory. We refer to [24] and [26] for such applications.

Theorem 6. Let $\mathcal{D} = \{k|(\mathrm{Im}k > 0) \cup |\mathrm{Arg}\, k| < \pi/2\} - s$ where $s =]-\infty, -\alpha^2[\cup \sigma_d(H) \cup \sigma_{res}(H) \cup \{0\}$. Then there exists a function $g(k, z)$ analytic in $\{(k, z)| k \in \mathcal{D}$ and $z \in E(k)\}$ where $E(k)$ is the ellipse with foci ± 1 and semi-major axis $\frac{|\alpha^2 + k^2|}{|k|^2} + \frac{\alpha^2}{|k|^2}$. For $k \in R^+$ one has $\lim_{z\to\cos\theta} g(k,z) = t(\underline{k},\underline{q})$.

Proof. For k and q reals $t(\underline{k}, \underline{q})$ is given by (36). Let us rewrite (37) as:

$$\phi_{k,\pm\Delta}(\underline{X}) = V(\underline{X}) \exp\left(i\sqrt{\frac{k^2(1+z)}{2}}\, \underline{e} \pm i \sqrt{\frac{k^2(1-z)}{2}}\, \underline{e}'\right)\cdot\underline{X} \tag{39}$$

where $z = \cos\theta$ and we have used formulas (30). Now let $\gamma(k) = (\underline{0}, \lambda(k))$ and

$$g_\gamma(k, z) = \lim_{\varepsilon\downarrow 0} (\Psi^+_{\bar{k},\bar{z}}(\bar{\gamma}), (H(\gamma) - k^2 - i\varepsilon)^{-1} \Psi^-_{k,z}(\gamma)) \tag{40}$$

where for $k \in R^+ - \{0\}$ one has $\Psi^\pm_{k,z}(\gamma) = u(\gamma) \phi_{k,\pm\Delta}$.

These vectors have analytic continuations simultaneously in k and z provided

$$\mathrm{Im}(\lambda(k)k\sqrt{1+z})^2 + \mathrm{Im}(\lambda(k)k\sqrt{1-z})^2 < 2\alpha^2(\mathrm{Re}\,\lambda(k))^2 \quad (41\text{-a})$$

$$|\mathrm{Arg}\,\lambda(k)| < \pi/2 \quad (41\text{-b})$$

Then the function g_γ provided by (40) defines an analytic continuation of $t(\underline{k}, \underline{q})$ in both k and z in this domain. One has apparently two branch points at $z = \pm 1$. However, turning around these points just amounts to changing $\underline{e}$ to $-\underline{e}$ or $\underline{e}'$ to $-\underline{e}'$; since V is spherically symmetric this does not change the amplitude. Now using the general relation:

$$|W|^2 = \mathrm{Re}(W^2) + 2(\mathrm{Im}W)^2 \qquad \forall W \in C.$$

one can rewrite (41-a) as

$$|1+z| + |1-z| < 2\left[\frac{\mathrm{Re}((\alpha^2+k^2)\lambda^2(k))}{|\lambda(k)|^2\,|k|^2} + \frac{\alpha^2}{|k|^2}\right]$$

Clearly the best choice of λ is

$$\lambda(k) = (\alpha^2 + k^2)^{-1/2}$$

It gives the stated analyticity domains for vectors in (40). In addition for $\mathrm{Im}k > 0$ the resolvent in (40) also is meromorphic since $\mathrm{Arg}(\alpha^2 + k^2) < \mathrm{Arg}\,k^2$. The investigation of the analytic continuation for $\mathrm{Im}k < 0$ requires an improvement of this proof which we don't develop here.

e) General two-body elastic reactions. We make the following simplifying assumptions:

i) Exponentially decreasing spherically symmetric potentials,

ii) E_α is the lowest threshold which implies in particular that the bound-states are non-excited,

iii) Bound-states have zero angular momentum; this is a technical assumption needed to have invariant amplitudes. Helicity amplitudes could be treated along the same lines.

Methods and results have then to be adapted as follows:

Forward scattering amplitudes. Use boosts in the direction of the relative momentum k between the bound-states. But Rollnick-like techniques are not available in N-particle systems; however, the exponential decay of potentials and bound-state wave-functions allows us to replace in the proof of theorem 4 $\gamma_1 = (-\underline{k}, 1)$ by $\gamma = (-\mu\underline{k}, 1)$ for some μ sufficiently close to 1.[1]

In fact one needs:

1) $\lim_{\varepsilon\downarrow 0} (H(\gamma) - (E_\alpha(k) + i\varepsilon))^{-1}$ exists and is meromorphic

2) $\phi_{\alpha,k}(\gamma) = u(\gamma)\ \phi_{\alpha,k}$ is analytic in k.

This gives the suitable limitations on μ.

Assuming now dilation analyticity for the potentials one can prove existence of a multi-sheeted meromorphic continuation through R^+, the poles belonging to $\sigma_{res}(H)$. Each E_α is a branch-point for this continuation as is generally expected for multiparticle scattering amplitudes.

Fixed momentum transfer analyticity. Hunziker's method applies exactly as for the two-particle scattering amplitude. However, the condition $|\Delta| < \alpha$, where a is the range of two-particle interactions has to be replaced here by $|\Delta| < \inf(\alpha, \alpha_1, \alpha_2)$ where

$\alpha_i = (\min\limits_{j\varepsilon C_i} \frac{m_j}{m_{C_i-\{j\}}})$. r_i, i = 1, 2 and r_i is the range of the bound-state wave-function $\phi^{b(C_i)}$. This limitation comes from the fact that $\phi_{\alpha,k,\Delta}$ defined as in (37) by substituting variables k and Δ for k in (28b) is analytic only for such values since bound-state wave-functions have a non-zero range.

Great Lehman Ellipse. The same result applies as long as Imk ≥ 0 with α replaced by $\inf(\alpha, \alpha_1, \alpha_2)$ as above and k^2 by $\omega_\alpha(k)$.

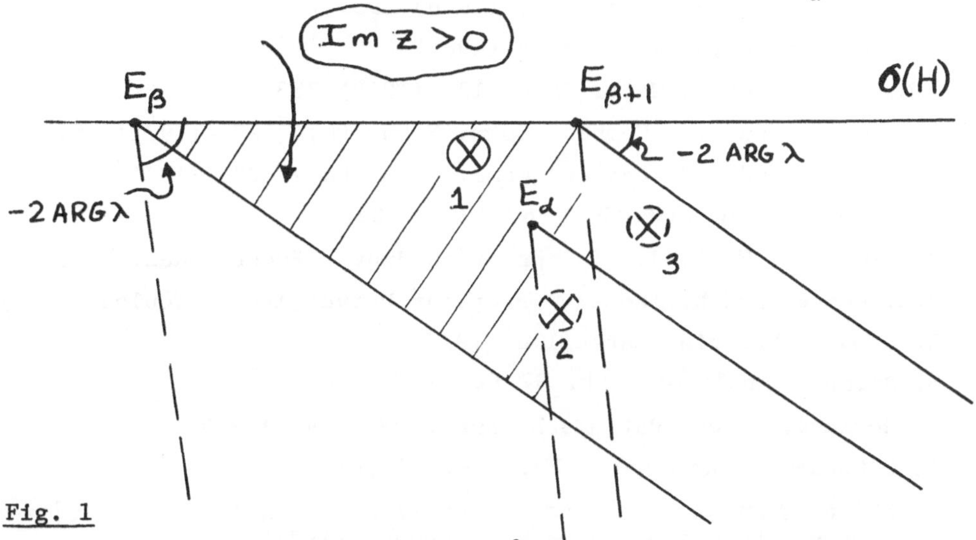

Fig. 1

Analytic continuation of $(\Psi,(H-z)^{-1}\Psi)$, $\Psi \varepsilon \mathcal{D}$, with threshold $E_r \varepsilon \Sigma^r$. Pole 1 has not been absorbed by the cut $E_\beta + \lambda^{-2}R^+$; pole 2 has been absorbed by the cut $E_\alpha + \lambda^{-2}R^+$, pole 3 by $E_{\beta+1} + \lambda^{-2}R^+$.

References:

1. T. Kato, *Perturbation theory for linear operators*. Springer-Verlag, N.Y.(1966).
2. Hack, Nuovo Cimento 13, (1959) 231
3. L. Faddeev, Mathematical aspects of the 3 body problem in the quantum scattering theory., Israel program for scientific translation, London, (1965).
4. K. Hepp, Helv. Phys. Acta 42, (1969) 425.
5. J. M. Combes, Nuovo Cimento, 64 A, (1969) 111.
6. S. Weinberg, Phys. Rev. 133,232 (1964). (Van Winter: Mat. Fys.skr. Dan Vid Selsk Nos. 8, 10 (1964, 1965)).
7. T. Ikebe, Arch. Rat. Mech. Anal 5, (1964) 1.
8. Nuttall, Jour. Math. Physics 4, (1967) 873.
9. Lovelace, *Strong interactions and High Energy Physics*, Moorhouse Oliver and Boyd, London (1964).
10. C. Dolph, AMS Bull. 67, (1961).
11. B. Simon, Comm. Math. Phys. 27, (1972) 1.
12. E. Balslev, Math. Scand. 19, (1966) 193.
13. W. Hunziker, Phys. Rev. 135 B (1964) 800.
14. J.M. Combes, Comm. Math. Phys. 12, (1969) 283.
15. J. Aguilar and J.M. Combes, Comm. Math. Phys. 22, (1971) 269.
16. E. Balslev and J.M. Combes, Comm. Math. Phys. 22, (1971) 280.
17. T. Ichinose, Nagoya Math J. 50(1973) 185.
18. B. Simon and M. Reed, To appear in Jour. Funct. Anal. 13 (1973).
19. J.M. Combes and L. Thomas, Preprint Université de Toulon (1973).
20. B. Simon, Preprint Marseille (1973).
21. B. Simon, Annals of Math. 97 (1973) 24 F.
22. J. Howland, Arch. Rat. Mech. Anal. 39, (1970) 323.
23. T. O'Connor, Comm. Math. Physics. 32 (1973) 319.
24. B. Simon, *Quantum mechanics for Hamiltonians defined as quadratic forms*, Princeton University Press, (1971).
25. W. Hunziker, Helv. Phys. Acta. (1961) 593.
26. J.R. Taylor, *Scattering theory*, John Wiley & Sons, (1972).

SPIN-WAVE SCATTERING

R. F. Streater

Bedford College

London

1. Introduction and Summary [1] [2].

We consider an infinite array of ions, each fixed at a point of the lattice $Z^\nu, \nu = 1,2$ or 3. Each ion has no degrees of freedom except its spin, 1/2, which is described by a *spinor*, that is a unit vector in C^2, $\binom{1}{0}$ denoting spin up, $\uparrow$, and $\binom{0}{1}$ spin down, $\downarrow$. A typical "product state" of all the ions can be pictured:

$$\ldots \otimes \uparrow \otimes \nearrow \otimes \nwarrow \otimes \uparrow \otimes \nwarrow \otimes \ldots$$

The "observables" associated with an ion $j \in Z^\nu$ are the spins $J_j^\alpha = 1/2\ \sigma^\alpha$ regarded as acting on the jth factor in the product; here, $\alpha = 1,2$ or 3 are the 3 spin (or isobaric spin) directions. For each ion j, its observables generate the 4 dimensional C^*-algebra $\mathcal{O}\!\mathcal{l}_j = \mathcal{B}(C^2)$ of 2×2 complex matrices. These are all independent observables, so the total system is described by $A = \bigotimes_{j \in Z^\nu} \mathcal{O}\!\mathcal{l}_j$, a C^*-algebra whose hermitian elements are called the quasi-local observables.

In a ferromagnetic system, the spins tend to align themselves, and we expect this to be the lowest energy state. In quantum mechanics, the energy operator, the Hamiltonian, is a function of the dynamical variables, in this case the spins J_j^α. Heisenberg suggested that, to get two spins $\vec{J}_j$ and $\vec{J}_k$ to attract each other in the parallel state, we should choose an interaction $\sum_\alpha fJ_j^\alpha J_k^\alpha = f\ \vec{J}_j \cdot \vec{J}_k$ where $f < 0$. Classically, the lowest value of $f\vec{s}_1 \cdot \vec{s}_2$ is $f|s_1||s_2|$

J. A. LaVita and J.-P. Marchand (eds.), Scattering Theory in Mathematical Physics, 273–298.

if $f < 0$, and this occurs when the spins are parallel. We therefore choose

$$H = \sum_{i,j \in Z^\nu} \sum_{\alpha=1}^{3} f_{ij} J_i^\alpha J_j^\alpha = \sum_{i,j \in Z^\nu} f_{ij} \vec{J}_i \cdot \vec{J}_i$$

with $f_{ij} \leq 0$. To make the theory invariant under space-translations we must have $f_{ij} = f_{i+k,j+k}$, $k \in Z^\nu$. This means that, even if each spin interacts only with its immediate neighbors, an infinite sum is involved in the definition of H. However, the Heisenberg equation of motion

$$\frac{d}{dt} A(t) = i[H,A(t)] \tag{0}$$

involves only those terms of H that do not commute with A, and, since J's at different points commute, this has a meaning if the range of the forces is not too long. Our first result is that this equation determines a unique *-automorphism group $\tau(t)$: $\mathfrak{A} \to \mathfrak{A}$, with $\tau(t_1)\tau(t_2) = \tau(t_1 + t_2)$, continuous in t if $\mathfrak{A}$ is furnished with the C^*-norm. Moreover, the space-translation automorphism $\sum(j)$, $j \in Z^\nu$, defined on J_k^α as $\sum(j)J_k^\alpha = J_{k+j}^\alpha$, commutes with $\tau(t)$ if $f_{ij} = f_{i-j}$. We also introduce *spin rotations* $\rho(R)$, $R \in SO(3)$, which act $\rho(R)J_k^\alpha = R_\beta^\alpha J_k^\beta$. These are not physical rotations since they do not rotate the position, k, of the spin, but they do define automorphisms $\rho(R)$ of $\mathfrak{A}$, which commute with $\sum(j)$ and with $\tau(t)$ if H is a scalar product $\sum \vec{J}_i \vec{J}_j$.

We thus have a *Heisenberg* or algebraic formulation of the quantum spin system. We recover the usual Hilbert space, or Schrödinger formulation, by considering a representation of $\mathfrak{A}$; a representation $(\pi,\mathcal{H})$ of a C^* - algebra $\mathfrak{A}$ is a *-homomorphism π: $\mathfrak{A} \to \mathcal{B}(\mathcal{H})$. Two representations, $(\pi_1,\mathcal{H}_1)$ and $(\pi_2,\mathcal{H}_2)$ are said to be *equivalent* if there exists a unitary map $V: \mathcal{H}_1 \to \mathcal{H}_2$ such that V intertwines π_1 and π_2 : $V\pi_1(A) = \pi_2(A)V$ for all $A \in \mathfrak{A}$. In systems of finitely many degrees of freedom (either spins, or canonical variables (p_i,q_i)) there exists only one equivalence class of irreducible representations, but for infinitely many, as here, there are many inequivalent representations. Another phenomenon occurs too, closely related to the first, namely, an automorphism need not be implemented in a given representation $(\pi,\mathcal{H})$; we say an automorphism γ of $\mathfrak{A}$ is *implemented* in $(\pi,\mathcal{H})$ if there exists a unitary operator $\Gamma: \mathcal{H} \to \mathcal{H}$, such that $\Gamma\pi(A)\Gamma^{-1} = \pi(\gamma(A))$ for all $A \in \mathfrak{A}$. Naturally, we are mostly interested in the representations in which $\tau(t)$ is implemented for each t, and in which the implementing operator T(t) say, can be chosen continuous in t.

In order to construct such representations we use the famous Gelfand-Naimark-Segal theorem connecting representations with states of $\mathfrak{A}$. A *state* ω is a functional $\mathfrak{A} \to C$ satisfying

(1) $\omega(1) = 1$

(2) $\omega(\lambda A + \mu B) = \lambda\omega(A) + \mu\omega(B), \quad \lambda,\mu \in C$

(3) $\omega(A^*A) \geq 0.$

If $(\pi,\mathcal{H})$ is a representation, and $\psi \in \mathcal{H}$, with $\|\psi\| = 1$, then the map $\omega_\psi(A) \equiv \langle\psi,A\psi\rangle$ is a state; a state of this form is called a *vector state* for the representation π. The famous GNS theorem says that, if ω is *any* state, there exists a Hilbert space $\mathcal{H}_\omega$, a representation π_ω and a vector $\Omega \in \mathcal{H}_\omega$, such that ω is the vector state ω_Ω relative to π_ω. Moreover, Ω is cyclic under $\pi_\omega(\mathfrak{A})$. If ω happens to be invariant under an automorphism γ, i.e. $\omega(\gamma(A)) = \omega(A)$ for all $A \in \mathfrak{A}$, then γ is implementable in $(\pi_\omega,\mathcal{H}_\omega)$; π_ω is irreducible if and only if ω is an extremal state (i.e. pure).

Product vectors define states of $\mathfrak{A}$; the product $\ldots\psi_1 \otimes \psi_2 \otimes\ldots$ defines, for $A = \ldots 1 \otimes A_1 \otimes A_2 \otimes \ldots 1 \otimes \ldots$ the state $\omega(A) = \prod_j \langle \psi_j, A_j\psi_j\rangle$, where $\otimes A_1 \otimes A_2 \otimes \ldots$ contains only finitely many factors not 1; and by linearity and continuity, it defines a unique state on $\mathfrak{A}$. We show that $\ldots \uparrow \otimes \uparrow \otimes \uparrow \ldots \otimes \uparrow = \Omega$ defines a pure state on $\mathfrak{A}$ and thus a representation $\pi^\uparrow$, invariant under space-time translations $\sum(j)\tau(t)$, and that the corresponding unitary group $T(t)$ is continuous. Moreover, the generator H is ≥ 0, and has Ω as its unique eigenvector: $H\Omega = 0$.

We have thus achieved a description of a ferromagnetic state. The state $\Omega^\downarrow = \ldots \downarrow \otimes \downarrow \otimes \ldots \downarrow \otimes \ldots$ is also space-time translation invariant, its representation $\pi^\downarrow$ being inequivalent to $\pi^\uparrow$. Indeed, we have $\Omega^\nearrow = \ldots \nearrow\otimes \nearrow \otimes \ldots \otimes\nearrow \ldots$, $\pi^\nearrow$ and $\mathcal{H}^\nearrow$ for each direction in space, each inequivalent, and with its own energy $H^\nearrow$ with groundstate $\Omega^\nearrow$. Each $\mathcal{H}^\nearrow$ is separable.

The spin rotations $\rho(R)$ are *not* implemented in $\mathcal{H}^\uparrow$ (except those rotations leaving the axis of spin invariant): in the concrete quantum mechanics $(\pi^\uparrow(\mathfrak{A}),\mathcal{H}^\uparrow,\Omega^\uparrow)$, there is no conserved quantum number corresponding to this symmetry; we say the symmetry is *spontaneously broken* [3]. The space $\mathcal{H}^\uparrow$ is obtained from $\Omega^\uparrow$ by applying the operators of the theory, $\pi^\uparrow(\mathfrak{A})$. In particular, the spin-lowering operator $\pi^\uparrow(J_j^-)$ applied n times (at different points) to $\Omega^\uparrow$ gives a state of n-"spin waves"

$$\ldots \uparrow \uparrow \underset{i_1}{\downarrow} \uparrow \uparrow \uparrow \underset{i_2}{\downarrow} \uparrow \ldots \uparrow \underset{i_n}{\downarrow} \uparrow \uparrow \uparrow \ldots$$

We may identify $\mathcal{H}^\uparrow$ with von Neumann's infinite tensor product $\bigotimes^{\Omega^\uparrow}_{j\in Z^\nu} C_j^2$; it is known that $\{\pi(J_{i_1}^- \ldots J_{i_n}^-)\Omega^\uparrow\}$ form an orthogonal basis in this space. We may therefore write $\mathcal{H}^\uparrow$ as a direct sum, $\mathcal{H}^\uparrow = \bigoplus_n \mathcal{H}_n$, where $\mathcal{H}_n$ is the span of product states with n spins down. Thus, $\mathcal{H}^\uparrow$ looks rather like a Fock space, except that no state has two spin-waves located at the same point. This is a consequence of $(J_j^-)^2 = 0$ for the spin 1/2 representation; here $J_j^\pm = J_j^1 \pm iJ_j^2$ are the spin raising and lowering operators. These almost behave as Boson creation and annihilation operators, except that

$$[\, J_j^+,\, J_k^- \,] = 2i\delta_{jk}\, J_j^3$$

The presence of J_j^3 on the right hand side leads to a kinematical "repulsion" of spin-waves.

The spin wave number, n, defines a selfadjoint operator N on $\mathcal{H}^\uparrow = \bigoplus \mathcal{H}_n$; in fact, it measures $\sum_j (J_j^3 - 1/2)$. Indeed, N is the generator of the representation of the group of rotations about the 3rd axis; this group, $SO(2)$, is implemented since $\Omega^\uparrow$ is invariant under it. Since H commutes with these rotations, N commutes with H that is, the number of spin-waves is conserved, and the symmetry is not spontaneously broken. It follows that H is reduced by $\bigoplus_n \mathcal{H}_n$. Clearly $H\Omega = 0$. We are able to prove that $H \geq 0$, and that H is bounded on each $\mathcal{H}_n$ with a bound proportional to n.

The one-particle space, $\mathcal{H}_1$, is spanned by vectors $J_k^-\Omega = \ldots \uparrow \otimes \uparrow \otimes \underbrace{\downarrow}_{k} \otimes \uparrow \ldots$; since there are mutually orthogonal, we see that $\mathcal{H}_1$ is isomorphic to $\ell_2(Z^\nu)$, and that S(k), the space translation group, becomes the regular representation when restricted to $\mathcal{H}_1$. This means that the group operators, $\{S(k),\ k \in Z^\nu\}$ form a maximal abelian set, so that any operator that commutes with them is a function of them, and of their generators, the momenta $\hat{p}$, which lie in the dual group to Z^ν, i.e. T^ν. The group T^ν is better known as the *first Brouillin zone*. Writing $S(j) = \exp i\hat{p}\cdot j$ where $p\cdot j = \sum_{\alpha=1}^{\nu} p_\alpha j_\alpha$, we see that $\hat{p}$ and the "position operator" $\hat{j}$ satisfy the Heisenberg commutation relations in the form proposed by Mackey:

$$e^{i\hat{p}\cdot j}e^{i\hat{j}\cdot q} = e^{i\hat{j}\cdot q}e^{i\hat{p}\cdot j}e^{ij\cdot q} \qquad j \in Z^\nu,\quad q \in T^\nu .$$

A localized state in $\mathcal{H}_1$, say $J_j^-\Omega^\uparrow$, moves around, because of the pull of its neighbors. In fact, since $[H, \hat{p}] = 0$, H is diagonalized by the p-representation, that is, $\mathcal{F} H \mathcal{F}^{-1}$ is a multiplication operator E on $L^2(T^\nu, d\mu)$, where $d\mu$ is Haar measure, and $\mathcal{F}: L^2(Z^\nu, n) \to L^2(T^\nu, d\mu)$ is the usual isomorphism given by the Fourier transform. Thus $E = \omega(p)$ for some function ω (known as the dispersion relation). A wave packet $\tilde{\psi} \in L^2(T^\nu, d\mu)$ evolves in time according to the pseudo-differential equation

$$i \frac{\partial}{\partial t} \tilde{\psi}(p,t) = \hbar\omega(p)\tilde{\psi}(p,t), \qquad p \in T^\nu$$

The solution of this, for a wide class of $\omega(p)$, is a function $\psi(i,t)$ which behaves as a quasi-particle with group velocity $\partial\omega/\partial p$. The function ω is explicitly determined as a function of the f_{ij} (for nearest neighbor interactions $\omega = f \sum_\delta (1 - \cos k\cdot\delta)$ where δ runs over the vectors εZ^ν describing the nearest neighbor of 0). Since $\psi(j,t) = \int_{T^\nu} e^{-i\omega(p)t+ip\cdot j} \tilde{\psi}(p)\, d\mu(p)$, the behavior of a wave-packet is determined for large t by the method of stationary phase.

The time-evolution of states containing 2,3 ... spin-waves, far apart from each other, might be expected to approximate that of free particles. This is proved to be the case for a wide class of ω, using Kato's theory of scattering in two Hilbert spaces. The asymptotic states are described in terms of states lying in the Fock space over $\mathcal{H}_1$ i.e. $C \oplus \mathcal{H}_1 + (\mathcal{H}_1 \otimes \mathcal{H}_1)_s + \ldots \equiv \mathfrak{F}$ in which the "free" time-evolution is defined to be

$$T_o(t) = 1 \oplus e^{-iH_1 t} \oplus e^{-iH_1 t} \otimes e^{-iH_1 t} \oplus \ldots$$

where $H_1 = \omega(\hat{p})$. These particles are quasi-free Bosons called *ideal spin-waves* by Dyson [4]. Clearly, $\mathcal{H}$ may be isometrically embedded in $\mathfrak{F}$ by the injection i:$i\, J^-_{j_1} \ldots J^-_{j_n}\, \Omega = |j_1\rangle \times\ldots\times |j_n\rangle \in \mathfrak{F}_n$ where $|j\rangle$ is the eigen-state of $\hat{j}$ in $\mathcal{H}_1$ with eigenvalue j.

The asymptotic condition in the form

$$\underset{t\to\pm\infty}{\text{s-lim}}\; T^*(t)\, i^*\, T_o(t)\Phi = \Omega_\pm\, \Phi \;, \qquad \Phi \in \mathfrak{F}$$

is proved, following Watts [5] and Hepp [6]. The convergence is faster than any inverse power of t on those states $\Phi \in \mathcal{H}_n$ representing particles whose group velocities do not overlap.

There are two concepts of the scattering operator; the first, $\tilde{S} = \Omega_+^*\Omega_-$, maps $\mathfrak{F}$ into $\mathfrak{F}$, through dymanics taking place in the physical space $\mathcal{H}$. Alternatively, $\Omega_\pm \mathfrak{F} \subset \mathcal{H}$ define subspaces $\mathcal{H}_\pm$ of scattering states of the physical space $\mathcal{H}$, and $S: \mathcal{H}_+ \to \mathcal{H}_-$ may be defined by $S\Omega_+\Phi = \Omega_-\Phi$, $\Phi \in \mathfrak{F}$. Using only the time-dependent method, we can establish the partial isometry $\Omega_\pm^*\Omega_\pm = 1$ and the intertwining property

$$T(t)\Omega_\pm = \Omega_\pm T_o(t)$$

One may express the presence of an impurity in the lattice, say at $j_o \in Z^\nu$, by relaxing the translation invariance of the f_{ij} for $j = j_o$. In this way, a spin-wave reaching a point i such that $f_{ij_o} \neq 0$ suffers a change in its equation of motion, and j_o

acts as a scattering center. We establish the existence and unitarity of the single-particle S-operator, using the time-dependent method, for scattering from a single impurity. We may then write $H = \sum f_{ij} \vec{J}_i \cdot \vec{J}_j + \sum v_{ij} \vec{J}_i \cdot \vec{J}_j$.

The time-independent method also leads to the existence and unitarity of the S-matrix, and shows that the T-matrix is a kernel operator on $L^2(T^\nu)$ of Hölder class. This part of the theory needs the usual assumptions, that the point spectrum for H_1 (i.e. bound states of one spin-wave and the impurity) does not lie in the continuum. The wave operator is given as the unique solution of the Lippmann-Schwinger equation (if there are no bound states)

$$\tilde{\Omega}^{\pm}(k,\ell) = \delta(k-\ell) \pm \lim_{\varepsilon\to+0} \int_{T^\nu} \frac{1}{\omega(\ell)-\omega(k)\pm i\varepsilon} \tilde{V}(k',\ell)\tilde{\Omega}(k,k')d\mu(k')$$

where $H_0 = \omega(k)$ is hte dispersion relation in the absence of the impurity, and $V = H - H_0$. Similarly, the T-matrix for complex energy z defined as $t(z) = V - V \frac{1}{H_0 - z} V$, satisfies

$$t(k,k',z) = V(k,k') - \int \frac{V(k,k'')}{\omega(k'')-z} t(k'',k',z)\, dk'' ,$$

and if there are finitely many bound states

$$t(k,k',z) = \sum_{i=1}^{N} \frac{\phi_i(k)\overline{\phi}_i(k)}{z-E_i} + V(k,k') + \int dq \frac{t(k,q,\omega(k)\pm io)t(q,k',\omega(k)\mp io)}{\omega(k) - z} \tag{1}$$

where $\phi_i(k)$ are the bound state wave-functions with energies E_i.

There are many open problems. Clearly, the 2 spin-wave scattering problem could be treated in the same way, since the kernel operation for t(z) in this case

$$t(k_1,k_2;k_1',k_2';z) = V(k_1,k_2;k_1',k_2') - \int \frac{V(k_1,k_2;k_1'',k_2'')t(k_1'',k_2'',k_1,k_2,z)}{\omega(k_1'') + \omega(k_2'') - z} dk_1'' dk_2''$$

can be reduced to an equation for the center of mass motion

$$t_\kappa(q,q',z) = V_\kappa(q,q') - \int \frac{V_\kappa(q,q')}{\omega_\kappa(q'')-z} t_\kappa(q'',q',z)\, dq''$$

where ω_κ is again an explicit function determined by the f_{ij}, the center of mass momentum κ and the relative momentum q:

$$\omega_\kappa(q) = 2 \sum_\delta f_{0\delta}(1 - \cos \kappa\cdot\delta \quad \cos q\cdot\delta \quad)$$

Because of its similarity with the impurity problem, the unitarity of S presumably goes through under the same assumptions, namely, absence of bound states and spurious solutions in the continuum. The explicit proof of this assumption might be quite hard.

Perez (E T H thesis, 1973) has attempted the scattering problem in the spirit of Faddeev. Presumably, the many spin-wave problem, in the presence of an impurity, can also be done.

An interesting case arises when the impurity potential is not rotation invariant (i.e. not of the form $\sum_{\alpha=1}^{3} v_{ij} J_i^\alpha J_j^\alpha$). For then spin-wave scattering from it will cause spin-wave production. There will be copious production of low energy spin waves and we will see an infra-red problem (a spin wave is the Goldstone particle corresponding to the spontaneous breakdown of rotation invariance).

Another interesting problem is scattering at finite temperature, in which case the spin-wave becomes unstable, and we no longer have a sharp dispersion relation.

Finally, one might study the dynamics of a "Bloch wall", that is the boundary between two magnetic domains magnetized in different directions. At least in one dimension this wall behaves as a particle, $\dots \uparrow\uparrow\underbrace{\uparrow\downarrow}_{j_o}\downarrow\downarrow \dots$ located at the break j_o. One can show that the space time translation automorphisms $\sum(j), \tau(t)$ are implemented in such a representation, but the detailed dynamics is hard to determine.

We now give some of the details.

2. Kinematics .

The algebra $\mathfrak{A}_j$ of observables associated with the ion $j \in Z^\nu$ is defined to be $\mathcal{B}(C^2)$, the C^* - algebra of 2×2 complex matrices, among which we recognize $\sigma^0 = 1$, $\sigma^1 = \begin{pmatrix} & 1 \\ 1 & \end{pmatrix}$, $\sigma^2 = \begin{pmatrix} & -i \\ i & \end{pmatrix}$, $\sigma^3 = \begin{pmatrix} 1 & \\ & -1 \end{pmatrix}$. These generate $\mathfrak{A}_j$, and $J^\alpha = \sigma^\alpha/2$ are interpreted as the spin operators in three orthogonal directions 1-2-3 in R^3. A system of n ions uses the algebra $\bigotimes_{j\in K} \mathfrak{A}_j \cong \mathcal{B}(C^{2n})$. Here, $K \subset Z^\nu$ is the subset containing the n ions under discussion. We denote $\bigotimes_{j\in K} \mathfrak{A}_j$ by $\mathfrak{A}(K)$, in which we identify $1 \otimes \dots \otimes \frac{\sigma^\alpha}{2} \otimes \dots \otimes 1$, $\alpha = 1,2,3$, with $J_j^\alpha \in \mathfrak{A}_j$.

More generally, if $K \subset L$ we identify $\mathfrak{A}_K$ with a subalgebra of $\mathfrak{A}_L$ by the C^* -homomorphism $i_{KL} : \mathfrak{A}_K \to \mathfrak{A}_L$. This is uniquely defined by its action on $\sigma^{\alpha_1} \otimes \ldots \otimes \sigma^{\alpha_{|K|}} \in \mathfrak{A}(K)$, which it takes to $\sigma^{\alpha_1} \otimes \ldots \otimes \sigma^{\alpha_{|K|}} \otimes \underbrace{1 \otimes \ldots \otimes 1}_{|L|-|K|}$. Here, $|K|$ denotes the number of points in K. Linear combinations of $\sigma^{\alpha_1} \otimes \cdots \otimes \sigma^{\alpha_{|K|}}$ are taken into linear combinations, and since these generate $\mathfrak{A}(K)$, $\alpha=0,\ldots,3$, this defines i_{KL} completely.

The system $\{ \mathfrak{A}(K) ;\ i_{KL} ;\ K, L \subset Z^\nu, |K| < \infty, |L| < \infty \}$ forms an *inductive system of C* - algebras* [7]; such a system defines a unique C* - algebra, called the *inductive limit*, $\uparrow_{K \to Z^\nu} \mathfrak{A}_K$, of which the $\mathfrak{A}_K$ may be identified as subalgebras. To obtain $\uparrow_{K \to Z^\nu} \mathfrak{A}_K$, we first form the union $\bigcup_K \mathfrak{A}_K$ of all the elements of any $\mathfrak{A}_K$. Then we say two elements, A_1, A_2 of this set, are *equivalent* if there exists a K such that $K \supset K_1$, $K \supset K_2$, where $\mathfrak{A}_{K_1} \supset A_1$, and $\mathfrak{A}_{K_2} \supset A_2$, and moreover $i_{K_1K} A_1 = i_{K_2K} A_2$. That is, $A_1 \simeq A_2$ iff they can be identified with a third element $i_{K_1K}A_1$ of a larger algebra. This relation is quickly proved to be an equivalence relation. The set of equivalence classes in $\bigcup_K \mathfrak{A}_K$ can be given the structure of a normed *-algebra, as usual by defining the product, sum, norm or * of equivalence classes as the class containing that operation on representatives. This is consistent, since the i_{KL} are injective homomorphisms. This *-algebra is called the algebra of local observables, and its completion (in the norm topology) is the algebra $\mathfrak{A}$ of quasi-local observables. It is simple and separable. We may regard each $\mathfrak{A}(K)$ as a subalgebra in the obvious way.

We remark that this C* - algebra is the same, that is, isomorphic, whatever the arrangemant of ions in space, i.e. in our case, whatever ν. These clearly different physical structures are distinguished by the *local structure* of $\mathfrak{A}$; namely the map from finite subsets K of Z^ν to subalgebras $\mathfrak{A}_K$ of $\mathfrak{A}$.

We now show that the lattice translation group Z^ν can be made to act on $\mathfrak{A}$, defining automorphisms: let $j \in Z^\nu$ and define $\Sigma(j): \mathfrak{A}(K) \to \mathfrak{A}(K+j)$ by the natural translation $\Sigma(j)\ \sigma_i^\alpha = \sigma_{i+j}^\alpha$, $\alpha = 1,2,3$, $i \in Z^\nu$, $j \in Z^\nu$. This can be extended to a unique C* - isomorphism between $\mathfrak{A}(K)$ and $\mathfrak{A}(K+j)$. These isomorphisms, for each $K \subset Z^\nu$, are compatible with the injections i_{KL}. Thus, if $\Sigma(j): \mathfrak{A}(K) \to \mathfrak{A}(K+j)$ is temporarily denoted $\Sigma_K(j)$, we have, if $K \subset L$, $i_{KL}\ \ \Sigma(j) = \Sigma_L(j)$. Because of this, the transformation $\Sigma(j)$ passes to the equivalence classes in $\bigcup_K \mathfrak{A}(K)$, defining a norm preserving *-algebraic map from $\uparrow \bigcup_K \mathfrak{A}(K)$ to itself. This can be extended to the norm completion, $\mathfrak{A}$, in a unique way, since the local algebra is dense in $\mathfrak{A}$. The group homomorphism law $\Sigma(j)\ \Sigma(k) = \Sigma(j + k)$ then also holds in the limit.

Each $\mathfrak{A}_j$ can be transformed onto itself by a spin rotation (not a real rotation since this would move j too). Let $\underline{\theta}$ denote a rotation in SO(3), of magnitude $|\underline{\theta}|$ and direction $\hat{\underline{\theta}}$. Let $U(\underline{\theta}) \in \mathcal{B}(C^2)$ denote the spin 1/2 representation, i.e. $U(\underline{\theta}) = e^{i\underline{\theta}\cdot\underline{J}}$. Then it is known that the J^α transform as a 3-vector, i.e. $U(\theta)J^\alpha U^{-1}(\theta) = R^{\alpha\beta}J^\beta$, where $R^{\alpha\beta}$, $\alpha\beta = 1,2,3$ is the rotation matrix of $\underline{\theta}$. Let us denote by $\rho(R)$ this automorphism of $\mathcal{B}(C^2)$. The simultaneous rotation of all the spins, each by its $\rho(R)$, in $\mathfrak{A}(K)$, defines an automorphism of each $\mathfrak{A}(K)$; again, this is compatible with the injections i_{KL}, and defines an automorphism group $R \to \rho(R)$ ρ: $SO(3) \longrightarrow \text{aut}\,\mathfrak{A}$. Clearly, $\sum(j)$ commutes with $\rho(R)$, $j \in Z^\nu$, $R \in SO(3)$.

So far, we have given the abstract formulation of the kinematics. We recover the usual form of quantum mechanics, with states and operators, by considering a representation $(\pi,\mathcal{H})$ of $\mathfrak{A}$. A class of representations with clear physical interpretations is obtained, via the GNS construction, from infinite tensor products of states, that is, positive linear forms of norm 1.

Let ω_j be a state on $\mathfrak{A}_j$, $j \in Z^\nu$. For example, $\omega_j(A)$ could be given by a vector ψ in C^2: $\omega_j(A) = \langle\psi_j, A\psi_j\rangle$, $A \in \mathcal{B}(C^2)$. The sequence $\{\omega_j\}$ defines an expectation value on any local operator: on the product $\sigma^{\alpha_1} \otimes \ldots \otimes \sigma^{\alpha_n}$, it gives $\omega_1(\sigma^{\alpha_1})\omega_2(\sigma^{\alpha_2})\ldots\omega_n(\sigma^{\alpha_n})$, and by linear extension , it gives a state, ω_K, on each $\mathfrak{A}(K)$, $K \subset Z^\nu$, $|K| < \infty$. Again, these states are compatible with the injections i_{KL}: $\omega_K(A) = \omega_L(i_{KL}A)$, and this allows us to define a unique state ω on $\mathfrak{A}$ such that $\omega\,\mathfrak{A}(k) = \omega_K$. Clearly, the ferromagnetic state $\ldots\uparrow\uparrow\ldots\uparrow\uparrow\ldots$ is a state of this kind, which we may write as $\omega^\uparrow$.

For any real α, $e^{i\alpha}\psi_j$ defines the same state as ψ_j in the sense of forms: $\langle e^{i\alpha}\psi_j, Ae^{i\alpha}\psi_j\rangle = \langle\psi_j, A\psi_j\rangle$. Similarly, a sequence $\{e^{i\alpha_j}\psi_j\}$, $j \in Z^\nu$, defines, in the inductive limit, the same state as the sequence $\{\psi_j\}$, $j \in Z^\nu$. By the GNS construction, a state such as $\omega^\uparrow$ defines a representation, $(\pi^\uparrow\mathcal{H}^\uparrow)$ and a vector $\Omega^\uparrow \in \mathcal{H}^\uparrow$(unique up to a phase) such that $\omega^\uparrow(A) = \langle\Omega^\uparrow, \pi^\uparrow(A)\Omega^\uparrow\rangle_{\mathcal{H}^\uparrow}$ for all $A \in \mathfrak{A}$. A dense set of vectors in $\mathcal{H}^\uparrow$ is obtained by operating on $\Omega^\uparrow$ with $\pi^\uparrow(A)$, as A runs over the local algebra. Such vectors differ from $\ldots\uparrow\uparrow\ldots\uparrow\ldots$ only in a finite region K, if $A \in \mathfrak{A}(K)$. More explicitly, the action of $\pi(\sigma_j^\alpha)$ on a product vector $\ldots\psi_1 \otimes \ldots\psi_j \otimes \ldots$ is

$$\pi(\sigma_j^\alpha)(\ldots\psi_1 \otimes\ldots\otimes \psi_j \times\ldots) = (\ldots\psi_1 \otimes\ldots\otimes \sigma^\alpha\psi_j \otimes\ldots) .$$

The representations obtained in this way from product vectors are not all equivalent. Klauder, McKenna and Woods [8] show that the product states defined by two sequences $\{\psi_j\}_{j\in Z^\nu}$ and $\{\phi_j\}_{j\in Z^\nu}$, lie in the same (or a unitary equivalent) representation if and only if the sequences are *weakly equivalent* in the sense of von Neumann's theory of infinite tensor products, i.e. $\sum_{j\in Z^\nu} \big||\langle\phi_j,\psi_j\rangle|-1\big| < \infty$. Two sequences can be weakly equivalent without being *equivalent* in von Neumann's sense $\sum_{j\in Z^\nu} |\langle\phi_j,\psi_j\rangle - 1| < \infty$; since unitary equivalence is the important thing; the concept of equivalent sequences [9] is a rather unnecessary complication.

Representations of this kind can sometimes be distinguished by a vector parameter, the *mean magnetization*, $\vec{M}$. Given a state ρ on $\mathfrak{A}$, $\vec{M}$ is defined, if it exists, by

$$\vec{M}(\rho) = \lim_{K\uparrow Z^{\nu}} \frac{1}{|K|} \rho\Big(\sum_{j\in K} \vec{J}_j\Big) .$$

Clearly, $\vec{M}(\omega^{\uparrow}) = \uparrow$, $\vec{M}(\omega^{\rightarrow}) = \rightarrow$ etc. Thus $\vec{M}$, a classical variable, takes continuous values and helps label the inequivalent representations of $\mathfrak{A}$, on which it takes constant values as is easily proved. Thus $\vec{M}$ behaves like a "Casimir operator" for $\mathfrak{A}$. We say that a superselection rule operates between states of different $\vec{M}$.

Haag and Kastler [10] have argued that all representations are physically indistinguishable, and that a "global" quantity like $\vec{M}$ is not an observable. I do not agree with this view, even though it is widely held. Their conclusion would be true of infinite ferromagnets, but is incorrect for actual magnets.

We now discuss whether the automorphisms $\rho(R)$ and $\sum(j)$ are implemented by unitary operators in the ferromagnetic representations, say $\pi^{\uparrow}$. The existence of unitary operators $S(j)$, $\mathcal{H}^{\uparrow} \to \mathcal{H}^{\uparrow}$, such that $\pi^{\uparrow}(\sum(j)A) = S(j)\pi^{\uparrow}(A)\,S^{-1}(j)$ for all $j\varepsilon Z^{\nu}$, $A\varepsilon\mathfrak{A}$ follows from the Gelfand-Naimark-Segal theorem, since $\omega^{\uparrow}$ is invariant under these automorphisms. For the same reason, if $\theta\varepsilon SO(3)$ represents a rotation about the 3rd axis, then there exists a unitary operator $U(\theta)$, such that $\pi^{\uparrow}(\sum(\theta)A) = U(\theta)\pi^{\uparrow}(A)U^{-1}(\theta)$ for all A. Its action on states differing from $\Omega^{\uparrow}$ only in a finite region is clearly continuous in θ, and as these form a dense set and U is unitary, we see that $U(\theta)$ is a continuous unitary representation of SO(2) on $\mathcal{H}^{\uparrow}$ Since SO(2) is compact, $\mathcal{H}^{\uparrow}$ splits into a direct sum of representations labelled by integers. The self-adjoint generator, N, of $U(\theta)$ is called *the spin-wave number operator*. Clearly $N\Omega^{\uparrow} = 0$; moreover, $N \geq 0$. To see this, note that an orthonormal basis in $\mathcal{H}^{\uparrow}$ consists of

$$\Omega;\ \Big\{\bigotimes_{j\neq j_1} |\uparrow\rangle_j \times |\downarrow\rangle_{j_1}\Big\}_{j_1\,\varepsilon\, Z^{\nu}};\ \Big\{\bigotimes_{j\neq j_1,j_2} |\uparrow\rangle_j \otimes |\downarrow\rangle_{j_1} \otimes |\downarrow\rangle_{j_2}\Big\}$$

etc.; these states are the 0,1,2... spin-wave states, and the spin-wave number, $N = \sum_j (J_j^3 - 1/2)$ defines a self-adjoint operator on $\mathcal{H}^{\uparrow}$ and generates rotations about the third axis. Thus we may write $\mathcal{H}^{\uparrow} = \bigoplus_{n=0}^{\infty}\mathcal{H}_n$, where $\mathcal{H}_n$ is the span of the spaces with n spins displaced. Since $S(j)$ and $U(\theta)$ commute, each $\mathcal{H}_n$ is invariant under space-translations (as can be seen too by their definition in terms of the basis).

The single particle space, $\mathcal{H}_1$, is naturally isomorphic to $\ell_2(Z^\nu)$, and, under the isomorphism, S(j) becomes the regular representation of Z^ν, and is therefore multiplicity free.

It is easy to prove that another rotation, $\sum(\underline{\theta})$, not leaving the vacuum invariant, is not implemented in $(\pi^\uparrow, \mathcal{H}^\uparrow)$. For, if it were, the states $\ldots\uparrow\uparrow\ldots\uparrow$, and $\ldots\nearrow\nearrow\ldots\nearrow\ldots$ would belong to the same representation. But, by the Klauder, McKenna and Woods criterion, $\prod_{j\epsilon Z^\nu} \left|\left|\langle\uparrow,\nearrow\rangle\right|-1\right| = \prod_j (\cos\theta - 1) = \infty$, a contradiction.

It is easy to prove that $\mathcal{H}_n$ contains no invariant vector under S(j) $(j \neq 0)$, and so $\Omega^\uparrow$ is the only invariant vector in $\mathcal{H}^\uparrow$. This property is called the uniqueness of the vacuum. This does not mean that $\omega^\uparrow$ is the only invariant state in the sense of positive linear forms; for example, $\omega^\downarrow$ or $\omega^\nearrow$ are also invariant. But they are not vector states of $\pi^\uparrow$. Each of the corresponding representations $\pi^\uparrow$, $\pi^\nearrow$, $\pi^\rightarrow$ ... has a vacuum, and only one.

The uniqueness of the translationally invariant vector $\Omega^\uparrow$ is equivalent to the *cluster decomposition property:* for $\phi,\psi \in \mathcal{H}^\uparrow$,

$$\lim_{j\to\infty} \langle\phi, U(j)\psi\rangle = \langle\phi,\Omega^\uparrow\rangle\langle\Omega^\uparrow,\psi\rangle$$

which can easily be proved independently.

We may use the uniqueness of the vacuum to prove that the spin automorphism $\sum(\underline{\theta})$ is not implemented, unless $\underline{\theta}$ leaves $\uparrow$ invariant. This is in fact the same statement, as saying that $\pi^\nearrow$ are inequivalent for different directions. We first observe that
$\langle\Omega^\uparrow,\vec{J}_i\Omega^\uparrow\rangle \neq \langle\Omega^\uparrow,\sum(\underline{\theta})(\vec{J}_i)\,\Omega^\uparrow\rangle$ for such $\underline{\theta}$. Now suppose, if possible, that there is a $V(\underline{\theta})$ such that $V(\underline{\theta})\, J_j^\alpha V^{-1}(\underline{\theta}) = R_{\alpha\beta}(\theta) J^\beta$. Then $V(\theta)S(j)V^{-1}(\theta)S^{-1}(j)$ commutes with $\pi^\uparrow(\mathfrak{A})$, since $\sum(j)$ and $\rho(\underline{\theta})$ commute. Since $\pi^\uparrow(\mathfrak{A})$ is irreducible, $V(\theta)S(j)V^{-1}(\theta)S^{-1}(j)$ is a multiple of the identity, λI say. Then

$$V(\theta)S(j)\Omega^\uparrow = \lambda S(j)V(\theta)\Omega^\uparrow,$$

i.e. $V(\theta)\Omega^\uparrow$ is an eigenstate of S(j). Because of uniqueness, $V(\theta)\Omega^\uparrow = e^{i\alpha}\Omega^\uparrow$, contradicting the non-invariance of $\Omega^\uparrow$ under $\rho(\underline{\theta})$.

The realization $\mathcal{H}_1 \simeq \ell_2(Z^\nu)$ is called the *Wannier representation*. We note that the position operator, $\hat{j}$, defined by $(\hat{j}\psi)(j)=j\psi(j)$, is conjugate to momentum $\hat{p}$: they satisfy the Heisenberg commutation relations in the form suggested by Mackey: if $j\epsilon Z^\nu$ and $k\epsilon \hat{Z}^\nu = T^\nu$, the dual group, define

$$(U(j)\psi)(i) = \psi(i + j)$$
$$(V(k)\psi)(j) = e^{ik\cdot j}\psi(j), \quad \text{where} \quad k\cdot j = \sum_{\alpha=1}^{3} k^\alpha j^\alpha .$$

Then

$$U(j)V(k) = e^{ik\cdot j}V(k)U(j).$$

Mackey's imprimitivity theorem [11] implies the uniqueness of the irreducible representation of these Weyl relations (analogue of the Stone - von Neumann theorem on the uniqueness of the Schrödinger representation).

3 Dynamics.

So far, we have given an interpretation of the operators of $\mathfrak{A}$ in terms of lattice spins at one time, say $t = 0$. We seek an automorphism group $\tau(t)$ of $\mathfrak{A}$, $t \in R$, such that $\tau(t)A$ has the interpretation as the measurement of $A \in \mathfrak{A}$ at a time t. If $\tau(t)$ is indeed as automorphism, then the operators $\tau(t)A$ form the same algebra, $\mathfrak{A}$.

In the Heisenberg model, one attempts to define the time-evolution by specifying the Hamiltonian

$$H = \sum_{i,j\in Z^\nu} f_{ij}\, \vec{J}_i\cdot\vec{J}_j \qquad (2)$$

For simplicity we shall assume that for given i, $f_{ij} = 0$ except for a finite number $\leq Q$, of j. This expresses that the range of the force is finite. Eq(2) has no immediate meaning if the sum is infinite. We would like time-evolution, and thus H, to commute with space-translations. This requires that $f_{i,j} = f_{i+k,j+k}$ for all k,i and j. Thus, an infinite sum in (2) is inevitable, and our first task is to give it a meaning.

For each i,j, the Hilbert space $C_i^2 \otimes C_j^2$ carries the representation $\mathcal{D}^{1/2} \otimes \mathcal{D}^{1/2}$ of SO(3). By the Clebsch-Gordan theorem, this is decomposable as $\mathcal{D}^0 \oplus \mathcal{D}^1$, the decomposition corresponding to the well known singlet (anti-symmetric) state, $|\uparrow\rangle \otimes |\downarrow\rangle - |\downarrow\rangle \otimes |\uparrow\rangle$, for $\mathcal{D}^0$, the scalar; and the triplet (symmetric) states $|\uparrow\rangle \otimes |\uparrow\rangle$, $|\uparrow\rangle \otimes |\downarrow\rangle + |\downarrow\rangle \otimes |\uparrow\rangle$ and $|\downarrow\rangle \otimes |\downarrow\rangle$, for $\mathcal{D}^1$, the vector.

The rotation invariant term $\vec{\sigma}_i\cdot\vec{\sigma}_j$ can be proved to be $P_{ij}^1 - 3P_{ij}^0$, where P^1 is the projection onto the triplet and P^0 the projection onto the singlet. The time-evolution, Eq. (0), uses the commutator, [H,A], and is unchanged if we add a multiple of the identity to H. We choose to add $-f_{ij}\, I_{\mathcal{H}}$; this is a simple "vacuum renormalization." Summing over i and j, and rearranging, leads to

$$H = \sum_{ij} \tfrac{1}{4} f_{ij}\, (\vec{\sigma}_i\cdot\vec{\sigma}_j - 1) = - \sum f_{ij}\, P_{ij}^0$$

This is ≥ 0 if $f_{ij} \leq 0$, the condition for ferromagnetism, since the energy is then *lowest* in the triplet state. We note that $P^o_{ij}\Omega^\uparrow = 0$, so $H\Omega = 0$, and H is defined on the dense set of states containing finitely many spin-waves. The existence of a self-adjoint extension is already assured, by Friedrich's theorem.

We note that H maps $\mathcal{H}_n$ (the space of n spin-waves) into itself. For, in considering $H\psi$ with $\psi \varepsilon \mathcal{H}_n$, a single term in the sum, $f_{ij}P^o_{ij}$, gives zero unless i and j have opposite spins (in one of the product vectors making up ψ); and if the spins are opposite at i and j,

$$P^o_{ij}\left(|\uparrow\rangle_i \times |\downarrow\rangle_j\right) = \frac{1}{\sqrt{2}}\left(|\uparrow\rangle \times |\downarrow\rangle - |\downarrow\rangle \times |\uparrow\rangle\right)$$

Each term on the right-hand side contains one spin-wave on the sites i,j. Collecting terms we see that $\mathcal{H}_n$ is mapped into itself by H. Tracing back the argument, we find that this is a consequence of the fact that $\vec{J}_i\cdot\vec{J}_j$ is invariant under spin rotations about the 3rd axis, so that the spin-wave number $\sum_j (J^3_j - 1/2)$ *should* be conserved in time.

Consider now nearest-neighbor interactions, $\nu = 1$, $f = 1$, on the one-particle space. If $\psi = \sum c_j|j\rangle$ is the Wannier representation of $\psi\varepsilon\mathcal{H}_1$, then

$$H\psi = f \sum c_j \left(|j\rangle - 1/2|j-1\rangle - 1/2|j+1\rangle\right)$$

$$= f\left(1 - 1/2\ S(-1) - 1/2\ S(+1)\right)\psi$$

where S(j) is the space translation operator. This expresses the 1-particle energy as a function of the translation operator. In fact, S(j) on $\mathcal{H}_1 \cong \ell_2(Z^\nu) = L_2(T^\nu, d\hat{p})$ form a maximal abelian set, and so $H|_{\mathcal{H}_1}$, which commutes with all S(j), must be a function of S(j).

In the Fourier transform space $L_2(T^\nu)$, U(j) becomes the multiplication operator $e^{ij\cdot k}$, where $k\varepsilon T^\nu$: $(U(j)\tilde{\psi})(k) = e^{ij\cdot k}\tilde{\psi}(k)$. The energy also is a multiplication operator

$$E(k) = \hbar\omega(k) = f\left(1 - \frac{e^{ik} + e^{-ik}}{2}\right) \quad \text{if } \nu = 1$$

$$= f(1 - \cos k)$$

Similarly, $E(k) = f(3 - \cos k - \cos k^2 - \cos k^3)$, $\nu = 3$. It is interesting to consider this equation for small k: $\omega(k) = \frac{f}{2}(k_1^2+k_2^2+k_3^2)$, so spin-waves behave as a particle of mass 1/f.

One can show that $H|_{\mathcal{H}_n}$ is a bounded operator, of norm $\leq fQn$, where Q is the number of neighbors. We may understand this in terms of spin-waves; if n spin displacements are far apart, each has energy $\leq fQ$. If they are closer together, so that they form triplet states with the range of the force, then the energy is less. This result implies at once that H is essentially self-adjoint on $\bigcup_n \mathcal{H}_n$, and that e^{-iHt} maps $\mathcal{H}_n$ into itself.

This discussion could equally well have been carried out in any other ferromagnetic representations $\mathcal{H}^{\nearrow}$, in which spin-waves are displacements in the opposite direction $\swarrow$.

However, we have not established yet that the map $\pi^{\uparrow}(A)\to T(t)\pi^{\uparrow}(A)T^{-1}(t)$ defines an automorphism group of $\mathfrak{A}$. For although $\pi^{\uparrow}(A_t) = T(t)\pi^{\uparrow}(A)T(-t) \in \mathcal{B}(\mathcal{H}^{\uparrow})$, it might not lie in $\mathfrak{A}$. If it were not, we would need to include all the A_t as observables, and enlarge $\mathfrak{A}$. We know that $\pi^{\uparrow}(\mathfrak{A}) \neq \mathcal{B}(\mathcal{H}^{\uparrow})$, since $\mathfrak{A}$ possesses some automorphisms, $\rho(R)$, that are not implemented in $\pi^{\uparrow}$, whereas it is known that every automorphism of $\mathcal{B}(\mathcal{H}^{\uparrow})$ (for any $\mathcal{H}$) is implemented.

The main step in proving that $T(t)\pi^{\uparrow}(a)T(-t) \in \pi^{\uparrow}(\mathfrak{A})$, thus defining an automorphism $\tau(t):\mathfrak{A}\to\mathfrak{A}$, is to prove that the perturbation expansion of Eq (1)

$$A(t) = A(0) + it[H,A(0)] + \frac{(it)^2}{2!}[H,[H,A(0)]] + \ldots$$

converges in norm. For finite range interactions, this was given in [1 This result was subsequently extended to wider and wider classes of interactions with rapidly falling tails, by Trottin and Manuceau, Robinson and Ruelle [16]. We give the simplest case

<u>Theorem.</u> Suppose $H = \sum_{j\in Z^\nu} X_j$, where X is a local operator and $X_j = S(j)XS^{-1}(j)$. Then for any local A the series

$$A + it[H,A] + \frac{(it)^2}{2!}[H,[H,A]] + \ldots$$

is norm convergent with a radius of convergence independent of A.

<u>Proof.</u> Let $\mathfrak{A}_r$ denote the algebra associated with a cube of side r and suppose $A \in \mathfrak{A}_r$, $X \in \mathfrak{A}_R$. The number of points in a cube of side r is r^ν. The nth term in the series can be written

$$\frac{(it)^n}{n!}\sum_{j_1\ldots j_n\in Z^\nu}[X_{j_1}[\ldots[X_{j_n},A]\ldots]$$

The norm of each term in the sum is $\leq 2^n \|X\|^{n-1}\|A\|\,|t|^n/n!$ It remains to estimate the number of non-zero terms. Now, $[X_{j_n},A] = 0$ unless $|j_n^\alpha| < R + r$, and $[X_{j_{n-1}}[X_{j_n},A]] = 0$ unless $|j_{n-1}^\alpha| < R + r$, or $|j_{n-1}^\alpha - j_n^\alpha| < 2R$, $\alpha = 1,2\ldots\nu$. Proceeding, we see that the multi-commutator is zero except in the intersection of cubes:

$$\{|j_n^\alpha| < R + r\} \cap$$
$$\cap \{|j_{n-1}^\alpha| < R + r \cup |j_{n-1}^\alpha - j_n^\alpha| < 2R\}$$
$$\cap \{|j_{n-2}^\alpha| < R + r \cup |j_{n-2}^\alpha - j_{n-1}^\alpha| < 2R \cup |j_{n-2}^\alpha - j_n^\alpha| < 2R\}$$
$$\cap \{j_1^\alpha < R + r \cup |j_1^\alpha - j_2^\alpha| < 2R \cup |j_1^\alpha - j_3^\alpha| < 2R \ldots, |j_1^\alpha - j_n^\alpha| < 2R\}.$$

This volume is the union of n! cubes, each of volum $(R + r')^{\nu n}$ where $r' = R$ or r. For large n, it can be shown [17] that the sum is dominated by $cn!\, R^{\nu n}$. Hence the series converges, independent of r.
[QED

This means that time evolution (obtained by iterating the time interval of convergence) can be defined purely algebraically, and is continuous in norm: $\|\tau(t)A - A\| \to 0$ as $t \to 0$.

Let us now return to the detailed behavior of a single spin wave. We have defined the position operator on $\mathcal{H}_1$ in the Wannier representation: $\hat{\underline{j}} = \sum_j \underline{j}\, J_j^- \cdot J_j^+$. The *velocity operator* is

$$\underline{v} = i[H, \hat{\underline{j}}]$$

Since any operator commuting with S(j) is a multiplication operator on $L^2(T^\nu)$, $\underline{v}$ is a function of $\underline{k}$. Since $\hat{\underline{j}} = -i\partial/\partial\underline{k}$ on $L^2(T^\nu)$ we see that $\underline{v}(\underline{k}) = \nabla\omega(\underline{k})$. Thus the velocity operator is multiplication by the group velocity. The momenta k_o at which $\nabla\omega(\underline{k}_o) = 0$, and the corresponding energies $\omega(\underline{k}_o)$, are called *critical*. Physically, they correspond to momenta at which the spin-wave ceases to propagate. We say a critical point k_o is *non-degenerate* if $\det \partial^2\omega/\partial k_i \partial k_j \neq 0$ at k_o. Then we may write, in suitable coordinates q on the manifold T^ν, in a neighborhood U of k',

$$\omega(k) = \omega(k') - \sum_{i=1}^{\lambda} q_i^2 + \sum_{i=\lambda+1} q_i^2 \quad , \quad k \in U \ .$$

If $\lambda = 0$ we get Schrödinger behavior. It is obvious that non-degenerate critical points are isolated.

It is obvious that, if H has finite range, $\omega(k)$ is analytic, and so the set of critical momenta forms a submanifold with Haar measure (on T^ν) zero. We shall consider this case here. More general functions ω, say C^∞, lead to the study of Morse theory [15].

Results on the density of critical points are important since they have a bearing on the spread and fall of a spin-wave packet. In the k-representation $(T(t)\tilde{\psi})(k) = e^{-i\omega(k)t}\,\tilde{\psi}(k)$, and so in the Wannier representation

$$(T(t)\psi)(j) = \frac{1}{(2\pi)^{\nu/2}} \int \tilde{\psi}(k) e^{-i\omega(k)t + i\underline{k}\cdot\underline{j}}\, d^\nu k$$

We know that the main contribution to the behavior at ∞ in t is given by the small interval around the critical points of ω. The rest of the integral contributes a function that is $o(t^{-N})$ for any N.

If $\omega = k^2/2m$ (Schrödinger case) then the behavior, $t \to \infty$, is governed by the usual "spread of the wave packet ",

$$|f(\underline{x},t)| < c(\underline{x})t^{-\nu/2} , \qquad t \to \infty .$$

This motivates the definition: a magnon(=spin-wave) in Z^ν is a *quasi-particle* if

$$|\psi(j,t)| < c(j)t^{-\nu/2} , \qquad |t| \to \infty.$$

For scattering theory it is sufficient for quasi-particles to be dense in $\ell^2(Z^\nu)$.

In a similar way, one can study the motion of spin-waves in a given external magnetic field $\{\vec{h}_j\}$ with the Hamiltonian $H = \sum f_{ij} \vec{J}_i \cdot \vec{J}_j + \sum \vec{h}_j \cdot \vec{J}_j$, where $h_j \in R^3$, $j \in Z^\nu$.

4. Impurity and Spin-Wave Scattering.

The Green's function for a spin-wave is

$$G(j,t) = {}_0\langle \downarrow | U(t) | \downarrow \rangle_j = \frac{1}{(2\pi)^{\nu/2}} \int e^{-i\omega(k)t} e^{i\underline{k}\cdot\underline{j}} \, d^\nu k .$$

For nearest neighbors, where $\omega = f \sum (1 - \cos k^\alpha)$, using $(2\pi)^{-1} \int_{-\pi}^{\pi} e^{in\theta} e^{ix\cos\theta} \, d\theta = -iJ_n(x)$, this becomes

$$G(j,t) = (-i)^\nu e^{i\nu ft} \prod_{\alpha=1}^{\nu} J_{j^\alpha}(ft) .$$

For large times we have Weber's inequality $|J_n(t)| \leq C(n)|t|^{-1/2}$ if $2|t| > |n| - 1/2$. Thus we do get quasi-particle behavior of a single spin-wave and any compact superposition, but the decay is not uniform in space. For $n \approx t$, it can be shown that $J_n(n) \sim \Gamma(1/3)/(2^{2/3}\, 3^{1/6}\, n^{1/3})$, so $G(n,n) \sim 1/n^{\nu/3}$, instead of $1/n$ $1/n^{\nu/2}$ as we might like.

An impurity can be represented by a Hamiltonian H of the same form, $\sum f_{ij} \vec{J}_i \cdot \vec{J}_j$, but where f_{ij} is not translation invariant; it could take different values if $i = i_o$, expressing the presence of an impurity at i_o. In this case we could define H_o, on $\mathcal{H}_1$, to be given by a translation invariant f_{ij} fixed by those i,j far from i_o. (If f_{ij} has not got finite range, the choice of H_o is more subtle, and is an open problem). The potential V is then defined as

$V = H - H_o$, and it has the form $V = \ldots 1 \otimes \ldots \otimes 1 \otimes \mathcal{V} \otimes \ldots \times 1 \ldots$ where $\mathcal{V} \varepsilon \mathcal{O}(K)$. V is bounded but not of finite rank. If the impurity is rotation invariant, as here, then V has a restriction to $\mathcal{H}_1$, and $V|_{\mathcal{H}_1}$ has finite range, so is zero on any state $\psi(j)$ with support outside K, so $V|_{\mathcal{H}_1}$ has finite rank. More generally, Watts [5] considers potentials such that $V|_{\mathcal{H}_1}$ has a tail, but is of trace-class.

If V is not invariant under rotations about the 3rd axis, then the spin-wave number is not conserved, and the scattering center produces spin-waves, like a fixed source in meson theory. The treatment of this case is an open problem.

We note that H_o, in $\mathcal{H}_1$, has no eigenvectors. $H = (H_o + V)|_{\mathcal{H}_1}$, might well have bound states, if V is attractive. Thus, an impurity can "capture" a spin-wave.

The time-dependent scattering theory of a spin-wave from a rotation-invariant impurity is easily disposed of. The existence of the four Moller operators

$$\Omega_\pm \quad (H_1, H_o) = \lim_{t\to\pm\infty} e^{iH_1 t} e^{-iH_o t}$$

$$\Omega_\pm \quad (H_o, H_1) = \lim_{\to\pm\infty} e^{iH_o t} e^{-iH_1 t} \qquad \text{on } \mathcal{H}_1$$

follows from a theorem of Kato [12], which uses only that V is compact.

The scattering of several spin-waves from an impurity is an open problem, as is the asymptotic completeness. Unitarity of $S = \Omega_+^* \Omega_-$ on $\mathcal{H}_1$ also follows from Kato [12].

The scattering of two spin-waves off each other (in the absence of an impurity) has been considered by Watts [5], and the n spin wave problem by Hepp [6], using " scattering in two Hilbert spaces." We wish to describe the ingoing and outgoing magnons in terms of free one-particle spin-waves. That is, we use the space $\mathcal{F} = C \oplus \mathcal{H}_1 \oplus (\mathcal{H}_1 \times \mathcal{H}_1)_s \oplus \ldots$, the Fock space over $\mathcal{H}_1$. However, the physical spin waves do not quite behave as Bosons, because of the kinematical repulsion caused by $(J_j^-)^2 = 0$, i.e. two spin-waves cannot lie at the same point. States in $\mathcal{F}$ are called *ideal spin-waves*. The free time-evolution $T_o(t)$ is defined in $\mathcal{F}$, and is taken as

$$T_o(t) = 1 \oplus e^{-iH_1 t} \oplus e^{-iH_1 t} \otimes e^{-iH_1 t} \oplus \ldots$$

To obtain this asymptotic description, we embed the physical space in the ideal space, $i: \mathcal{H}_n \to \mathcal{H}_1 \otimes \ldots \otimes \mathcal{H}_1$, by the isometric map

$$i(J_1^- \ldots J_n^- \Omega) = J_1^- \Omega \otimes \ldots \otimes J_n^- \Omega$$

where of course, all 1, ...n are different points.

As usual, we compare e^{-iHt} on $\mathcal{H}_n$ with $T_o(t)$ on $\mathfrak{F}$. Thus we wish to show

Theorem. (Watts, Hepp). For nearest neighbor coupling:

$$\Omega(t) = e^{iHt} i^* e^{-iH_ot} \xrightarrow[\text{s-lim}]{} \Omega_\pm , \quad t \to \pm\infty.$$

$$\Omega^*_\pm \Omega_\pm = 1 \quad \text{and } e^{iHt}\,\Omega_\pm = \Omega_\pm e^{iH_ot}$$

Remarks. $\Omega_\pm$ maps $\otimes^n \mathcal{H}_1$, into $\mathcal{H}_n$, and so $\mathcal{H}^\pm = \Omega^\pm \mathfrak{F} \subset \mathcal{H}^\uparrow$ is a pair of subspaces of the physical space, that represent asymptotic states. The S operator, $\Omega^*_+\Omega_-$, maps $\mathfrak{F}$ into $\mathfrak{F}$. On the two particle space, the effective potential is

$$V_2 = iHi^* - H_o \otimes 1 - 1 \otimes H_o.$$

Here, i, i^* eliminate states $|j\rangle \otimes |j\rangle$, i.e. act as a kinematic repulsion.

Proof. It suffices to prove convergence on a dense set of $\Phi \in \mathfrak{F}$, and for this, it suffices to prove it for $\Phi \in \otimes^n(\mathcal{H}_1)_s$ for each n, with wave-function $\Phi(k_1,\ldots k_n)$, with non-overlaping velocities. Then

$$\frac{d}{dt} e^{iHt} i^* e^{-iH_ot}\,\Phi = \int dk_1\ldots dk_n\, \Phi(k_1\ldots k_n) e^{-it\sum^n \omega(k_i)}$$
$$\times\, e^{iHt}(H - \sum^n \omega(k_i))\, \tilde{J}_-(k_1)\ldots\tilde{J}_-(k_n)\Omega \quad ,$$

where $\tilde{J}_-(k) = \sum_{j\in Z^\nu} J^-_j\, e^{i\underline{k}\cdot\underline{j}}$ which we may write as

$$\frac{d}{dt} e^{iHt} i^* e^{-iH_ot}\Phi = ie^{iHt} i^* \,|\Phi_t\rangle$$

where $\Phi_t(k_1,\ldots k_n) = \sum_{1\le i\le j\le n} \int d^\nu \ell\, V(k_i,k_j,\ell) \;\times$

$$\times\, \Phi(\ldots,k_i + \ell/2,\ldots,k_j - \ell/2) e^{it[\omega(k_1)+\ldots+\omega(k_i+\ell/2)+..\omega(k_j-\ell/2)]}$$

and $V(p,q,\ell) = f/(2\pi)^3 \sum_\delta \cos(\delta\cdot(p-q)/2)(\cos\delta(p+q) - \cos\delta\cdot\ell)$.

Here, δ are the vectors from a point to its neighbors, in the sense that $f_{ij} \neq 0$ if $\underline{i}-\underline{j} = \underline{\delta}$. For states with non-overlapping velocities, the norm of this is $o(t^{-N})$ for any N. Hence

$$\left\| \frac{d}{dt} (e^{iHt} i^* e^{-iH_ot})\Phi \right\|$$

is integrable, and so the wave-operators converge, by Cook's criterion.

To show that $\Omega_\pm$ are isometric, we follow Hepp [6]. While i is isometric, i* is only partially isometric. Hepp shows that it is asymptotically isometric, that is

Lemma.

$$\lim_{t\to\pm\infty} \langle \Phi_t | ii^*\Phi_t \rangle = \langle \Phi, \Phi \rangle \tag{3}$$

where $\Phi \in \mathfrak{F}$, $\Phi_t = e^{-iH_0 t}\Phi$.

Proof. Since $i^* e^{-iH_0 t}$ is a contraction, $\mathfrak{F}_n \to \mathcal{H}_n$, it is sufficient to prove (3) for a dense set $\Phi \in \mathcal{S}_n$. Now

$$\langle \Phi, \Phi \rangle - \langle \Phi^t, ii^*\Phi^t \rangle = \sum_{j_1 \ldots j_n} |\tilde{\Phi}^t(j_1, \ldots j_n)|^2 \tag{4}$$

where the sum extends over all $j_1, \ldots j_n \in Z^\nu$ with $j_i = j_k$ for some $i \neq k$, since, if all $j_i \neq j_k$, i acts as an embedding, $\Phi \to \Phi^t$ is unitary, and the term cancels. Now (4) can be majorized by

$$C \sum_{j_1 \ldots j_n} |\tilde{\Phi}^t(j, j, j_3, \ldots j_n)|^2$$

$$= C \int_{T^{n-2}} d\check{k}_3 \ldots d\check{k}_n \int_D d\check{k}\, d\check{\ell}\, d\check{m}\, e^{itE(k_1 \ell, m)}$$

$$\times \Phi(k/2+\ell, k/2-\ell, k_3, \ldots k_n)\Phi(k/2+m, k/2-m, k_3 \ldots k_n)$$

where $D = \{k, \ell, m \in R^9; (k/2 \pm \ell) \leq 2\pi,\ k/2 \pm m < 2\pi\}$ and

$$E(k, \ell, m) = \omega(k/2 + \ell) + \omega(k/2 - \ell) - \omega(k/2 + m) - \omega(k/2 - m)$$

$$= 2f \sum_{\alpha=1}^{\nu} \cos k^\alpha/_2 \{\cos \ell^\alpha - \cos m^\alpha\}$$

for nearest neighbor interactions. Thus, if $\Phi(k/_2 + \ell, k/_2 - \ell, k_3 \ldots k_n)$ is symmetric and zero in a neighborhood of $\nabla_\ell E(k, \ell, m) = 0$, (4) is $O(t^{-N})$. Since such Φ form a dense set, we have the lemma. As a corollary, $\Omega_\pm^*\Omega_\pm = 1$ and $e^{iHt}\Omega_\pm = \Omega_\pm e^{iH_0 t}$ (see Jauch [13]). This proves the theorem.

An isometric scattering operator S is defined in $\mathcal{H}^+ \to \mathcal{H}^-$ by

$$S\Omega_+\Phi = \Omega\Phi, \quad \Phi \in \mathfrak{F}.$$

However, it is known that $\mathcal{H} \neq \mathcal{H}_\pm$, because of the existence of spin-wave bound states, and so, of many scattering channels.

This question has been studied by Perez by the Faddeev method [14]. We now give an introduction to the usual time independent scattering theory, with a few surprises.

5. The Time Independent Theory of Impurity Scattering.

The potential V for spin-wave scattering from an impurity, is clearly a kernel operator on $L^2(T^\nu)$

$$(V\tilde{\psi})(k) = \int_{T^\nu} V(k_1,k_2)\psi(k_2)\, d^\nu k_2$$

where $V(k_1,k_2) = \sum d_{j,j+\delta}(e^{-i\underline{k}_1\cdot\delta}-1)(1- e^{i\underline{k}_2\cdot\underline{\delta}})$.
It turns out to be convenient to consider potentials of Hölder class:

$$|V(k,k') - V(k + h,k' + h')| \leq C(|h|^{\mu_o} + |h'|^{\mu_o'})$$

where $\mu_o,\mu_o' > 1/2$ and addition is on T^ν,i.e. *mod* 2π. Clearly, our V satisfies these conditions.

We note that the resolvent $r(z) = (H-z)^{-1}$ and the free resolvent, $r_o(z) = (H_o-z)^{-1}$ satisfy the equation

$$r(z) = r_o(z) - r_o(z)Vr(z) \qquad \text{if } \operatorname{Im}z \neq 0 .$$

Putting $S = 1 + iT$, and defining

$$t(z) = V - Vr(z)V$$

we know that $T = \lim_{\substack{\operatorname{Im}z\to 0\\ z=k^2}} t(z)$.

Our task is to prove that t(z) is a kernel operator, whose limit, T, is too. The operator t(z) satisfies

$$t(z) = V - Vr_o(z)t(z) \tag{5}$$

We prove that t is a kernel operator of Hölder type, and that it satisfies unitarity.

We start by considering (5) in kernel form

$$t(k,k'z) = V(k,k') - \int \frac{V(k,k'')}{\omega(k'')-z}\, t(k'',k',z)\, dk''$$

and show that this has a unique solution, $t(\cdot,k',z)$ in the class $b(\mu)$ for some μ:

$$b(\mu) = \{f: T^\nu \to \mathbb{R};\ |f(k+h) - f(k)| \leq C|h|^\mu\}$$

with norm

$$\|f\|_\mu = \sup_{h,k}\left[\frac{|f(k+h) - f(k)|}{|h|^\mu} + |f(k)|\right].$$

Definition. Let $f \in b(\mu)$ and define the operator $a(z)$

$$(a(z)f)(k) = \int \frac{v(k,k'')}{\omega(k')-z} f(k'')\, dk''$$

Then the equation (5) can be written $a(z)t(\cdot,k';z)+\mathbb{1}\, t(\cdot,k';z)=\mathbb{1}\, v(\cdot,k')$. This is the inhomogeneous equation (IHE). The corresponding homogeneous equation (HE) is $a(z)s(\cdot,k',z) + \mathbb{1}\, s(\cdot,k',z) = 0$.

We can thus apply the Fredholm alternative: I.H.E. has a unique solution if and only if H. E. has no non-zero solution.

Our first task is to prove $a(z)$ is compact, and then to discuss the homogeneous equation. The proof that $a(z)$ is a compact operator, with domain $b(\mu)$ and range $b(\mu')$, where $\mu' > \mu$, follows the same method as found, for example, in Faddeev's book [14]. It remains to discuss the non-existence of solutions to HE .

Lemma. If $\text{Im } z \neq 0$, HE has no non-trivial solutions in $b(\mu)$, for $\mu < \mu_0$.

Proof. Suppose HE has a solution $\phi_0 \neq 0$ for z_0. Then

$$\phi_0(k) + \int \frac{V(k,k')}{\omega(k')-z_0} \phi_0(k')\, dk' = 0 .$$

Let $\psi_0(k) = \dfrac{\phi_0(k)}{\omega(k)-z_0}$. Since $\text{Im } z_0 \neq 0$, $\psi_0 \in L^2(T^\nu)$. HE can be rewritten $[\omega(k)-z_0]\psi_0(k) + \int V(k,k')\psi_0(k')\, dk' = 0$, i.e. $-H\, \psi_0 = z_0\psi_0$, i.e. ψ_0 is a normalizable eigenstate for complex eigenvalue, impossible since $H = H^*$.

Lemma. For z sufficiently large, HE has no solution $\neq 0$. For sufficiently large $z, k = \|a(z)\| < 1$; then $\phi = -a(z)\phi$ implies $\|\phi\| \leq k\, \|\phi\| < \|\phi\|$ if $\phi \neq 0$, impossible.

Let us now consider solutions $\phi_{\xi\pm} \in b(\mu)$ to HE in the following sense of distributions:

$$\phi_{\xi\pm}(k) = -\lim_{\varepsilon\to 0} \int \frac{V(k,k')\phi_{\xi\pm}(k')}{\omega(k')-\xi^2 \mp i\varepsilon}\, dk' . \tag{6}$$

Lemma. If $\phi_{\xi\pm} \in b(\mu)$, satisfying (6), and ξ^2 not a critical energy, then $\phi_{\xi\pm}(k) = 0$ a.e. on the energy surface $\omega(k) = \xi^2$.

Proof. Take the + sign. Let $d(\varepsilon) = [a(\xi^2+i\varepsilon) - a(\xi^2-i\varepsilon)]\, \phi_{\xi+}$. Now, the usual methods used to prove $a(z)$ is compact, lead to the estimate $\|a(z+\Delta) - a(z)\|_\mu \leq C(1+|z|)^{-1}\, |\Delta|^\nu$ where $\nu \leq \mu$. From this it follows that $\|d(\cdot,\varepsilon)\|_\mu \to 0$ as $\varepsilon \to 0$. Both the functions $\phi_{\xi+}$ and $d(\cdot,\varepsilon)$ are square integrable. The equation can be rewritten $\phi_{\xi+} + a(\xi^2+io)\phi_{\xi+} = 0$, so that $\phi_{\xi+} + Vr_0(\xi^2+ i\varepsilon)\phi_{\xi+} = d(\varepsilon)$. Since V is symmetric $\langle r_0(\xi^2+i\varepsilon)\phi_{\xi+}, V\, r_0(\xi^2+i\varepsilon)\phi_{\xi+}\rangle$ is real, so

$$\mathrm{Im}\left\langle r_o(\xi^2+i\epsilon)\phi_{\xi+}\ ,\ \phi_{\xi+}\right\rangle = \mathrm{Im}\left\langle r_o(\xi^2+i\varepsilon)\phi_{\xi+}\ ,\ d(\varepsilon)\right\rangle .$$

Thus

$$\frac{1}{2i}\int \phi_{\xi+}(k)\left[\frac{1}{\omega(k)-\xi^2-i\varepsilon}-\frac{1}{\omega(k)-\xi^2+i\varepsilon}\right]\phi_{\xi+}(k)\ d\check{k}$$

$$= \mathrm{Im}\int \frac{\phi_{\xi+}(k)}{\omega(k)-\xi^2-i\varepsilon}\ \overline{d(k,\varepsilon)}\ d\check{k}\ .$$

Now let $\varepsilon \to 0$. Since $\|d(\cdot,\varepsilon)\|_\mu \to 0$, a singular integral estimate shows r.h.s. $\to 0$, as $\mu > 1/2$. The ℓ.h.s. tends to

$$\frac{1}{\pi}\int |\phi_{\xi+}(k)|^2\ \delta(\omega(k)-\xi^2)\ \ d\check{k}.$$

If ω is not a critical point, this gives $\phi_{\xi+} = 0$ a.e.

<u>Definition.</u> Those real values, $z = x$, for which HE has a solution, are called *singular* points of $a(z)$, *provided* x is not a critical energy level.

<u>Theorem.</u> Any singular point of $a(z)$ belongs to the discrete spectrum of H, and conversely.

<u>Proof.</u> When $x = -\chi^2$, the proof is trivial: Let ϕ be a solution and set $\psi(k) = \phi(k)/(\omega(k)-x) = \phi(k)/(\omega(k)+\chi^2)$. Clearly, $H\psi = -\chi^2\psi$ and $\psi \in L^2(T^\nu)$.

We need not consider the case $x = 0$ since $\omega = 0$ is a critical energy level.

Now let $x = \chi^2$, $\psi(k) = \phi(k)/(\omega(k)-\kappa)$. We show $\psi \in L^2$ if κ is not critical. The only difficulties arise when $\omega(k) = k$, as ψ is singular there. On the other hand, $\phi(k) = 0$ on this surface (previous lemma). Since κ is not critical, $\omega(k) - \kappa = \nabla\omega|_{k_o}\cdot(\underline{k}-k_o) + O|k-k_o|^2$ where $\omega(k_o) = \kappa$, so $\phi(k)/(\omega(k)-\kappa)$ behaves as $\psi \lesssim \dfrac{|k-k_o|^\mu}{\nabla\omega|_{k_o}\cdot|k-k_o|}$, square integrable in all dimensions if $\mu > 1/2$. The converse, that $\psi \in L_2 \Rightarrow \phi \in b(\mu)$, is not hard!

It is difficult to be specific about V, such that HE possesses no solution at critical points, but in the *generic* case this will be the case, since *Pontrjugin's theorem* assures us that if $\omega \in C^\infty$, critical energies C_ω are *meagre*, and more, as T^ν is compact, the critical energies are *nowhere dense*.

At this stage, the theory becomes postulational, i.e. we assume that there is no solution to HE in $b(\mu)$ except for negative ω.

In the usual time-independent theory with $\omega = k^2/2\omega$, $\vec{k} = 0$ is the only critical point, and "spurious" solutions at $\vec{k} = 0$ have to be ruled out.

We would expect asymptotic completeness to hold. We now sketch a proof of this [5], under the assumption that HE has no spurious solutions.

We first notice that $t = -i(S - 1)$ satisfies the equation

$$t(k,k',z_1) - t(k,k',z_2) = (z_2-z_1) \int \frac{t(k,k'',z_1)\,t(k'',k',z_2)}{(\omega(k'')-z_1)(\omega(k'')-z_2)}\,dk'' \tag{7}$$

This actually follows from Hilbert's identity $r(z_1) - r(z_2) = (z_1-z_2)r(z_1)r(z_2)$ and the intertwining properties

$$t(z)r_o(z) = Vr(z) \quad ; \quad r_o(z)t(z) = r(z)V.$$

For, using the definition of $t(z)$,

$$t(z_1) - t(z_2) = Vr(z_2)V - Vr(z_1)V = (z_2-z_1)Vr(z_1)r(z_2),$$

$$= (z_2-z_1)t(z)r_o(z_1)r_o(z_2)\,t(z_2)$$

whose kernel form is (7).

In addition to obtaining a solution $t(\cdot,k',z) \in b(\mu)$, Watts proves the following estimates by the usual methods:

(i) $T(k,k',z)$ is bounded uniformly in k,k',z

(ii) $|t(k+h,k'+h',z+\Delta)-t(k,k',z)| \leq C(|h|^{\mu}+|h'|^{\mu}+|\Delta|^{\lambda})$

(iii) $|t(k,k',z)-V(k,k')| \leq C(1+|z|)^{-1}$

with μ,λ arbitrarily close to μ_o and $1/2$ resp. . Because of (i), the integral

$$\int \frac{t(k,q,z)\,t(q,k',z)}{[\omega(q) - z]^2}\,dq \qquad \text{exists uniformly, if } \operatorname{Im} z \geq \delta \neq 0,$$

and hence $t(k,k',z)$ is holomorphic with this as derivative, by taking the limit of (7) as $z_1 \to z_2$. We may therefore write a Cauchy contour integral for t round its singularities. Because of (iii), $t - V$ is zero at ∞, and so an infinite contour can be neglected if we write a contour integral for $t - V$ instead of for t.

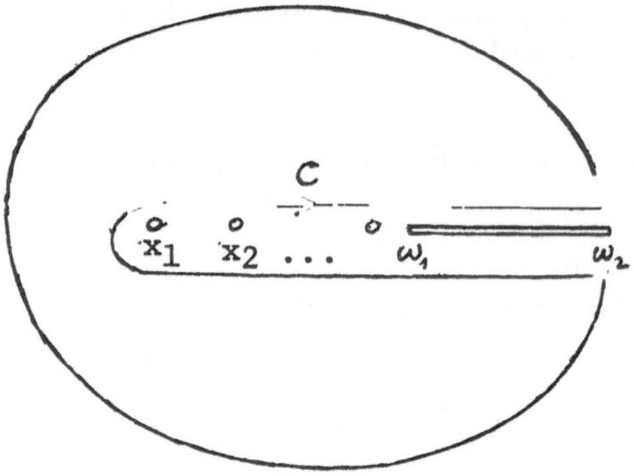

$$t(z) - V = \frac{1}{2\pi i} \int_C \frac{t(z') - V}{z' - z} dz'$$

As usual, the contribution from the bound states gives rise to poles in t; to see this, note that the resolvent r(z) is analytic except on its spectrum, and it has poles at the bound states

$$r(z) = \sum_i \frac{-p_i}{(z-\kappa_i)} + \hat{r}(z) ,$$

where $\hat{r}$ is analytic for $z < 0$, and $p_i = |\psi_i\rangle \langle\psi_i|$ is the projection onto the eigen-space i,ψ_i being the energy eigenfunction. So, from the definition $t = V - Vr(z)V$ we get the representation

$$t = \sum_i \frac{-Vp_iV}{z-\kappa_i} + V + \hat{t}(z) ,$$

where $\hat{t}(z)$ is analytic if $z < 0$. Putting $\phi_i = -V\psi_i$ we get, from HE $\tilde{\phi}_i(k) = (\omega(k) - \kappa_i)\tilde{\psi}_i(k) \in b(\mu)$, so

$$t = \sum \frac{\tilde{\phi}_i(k)\tilde{\phi}_i(k')}{\omega(k) - \kappa_i} + V(k.k') + \hat{t}(z) .$$

Define the distribution

$$\begin{aligned} \text{disc } t(k,k',\kappa) &= \lim_{\varepsilon\to 0} [t(k,k',\kappa+i\varepsilon) - t(k,k',\kappa-i\varepsilon)] \\ &= \lim_{\varepsilon\to 0} \int t(k,q,\kappa+i\varepsilon)\frac{2i|\varepsilon|}{(\omega(k)-\kappa)^2+\varepsilon^2} t(q,k',\kappa-i\varepsilon)\, dq \end{aligned}$$

Watts has proved, that if $\kappa \notin C_\omega$, the critical set,

$$\lim_{\epsilon\to 0} \frac{|\varepsilon|}{(\omega(k)-\kappa)^2+\varepsilon^2} = \pi\delta(\omega(k) - \kappa) \qquad (8)$$

in the sense that one integrates both sides with test-functions with support outside the critical set. To be a non-trivial statement, we must use the Pontrjagin result that C_ω is nowhere dense, so that (8) is true on intervals.

If HE has no spurious solutions at critical points, then IHE has a unique solution at all positive κ. Since the critical points have measure zero, and t is finite, we can ignore the critical points and take the limit, $\varepsilon \to 0$, to get (1). This equation immediately implies asymptotic completeness i.e. for $o \leq \omega \leq \omega_{max}$

$$t(k,k',\omega+io) - t(k,k',\omega-io) = 2\pi i \int_S t(k,q,\omega(q)\pm io)\, t(q,k',\omega(q)\mp io)\, d\check{q}$$

where the integration is over the energy-surface $S = \{q : \omega(q) = \omega\}$. If HE has spurious solutions, it does not appear possible to establish this result.

It would be interesting to find an interpretation of such a situation.

References.

1. R. F. Streater, The Heisenberg Ferromagnet as a Quantum Field Theory, Commun. Math. Phys. 6 (1967), 233.
2. R. F. Streater, *Lectures at Karpacz Winter School (1968)*; published by Institute of Theoretical Physics, Wroclaw, 1968.
3. R. F. Streater, Spontaneous Breakdown of Symmetry in Axiomatic Theory, Proc. Roy. Soc. A287 (1965), 510.
4. F. J. Dyson, General Theory of Spin-Wave Interactions Phys. Rev. 102 (1956), 5.
5. G. J. Watts, Theory of Spin-Wave Scattering, Ph.D. Thesis, Mathematics, Bedford College London, 1973.
6. K. Hepp, Scattering Theory in the Heisenberg Ferromagnet, Phys. Rev B, 5, (1972) 95.
7. Z. Takeda, Tohuku Math, J. 6, (1954), 212.
8. J. R. Klauder, J. McKenna and E. J. Woods, J. Mathematical Phys, 7, (1966) 5.
9. J. von Neumann, Infinite Tensor Products, Compositio Mathematica 6, (1938) 1.
10. R. Haag and D. Kastler, An Algebraic Approach to Quantum Field Theory, J. Mathematical Phys., 5, (1964) 848.

11. G. W. Mackey, *Induced Representations of Groups and Quantum Mechanics*, Benjamin N. Y. (1968).
12. T. Kato, *Perturbation Theory for Linear Operators*, Springer, 1966.
13. J. M. Jauch, Helv. Phys, Acta. 31, (1958) 136.
14. L. D. Faddeev, Mathematical Aspects of the Three Body Problem in Quantum Scattering Theory; Israel Programme for Scientific Translations, 1965.
15. J. W. Milnor, *Morse Theory*, Princeton U. Press.
J. Eels, Singularities of Smooth Maps, Nelson.
16. J. Manuceau and J. C. Trottin, Ann. Inst. Henri Poincaré, Vol. X4 (1969).
D. W. Robinson, Commun. Math. Phys. 7, (1968), 337.
D. Ruelle, *Statistical Mechanics*, Benjamin, N.Y. 1969.
17. R. F. Streater and I. F. Wilde, Time Evolution of Quantum Fields with Bounded Quasi-local Interaction Density, Commun. Math. Phys. 17, (1970) 21.

AN INTRODUCTION TO SOME MATHEMATICAL ASPECTS OF SCATTERING THEORY IN MODELS OF QUANTUM FIELDS

Sergio Albeverio

Università di Napoli

Universitet i Oslo

ABSTRACT

We give an elementary introduction to some results, problems and methods of the recent study of scattering in models developed in connection with constructive quantum field theory.

As far as the main core of these lectures is concerned, we have made a deliberate effort to be understandable also for mathematicians having some notions of non-relativistic quantum mechanics but no specific previous knowledge of quantum field theory.

We introduce the Fock space, the free fields and the free Hamiltonian and discuss the singular perturbation problem posed by local relativistic interactions. Scattering theory is first discussed for the simplified cases of space cut-off interactions and of translation invariant interactions with persistent vacuum. We give then the Wightman-Haag-Ruelle axiomatic framework as a guide for the construction of models with local, relativistic interactions and of the corresponding scattering theory. The verification of the axioms is carried through in a class of models with local relativistic interactions in two-dimensional space-time.

INTRODUCTION

Quantum field theory has a long and tortuous history. In a sense it started already before the advent of quantum mechanics, with the study of the electromagnetic field enclosed in a cavity, which led Planck to the introduction of his universal quantum of action.[1] It was thus quite natural that, after the quantum mechanics of one and n non-relativistic particles was developed to a new satisfactory physical theory ($\approx$1925-26), the attention came back to the study of the original problem of the electromagnetic

J. A. LaVita and J.-P. Marchand (eds.), Scattering Theory in Mathematical Physics, 299–381.

field, in particular to understand its influence on the motion and radiation of an atom ([Dir])

Considering the classical electromagnetic field in a cubic box Λ and expanding the vector potential $\vec{A}(t, \vec{x})$, which satisfies the Maxwell-Lorentz equations

$$\left(\frac{\partial^2}{\partial t^2} - \Delta \right) \vec{A}\,(t, \vec{x}) = 0, \quad \mathrm{div}\, \vec{A} = 0$$

with zero boundary conditions on the walls, in terms of the complete orthonormal functions of the Laplacian Δ in Λ, with zero boundary conditions on the walls, one obtained equations for the expansion coefficients (normal coordinates) of the same type as those of classical harmonic oscillators, so that the classical field could be looked upon as an infinite assembly of independent harmonic oscillators, whose sum of energies gives the total classical energy integral associated with the the electromagnetic field. [2] It was then natural to proceed to the quantization of the classical field as a limit for $n \to \infty$ of the case of n classical harmonic oscillators. The latter are quantized by replacing the classical position x_j and momentum variables P_j by operators on the Hilbert space of states, satisfying the canonical commutation relations

$$[P_j, x_\ell] = i^{-1}\, \delta_{j\ell}, \quad [P_j, P_\ell] = 0 = [x_j, x_\ell] \quad (i \equiv \sqrt{-1}),$$

where $[A, B] \equiv AB - BA$.

So, by analogy, the electromagnetic field was quantized by replacing each classical field component $A_r(t, \vec{x})$ of the vector $\vec{A}(t, \vec{x})$ by a "field operator" $A_r(t, \vec{x})$, satisfying, with its conjugate "momentum operator" $\pi^r(t, \vec{x})$ (corresponding to the classical field momentum $\frac{\partial}{\partial t} A_r(t, \vec{x})$), the singular equal-time commutation relations

$$[\pi^r\,(t, \vec{x}), A_s(t, \vec{y})] = i^{-1}\, \delta_{rs}\, \delta(\vec{x} - \vec{y}),$$

$$[\pi^r\,(t, \vec{x}), \pi^s(t, \vec{y})] = 0 = [A_r(t, \vec{x}), A_s(t, \vec{y})],$$

$$r,s = 1, 2, 3.$$

The singular nature of these commutation relations became quite early the source of troubles in the computation of physical quantities using the usual quantum mechanical perturbation series (divergent or meaningless results order by order).

Despite its shortcomings, the formalism of quantized fields became, since the 30's, a major method to describe at least formally and qualitatively high energetical quantum mechanical particles (with velocities comparable with the one of the light) and their interactions.

The method was extended from the original case of the electromagnetic field to fields defined first as solutions of classical relativistic invariant equations of motions (e.g. Klein-Gordon equation, Dirac equation) and then quantized according to the rules developed originally for the electromagnetic field.

In particular Fermi developed a theory of β-decay based on the idea of weak interactions described in terms of fields and the nuclear interactions were also described in terms of quantized fields. However here the hope for quantitative predictions going beyond the (sometimes bad) first order of perturbation theory had to be abandoned.

In the 40's and 50's however, renormalization theory was developed as a systematic way of extracting finite results from the divergent and/or meaningless formal perturbation series of quantum electrodynamics. But again, despite considerable computational success, the mathematical and theoretical foundations of the rules set up by renormalization theory remained obscure.

To get out of such a confused situation, in the early 50's some mathematicians and physicists started seeking for structural results, of mathematical and theoretical nature, of sufficient generality and strength as to gain insights into the problems of quantum field theory, taking seriously the fact that it is concerned with infinitely many degrees of freedom.

It was realized e.g. [3] that, roughly speaking, whereas the commutation relations $[P_j, x_\ell] = i^{-1} \delta_{j\ell}$, etc. given above have, under suitable assumptions, a unique (up to unitary equivalence) irreducible representation as operators on a Hilbert space [4] whenever the index set I over which j, ℓ run is a finite set, this is not true any more if the index set I is infinite. A whole lot of inequivalent representations can be easily exhibited.

Since when working in a Hilbert space quantum field theorists had been using exclusively a special representation of the canonical commutation relations, namely the Fock representation, [5] the question arose whether in quantum field theory all other representations can actually be excluded on the basis of some general principle.

This was answered in the negative by "Haag's Theorem", of which we will get to see effects in Chapter I.

Roughly from the mid-fifties, structural analysis of relativistic quantum mechanical system with infinitely many degrees of freedoms was carried forward in a systematic way by the so called axiomatic approaches.

These approaches (and we refer mainly, in view of Chapter IV, to the Wightman's axiomatic approach, see bibliographical note) [6,7] had the aim of clarifying the conceptual and mathematical basis of relativistic local quantum field theory by postulating a reasonably small number of selected general principles ("axioms"), formulated in mathematically precise terms and leading to interesting physical deductions, including the existence of scattering states and an S-matrix. No use of field equations, canonical commutation relations and Hamiltonian formalism was made in these general formalisms.

However the question of the inner consistency of such frameworks with (the only physically interesting case) $S \neq I$ could not be solved (and is still open for the physical case of a 3-dimensional space). In order to solve this question at least in a model world with less space dimensions and to gain further insight for future work on more realistic cases, a return to the Hamiltonian formalism[8] was strongly advocated by Wightman [Wi 1] and the program was carried forward since about '64 with full hard analytic equipment by a number of mathematicians and physicists.

It is called "constructive quantum field theory" (CQFT) and has led to a number of interesting results, including the proof of above consistency by constructing models in two space-time dimensions.[9]

In these lectures we shall discuss some scattering problems as they arise in connection with some models of CQFT. We shall give for this a short introduction to some tools and procedures of CQFT, without presupposing any previous knowledge of the subject and having always as a guideline applications to scattering theory. We shall actually stress more what can be done rather than what might not work in more complicated (higher dimensional) cases.[10]

We shall establish the connection with the general axiomatic theories of quantized fields, and in particular with Wightman's theory, in section IV, where we will see that certain models constructed in 2 space-time dimensions satisfy the Wightman axioms and further postulates for the corresponding Haag-Ruelle scattering theory.

We shall not however enter into details concerning those consequences of this fact, that follow from the general results proved in axiomatic quantum field theory.

This is because most interesting results of axiomatic quantum field theory (mostly proved a number of years ago) are available in excellent book form accounts to which we can refer the interested reader.[11] [12] The main line of these lectures is the following.

In Chapter I we introduce the Hilbert space (Fock space) suitable for the description of countably many identical particles and then discuss both the free energy operator in this space as well as the perturbations. These are discussed from the point of view of general sums of Wick monomials as well as from the point of view of local relativistic invariant interactions. The basic difficulties pertaining to the latter are discussed.

In Chapter II we discuss interactions with smooth kernels, which correspond e.g. to space cut-off local relativistic interactions, with vacuum polarization problems.

Such interactions appear as approximations to the local relativistic interactions discussed later in Section IV. However they present also interesting problems on their own, which are discussed in Chapter II. In particular we discuss existence of scattering quantities, location of the absolutely continuous spectrum of the Hamiltonian and unitarity of the S-operator.

In Chapter III we make a brief digression on another class of models, which present, in a sense, problems complementary to those of Chapter II.

These models have translation invariant interactions but do not have vacuum polarization. The existence of scattering quantities requires here a modified construction (as compared to the one of Chapter II). Results on other "intermediate models" (translation-rotation invariant interactions in higher space dimensions with vacuum polarization, but without Lorentz invariance) will be discussed shortly in a remark in Chapter IV. [13] This Chapter is divided into two sections. In the first one we give the Wightman-Haag-Ruelle framework for local relativistic scattering theory (realized, as we shall see in the second section, by models which have interactions that are in particular both translation invariant and have vacuum polarization, a synthesis of the properties of those of Chapter II and III).

In the second section of Chapter IV we mention two types of models for which all Wightman's axioms have been proved, as well as all or almost all additional assumptions for scattering theory.

For a type of models we shall sketch the whole construction by which, starting from a perturbation problem in Fock space (a space cut-off approximation of the true local relativistic interaction) one arrives at a new Hilbert space, carrying a new representation of the inhomogeneous Lorentz group and fields, which satisfy all Wightman axioms.

An hasty reader, mainly interested in the direct line carrying from Fock space to above local relativistic models, can jump from Section I directly to Chapter IV, since the problems discussed in Chapter II and in Chapter III have a more special character. The problems discussed in these lectures involve systems with infinitely many degrees of freedom. Other problems for other systems of such type are discussed in these lecture series by W. Strauss and R. Streater and we advise the reader to look at the connections.

Formulae are numbered consecutively in each section and footnotes as well as bibliographical comments are given separately for each section. The latter, as well the references, have no pretention of completeness and I apologize for omissions. It is my hope that I have not outbalanced the advantages of a fast publication (as has been urged for these proceedings) by inaccuracies in the final preparation of these lecture notes.

I. Fock Space Formulation of the Perturbation Problem.

I.1 Symmetric Fock Space (particle representation).

We shall introduce a separable Hilbert space, called symmetric Fock space[1], as a mathematical structure suitable on one hand for the description of a quantum mechanical system consisting of countably many non-interacting identical particles and, on the other hand, for the formulation of the perturbation (interactions of the particles) as a (singular)[2] perturbation of the free motion of the system. In this section we shall give one realization of this mathematical structure which stresses the "particle aspects".

Another realization will be given in section I.4, in which the "field aspects" are stressed.[3] Let $\mathcal{H}$ be a given separable Hilbert space over the field $\mathbb{C}$ of complex numbers. Form the symmetric n-fold Hilbert tensor product $S\mathcal{H}^{\otimes n}$ of $\mathcal{H}$ with itself.[4]
Set $\mathcal{F}^{(n)} \equiv S\mathcal{H}^{\otimes n}$. Then $\mathcal{F}^{(o)} \equiv \mathbb{C}$ is called the zero-particle space;
$\mathcal{F}^{(1)} \equiv \mathcal{H}$ is called the one-particle space;
and $\mathcal{F}^{(n)}$ is called the n - particle space.
$\mathcal{F}^{(n)}$ is a Hilbert space with scalar product $(.,.)_{\mathcal{F}^{(n)}}$, induced, for $n \geq 1$, in the natural way from the one given in $\mathcal{H}$

$$(\psi_1 \otimes \cdots \otimes \psi_n, \chi_1 \otimes \cdots \otimes \chi_n)_{\mathcal{F}^{(n)}} = \prod_{i=1}^{n} (\psi_i, \chi_i)_{\mathcal{H}},$$

S being the projector from $\mathcal{H}^{\otimes n}$ to $\mathcal{F}^{(n)}$

We look upon $\mathcal{F}^{(o)} = \mathbb{C}$ as a one-dimensional Hilbert space with scalar product $(z, z') \equiv \bar{z} z'$, where — means complex conjugation. Form $\mathcal{F} \equiv \bigoplus_{n=0}^{\infty} \mathcal{F}^{(n)}$ as the symmetric Hilbert tensor algebra[5] over $\mathcal{H}$.
$\mathcal{F}$ is called the symmetric Fock space (bosons Fock space). The elements of $\mathcal{F}$ are sequences

$$\psi = \{\psi^{(n)} \mid n = 0, 1, 2, \ldots; \psi^{(n)} \in \mathcal{F}^{(n)}\}.$$

ψ is said to have the component $\psi^{(n)}$ in $\mathcal{F}^{(n)}$. The scalar product in $\mathcal{F}$ is the one induced by the scalar product in $\mathcal{H}$:

$$(\psi, \chi) \equiv \sum_{n=0}^{\infty} (\psi^{(n)}, \chi^{(n)})_{\mathcal{F}^{(n)}}.$$

A basic vector in $\mathcal{F}$ is the so called Fock vacuum $\Omega_0 \equiv \{1, 0, 0, \ldots\}$, whose only non-vanishing component is the one in $\mathcal{F}^{(o)}$, which is set equal to 1. If A is a self-adjoint operator from a dense subset of $\mathcal{H}$ into $\mathcal{H}$, so that e^{isA} for real s is a one-parameter strongly continuous unitary group on $\mathcal{H}$, then there exists a uniquely defined unitary group $\Gamma(e^{isA})^{(n)}$ on $\mathcal{F}^{(n)}$ such that

$$\Gamma(e^{isA})^{(n)} \equiv e^{isA} \otimes \ldots \otimes e^{isA}.$$

Define $\Gamma(e^{isA})$ on $\mathcal{F}$ by

$$\Gamma(e^{isA})\psi = \{\Gamma(e^{isA})^{(n)} \psi^{(n)} \mid n = 0, 1, 2, \ldots\}.$$

$\Gamma(e^{isA})$ is a strongly continuous group of unitary transformations, uniquely defined on $\mathcal{F}$. Call $d\Gamma(A)$ its infinitesimal generator. Then

$$d\Gamma(A)\psi = \{d\Gamma(A)^{(n)} \psi^{(n)}\},$$

where

$$d\Gamma(A)^{(n)} \equiv A \otimes I \otimes \ldots \otimes I + I \otimes A \otimes I \otimes \ldots \otimes I + \ldots + I \otimes \ldots \otimes I \otimes A,$$

and I is the identity operator in $\mathcal{H}$. In particular we define, taking A to be I, $d\Gamma(I) \equiv N$, where N is the so called number operator on $\mathcal{F}$. N is the self-adjoint operator defined to be multiplication by n on $\mathcal{F}^{(n)}$, $n = 0, 1, 2, \ldots$.

We shall be particularly interested in taking for $\mathcal{H}$ the space of states of a single particle with spin zero and no charge, moving in (s + 1) - dimensional space-time, with s dimensions of space and one dimension of time: s is a fixed natural number (i.e. s = 0, 1, 2, ...).[6] In this case $\mathcal{H}$ can be taken to be (momentum space representation of quantum mechanics) the space of all functions, defined on "momentum space" $\hat{R}^s$,[7] which are square integrable with respect to the Lebesgue measure dk in momentum space. Thus we take $\mathcal{H} = L_2(\hat{R}^s)$. $\mathcal{F}^{(n)}$ is in this case the space $L_2^{sym}(\hat{R}^{sn})$ of all square integrable functions $\psi^{(n)}(k_1 \ldots k_n)$ of the n momenta $k_1 \ldots k_n$, such that moreover

$$\psi^{(n)}(k_{\pi_1} \ldots k_{\pi_n}) = \psi^{(n)}(k_1 \ldots k_n),$$

for any permutation $(1 \ldots n) \to (\pi_1 \ldots \pi_n)$ of the numbers $(1 \ldots n)$. $\mathcal{F}^{(n)}$ is thus the usual space of momentum space wave functions characterizing the states of a quantum mechanical system consisting of n identical particles.

Consider now a system consisting of a single free relativistic particle with momentum k and rest mass $m_0 \geq 0$. Its relativistic energy is given in terms of k and m_0 by $\mu(k) = +\sqrt{m_0^2 + k^2}$, where k^2 is the square of the euclidean length of k in $\hat{R}^s$. Let f be a function of k lying in $\mathcal{S}(\hat{R}^s)$, where $\mathcal{S}(\cdot)$ is the Schwartz space of C^∞ functions decaying at infinity together with all their derivatives stronger than any inverse power. Then $e^{-it\mu(k)} f(k)/\sqrt{\mu(k)}$ has, for each $t \in R$, the Fourier transform over R^s

$$F(t,x) \equiv (2\pi)^{-s/2} \int e^{-ikx} e^{-it\mu(k)} \frac{f(k)}{\sqrt{\mu(k)}} dk,$$

where kx is the natural pairing of $\hat{R}^s$ (momentum variables) and its dual space R^s (position variables).

One sees immediately that $F(t,x)$ satisfies the Klein-Gordon equation $(\Box + m_0^2)\, F(t,x) = 0$ with

$$\Box \equiv \frac{\partial^2}{\partial t^2} - \Delta,$$

Δ being the Laplacian on R^s, and t the time variable. Moreover the inhomogeneous orthochronous Lorentz group has a natural unitary irreducible representation $U_0(a,\Lambda)$ in $\mathcal{H}$, given by $F(\cdot) \to (U_0(a,\Lambda)f)(\cdot)$, with $U_0(a,\Lambda)f$ such that $(2\pi)^{-s/2} \int e^{-ikx}\, e^{-it\mu(k)} (U_0(a,\Lambda)f)(k) \frac{dk}{\sqrt{\mu(k)}} = F(\Lambda^{-1}((t,x)-a))$, and $a \in R^{s+1}$ are translations, Λ homogeneous Lorentz transformations on R^{s+1}. Thus we see that $e^{-it\mu} f/\sqrt{\mu}$ and F have a natural interpretation as wave functions describing the free motion of a single relativistic particle of momentum k and energy $\mu(k)$.

Define now $H_0 \equiv d\Gamma(\mu)$. This is a self-adjoint operator on its natural domain $D(H_0)$, induced on $\mathcal{F}$ by the natural domain of μ as multiplication operator in $\mathcal{H}$. $\mathcal{F}$ is a "spectral representation space" for H_0.

It is immediate to see that $e^{-itH_0} \upharpoonright \mathcal{F}^{(n)}$ describes the motion of n free particles with relativistic energy-momentum dependence.

H_0 is the so called free energy operator or free Hamiltonian. The Fock space, as symmetric closed tensor algebra over a space in which the free energy is "diagonal", solves in a natural way the problem of the formulation of the free motion of particles. In the next sections we shall examine the problem of introducing interactions between the particles.

Remark: The representation $U_0(a,\Lambda)$ has a natural (reducible) extension to $\mathcal{F}(\mathcal{H})$ given by $\Gamma(U_0(a,\Lambda))$.

I.2 Annihilation-creation operators.

Since generally in the processes between relativistic particles the number of particles is not conserved[1], it is quite natural to think of introducing shift operators in the Fock space, carrying from $\mathcal{F}^{(n)}$ to $\mathcal{F}^{(m)}$, with $m \neq n$. It turns out that it is appropriate to introduce certain suitably normalized elementary shifts $a(k)$ and $a^*(k)$, destroying and creating one particle of sharp momentum k (in the sense of bilinear forms, to be made precise below), in terms of which the interactions can be expressed. The introduction of the elementary shifts $a(k)$, $a^*(k)$ is suggested by a number of related facts which can be somewhat summarized by saying that the structure $(\mathcal{F}, H_0)$ consisting of Fock space $\mathcal{F}$ and the free energy operator H_0 is a limiting case (for $n \to \infty$) of the familiar structures associated with the study of the n-dimensional harmonic oscillator. In section I.4 we shall discuss this correspondence. For the moment we propose to accept this fact (which is incidentally an essential part of the physical background for quantum field theory)[2] and take it as a starting point for the introduction of the elementary shifts $a(k)$, $a^*(k)$. In fact in the theory of n uncoupled one-dimensional harmonic oscillators, a well known method to find the spectral representation for the correspondent free energy operator is based on a study of the algebraic commutation relations

$$[a_j, a^*_\ell] = \delta_{j\ell}, \ [a_j, a_\ell] = 0 = [a^*_j, a^*_\ell], \ \ell, j = 1, \ldots, n$$

and the representation on the relevant Hilbert space of states by operators

$$a_\ell \equiv \frac{1}{\sqrt{2}}(x_\ell + i\, p_\ell), \ a^*_\ell \equiv \frac{1}{\sqrt{2}}(x_\ell - i\, p_\ell).$$

Moreover the a_ℓ, a^*_ℓ act as shift operators in a suitable decomposition of the Hilbert space of states, since they shift the energy of the ℓ - th one-dimensional harmonic oscillator.[3]

We are lead similarly to the introduction of algebraic (formal) quantities $a(k)$, $a^*(k)$ as candidates for elementary shifts on $\mathcal{F}$, such that

$$[a(k), a^*(k')] = \delta(k - k'), \ [a(k), a(k')] = 0 = [a^*(k), a^*(k')]. \quad (1)$$

The appropriate realization of $a(k)$, $a^*(k)$ is through densely defined bilinear forms on $\mathcal{F} \times \mathcal{F}$. As a matter of fact, $a(k)$ can be even realized as a densely defined operator on Fock space $\mathcal{F}$ and we start with its description. Let D_0 be the linear span in $\mathcal{F}$ of Ω_0 and $(S f_1 \otimes \ldots \otimes f_n)(k_1, \ldots, k_n) \equiv \sum_{\pi_1,\ldots,\pi_n} \frac{1}{n!} f_{\pi_1}(k_1) \ldots f_{\pi_n}(k_n)$, where the sum extends over all permutations $(1, \ldots, n) \to (\pi_1, \ldots, \pi_n)$ of the numbers $1, \ldots, n$ and $f_i \in \mathcal{S}(\hat{R}^s)$ for all $i = 1, \ldots, n$ and all finite $n = 1, 2, \ldots$. Then D_0 is dense in $\mathcal{F}$. Define, for $k \in \hat{R}^s$ and $\psi \in D_0$, $a(k)\psi$ to be the vector with component $(a(k)\psi)^{(n)}$ in $\mathcal{F}^{(n)}$ given by $(a(k)\psi)^{(n)}(k_1, \ldots, k_n) \equiv \sqrt{n+1}\, \psi^{(n+1)}(k_1, \ldots, k_n, k)$, for $n = 1, 2, \ldots$; $a(k)\, \Omega_0 \equiv 0$.

Clearly $a(k)\, D_0 \subset D_0$ and so for $k_1, \ldots, k_\ell \in \hat{R}^s$ we can define the product $a(k_1) \ldots a(k_\ell)$ as a linear map from D_0 into D_0 . Note that $D_0 \cap \mathcal{F}^{(n)}$ is mapped into $D_0 \cap \mathcal{F}^{(n-\ell)}$. One has $[a(k), a(k')] = 0$ on D_0. $a(k)$ is called the annihilation operator for a particle of momentum k. In order to define the object $a^*(k)$ we spoke about before, we look at $a(k)$ as a bilinear form[4] $a(k)\{\psi, \psi'\}$ which associates to the pair of vectors ψ, ψ' in D_0, i.e. to the element $\psi \times \psi'$ in $D_0 \times D_0$, the complex number $a(k)\{\psi, \psi'\} \equiv (\psi, a(k)\psi')$, where $(\cdot, \cdot)$ is the scalar product in $\mathcal{F}$, so that $a(k)\{\cdot, \cdot\}$ is linear in the second argument and antilinear in the first. Define now $a^*(k)\{\cdot, \cdot\}$ as the bilinear form mapping $\psi \times \psi' \in D_0 \times D_0$ into the complex number $a^*(k)\{\psi, \psi'\} = (a(k)\psi, \psi')$. The bilinear form $a^*(k)\{\cdot, \cdot\}$ is thus determined on $D_0 \times D_0$ by $a(k)$ and is adjoint to $a(k)$, $\{\cdot, \cdot\}$ in the sense that $a^*(k)\{\psi, \psi'\} = \overline{a(k)\{\psi', \psi\}}$, where $\overline{}$ stands for complex conjugate of a number. We call $a^*(k)\{\cdot, \cdot\}$ the creation bilinear form for a particle of momentum k. We shall write $a^{\#}(k)\{\cdot,\cdot\}$ for both possibilities, $a^*(k)\{\cdot,\cdot\}$ or $a(k)\ \{\cdot,\cdot\}$. One has $a^{\#}(k)\{\psi,\psi'\} \in \mathcal{S}(\hat{R}^s)$ as a function of k, for all ψ,ψ' in D_0. Hence $\int w(k)\ a^{\#}(k)\{\psi,\psi'\}\,dk$ exists as the tempered distribution $w \in \mathcal{S}'(\hat{R}^s)$, tested against $a^{\#}(k)\{\psi,\psi'\}$ (considered as test function in $\mathcal{S}(\hat{R}^s)$).

We are going to define now for any $h \in \mathcal{S}(\hat{R}^s)$ operators $a^{\#}(h)$ and will then show that $a^{\#}(h) = \int h(k)\ a^{\#}(k)dk$ in the sense of bilinear forms. Define namely, for $h \in \mathcal{S}(\hat{R}^s)$, $a(h)$ as the linear operator on D_0 such that

$$(a(h)\psi)^{(n)}(k_1, \ldots, k_n) \equiv \sqrt{n+1}\int \psi^{(n+1)}(k_1,\ldots, k_n,k)h(k)dk$$

$$a(h)\Omega_0 \equiv 0.$$

Then $a(h)$ maps D_0 into D_0 and $D_0 \cap \mathcal{F}^{(n)}$ into $D_0 \cap \mathcal{F}^{(n-1)}$. It is called the annihilation operator for a particle of wave function h. One has the relation with the corresponding annihilation bilinear form for a particle of momentum k:

$$(\psi,\ a(h)\psi') = \int h(k)\ a(k)\{\psi,\psi'\}\,dk$$

for all ψ, $\psi' \in D_0$.[5] The corresponding creation operator for a particle of wave function h is defined for all $h \in \mathcal{S}(\hat{R}^s)$ as the linear operator $a^*(h)$ acting on D_0 as

$$(a^*(h)\psi)^{(n)}(k_1,\ldots,k_n) \equiv \frac{1}{\sqrt{n}} \sum_{i=1}^{n} h(k_i)\ \psi^{(n-1)}(k_1,\ldots,\not{k}_i,\ldots k_n),$$

where $\not{k}_i$ means that the argument k_i is missing. $a^*(h)$ maps D_0 into D_0 and $D_0 \cap \mathcal{F}^{(n)}$ into $D_0 \cap \mathcal{F}^{(n+1)}$ and one has the relation with the correspondent creation bilinear form of a particle of momentum k: $(\psi,a^*(h)\psi') = \int h(k)\ a^*(k)\ \{\psi,\ \psi'\}\,dk$, for all ψ, $\psi' \in D_0$.[6] The $a(h)$, $a^*(h)$ have closures, which we shall denote by the same symbols. From the definition one proves easily the estimates $||a^{\#}(h)\psi|| \leq ||h||_2\ ||(N + 1)^{1/2}\ \psi||$, first for all $\psi \in D_0$, $h \in \mathcal{S}$ and then, by extension, for all $\psi \in D((N + 1)^{1/2})$ and $h \in L_2(\hat{R}^s)$ (N being the number operator introducted in section I.1); here and in the following we write $a^{\#}(h)$ for $a^*(h)$ or $a(h)$ and in a product, e.g. $a^{\#}(h_1)\ a^{\#}(h_2)$, the label # is allowed to have a different meaning in different factors (thus $a^{\#}(h_1)\ a^{\#}(h_2)$ stands for $a^*(h_1)\ a^*(h_2)$, $a^*(h_1)\ a(h_2)$, $a(h_1)\ a(h_2)$, $a(h_1)\ a^*(h_2)$).

Basic properties of the annihilation-creation operators are given by the following:

Proposition 1. For any wave function $h \in \mathcal{H}$, $a^{\#}(h)$ is a linear, unbounded[7], densely defined closed operator on Fock space $\mathcal{F}$, and linear in h (thus a linear map of $\mathcal{H}$ into operators on $\mathcal{F}$). The natural domains of definition of $a(h)$ and $a^*(h)$ are equal. Moreover

1) $(a(h))^* = a^*(\bar{h})$;

2) $||a^{\#}(h_1)\ldots a^{\#}(h_j)\psi|| \leq ||h_1||_2 \cdots ||h_j||_2\ ||(N + 1)^{j/2}\ \psi||$, for all $\psi \in D\ ((N + 1)^{j/2})$, $h_1,\ldots, h_j \in \mathcal{H}$;

3) On $D(N + 1)$

$$[a(h), a(f)] = 0 = [a^*(h), a^*(f)]$$

and $$[a(h), a^*(f)] = (\bar{h}, f)_2,$$

for any $h, f \in \mathcal{H}$, where $[A,B] \equiv AB - BA$ on $D(A) \cap D(B)$ for any two operators in a Hilbert space, $D(\cdot)$ denoting domains. These are the so called (strong) canonical commutation relations.

4) a) $e^{-itH_0}\, a^\#(h)\, e^{itH_0} \supset a^\#(e^{\mp it\mu}h)$, where $A \supset B$ between operators means A extends B. On the right hand side - goes with a^*, + goes with a.

b) $[H_0, a^\#(h)]^{**} = a^\#(\pm\mu h)$, for all h in the domain $D(\mu)$ of the multiplication operator μ in $\mathcal{H}$ and $[,]^{**}$ means double adjoint i.e. closure of $[,]$.

5) The closed linear span of all vectors of the form $a^*(h_1)\ldots a^*(h_n)\Omega_0$, for all h_i, $i = 1,\ldots, n$ in some dense subset of $\mathcal{H}$, is the subspace $\mathcal{F}^{(n)}$ of $\mathcal{F}$.

Proof. The proof is easy: see [Co 1]. ∎

The operators $a^\#(h)$ are not self-adjoint, so they do not correspond to any physical observables. However it is easy to construct from the $a^\#(h)$ self-adjoint operators which (will be called free fields and) are the quantization of classical fields, as we shall see in section I.4. Define

$$\phi(g) \equiv \frac{1}{\sqrt{2}}\,[a^*(\hat{g}/\sqrt{\mu}) + a(\hat{g}/\sqrt{\mu})\,]^{8} \tag{2}$$

on D_0, for *any* $\hat{g}$ such that $\hat{g}/\sqrt{\mu} \in L_2(\hat{R}^s)$ and $\hat{g}(-k) = \overline{\hat{g}(k)}$ and where g is the function on R^s given by

$$g(x) \equiv (2\pi)^{-s/2} \int e^{-ikx}\ \hat{g}(k)\ dk.$$

Note that $g(x)$ is a real-valued function of the position variable $x \in R^s$.[9]

(The reason for the introduction of the energy denominators $\sqrt{\mu}$ will become apparent below). $\phi(g)$ is essentially self-adjoint on D_0[10] and we shall denote its self-adjoint closure again by $\phi(g)$. This is by definition the free time zero quantum field operator with real valued wave function (test function) g. The time t free field is by definition

$$\phi(t,g) \equiv e^{itH_0}\,\phi(g)e^{-itH_0} = \frac{1}{\sqrt{2}}\,[a^*(e^{it\mu}\hat{g}/\sqrt{\mu}) + a(e^{-it\mu}\bar{\hat{g}}/\sqrt{\mu})]$$

and is essentially self-adjoint on D_0 . For every fixed t, the family of maps from the set of test functions $g \in \mathcal{S}(R^s)$ to operators $\phi(t,g)$ on D_0 is a tempered operator-valued distribution, denoted by $\phi(t,x)$. We denote this distribution for $t = 0$ by $\phi(x)$. $\phi(x)$ is called the free time zero field at point x. It can be looked upon as the bilinear form

$$\psi \times \psi' \in D_0 \times D_0 \to \frac{1}{\sqrt{2}}\int [a^*(k)\{\psi,\psi'\} + a(-k)\{\psi,\psi'\}]\ \frac{e^{-ikx}}{\sqrt{\mu(k)}}\ dk. \tag{3}$$

Note that $\phi(g)$, looked upon as a bilinear form on $D_0 \times D_0$, is, for $g \in \mathcal{S}(R^s)$, the weak integral of the bilinear form $\phi(x)$ tested with g. $\phi(t,x)$, which can be looked upon as the bilinear form

$$\psi \times \psi' \to \frac{1}{\sqrt{2}} \int [a^*(k)\{\psi,\psi'\}\, e^{it\mu} + a(-k)\{\psi,\psi'\}\, e^{-it\mu}]\, \frac{e^{-ikx}}{\sqrt{\mu(k)}}\, dk, \quad (4)$$

is called the free field at point x and time t. The reason for the introduction of the energy denominator $\sqrt{\mu}$ in the definition (2) of $\phi(g)$, which implies the energy denominators in (3), (4), is that, so defined, $\phi(t,x)$ transforms covariantly under the unitary representation $\Gamma(U_0(\underline{a},\Lambda))$ of the inhomogeneous proper orthochronous Lorentz group given in $\mathcal{F}$ (see section I.1), in the sense that under $\Gamma(U_0(\underline{a},\Lambda))$ the bilinear form $\phi(\underline{x})\{\psi,\psi'\}$ transforms to the bilinear form $\phi(\Lambda\underline{x} + \underline{a})\{\psi,\psi'\}$ where $\underline{x}$ stands for (t,x).[11]

Finally we note for later reference that ϕ has the important property $[\phi(g_1), \phi(g_2)] = 0$ on D_0, for all $g_1, g_2 \in \mathcal{S}(R^s)$. This is an expression of the "locality" of the free time zero fields.[12]

I.3 The Perturbation problem in the particle picture.

In section I.2 we have introduced annihilation-creation bilinear forms. We shall now show that these operations on Fock space are suitable for the description of interactions not conserving (in general) the number of particles. We shall define general classes of such perturbations and then in the next Sections I.5, I.6 examine certain restrictions adequate for the eventual description of the special but most interesting class of local relativistic interactions given in the Hamiltonian formalism as quantized versions of classical relativistic field interactions. A general[1] perturbation of the free Hamiltonian H_0, which creates i particles and destroys j particles, can be given in Fock space by a bilinear form $W(w_{ij})$, where $W(w_{ij})$ is the bilinear form which associates to the pair of vectors ψ,ψ' in D_0 the complex number

$$W(w_{ij})\{\psi,\psi'\} = \int (a(k_i^*)\dots a(k_1)\psi, a(-p_1)\dots a(-p_j)\psi')$$
$$w_{ij}(k_1,\dots,k_i;\, p_1,\dots,p_j)dk_1\dots dk_i\, dp_1\dots dp_j \quad (1)$$

where $w_{ij}(\cdot)$ is a complex-valued function which belongs to the space of tempered distributions $\mathcal{S}'(\hat{R}^{s(i+j)})$. That $W(w_{ij})$ is indeed a bilinear form over $D_0 \times D_0$ follows from the definition of $a(k)$ and proposition 1, point 2) of Section I.2.

$W(w_{ij})$ is called the Wick ordered monomial, with i creation and j annihilation "operators" and numerical kernel w_{ij}. Note that the bilinear form adjoint to $W(w_{ij})$ is $W(w_{ji})$, where

$$w_{ji}(k_1,\dots,k_j;\, p_1,\dots,p_i) = \bar{w}_{ij}(-p_i,\dots,-p_1;\, -k_j,\dots,-k_1). \quad (2)$$

General candidates of symmetric perturbations[2] to H_0 are therefore of the form

$$V = \sum_{\substack{i,j=0 \\ i\leq j}}^{M} (W(w_{ij}) + W(w_{ji})), \tag{3}$$

with w_{ji} given in terms of w_{ij} by (2), w_{ij} being arbitrary kernels in $\mathcal{S}'(\hat{R}^{s(i+j)})$. Of course in order that to the symmetric bilinear form $H_0 + V$ (+ being here understood in the sense of bilinear forms) there is indeed associated a self-adjoint operator H which can be interpreted as the Hamiltonian of a physical system, certain additional conditions have to be satisfied. First of all H and $H_0 + V$ should be bounded from below and H should be "uniquely determined" by V (in a suitable sense). A most favorable case is the one[3] in which V is itself a symmetric operator on Fock space, not only a bilinear form. A sufficient condition for this to happen follows easily as a corollary from the following:

Proposition 1. Consider the general Wick monomial $W(w_{ij})$ of (1) and suppose $w_{ij} \in L_2(\hat{R}^{s(i+j)})$. Then to the bilinear form $W(w_{ij})$ there is associated in a unique way a densely defined operator, denoted again by $W(w_{ij})$, such that

$$W(w_{ij})\{\psi,\psi'\} = (\psi, W(w_{ij})\psi'),$$

for all $\psi, \psi' \in D_0$. Moreover $W(w_{ij})$ has a natural extension to a domain containing $D((N+1)^{(i+j)/2})$ and satisfies the estimate

$$||W(w_{ij})\psi|| \leq c(i,j)\, ||w_{ij}||_2\, ||(N+1)^{(i+j)/2}\psi||$$

for all $\psi \in D((N+1)^{(i+j)/2})$, where $c(i,j)$ is independent of w_{ij}, ψ and $||\cdot||_2$ is the $L_2(\hat{R}^{s(i+j)})$ norm.

Corollary: A sufficient condition for a perturbation V of the form (3) to be a symmetric operator defined on $D((N+1)^M)$ is that all kernels w_{ij} are square integrable with respect to all their variables.

Proof: From the definition of the bilinear form $W(w_{ij})\{\cdot,\cdot\}$ it is easy to verify that, for any $\psi \in \mathcal{F}^{(m)} \cap D_0$, $\psi' \in \mathcal{F}^{(n)} \cap D_0$

$$\begin{aligned} |W(w_{ij})\{\psi, \psi'\}| &= (m!\, n! [(m-i)!(n-j)!]^{-1})^{1/2}\, |(\psi, w_{ij}\, \psi')_2| \leq \\ &\leq a\, b^{(i+j)}\, (n-j)^{(i+j)/2} ||w_{ij}||_2\, ||\psi||\, ||\psi'||, \end{aligned} \tag{4}$$

where a, b are independent of i, j, m, n. From this we see that $W(w_{ij})$, as a bilinear form on $\mathcal{F}^{(m)} \times \mathcal{F}^{(n)}$ is the bilinear form of an operator X, bounded from $\mathcal{F}^{(n)}$ to $\mathcal{F}^{(n+i-j)}$. Moreover $X(N+1)^{-\beta}$ extends to a bounded operator X' on $\mathcal{F}$, for all $\beta \geq (i+j)/2$. This bounded operator gives a bilinear form $\psi \times \psi' \to (\psi, X'\psi')$ which is equal to the bilinear form on $D_0 \times D_0$

$$\psi \times \psi' \to W(w_{ij})\{\psi, (N+1)^{-\beta}\chi'\}$$

and then from (4) we deduce that $||X'|| \leq c(\beta, i, j)||w_{ij}||_2$

The proposition then follows taking $\beta = (i + j)/2$. ■ Even in the case in which the assumptions of above Corollary are satisfied, we are still not assured that $H_0 + V$ is bounded below, and in fact this is not generally so[4] unless further restrictions on V are made. Also the self-adjointness of $H_0 + V$ is not assured in general. In handling these latter problems, further smoothness assumptions on the different kernels of the Wick monomials forming V seem appropriate. Even though these smoothness assumptions simplify considerably the problem of the definition of a suitable Hamiltonian, and thus of a dynamics, it is essential to point out already here that specific scattering problems have, as we shall see later, their own difficulties not only in these definition problems for the Hamiltonian but also in the fact that the perturbation causes certain particular shifts of the spectrum. Smoothness assumptions can only make these shifts finite. If $H_0 + V$ is bounded from below, the assumptions $w_{io} = w_{oi} = 0$ and possibly $w_{i1} = w_{1i} = 0$, for all i (depending on how smooth the kernels are) would take care of the above shifts.[5] These assumptions are however, in contrast to the previous smoothness ones, too strong for the study of suitable "removable" approximations of local relativistic interactions.

To analyze these points and to recognize the physical meaning of the different assumptions, we shall discuss in Sections I.5, I.6 the form of potential candidates for local relativistic perturbations, looking at quantum field theory as a quantized version of a classical field theory. To do this we shall introduce in the next Section I.4 the so called wave picture (or Schrödinger representation) of the space of states of the quantum field theoretical systems, described before by the Fock space $\mathcal{F}$. In this representation space, which is a spectral representation space for the free time zero field $\phi(g)$ defined above, the perturbation problem ($H_0 \to$ "$H_0 + V$") appears as a limit for $n \to \infty$ of a quantum mechanical local potential problem in R^n and as a perturbation problem of an elliptic partial differential operator in infinitely many variables by an operator multiplication by a measurable function.

I.4 The Schrödinger Wave Representation of the Symmetric Fock Space.

We shall now give an isomorphism of the Fock space $\mathcal{F}$, as introduced in Section I.1, and an L_2-space with respect to a gaussian measure. The elements of this L_2-space can be interpreted as functions over a "configuration space" of the classical field associated with the time zero free field. Thus the isomorphism gives us a bridge between the particle picture of the preceding sections and classical field theory[1] permitting the introduction, by a correspondence principle, of candidates for local relativistic interactions.

Consider the set $\mathring{K}$ of all real-valued functions g in $\mathcal{S}(R^s)$.

Then their Fourier transforms

$$\hat{g}(k) \equiv (2\pi)^{-s/2} \int e^{ikx} g(x)\, dx$$

satisfy $\hat{g}(-k) = \overline{\hat{g}(k)}$, where —— stands for complex conjugate. Introduce in $\hat{K}$ the scalar product $\langle g_1, g_2\rangle \equiv \int \hat{g}_1(k)\, \hat{g}_2(k)\, dk/\mu(k)$. Form the closure K of $\hat{K}$ with respect to the norm defined by $\langle\cdot,\cdot\rangle$. K is a Sobolev Hilbert space of distributions[2], with scalar product $\langle\cdot,\cdot\rangle$. Let ξ be the (unique up to equivalence) real unit Gaussian generalized stochastic process indexed by K, with mean zero and covariance given by the scalar product in K, i.e. for every $g \in K$ one has a map to a random variable[3] $\xi(g)$ on a probability space $(Q, \sum, \mu_0)$, such that $\xi(g)$ depends real-linearly on g, the σ-algebra $\sum$ is generated by the $\xi(g)$ with $g \in K$ and

$$\int e^{i\xi(g)}\, d\mu_0 = e^{-1/2 \langle g,g\rangle} .$$

One has then in particular $\xi(g) \in L_p(Q, \sum, \mu_0)$, for all $1 \leq p < \infty$ and $\int \xi(g)\, d\mu_0 = 0$, $\int \xi(g_1)\, \xi(g_2) d\mu_0 = \langle g_1, g_2\rangle$, for all g, g_1, g_2 in K. Moreover for any orthonormal system $g_1,\ldots, g_n$ in K we have the joint distribution function

$$\mu_0\{\xi(g_1) < \alpha_1,\ldots, \xi(g_n) < \alpha_n\} = \prod_{i=1}^{n} \int_{-\infty}^{\alpha_i} (2\pi)^{-1/2} e^{-1/2 x_i^2}\, dx_i$$

i.e. for every bounded continuous function $F(\cdot)$ on R^n:

$$\int F(\xi(g_1),\ldots,\xi(g_n))\, d\mu_0 = (2\pi)^{-n/2} \int_{R^n} F(x) e^{(-1/2)x^2}\, dx.$$

A canonical choice[4] for $(Q, \sum, \mu_0)$ is the following. Take a complete orthonormal system $\{e_n\}$ in K and consider correspondently the infinite tensor product of copies of $(R, \mathcal{B}, \nu)$ indexed by n, where R is the real line, $\mathcal{B}$ are the Borel sets of the real line and ν is the unit Gaussian measure $(2\pi)^{-1/2} e^{-(1/2)x^2}dx$ on the line. In this choice points q of Q are of the form $q = q_1 \times q_2 \times \ldots$, where q_n is a point of the n-th copy of R (the "n-th component of q"), and $\xi(e_n)(q) = q_n$, thus $\xi(e_n)$ as a function on $L_p(Q)$ [5] acts as multiplication by the n-th coordinate. Of particular importance for us will be the (complex) $L_2(Q)$-space. A standard Hermite polynomials decomposition of $L_2(Q)$ used in probability theory[6] is the following. Let $L_2^{(n)}(Q)$ be defined as the orthogonal complement in $L_2(Q)$ of the subspace $L_2^{(\leq(n-1))}(Q)$ with respect to the subspace $L_2^{(\leq n)}(Q)$, where $L_2^{(\leq j)}$ denotes, for all $j = 0, 1, 2,\ldots$, the smallest closed subspace containing 1 and all monomials $\xi(g_1)..\xi(g_i)$, for all $i \leq j$, with $g_i \in K$. Then one proves that the algebraic direct sum

$$\bigoplus_{n=0}^{\infty} L_2^{(n)}(Q)$$

is dense in $L_2(Q)$, in much the same way as one proves that the Hermite polynomials are dense in $L_2(R, \mathcal{B}, \nu)$.[7] This is the key for the proof of the following Proposition, giving the connection between the Fock space of Section I.1 (defined as the Hilbert symmetric tensor algebra over $\mathcal{H}$) and the $L_2(Q)$ space and realizing the free field $\phi(g)$ as a multiplication operator.

Proposition. The Fock space $\mathcal{F}(\mathcal{H})$, where $\mathcal{H} = L_2(\hat{R}^s)$ is isomorphic to the (complex) space $L_2(Q, \sum, \mu_0)$ of square integrable functions over a probability space $(Q, \sum, \mu_0)$. $(Q, \sum, \mu_0)$ can be realized as described above. The isomorphism is given by the unitary operator U which maps $\mathcal{F}(\mathcal{H})$ onto $L_2(Q, \sum, \mu_0)$, in such a way that

1) $U\Omega_0 = 1$, where on the right hand side 1 stands for the random variable equal identically to the constant one on $(Q, \sum, \mu_0)$;

2) $U\phi(g)U^* = \xi(g)$, for any $g \in K$, where ξ is the unit Gaussian (generalized) stochastic process described above;

3) $U\, S_n(\frac{\hat{g}_1}{\sqrt{\mu}} \otimes \ldots \otimes \frac{\hat{g}_n}{\sqrt{\mu}}) = (n!)^{-1/2} E^{(n)}(\xi(g_1)\ldots\xi(g_n))$, where S_n is the symmetrization operator, $E^{(n)}$ is the orthogonal projection of $L_2(Q, \sum, \mu_0)$ onto the subspace $L_2^{(n)}(Q, \sum, \mu_0)$ and $g_1,\ldots,g_n \in K$

Proof. The unitarity of U follows from (3) and the facts that $\mathcal{F} = \oplus \mathcal{F}^{(n)}$ and, as mentioned above, $L_2(Q) = \oplus L_2^{(n)}(Q)$. 2) follows by an explicit computation on corresponding natural basis for $\mathcal{F}$ and $L_2(Q)$ (See [Se 1]) ■ The isomorphism given by the above proposition shows that we can interpret $L_2(Q, \sum, \mu_0)$ as a spectral representation for the free time zero field $\phi(g)$, since it is represented in $L_2(Q)$ by the operator multiplication by a coordinate function $\xi(g)$ in $L_2(Q, \sum, \mu_0)$. Moreover the structure $\{L_2(Q, \sum, \mu_0), \xi(g), g \in K\}$ is the "limit as $n \to \infty$", as described above, of the structure

$$\{L_2(\bigotimes_{j=1}^{n} (R, \mathcal{B}, \nu), \xi(x_i), x_i \in R\ (i=1,\ldots,n)\},$$

where $\xi(x_i)$ acts as multiplication by the coordinate x_i on

$$L_2(\bigotimes_{j=1}^{n} (R, \mathcal{B}, \nu)) = L_2(R^n, \mathcal{B}(R^n), \mu_0^{(n)}),$$

where $\mathcal{B}(R^n)$ are the Borel sets of R^n and $\mu_0^{(n)} = \nu\times\ldots\times\nu$ is the Gaussian measure $(2\pi)^{-n/2} \prod_{j=1}^{n} e^{(-1/2)x_j^2} dx_j$, on R^n. But

$L_2(R^n, \mathcal{B}(R^n), \mu_0^{(n)})$ is unitary equivalent to the Lebesgue measure space $L_2(R^n, \mathcal{B}(R^n), d^n x)$, where the unitary equivalence is given by the unitary map $\mho : f(x) \to (2\pi)^{-n/4} e^{(-1/4)x^2} f(x)$, from $f \in L_2(\mu_0^{(n)})$ to $L_2(d^n x)$. Hence we can interpret $L_2(\bigotimes_{j=1}^{n} (R, \mathcal{B}, \nu))$ as the Hilbert space of states of a non-relativistic quantum mechanical system with n degrees of freedom, in particular of a harmonic oscillator in n space dimensions. Consider now such a harmonic oscillator with energy operator

$$T = \frac{1}{2}\,[-\Delta + \sum_{\ell=1}^{n} \mu_\ell^2\, x_\ell^2 - \mu_\ell] = \frac{1}{2}\sum_{\ell=1}^{n} [p_\ell^2 + \mu_\ell^2 \quad x_\ell^2 - \mu_\ell],$$

where Δ is the Laplacian in $L_2(d^n x)$, the μ_ℓ are strictly positive numbers and $P_\ell = i^{-1}\, d/dx_\ell$ is the conjugate momentum operator to the position operator x_ℓ. Call K the R^n space with scalar product given by

$$<x,\ y> \equiv \sum_{\ell=1}^{n} x_\ell\ \mu_\ell\ y_\ell,$$

and consider the space $L_2(R^n,\ \delta)$ where δ is the Gaussian measure defined by the above scalar product

$$d\delta \equiv (\prod_{\ell=1}^{n} \int e^{-\mu_\ell\, x_\ell^2}\, dx_\ell)^{-1} \prod_{\ell=1}^{n} e^{-\mu_\ell \quad x_\ell^2}\, dx_\ell.$$

One has $L_2(d^n x) \cong L_2(\mu_0^{(n)}) \cong L_2(\delta)$, where $\cong$ stands for unitary equivalence. On the other hand, similarly as above, $L_2(\delta) \cong \mathcal{F}(\bar{K})$, where $\mathcal{F}(\bar{K})$ is the Fock space constructed on the complexification $\bar{K}$ of K. T acts on $\mathcal{F}(\bar{K})$ as $d\Gamma(\mu)$, where μ is the nxn matrix $((\mu_{ij}))=\delta_{ij}\mu_i$ on $\bar{K}$. We thus see that the free energy operator H_0 on the Boson Fock space and the Boson Fock space itself are the generalization of the energy operator and the space of states for an n-dimensional harmonic oscillator. In particular we see that H_0 in $L_2(Q)$ appears as an elliptic partial differential operator in infinitely many variables.

In the next section we shall use the Schrödinger representation of the Fock space in the introduction of the form of relativistic perturbations.

I.5 Local Relativistic Perturbations.

The time t free field $\phi(t,x)$ introduced in I.2 satisfies (in the operator distributional sense) the Klein-Gordon equation

$$(\frac{\partial^2}{\partial t^2} - \Delta)\ \phi\ (t,x) + m_0^2 \quad \phi(t,\ x) = 0,$$

where Δ is the Laplacian on R^s. This equation, looked upon as a classical equation, can be derived from a variational principle with Lagrangian

$$L = \frac{1}{2}\int [(\frac{\partial}{\partial t}\ \phi^{c\ell}\ (t,x))^2 - (\nabla\ \phi^{c\ell}\ (t,x))^2 - m_0^2\ (\phi^{c\ell}\ (t,x))^2]\ dx,$$

where ∇ is the gradient and we denote by $\phi^{c\ell}$ the classical field. Its classical energy integral is (cfr. lectures by Strauss)

$$\mathcal{E}(\phi^{c\ell},\ \pi^{c\ell}) = \frac{1}{2}\int [(\pi^{c\ell}(x))^2 + (\nabla\ \phi\ ^{c\ell}(x))^2 + m_0^2\ (\phi^{c\ell}(x))^2\ ]\ dx,$$

where $\phi^{c\ell}(x) \equiv \phi^{c\ell}\ (0,\ x)$ and where $\pi^{c\ell}(x)$ is the value at $t = 0$ of the classical conjugate momentum $\pi^{c\ell}(t,x) = \frac{\partial\phi^{c\ell}}{\partial t}\ (t,x)$.

If we introduce Fourier transforms

$$\hat{\phi}^{c\ell}(k) \equiv (2\pi)^{-s/2} \int e^{ikx} \phi^{c\ell}(x)\, dx,$$

$$\hat{\pi}^{c\ell}(k) \equiv (2\pi)^{-s/2} \int e^{ikx} \pi^{c\ell}(x)\, dx,$$

we obtain $\mathcal{E}$ in the form

$$\mathcal{E} = \frac{1}{2} \int [(\mathrm{Re}\ \hat{\pi}^{c\ell}(k))^2 + (\mathrm{Im}\ \hat{\pi}^{c\ell}(k))^2]\, dk + \frac{1}{2} \int \mu(k)^2 [(\mathrm{Re}\ \hat{\phi}^{c\ell}(k))^2 + (\mathrm{Im}\ \hat{\phi}^{c\ell}(k))^2]\, dk,$$

with $\mu(k) \equiv \sqrt{m_0^2 + k^2}$. From the preceding section we know on the other hand that the harmonic oscillator coordinate x_ℓ as an operator on L_2 corresponds to the multiplication operator $\xi(e_\ell)$ on Schrödinger space $L_2(Q)$, e_ℓ being a basis vector in K. Let us make now a formal argument to suggest a relation between $\mathcal{E}$ and the free Hamiltonian H_0 in Fock space. Consider the n-dimensional harmonic oscillator Hamiltonian

$$\frac{1}{2} \sum_{\ell=1}^{n} (P_\ell^2 + \mu_\ell^2\ x_\ell^2)$$

and take formally the limit to continuous variables $\mu_\ell \to \mu(k)$, $e_\ell \to e^{ikx}$. Then the limit is formally $\mathcal{E}$ with $\hat{\phi}^{c\ell}$, $\hat{\pi}^{c\ell}$ replaced by corresponding operators $\hat{\phi}$, $\hat{\pi}$ (Fourier-transforms of $\phi(x)$ and $\pi(x)$ which should satisfy $[\pi(y), \phi(x)] = i^{-1}\ \delta(y-x)$, $[\pi(y), \pi(x)] = 0$). On the other hand such a limit is, by the preceding section (modulo the additive constant $\sum \mu_\ell$) precisely H_0. So we suspect that H_0 has an expression of the form $\mathcal{E}$ with the classical quantities replaced by corresponding quantized ones. In fact define $\pi(x)$ as the bilinear form on Fock space

$$\psi \times \psi' \in D_0 \times D_0 \to \frac{i}{\sqrt{2}} (2\pi)^{-s/2} [\int a^*(k) \{\psi, \psi'\} e^{-ikx} \sqrt{\mu(k)}\, dx - \int a(k) \{\psi, \psi'\} e^{ikx} \sqrt{\mu(k)}\, dk]. \quad (3)$$

(Formally then indeed $[\pi(y), \phi(x)] = i^{-1}\ \delta(x-y)$, $[\pi(y), \pi(x)] = 0$). Moreover for $\psi, \psi' \in D_0$ one has

$$(\psi, H_0 \psi') = \frac{1}{2} \int [:(\pi(x))^2: + :(\nabla \phi(x))^2: + m_0^2\ :(\phi(x))^2:]\{\psi\ \psi'\}dx, \quad (4)$$

where $:\phi(x)^2:$, $:\nabla \phi(x)^2:$, $:\pi(x)^2:$ are the bilinear forms

$$\frac{1}{2} (2\pi)^{-s} \sum_{j=0}^{2} W(w_{j\ 2-j}^{(\alpha)}(x)),$$

where the kernels are equal respectively to

$$w_{j\ 2-j}^{(1)}(x) \equiv \binom{2}{j} e^{-i(k_1 + k_2)x} (\mu(k_1)\ \mu(k_2))^{-1/2}$$

$$w_{j\ 2-j}^{(2)}(x) \equiv \binom{2}{j}(-1)\ e^{-i(k_1 + k_2)x}\ k_1 k_2 (\mu(k_1)\ \mu(k_2))^{-1/2}$$

$$w_{j\ 2-j}^{(3)}(x) \equiv \binom{2}{j}(-1)(-1)^{j}\ e^{-i(k_1 + k_2)x} (\mu(k_1)\ \mu(k_2))^{1/2}.$$

The passage from the formal expressions $(\phi(x))^2$, $(\nabla\,\phi(x))^2$, $(\pi(x))^2$, not definable as bilinear-forms, to the bilinear forms appearing in (4) is a standard renormalization procedure, which can be described in the following terms. We do it for the quantity $(\phi(x))^n$, the others being treated similary (n can be any positive number). Write the expression for $\phi(x)$ in terms of annihilation bilinear forms formally as $(\sqrt{2})^{-1}(2\pi)^{-s/2} \int [a^*(k) + a(-k)] \frac{e^{-ikx}}{\sqrt{\mu(k)}}\, dk$. Form formally $(\phi(x))^n$ as $(\sqrt{2})^{-n}(2\pi)^{-ns/2} \int \prod_{j=1}^{n} [a^*(k_j) + a(-k_j)] \frac{e^{-ik_jx}}{\sqrt{\mu(k_j)}}\, dk_j$, multiply out in $\prod_{j=1}^{n} [a^*(k_j) + a(-k_j)]$ and then rewrite each term by writing all symbols a^* before the symbols a (the order of the a^* among themselves and of the a among themselves is immaterial). Then one obtains

$$(\sqrt{2})^{-n}(2\pi)^{-ns/2}\sum_{j=0}^{n} \int w_{j\,n-j}(x)(k_1,\dots,k_n)\,a^*(k_1)\dots a^*(k_j)\,a(-k_{j+1})\dots a(-k_n)\, dk_1\dots dk_n$$

with $w_{j\,n-j}(x)(k_1,\dots,k_n) \equiv \binom{n}{j} e^{-i\sum_{\ell=1}^{n} k_\ell x}(\mu(k_1)\dots\mu(k_n))^{-1/2}$ (5)

This has then a natural interpretation as the bilinear form

$$(\sqrt{2})^{-n}\ (2\pi)^{-n\,s/2}\ \sum_{j=0}^{n} W\,(w_{j\,n-j}(x)),$$

which is denoted $:\phi(x)^n:$ and called the n-th Wick power of $\phi(x)$.[1] Summarizing, we have thus seen that H_0 is given in terms of the fields as the classical energy integral for a Klein-Gordon field, with the classical quantities $\phi^{c\ell}(x)$, $\pi^{c\ell}(x)$ replaced by $\phi(x)$, $\pi(x)$ and the squares by Wick squares. This gives us then a hint on how to introduce local relativistic perturbations. For the classical field equation this is done (cfr. Strauss lectures) by adding a non-linear term, say $\sum_n c_n \int \phi^{c\ell}(x)^n\, dx$ to the energy integral, which corresponds to adding the non linear term $\sum_n n c_n \phi^{c\ell}(x)^{n-1}$ to the Klein-Gordon equation

$$\left(\frac{\partial^2}{\partial t^2} - \Delta + m_0^2\right) \phi^{c\ell}(t,x) + \sum_n n\, c_n\, \phi^{c\ell}(x)^{n-1} = 0.$$

The analogy with what has been done to identify $\mathcal{E}(\phi,\pi)$ with H_0 suggests taking as perturbation on Fock space an expression of the form

$$V = \sum c_n \int :\phi(x)^n:\, dx. \qquad (6)$$

We shall now motivate a little the replacement of the non-existing expression $(\phi(x))^n$ by $:\phi(x)^n:$ in the above choice. This is essentially a mathematical device to renormalize the non-existing power of a distribution, but preserves, as we shall see, physically desirable properties of the classical power $\phi^{c\ell}(x)^n$ of a field, namely euclidean covariance, Lorentz covariance of $\phi^{c\ell}(t,x)^n$ and locality.[2] To get a better insight into this and also for later use, let us look at the realization of $:\phi(x)^n:$ in the Schrödinger representation $L_2(Q)$ of Fock space (see Section I.4). Define for any smooth g (e.g. in $C_0^\infty(R^s)$) the element $:\xi^n:(g)$ of $L_2^{(n)}(Q)$ (see Section I.4) for the definition of $L_2^{(n)}(Q)$) as the function in $L_2^{(n)}(Q)$ such that

$\int :\xi^n:(g)E^{(n)}\xi(h_1)\dots\xi(h_n)=n!\int g(x)\prod_{j=1}^{n} G(x-y_j)h_j(y_j)dy_j dx$, for all h_i e.g. in $C_0^\infty(R^s)$, where G(z) is the kernel of $(-\Delta + m_0^2)^{-1/2}$; Δ being the Laplacian in R^s. It is not difficult to see, by explicit computation,[3] that the preimage of $:\xi^n:(g)$ under the unitary map U of $\mathcal{F}$ onto $L_2(Q)$ is the sum of Wick monomials

$$(\sqrt{2})^{-n}(2\pi)^{-(n-1)s/2}\sum_{j=0}^{n} W(w_{j\,n-j}(g)) \tag{7}$$

with kernels $w_{j\,n-j}(g)=(2\pi)^{-s/2}\int w_{j\,n-j}(x)g(x)dx$,

where $w_{j\,n-j}(x)$ are the kernels appearing in $:\phi(x)^n:$ and given by (5). Thus, setting: $:\phi^n:(g)=U^*:\xi^n:(g)U$ we have

$$:\phi^n:(g)\,\{\psi,\psi'\} = \int g(x) : \phi(x)^n : \{\psi,\psi'\}\,dx,$$

in the sense of bilinear forms. Thus under U, $:\phi(x)^n:$ is mapped into the random-variable valued distribution $:\xi(x)^n:$, which tested with $g \in C_0^\infty(R^s)$ gives $:\xi^n:(g)$ as an element of $L_2^{(n)}(Q) \subset L_2(Q)$. One calls $:\phi^n:(g)$ the n-th Wick power of ϕ with test function g and the same denomination is carried over to $:\xi^n:(g)$. $L_2(Q)$ is a spectral representation space for the Wick powers $:\phi^n:(g)$ and interactions of the form

$$V_g = \sum c_n : \phi^n : (g), \tag{8}$$

which all are realized in $L_2(Q)$ by operators multiplication by $L_2(Q)$ functions. In particular this shows that $:\phi^n:(x)$ has indeed the same locality property as $\phi(x)$, since $:\xi^n:(g)$, $:\xi^n:(g')$ are real valued functions and commute. Also the covariance properties of $:\phi(x)^n:$, resp. $\exp(itH_0) : \phi(x)^n : \exp(-itH_0)$ follow easily from the re presentation. We shall now look again at interactions of the form (6). We have already seen above that, by construction, $:\phi(x)^n:$ is a bilinear form on $D_0 \times D_0$. It is easy to see that $\int :\phi(x)^n: dx$ is the bilinear form $\sum_{j=0}^{n} W(w_{j\,n-j})$, where the kernels $w_{j\,n-j}$ of the Wick monomials are tempered distributions

$$(\sqrt{2})^{-n}(2\pi)^{-\frac{s}{2}(n-1)}\binom{n}{j}\,\delta\left(\sum_{\ell=1}^{n} k_\ell\right)(\mu(k_1)\dots\mu(k_n))^{-1/2}. \tag{9}$$

Thus we have from (5) as candidates[2] of local relativistic perturbations:

$$V = \sum_{n=0}^{M} c_n \sum_{j=0}^{n} W(w_{j\,n-j}). \tag{10}$$

If M is finite we know already that these interactions V are bilinear forms on $D_0 \times D_0$: they are called polynomial interactions in s space dimensions and denoted by $:P(\phi):_{s+1}$. In the case $M = \infty$ and for suitable coefficients c_n "non polynomial interactions" exist in certain cases and more specific examples will be considered later. We shall call simply "local relativistic interactions" perturbations V of the form (9), (10) (with $M \leq \infty$). In the next section we shall discuss some implications of the particular form of the kernels $w_{j\,n-j}$ entering in local relativistic perturbations (10), in relation to the discussion of the general perturbation by annihilation-creation bilinear forms introduced in Section I.3.

I.6 Some Problems Posed by the Perturbations.

We shall now discuss some implications of the particular form (10) (Section I.5) of local relativistic perturbations in relation with the perturbations of Section I.3 (formula (3)) and fix some terminology.

A. The tempting assumptions $w_{io} = 0 = w_{oi}$ (and $w_{1i} = 0 = w_{i1}$) for all i, mentioned in Section I.3 (and arising from a discussion of possible spectral shifts, along the lines of e.g. [Fr 2]), are certainly not satisfied for these interactions: one has namely from (9) $w_{j\,n-j} \neq 0$ for all $j = 0, 1, \ldots, n$. The fact that $w_{no} \neq 0$ (and then also $w_{on} \neq 0$!) i.e. the fact that V does not leave invariant the vacuum (or contains "pure creation terms") (in the sense that, as a bilinear form, $V\{\psi, \Omega_0\} \neq 0$ for some ψ, is called the presence of "vacuum polarization"[1]. In fact all local relativistic interactions V have the property of not conserving the number of particles (in the sense $V\{\psi, \psi'\} \neq 0$ for some ψ, ψ' not in the same $\mathcal{F}^{(n)}$, n arbitrary); this feature is sometimes called (to stress the difference with the n particle non-relativistic case or certain particular models[2] which, from other points of view, show some partial quantum field theoretical features) the problem of "number of particle divergences". The interactions V of the form of Section I.3 "without vacuum polarization" i.e. with $w_{io} \equiv 0 \equiv w_{oi}$ for all i (also called "interactions with persistent vacuum"), even in the case where they are obtained from a V of the form (10) (I.5) simply dropping the terms $W(w_{M0})$ and $W(w_{0M})$, cannot properly be regarded as mathematically suitable approximations for the passage to local relativistic interactions, since the terms dropped are "the most singular" ones. However, such interactions have in some cases physical interest in their own inasmuch as they correspond to a low energy approximation of the local relativistic interactions. Moreover, they present interesting mathematical problems. We shall come back to such interactions in Chapter III, where we shall also discuss the consequences of an assumption of the form $w_{1i} \equiv w_{i1} \equiv 0$ for all i (also not satisfied of course for local relativistic interactions).

B. The kernels $w_{j\,n-j}$ as given by (9) (Section I.5) are not square integrable in all their variables and therefore do not satisfy the assumption $w_{j\,n-j} \in L_2$ of Section I.3, which was sufficient for making V a symmetric operator on Fock space. There are two factors which make $w_{j\,n-j} \notin L_2$: we shall discuss them separately under B.1 and B.2.

B.1 Presence of the δ-distribution in (9) (Section I.5): this δ-distribution comes from the dx integration of $:\phi(x)^n:$ over the infinite volume space R^s. It is due uniquely to the translation invariance of the perturbation and the presence of vacuum polarization [3,4] and is independent of the number of space dimensions and of the locality and Lorentz invariance of the interaction (It arises for interactions of the general form (3) of Section I.3, whenever translation invariance and $w_{M0} \neq 0$ is required.) It is called the "infinite volume problem". Even if instead of the factors $(\mu(k_1)\ldots\mu(k_n))^{-1/2}$ the rest of the kernels of $w_{j\,n-j}$ would consist of very nice smooth functions, it is easy to see that the term $W(w_{M0})$, always present in a local relativistic perturbation, could never be an operator defined on Ω_0 and on vectors with finitely many non vanishing components.[5] A way to get around this problem is to first introduce an approximation V_g of V, so called "space cut-off interactions", such that the $\delta(\cdot)$-distribution gets replaced by a smooth function $\hat{g}(\cdot)$.[6] E.g. in the case of the local relativistic interactions (10) of Section I.5, the space cut-off interaction is

$$V_g = \sum_{n=o}^{M} c_n \sum_{j=o}^{n} W(w_{j\,n-j}(g)) \tag{1}$$

with $w_{j\,n-j}(g) \equiv (\sqrt{2})^{-n}(2\pi)^{-(n-1)s/2} \binom{n}{j} \hat{g}\left(\sum_{\ell=1}^{n} k_\ell\right)(\mu(k_1)\ldots\mu(k_n))^{-1/2}$ (2)

Note that (1) can be written, in the sense of bilinear forms, as

$$V_g = \sum c_n \int :\phi(x)^n: g(x)\,dx,$$

where

$$g(x) \equiv (2\pi)^{-s/2} \int e^{-ikx}\,\hat{g}(k)\,dk.$$

We shall write shortly: $P(\phi,g):_{s+1}$ for the space cut-off $:P(\phi):_{s+1}$ interactions. The replacement of V by V_g is of course in order to associate in a satisfactory way to $H_0 + V_g$, which exists to begin with as a bilinear form, a self-adjoint lower bounded operator on Fock space and then proceed to the "infinite volume limit" $g \to 1$ (i.e. $\hat{g}(k) \to \delta(k)$) (removal of the space cut-off), done, as we shall see in Section IV.2, by a study of suitable measures defined on Fock space. But even before the removal of the space cut-off the study of the perturbations $H_0 \to H_0 + V_g$ presents many interesting mathematical problems in itself and will be discussed in Section II.

B.2 Consider now a space cut-off perturbation V_g of the form (1), with kernels of the form (2). Suppose m_0 (the mass of the particles in Fock space) is strictly positive. Then the kernels (2) are locally square integrable over momentum space $\hat{R}^{sn}$, but they are not square integrable at infinity in momentum space, for $s > 1$. For $s = 1$ the logarithmic divergence of the L_2-norm of the energy term $(\mu(k_1)\dots\mu(k_n))^{-1/2}$ is compensated by the decay of $\hat{g}(\Sigma k_\ell)$ and in fact the kernels are in $L_2(\hat{R}^{sn})$ ([Ros1]). For $s = 1$ we can thus apply Proposition 1 of Section I.3 to obtain V_g as a densely defined operator on Fock space, but this procedure cannot be applied for $s > 1$. This is a manifestation of the so called problem of "ultraviolet divergences", which depends strongly on the form of interaction and on the number of space dimensions (worse divergences for higher s, no divergences for $s = 1$, in these models). For $s > 1$, in order to handle "ultraviolet divergences" a first step is the introduction of a "regularization" or "ultraviolet cut-off", so that the kernels become square integrable. This is done by replacing all energy factors $1/\sqrt{\mu(k_j)}$ in (2) by smooth functions $\hat{\chi}_\varepsilon(k_j)/\sqrt{\mu(k_j)}$, decreasing sufficiently quickly for $|k_j| \to \infty$ (e.g. $\hat{\chi}_\varepsilon(\cdot) \in \mathcal{S}(\hat{R}^s)$) and such that $\hat{\chi}_\varepsilon(\cdot) \to 1$ as $\varepsilon \to 0$. This corresponds to replacing the free field $\phi(x)$ appearing in the expression (3) for V_g by its regularized version $\phi_\varepsilon(x) \equiv \phi(\chi_\varepsilon(x-\bullet))$, where $\chi_\varepsilon(\cdot)$ is the Fourier transform of $\hat{\chi}_\varepsilon(\cdot)$ (such that $\chi_\varepsilon(\cdot) \to \delta(\cdot)$ as $\varepsilon \to 0$) and $\phi(\chi_\varepsilon(x-\bullet))$ is the free time zero field with test function depending on x as a parameter and equal to $\chi_\varepsilon(x-y)$, when y is the variable running over R^s. Thus we obtain

$$V_{g,\varepsilon} = \sum_{n=0}^{M} c_n \int :\phi_\varepsilon(x)^n: g(x)dx = \sum_{n=0}^{M} c_n \sum_{j=0}^{n} W(w_{j\,n-j}(g,\varepsilon)),$$

where **(3)**

$$w_{j\,n-j}(g,\varepsilon) \equiv (\sqrt{2})^{-n}(2\pi)^{-(n-1)s/2} \binom{n}{j} \hat{g}\Big(\sum_{\ell=1}^{n} k_\ell\Big) \prod_{\ell=1}^{n} \frac{\hat{\chi}_\varepsilon(k_\ell)}{\sqrt{\mu(k_\ell)}},$$

and the Wick power $:\phi_\varepsilon(x)^n:$ can be defined from the usual power $(\phi_\varepsilon(x))^n$ (which exists for $\varepsilon > 0$ as a self-adjoint operator on a natural domain containing D_0), in the same way as $:\phi(x)^n:$ was defined from the formal power $\phi(x)^n$.[7] $V_{g,\varepsilon}$ is called a space cut-off, ultraviolet cut-off interaction and is, by Proposition 1 of Section I.3, a symmetric operator on Fock space. We denote by $:P(\phi_\varepsilon)(g):_{s+1}$ the ultraviolet cut-off $:P(\phi)(g):_{s+1}$ interactions. As in B.1, the first problem to be solved now is to associate to the bilinear form $H_0 + V_{g,\varepsilon}$ a satisfactory self-adjoint lower bounded operator on Fock space and then to take the limit $\varepsilon \to 0$. This has been done for $:P(\phi_\varepsilon)(g):_{s+1}$ models for $s \leq 2$, with g fixed [G1-J4][8]. The case $m_0 = 0$, $s \leq 2$ presents further problems, namely lack of square integrability of the kernels at the origin in momentum space. This is a manifestation of the "problem of infrared divergences", which again depend strongly on the form of interaction and on the number of space dimensions. From the point of view of scattering problems, the case $m_0 = 0$ presents, in addition to the general difficulties, perculiar difficulties which correspond to the situation presented by long range forces as compared with short range ones in non-relativistic quantum (and classical) mechanics. Since we are only giving an introduction to the subject, we shall not be able to say much, unfortunately, on the $m_0 = 0$ problem (see however, remarks in Section III). From now on we shall always assume $m_0 > 0$.

Summarizing the examination of local relativistic perturbations and coming back to the general perturbations by annihilation-creation operators of Section I.3, we see that assumptions of "smoothness of the kernels" correspond to space cut-off and/or ultraviolet cut-off models whereas $w_{io} \equiv 0 \equiv w_{oi}$ and $w_{i1} \equiv 0 \equiv w_{1i}$ assumptions correspond to absence of vacuum polarization or one particle coupling terms models. If one is interested in deriving results for local relativistic interactions themselves, then above assumptions have to be considered as approximations to be removed at a later stage. Call now V any of the interactions discussed above, such that we are able to prove that V, given first as a bilinear form, is indeed a densely defined symmetric operator on Fock space (this is e.g. the case, by Proposition 1 of Section I.3, if all kernels are square integrable). Some problems we can then ask are:
1. Associate "uniquely to "$H_0 + V$" (defined suitably) a self-adjoint operator H, bounded from below.[9]
2. Study the qualitative spectral properties of H (existence of an isolated eigenvalue at the infimum of the spectrum, discrete spectrum, essential spectrum, continuous spectrum, absolutely continuous spectrum, eigenvalues in the continuum, etc.).
3. Establish the existence of scattering quantities (asymptotic states, fields, S-matrix) associated with H,[10] study related spectral problems, unitarity of the S-matrix.
4. If V is a space cut-off approximation of a translation invariant interaction remove the space cut-off on suitable quantities constructed using the information on H.

5. Study the limiting quantities and construct an associated scattering theory. In particular remove the space cut-off (and untraviolet cut-off) approximations for local relativistic interactions and study the limiting quantities (scattering theory for local relativistic quantum field theoretical models). Point 1) and at least part of 2) are, by the present methods, important steps to tackle the problems 4). The study of 3) presents interesting physical and mathematical problems in its own. Moreover, in some models a connection between the scattering quantities constructed in 3) and the similar quantities for 5) can be established.[11]

In the next chapters we shall examine some of these questions, particularly those more directly connected with scattering theory.

II. Scattering Theory: Space Cut-Off Models.

II.1 Introduction.

As mentioned before (problem 1 of Section I.6) the first problem to be solved in order to formulate scattering theory is[1] (as in non relativistic quantum mechanics) to find a suitable self-adjoint lower bounded operator H as "sum" of H_0 and the perturbation. In I. we have given a sketch of the main problems (infinite volume, ultraviolet divergences) encountered in trying to do so for quantum field theoretical models with local relativistic interactions. In this chapter we shall content ourselves with discussing cases in which the above problem can be solved on Fock space in the sense that H is equal to an extension of $H_0 + V$.[2] By what has been said in Section I.6, this already excludes the case of translation invariant interactions V such that $V\Omega_0 \neq 0$. Thus the discussion of this section will be basically limited to space cut-off interactions, like e.g. $:P(\phi)(g):_2$ or $:P(\phi_\varepsilon)(g):_{s+1}$. However, except for remarks, we shall not make any assumption, besides possibly ultraviolet cut-off, which would break the formal potential relativistic invariance of the interaction if the space cut-off could be removed.

In particular we do not make the "absence of vacuum polarization" assumption. The results have interest on their own, since they provide non-trivial scattering models with vacuum polarization as well as intermediate steps for the study of the limit of no space cut-off (e.g. by the information they give on the location of the absolutely continuous spectrum of $H_0 + V_g$). Instead of discussing in general the problem of the definition of H, we shall assume $H_0 + V_g$ essentially self-adjoint on $D(H_0) \cap D(V_g)$ and then prove a general theorem on the existence of scattering quantities. We shall however mention the actual solution of the problem of the essential self-adjointness of $H_0 + V_g$ in some specific models, to which the general theorem will be applied. The general theorem itself is formulated in such a way that both the statement and the proof can be adapted directly to other situations, particularly to the case of the translation invariant persistent models, to be mentioned later (which present difficulties of a different kind compared with those encountered in space cut-off models.) We shall now make a series of preliminary observations:

Remark 1. Since we want to discuss interactions V_g for which $V_g\Omega_0 \neq 0$, we cannot expect H to have zero as the bottom E_g of its spectrum (as it is the case for H_0).

In fact e.g. in $:P(\phi)(g):_2$ interactions $E_g \to -\infty$ as $g \to 1$. ([Guer 2]). Because of this reason $W(t) \equiv \exp(itH)\exp(-itH_0)$ will not converge strongly, as $t \to \pm\infty$, on the subspace $\mathcal{F}^{(0)}$ of $\mathcal{F}$. It is not difficult to convince oneself, using the "particle structure" of the spectrum of H_0, that in general the operators W(t) will not converge strongly in these models[3] (and in fact in no models with a shift of the ground state energy and/or, for translation invariant V, shifts of the infimum of the spectrum of the restriction to a subspace of fixed total momentum: see SectionIII). It turns out that a field theoretical substitute for the ordinary wave operators of quantum mechanics and the relative Cook's method for their construction are the so called "asymptotic creation-annihilation operators", studied first in details by Kato-Mugibayashi and Höegh-Krohn.

Remark 2. Again, as in I, we restrict ourselves to the discussion of models defined in the symmetric Fock space $\mathcal{F}$, which describes, in physical language, a single sort of scalar bosons. The same kind of methods extend however to the cases of different sorts of boson particles interacting with each other (such a case will be discussed quickly later). For the case in which the particles are fermions the Fock space is not the symmetric Fock space but rather an analogue construct in which $S\,H^{\otimes n}$ is replaced by $A\,H^{\otimes n}$, A being the projector on the antisymmetric subspace of $H^{\otimes n}$

The creation-annihilation operators are bounded in this case and this brings in a considerable amount of simplification in some of the proofs. On the other hand local relativistic interactions in the antisymmetric case have more singular kernels than boson self-interactions and a control on the infinite volume limit as the one we shall describe in Chapter IV is still lacking for models involving fermions.[4]

Remark 3. We are primarily concerned in these lectures with the basic formulation of the scattering problem and the general construction of scattering quantities. In this optics the problem of ultraviolet divergences and their renormalization is concentrated in the definition of the Hamiltonian H but, besides this, somewhat subordinate. In fact in the cases in which H has been defined by solving first a (non trivial) ultraviolet problem, the construction of scattering quantities procedes essentially along the same lines as in the correspondent ultraviolet cut-off models. For this reason we shall limit ourselves to mention (later) scattering results for models with ultraviolet divergences, but shall not enter into details

II.2 A general proposition on the existence of asymptotic fields.

In this section we shall assume that the interaction V is such that it has a common dense domain with H_0 and that there exists a self-adjoint operator H, bounded from below, which is a self-adjoint extension of $H_0 + V$. We shall study the strong limits as time t goes to $\pm\infty$ of operators of the form

$$a_t^{\#}(h) \equiv e^{itH}\ a^{\#}\ (e^{\mp it\mu} h)e^{-itH},$$

where $a^{\#}$ are the creation-annihilation operators, h is a wave function,[1] μ the one-particle energy. We shall look for a general statement holding whenever there is a control on V essentially in terms of "a priori estimates", not involving detailed features of V. Roughly the statements will hold e.g. whenever V is given as an expression in creation-annihilation operators, with smooth kernels (space cut-off).

A general strategy to the proofs, which we will only sketch, is to prove first convergence on a suitable dense set of vectors and for h in a suitable dense subset of $\mathcal{H}$ and then to extend it to larger domains and to $\mathcal{H}$ by using uniform fixed time estimates.

We shall now give the formal procedure of the proof of the strong convergence of $a_t^{\#}(h)$, which also suggests the convenient assumptions to be made. We just handle the case $a_t(h)$ and $t \to +\infty$, the other cases (a^* instead of a and/or $-\infty$ instead of $+\infty$) being entirely similar. One proves first for suitable ψ and h that

$$a_t(h)\,\psi = a_{t_0}(h)\,\psi + \int_{t_0}^{t} \frac{d}{d\tau}\, a_\tau(h)\,\psi\, d\tau \tag{1}$$

and that $\int_{t_0}^{\infty} \frac{d}{dt}\, a_t(h)\,\psi\, dt$ exists. (2)

Then it follows of course that $\underset{t \to \infty}{s-\lim}\, a_t(h)\psi$ also exists.

To prove (2) we compute (formally)

$$\frac{d}{dt}\, a_t(h)\psi = iU_t[H, a(h_t)]U_t^* + U_t\left(\frac{d}{dt}\, a(h_t)\right)U_t^* = iU_t[V, a(h_t)]U_t^*,$$

where

$$U_t \equiv e^{itH}, \quad h_t(k) \equiv e^{it\mu(k)}\, h(k)$$

and we have used $H = H_0 + V$ and $\frac{d}{dt}\, a(h_t) = i[H_0, a(h_t)]$. We would like to show that on suitable vectors ψ and for suitable h

$$\left\|\frac{d}{dt}\, a_t(h)\,\psi\right\| \le \beta(t),$$

where $\beta(t)$ is integrable for all $t \ge t_0$. i.e. we want

$$\|[V, a(h_t)]\, U_t^*\,\psi\| \le \beta(t).$$

Take as an example the case in which V is a space cut-off interaction of the form (1) in Section I.6, where we take $M < \infty$. V is then a finite sum of Wick monomials and it is therefore sufficient to estimate

$$\|[W(w_{j\,n-j}),\, a(h_t)]\, U_t^*\,\psi\|\,. \tag{3}$$

Using the canonical commutation relations one can see that the commutator in (3) can be written again as a sum of Wick monomials $W(w'_{(j-1)n-j})$ with kernels of the form

$$w'_{(j-1)n-j}(t;\, \underline{k}';\, \underline{p}) \equiv \int w_{j\,n-j}(\underline{k}'\, q;\, \underline{p})\, e^{it\mu(q)}\, h(q)\, dq,$$

where $\underline{p} \equiv (p_1,\ldots,p_{n-j})$, $\underline{k} \equiv (k_1,\ldots,\not{k}_\ell,\ldots,k_j)$, $\not{k}_\ell$ meaning that k_ℓ is missing. Then the L_2-norm $||w'_{(j-1)n-j}||_2$ of w' with respect to all the variables is bounded by

$$c_j \int \left| \int w_{j\,n-j}(\ldots q\ldots)\, h_t(q)\, dq \right| d\underline{k}' d\underline{p} \ .$$

Take $h \in \overset{\circ}{C}{}_0^\infty\ (R^s)$, where $\overset{\circ}{C}{}_0^\infty$ is the (dense) set of all C^∞ functions $h(k)$ of compact support in k space and vanishing identically in some neighborhood of the origin $k = 0$. Use

$$e^{it\mu(q)} = - t^{-2} \left(\frac{\mu}{|q|}\right)^2 \frac{d^2 e^{it\mu(q)}}{d|q|^2},$$

integrate by parts twice using that the kernels are smooth enough and

$$\int \left| \frac{d}{d|q|^2} \left(w_{j\,n-j} \left(\frac{\mu}{|q|}\right)^2 h\right)\right| d\underline{k}' d\underline{p}\ \ dq < \infty.$$

($\hat{g}$ was taken to be smooth, e.g. in $\mathcal{S}(\hat{R}^s)$). Then $||w'_{(j-1)n-j}||_2 \leq$ const. $(1 + t^2)^{-1}$ is proved and using the bound on Wick monomials given by Proposition 1, Section I.3, we have

$$||[W(w_{j\,n-j}),\ a(h_t)]\ (N+1)^{-(i+j+1)/2}|| \leq$$

const. $(1+t^2)^{-1}$, which implies $||[V,\ a(h_t)]\ (N+1)^{-\alpha}|| \leq$ const. $(1 + t^2)^{-1}$ for some suitable α.

Remark: This argument is depending on the presence of the space cut-off in the sense that if g were replaced by 1, thus $\hat{g}(k)$ by $\delta(k)$, we would have distributional kernels and we would have $|\int w_{j\,n-j}(\ldots q\ldots)\ h_t\ (q)\ dq| = |w_{j\,n-j}(\ldots 0\ldots)|\ |h(Q)|$, which is independent of t, thus invalidating the whole argument. Thus smoothness properties in momentum space for the kernels of the Wick monomials out of which the interaction is built are essentially used.

To conclude now the (formal) proof of (2) it suffices to have an estimate of the form that $||(N+1)^\alpha\ U_t{}^*\ \psi||$ be bounded by a constant independent of t. For this it is sufficient in turn to have a bound of the form $||(N+1)^\alpha\ \chi|| \leq ||f(H)\ \chi||$, for all $\chi \in D(f(H))$, where $f(H)$ is a suitable function of the self-adjoint operator H and $D(\cdot)$ denotes the domain of the operator.

We have so seen the motivation for an "abstract" Theorem to be formulated in the following. First we need some definitions. Let T be a self-adjoint operator in $\mathcal{F}$, $\int \lambda\ d\ E_\lambda$ its spectral decomposition. We denote by M_T the set of all real valued Borel measurable functions $g(\cdot)$ on the real line R which are finite almost everywhere with respect to $d\ E_\lambda$. For $g \in M_T$, $g(T)$ is defined in the usual way of the functional calculus.

We shall use operators in M_T, for suitable T, to bound other operators. In particular we use $T = N_\theta$, where $0 \leq \theta$, $N_\theta \equiv d\Gamma(\mu^\theta)$ [2].

Note that $N_0 \equiv N$, $N_1 \equiv H_0$.

We shall also use the notation $A \leq B$ for any quadratic forms associated with positive self-adjoint operators A, B to say

$$(\psi, A\psi) \leq (\psi, B\psi)$$

for all $\psi \in D(B^{1/2})$.

Proposition 1: Assume V is a symmetric operator on Fock space $\mathcal{F}$, and such that $H_0 + V$ has a self-adjoint extension H which is bounded from below. Assume furthermore there exist $g_1, g_2 \in \mathcal{M}_H$ such that a) $N + 1 \leq g_1(H)^2$ b) $D(H_0) \supset D(g_2(H))$ c) Moreover assume there exists $f_0 \in \mathcal{M}_K$ where $K = N_\theta$ (for some θ) or $K = H$, such that $f_0 \neq 0$ almost everywhere with respect to the spectral measure of K and, in the case $K = N_\theta$, $f_0(K)^2 \leq f(H)^2$ for some f in $\mathcal{M}_H$. In all cases

$$||[V, a^*(e^{-it\mu}h)]\ f_0(K)^{-1}\ \psi|| \leq \alpha_f(t)\ ||\psi||$$

for all $\psi \in D(f(H))$, where $\alpha_f(t)$ is bounded continuous and $\int_{t_0}^{\infty} \alpha_f(t)\, dt < \infty$ for some finite t_0. Assume finally

$$[V, a^*(e^{-it\mu}h)]\ f_0(K)^{-1}\ \psi$$

is strongly continuous in t for all h in a dense linear subset $\mathcal{S}$ of $L^2(\hat{R}^s)$. Then:

1) $\text{s-}\lim_{t\to+\infty} e^{itH}\ a^*\ (e^{-it\mu}h)\ e^{-itH}$ exists for all $h \in L_2(\hat{R}^s)$ on $D(g(H))$, where $g(\beta) \equiv \sup\ (\beta\, g_1(\beta), \beta g_2(\beta), \beta f(\beta))$.

2) Call $a^*_+(h)$ the above limit. Then $||a^*_+(h)\psi|| \leq ||h||_2\ ||g_1(H)\psi||$ for all $\psi \in D(g_1(H))$.

3) If $g_1(H)^{-1}$ exists as a bounded everywhere defined operator, then convergence 1) and the estimate 2) extend to all $D(g_1(H))$.

4) Completely similar statements under accordingly modified assumptions hold for $\exp(itH)\ a(\exp(it\mu)h)\ \exp(-itH)$ in the limit $t\to+\infty$, as well as for $\exp(itH)\ a^\#\ (\exp(\mp it\mu)h)\ \exp(-itH)$ in the limit $t\to-\infty$.

Remark 1: This proposition is an abstract formulation of methods used by Hoegh-Krohn and Kato-Mugibayashi in a series of models (see bibliographical note).

Remark 2: The condition a) is already used in order to define $a_t^\#(h)$ on a dense domain, using Proposition 1, Section I.2:

$$||a_t^\#(h)\psi|| \Rightarrow ||a^\#(e^{\mp it\mu}h)(N+1)^{-1/2}(N+1)^{1/2}\ e^{-itH}\ \psi|| \leq$$
$$\leq ||h||_2\ ||g_1(H)\psi|| \qquad (4)$$

which shows that $a_t^\#(h)$ is defined on $D(g_1(H))$. The same condition leads also to the estimate 2), which gives information on the domain of the asymptotic creation-annihilation operators (also called asymptotic fields, for brevity). Later on we shall use a stronger condition than a), namely $(N+1)^2 \leq g_1(H)^2$ for some $g_1 \in \mathcal{M}_H$, in order to discuss asymptotic Fock spaces.

As discussed before the statement of the Proposition, the condition c) is quite naturally fulfilled when V is a sum of Wick monomials with smooth kernels.

Proof of Proposition 1: The proof is easily achieved using the sketch given before the statements of the Proposition. Remark that although in condition c) only the behaviour for h in a dense subset δ of $L_2(\hat{R}^s)$ enters, the statements first proven only for such h, extend to the whole $L_2(\hat{R}^s)$ using the uniform bound 2). Similarly 3) follows from 1), 2) using the uniform bound and the closedness of H. ∎

Applications: We shall now give some examples of models in which all the assumptions made above are satisfied and for which therefore the existence of asymptotic creation-annihilation operators is given by Proposition 1.

Example 1: $:P(\phi)(g):_2$ models. The interactions here are of the form

$$V = V_g = \sum_{n=0}^{2p} c_n : \phi^n :(g) = \sum_{n=0}^{2p} c_n \int : \phi(x)^n : g(x)\, dx,$$

$c_{2p} > 0$, $g(\cdot) \geq 0$, i.e. sums of Wick monomials with kernels

$$\hat{g}\,(k_1 + \ldots + k_n)(\mu(k_1)\ldots\mu(k_n))^{-1/2},$$

where e.g. $g(\cdot) \in C_0^\infty(R^s)$. As remarked before (Section I.6, B.2) the kernels are square integrable as functions of all their variables and hence V is a densely defined symmetric operator on $\mathcal{F}$, by Proposition 1 of Section I.3, with domain containing $D((N+1)^p)$. Hence $H_o + V$ is also a symmetric operator. There exists several proofs for its essential self-adjointness. Since the proofs are rather involved and we are lacking time, we shall limit ourselves to refer to the literature. The same as regards the proof of the lower boundedness of this operator. Let us mention however that in some of these proofs the wave representation of the Fock space as $L_2(Q)$ is used. We know that $:\phi^n(g):$ is mapped into the operator multiplication by the L_2 function $:\xi^n:(g)$, self-adjoint on a natural domain, as well as $V_g^W = \sum c_n : \xi^n :(g)$, the image of V_g under the same map. By a basic estimate $e^{-V_g^W} \in L_p(Q)$, for all $1 \leq p < \infty$. On the other hand $\exp(-tH_0)$ is mapped into the Hermite operator $\exp(-tH_0^W)$, which is a contraction from $L_p(Q)$ to $L_q(Q)$, for any given p, q > 1 and t > 0 sufficiently big. The lower bound follows then rather simply from these facts, which are ingredients also for a proof of the essential self-adjointness (another essential ingredient is the Lie-Trotter-Kato product formula.)

Remark: In these models the infimum of the spectrum of $H=(H_0+V)^{**}$ is an isolated simple eigenvalue E_g (ground state energy) and the spectrum in $[E_g, E_g + m_0)$ is discrete. The assumptions a), b), c) and the one in point 3) of the above proposition are satisfied for H, because of the following estimates ([Ros 2]).

$$N^j \leq a\,(H + b)^j$$
$$H_0^{3-\varepsilon} N^j \leq a(H + b)^{\alpha} \quad , \tag{5}$$

for all j, $0 < \varepsilon \leq 3$, some finite constants a, b, α (which might depend on j,ε), b such that $H + b > 0$. Hence $g_1(H)$ can be taken to be $a(H+b)$, $f(H)$ to be $a(H+b)^{\alpha/2}$. Thus the proposition applies in the strongest form given by 3), which shows that

$$s - \lim_{t\to\pm\infty} e^{itH}\, a^{\#}(e^{\mp it\mu}h)\, e^{-itH}$$

exist on $D(H) \supset D(H_0) \cap D(V)$. This result was obtained by Høegh-Krohn in the general case, and by Kato-Mugibayashi in the particular case p = 2. ([Ho-K 12], [Ka-M 2]).

Example 2: $:P(\phi_\varepsilon)(g):$ models. The interactions are of the same form as in Example 1, but with the ultraviolet cut-off field ϕ_ε replacing ϕ (and $s \geq 1$). For these models essentially the same results hold as for the models of Example 1, and hence proposition 1 applies. ([Ho-K 11]).

Example 3: Space cut-off bounded interaction models in s + 1-space time dimensions.[+] In these models the interaction has the form

$$\lambda V = \lambda\, V_g = \lambda \int e^{is\phi_\varepsilon}\ (g)\ d\mu(s),$$

where λ is any real number (coupling constant), g is e.g. the characteristic function of the sphere $|x| \leq r$ in R^s and $d\mu(s)$ is any complex finite measure on the real line, such that $d\mu(-s) = d\mu(s)$ (in order that V be symmetric) and $\int |s|\ d|\mu|\ (s) < \infty$. Note that $e^{is\phi_\varepsilon(x)}$ is essentially self-adjoint in D_0. In this case V is a bounded self-adjoint operator on $\mathcal{F}$ and thus we have a regular perturbation of H_0. The assumptions a), b), c) are easily verified in this case. $a_t^{\#}(h)$ converge in this case on $D_0(H_0^{1/2})$. All properties of Proposition 1 hold. ([Ho-K 8]).

II.3 General properties of the asymptotic creation-annihilation operators. The statements in this section are partly independent of the particular assumptions made in proposition 1 of the preceding Section II.2 to prove the existence of asymptotic creation-annihilation operators $a_{\pm}^{\#}(h)$ and follow solely from the existence of $a_{\pm}^{\#}(h)$ on some dense domain. This will be evident from the proof of the different statements. Nevertheless we do not seek maximal generality, would like much more to give just some insight into the problems, thus we shall assume for simplicity, throughout this section, that all the assumptions leading to the statements 1) to 4) of proposition 1 (Section II.2) are satisfied. Let D be a common (dense) convergence domain for $a_t^{\#}(h)$ as $t \to +\infty$, coming from proposition 1 (Section II.2).

Proposition 1: The asymptotic creation-annihilation operators $a_+^{\#}(h)$ constructed in proposition 1 of Section II.2 are, for any $h \in H$, linear, densely defined on a domain $D_{\#}$ and closable on this domain. We shall denote their closures by $a_+^{\#}(h)$. One has:
1) $(a_+(h))^* \supset a_+^*(\bar{h})$, $(a_+^*(h))^* \supset a_+(\bar{h})$;

2) $e^{-itH} a_+^{\#}(h) e^{itH} \supset a_+^{\#}(e^{\mp it\mu} h)$, where $-$ goes with a^*, $+$ with a. Moreover as bilinear forms on D x D:

3) $[H, a_+^{\#}(h)] = a_+^{\#}(\pm \mu h)$, $h \in D(\mu)$.

4) $[a_+(h), a_+^*(f)] = (\bar{h},f)$, $[a_+(h), a_+(f)] = 0$, $[a_+^*(h), a_+^*(f)] = 0$, for all $h, f \in H$, where [A,B] is to be understood in the sense of bilinear forms as the bilinear form:

$$\psi \times \psi' \in D \times D \to (A\psi, B\psi') - (B\psi, A\psi').$$

5) i) For any eigenvector ψ of H (i.e. any $\psi \neq 0$ in Fock space $\mathcal{F}$, such that $H\psi = E\psi$ for some (real) number E) one has $a_+(h)\psi = 0$ for all $h \in H$;
  ii) For any vector $\psi \in D$ such that $a_+(h)\psi = 0$ one has

$$\|a_+^*(h)\psi\| = \|h\|_2 \|\psi\|$$

6) The statements 1) to 5) hold with all lower labels + replaced by $-$.

Remark 1: The statement 3) will be called the weak commutation relations of $a_+^{\#}$ with H. 4) will be called the weak (canonical) commutation relations for the asymptotic creation-annihilation operators $a_+^{\#}$. 5) is non-void, at least in any physically interesting case, since H has at least the bottom of its spectrum (ground state energy) as an eigenvalue. In particular this is the case for all the examples given in the preceding section.

Proof: It is easy and we just mention the ingredients for the proof of the different statements:
1) Use $(\chi, a_t(h)\psi) = (a_t^*(\bar{h})\chi, \psi)$ for all $\chi, \psi \in D$, together with $a_t^{\#}(h) \to a_+^{\#}(h)$ strongly on D;
2) Use $U_t^* a_s^{\#}(h) U_t = a_{s-t}^{\#}(e^{\mp it\mu} h)$ on D and make $s \to \infty$.

3) Use the estimate $\|a_+^{\#}(h)\psi\| \leq \|h\|_2 \|g_1(H)\psi\|$ from proposition 1 and differentiate 2) with respect to t.

4) Note that $(a_t(h)\psi, a_t^*(f)\psi')$ converges to $(a_+(h)\psi, a_+^*(f)\psi')$ as $t \to \infty$, using the strong convergence of $a_t^{\#}$ and the uniform bound. For $a_+^{\#}$ replaced by $a_t^{\#}$, on the other hand, the commutation relations in 4) hold trivially, by the unitary equivalence with $a^{\#}$.

5) i) Suppose ψ is an eigenvector of H belonging to the eigenvalue E. Then $a_t(h)\psi = U_t a(e^{it\mu} h) U_t^*\psi = e^{it(H-E)} a(e^{it\mu} h)\psi$.

Since $a_t(h)\,\psi \to a_+(h)\,\psi$ as $t \to \infty$, we have then that

$$||a_+(h)\,\psi|| = \lim_{t\to\infty} ||a(e^{it\mu}h)\,\psi\,||$$

Because of the uniform estimate

$$||a(e^{it\mu}h)\,\psi|| \leq ||h||_2\ ||(N+1)^{1/2}\,\psi||$$

and the closedness of the operator $a(\cdot)$ it is now enough to prove

$$\lim_{t\to\infty} ||a(e^{it\mu}h)\,\psi|| = 0$$

where h is in $\overset{\circ}{C}{}_0^\infty(\hat{R}^s)$ and ψ is in the dense subset D_0 of finite vectors with components in $\mathcal{S}$. An estimate similar to the one made in the preceding section then gives this. 5) ii) follows then from 4) and 5) i). 6) is trivial ∎

Proposition 2: If the infimum of the spectrum $\sigma(H)$ of H is an eigenvalue E^0, then we have $\sigma(H) \supseteq [E^0 + m_0, \infty)$.

Proof: We follow [Di] and remark that it is enough to prove that, given $\delta > 0$ and $E^* \in [m_0, \infty)$, there exists $\psi \equiv \psi_{\delta,E^*} \in D(H)$ such that $||(\bar{H} - E^*)\,\psi|| \leq \delta\ ||\psi||$, where $\bar{H} \equiv H - E^0$. Let Ω be the eigenvector of H corresponding to the eigenvalue E^0 of H. One has $\Omega\in D(H)$ thus $a_+(h)\,\Omega \in \mathcal{F}$. By the commutation relation between H and $a_+^*(h)$ given in point 3) of proposition 1 we get easily $(\bar{H} - E^*)\ a_+^*(h)^+\ \Omega = a_+^*((\mu - E^*)h)\Omega$. Given $\delta > 0$, E^*, we can always choose h such that $|E^* - \mu(k)| \leq \delta$, for all k in the support of h ($E^* - \mu(k)$ being a bounded continuous function of k which vanishes for some k). Using proposition 1, point 5), ii) we then have $||(\bar{H} - E^*)\ a_+^*\ (h)\ \Omega|| \leq \delta\ ||h||\ ||\ a_+^*(h)\ \Omega||$. ∎

Remark 2: All the statements of this section hold, in particular, for the examples discussed in Section II.2.

II.4 Asymptotic Fock spaces. The asymptotic creation-annihilation operators $a_\pm^\#(h)$ which satisfy the weak canonical commutation relations have already given us some information on the spectrum of the Hamiltonian, namely that it contains the whole interval $[E^0 + m_0, \infty)$, where E^0 is the infimum of the spectrum of H. It is natural to try to get information on the absolutely continuous spectrum, by solving, partially at least, the diagonalization problem for H, as was done, by the very introduction of Fock space, for H_0. The commutation relations of H_0 and $a^*(h)$ give

$$H_0\ a^*(h_1)\ldots a^*(h_n)\ \Omega_0 = \sum_{j=1}^{n}\ a^*(h_1)\ldots a^*(\mu h_j)\ldots a^*(h_n)\ \Omega_0,$$

which of course is the same as saying that $H_0 = d\Gamma(\mu)$, since $\mathcal{F}$ as a carrier of an irreducible representation of the canonical commutation relations is spanned by all vectors of the form Ω_0, $a^*(h_1)\ldots a^*(h_n)\ \Omega_0$.

We would like of course to have similary a Fock representation of the commutation relations for the $a_+^\#(h)$, with some cyclic vector

Ω in such a way that then by the commutation relations of H and $a_+\#(h)$, which we know already are (in the weak sense) the same as those of H_0 and $a\#(h)$, we could have H acting as a free energy operator in this space i.e. as a $d\Gamma(\cdot)$ of a multiplication operator on the asymptotic one particle space $\{a_+^*(h)\ \Omega,\ h \in H\}$.

To fulfill this program it is useful to have information on the domain of products of $a_+\#(h)$ operators. Again we do not seek maximal generality and simply make the following additional assumptions, also satisfied in all previously given examples:

1) $(N + 1)\ H_0{}^2 \leq g_3(H)^2$ for some $g_3 \in M_H$;

2) $(N + 1)^2 \leq g_1(H)^2$

3) $C^\infty(H) \subset D\ (H\ g_1(H)) \cap D(H\ g_3(H))$

Remark 1: The condition 2) absorbs the weaker assumption we used thoughout before (from II.2 on): $(N + 1) \leq g_1(H)^2$.

Proposition 1: Under the assumptions of proposition 1 of Section II.2 and moreover under the assumptions 1) and 3) above we have

i) $[H,a_+\#(h)] = a_+\#(\pm\mu h)$ on $D(Hg_1(H)) \cap D(Hg_3(H))$, $h \in D(\mu)$

ii) For all $n = 1, 2,\ldots,h_i \in \mathcal{S}(\hat{R}^s)$ $\prod_{i=1}^{n} a_+\#(h_i)$ is defined on $C^\infty(H)$.

iii) Moreover under the additional condition 2) above the following strong commutation relations hold:

$$[a_+(h),\ a_+^*(f)] = (\bar{h},\ f)$$
$$[a_+(h),\ a_+(f)] = 0$$
$$[a_+^*(h),\ a_+^*(f)] = 0,$$

on $C^\infty(H)$, for all $h,\ f \in H$.

Furthermore $||a_+\#(h)\ a_+\#(f)\ \psi|| \leq ||h||_2\ ||f||_2\ ||g_1(H)\ \psi||$

iv) All statements i), ii), iii) hold also with + replaced everywhere by −.

Proof: i) is elementary but somewhat lengthy. To prove ii) we first remark that it is enough to show $a_+\#(h)\ C^\infty(H) \subset C^\infty(H)$, because then by assumption 3)

$$a_+\#(h)\ C^\infty(H) \subset D(H\ g_1(H)) \cap D(H\ g_3(H)) \subset D(g_1(H)) \subset D(a_+\#(f)).$$

Let $\psi \in C^\infty\ (H)$. Then $a_+\#(h)\ \psi \in D(H)$, by i), and thus $H\ a_+\#(h)\psi \in \mathcal{F}$. On the other hand $H\ \psi \in C^\infty\ (H) \subset D(a_+\#(f))$, thus $a_+\#(f)\ H\ \psi \in D(H)$. Finally $a_+\#(\pm\mu h)\ \psi \in D(H)$, again by i). From the weak commutation relations of $a_+\#(h)$ with H (given by proposition 1, 3) Section II.3) we deduce then $H\ a_+\#(h)\psi = a_+\#(h)H\ \psi + a_+\#(\pm\mu h)\psi$, where we know already by the above considerations that the right hand side is in D(H), hence also the left hand side. It is then easy to prove by induction, using the same kind of arguments, that $a_+\#(h)C^\infty(H) \subset D(H^n)$ for all $n = 1, 2, \ldots$. iii) This is a consequence of the assumption 2) above, yielding the t-uniform bound

$$||a_t\#(h)\ a_t\#(f)\ \psi|| \leq ||h||_2\ ||f||_2\ ||g_1(H)\ \psi||,$$

Proposition 1.2 of Section I.2 and e.g. the weak canonical commutation relations of the $a_{+}\#(h)$. iv) is trivial. ∎

We can now proceed to the construction of asymptotic Fock spaces i.e. to the Fock space representations of the $a_{\pm}\#(h)$. Assume there is at least a vector Ω_+ in $\mathcal{F}$ such that $a_+(h)\,\Omega_+ = 0$ and a vector Ω_- in $\mathcal{F}$ such that $a_-(h)\,\Omega_- = 0$, for all $h \in \mathcal{H}$. Such vectors certainly exist when H has at least an eigenvalue, e.g. the infimum of its spectrum (which is verified in all physical examples). Then, since we have the commutation relations for the $a_{\pm}\#(h)$ on an invariant domain $D = C^{\infty}(H)$ and $\Omega_{\pm} \in D$, we can find Fock representations $\mathcal{F}_{\pm}(\Omega_{\pm})$ with cyclic vectors Ω_+, such that $\mathcal{F}_+$ is the closed linear span of all vectors Ω_+, $a_+{*}(h_1)\ldots a_+{*}(h_j)\,\Omega_+$, for all $j = 1, 2, \ldots$ and $h_i \in \mathcal{S}(\hat{R}^s)$(and similarly for $\mathcal{F}_-$). Note that $\mathcal{F}_{\pm}(\Omega_{\pm})$ coincide here by construction with, in general proper and different, subspaces of $\mathcal{F}$, but it is often more instructive to think of them abstractly as separate "asymptotic spaces" constructed with the asymptotic creation-annihilation operators. If e.g. there exist two orthogonal vectors $\Omega_+^{(1)}$, $\Omega_+^{(2)}$ such that $a_+(h)\,\Omega_+^{(i)} = 0$ for $i = 1, 2$ and all $h \in \mathcal{H}$, then $\mathcal{F}_+(\Omega_+^{(1)})$ is orthogonal to $\mathcal{F}_+(\Omega_+^{(2)})$, with respect to the scalar product in $\mathcal{F}$. In particular this holds whenever $\Omega_+^{(1)}$, $\Omega_+^{(2)}$ are two orthogonal eigenvectors of H (if they exist). Suppose now that H has an eigenvalue E, with correspondent eigenvector Ω. Define a mapping $W_+(\Omega)$ from $\mathcal{F}$ to $\mathcal{F}_+(\Omega)$ by setting $W_+(\Omega)\,\Omega_0 \equiv \Omega$ and $W_+(\Omega)a{*}(h_1)\ldots a{*}(h_n)\Omega_0 \equiv a_+{*}(h_1)\ldots a_+{*}(h_n)\,\Omega$, for all $h_i \in \mathcal{S}(\hat{R}^s)$ $i = 1, \ldots, n$; $n = 1, 2, \ldots$ and extending then $W_+(\Omega)$ by linearity.

Then $W_+(\Omega)$ is a linear operator on $\mathcal{F}$, densely defined, and by the canonical commutation relations we have that $W_+(\Omega)$ is isometric with range the linear span of Ω, $a_+{*}(h_1)\ldots a_+{*}(h_n)\,\Omega$. Thus $W_+(\Omega)$ extends uniquely to a partial isometry, with initial domain $\mathcal{F}$ and final domain $\mathcal{F}_+(\Omega)$. We call $W_+(\Omega)$ "generalized wave operators." One has, using the definition of $W_+(\Omega)$ together with the commutation relations of $\exp(-itH_0)$ with $a{*}(h)$ (proposition 1,4) in Section I.2) $W_+(\Omega)\exp(-itH_0)\,a{*}(h_1)\ldots a{*}(h_n)\,\Omega_0 =$ $W_+(\Omega)\,a{*}(\exp(-it\mu)h_1)\ldots a{*}(\exp(-it\mu)h_n)\,\Omega_0$ and using the commutation relations of $\exp(-itH)$ and the $a_+{*}(h)$, this is equal to

$$U_t{*}\;a_+{*}(h_1)\ldots a_+{*}(h_n)\,U_t\,\Omega = e^{-it(H-E)}\,a_+{*}(h_1)\ldots a_+{*}(h_n)\,\Omega.$$

Thus we have proven $W_+(\Omega)\,e^{-itH_0} = e^{-it(H-E)}\,W_+(\Omega)$, hence the unitary equivalence of H_0 on $\mathcal{F}$ and the restriction $H-E \restriction \mathcal{F}_+(\Omega)$ of $H - E$ to the subspace $\mathcal{F}_+(\Omega)$ of $\mathcal{F}$. Since the absolute continuous spectrum of H_0 is $[m_0, \infty)$ and everything said holds with + replaced everywhere by −, we have proven the following:

<u>Proposition 2</u>: Under the same assumptions as for proposition 1, and supposing that H has at least one eigenvalue E, one has that the absolutely continuous spectrum of H contains the interval $[E + m_0, \infty)$. $H - E$ is unitary equivalent H_0 on the invariant subspaces $\mathcal{F}_{\pm}(\Omega)$ of $\mathcal{F}$. $\mathcal{F}_{\pm}(\Omega)$ is the closed linear span of the eigenvector Ω to the

eigenvalue E and all vectors of the form $a_{\pm}^*(h_1)\ldots a_{\pm}^*(h_n)\,\Omega$, for all n = 1, 2, ..., $h_i \in H$, hence it has the structure of a Fock space. ■

Høegh-Krohn has proved the following proposition, which uses in an essential way the lower boundedness of H ([Ho-K 5]):

Proposition 3: Under the same assumptions as for proposition 1, $\mathcal{F}$ is the closure of $\cup\,\mathcal{F}_+(\phi_0)$ as well as of $\cup\,\mathcal{F}_-(\phi_0)$, where the union is taken over all ϕ_0 in $V_+^{(0)}$ resp. $V_-^{(0)}$ where

$$V_{\pm}^{(0)} \equiv \{\, \psi \in \mathcal{F} \mid a_{\pm}(h)\,\psi = 0 \text{ for all } h \in H \}.$$

Remark: This proposition should not be confused with asymptotic completeness. It is just the statement that $\mathcal{F}$ is the closure of the union of a part $\mathcal{F}_{+b}$ which is the direct sum $\oplus_i \mathcal{F}_+(\Omega^i)$ over the asymptotic spaces constructed with all eigenvectors Ω^i of H and a remainder which is the union of all $\mathcal{F}_+(\phi_0')$, where ϕ_0 is any vector annihilated by all $a_+(h)$ and which is not an eigenvector of H.

Proof: It consists of 2 steps. Let F_λ be the spectral decomposition of H.

1) One has $a_+(h)F_\lambda \mathcal{F} \subset F_{\lambda-m_0}\mathcal{F}$ (this follows observing that $a_+(h)$ is bounded on $F_\lambda \mathcal{F}$ and, by the commutation relations,

$$\| e^{tH} a_+(h) F_\lambda \psi \| \leq e^{t(\lambda - m_0)},$$

hence by the spectral theorem $a_+(h)F_\lambda \psi = F_{\lambda-m_0} \psi$).

2) H is lower bounded, hence $F_{E^0-\varepsilon} = 0$ for all $\varepsilon > 0$, with $E^0 \equiv \inf\sigma(H)$. Then from 1) one has $a_+(h)F_{E^0-\varepsilon+m_0} = 0$, hence $F_{E^0-\varepsilon+m_0}\mathcal{F} \subset V_+^{(0)}$. Similarly one proves $F_{E^0-\varepsilon+jm_0}\mathcal{F}^0 \subset A_+^{(j-1)}$, where $A_+^{(j)}$ is the maximal closed subspace annihilated by all $a_+(h_1)\ldots a_+(h_j)$. Hence $V_+^{(0)} \bigcup_{j=1}^{\infty} A_+^{(j)} = \bigcup \mathcal{F}_+(\phi_0)$ is dense in $\mathcal{F}$. ■ From the above proposition 3 we have now the following tensor decomposition of $\mathcal{F}$ and H. Suppose H has the eigenvalue E with eigenvector Ω. Let $\mathcal{J}_+$ be the map defined by $\phi_0 \to \Omega \otimes \phi_0$, $a_+^*(h_1)\ldots a_+^*(h_n)\,\phi_0 \to a_+^*(h_1)\ldots a_+^*(h_n)\,\Omega \otimes \phi_0$, for all $h_i \in H$, n = 1, 2, ..., $\phi_0 \in V_+^{(0)}$.

This map extends by linearity and proposition 3 to a densely defined map, and the range of $\mathcal{J}_+$ is by construction dense in $\mathcal{F}_+(\Omega) \otimes V_+^{(0)}$. Moreover $\mathcal{J}_+$ preserves scalar products. Hence it extends to a unitary map from $\mathcal{F}$ onto $\mathcal{F}_+(\Omega) \otimes V_+^{(0)}$. Under this map H goes to $(E + H_0^+) \otimes 1 + 1 \otimes (H \restriction V_+^{(0)})$, where H_0^+ is the free energy operator $d\Gamma(\mu)$ in the Fock space $\mathcal{F}_+(\Omega)$. Similarly with − replacing + everywhere. We close this section with some remarks and problems.

1. As in the preceding section the propositions of this section hold for the models of Examples 1), 2), 3) of Section II.2.

Moreover they hold also ([Ho-K 16]), in a modified form, for the space cut-off version of another type model we shall discuss later, in Section IV, namely the so called "exponential interactions in 2 space-time dimensions." ([Ho-K 15]) These space cut-off interactions have the form $V_g = \int :e^{\alpha\phi}:(g)\, d\nu(\alpha)$, $g \in L_2(R^s) \cap L_1(R^s)$, $g \geq 0$, (1) where $d\nu(\alpha)$ is any positive, bounded measure on the real line with support in the interval

$$\left(\frac{-4}{\sqrt{\pi}}, \frac{4}{\sqrt{\pi}}\right),$$

and where $:e^{\alpha\phi}:(g)$ is defined as $\sum_{n=0}^{\infty} \frac{\alpha^n}{n!} :\phi^n:(g)$ and the series is strongly convergent: see Chapter IV.

2. What information is available on the possible eigenvalues of H (bound states)? For all examples mentioned H is proved to have at least one eigenvalue, E_g namely the infimum of its spectrum (ground state energy). In general, as known from special $:P(\phi)(g):_2$ examples, H can have other eigenvalues besides the ground state energy. For the exponential interactions of the form (1) in point 1 above, with $d\nu(-\alpha) = d\nu(\alpha)$ the non-existence of any eigenvalues and moreover of any spectral point in the whole interval $(E_g, E_g + m_0)$ is proven. ([Al-H.K. 5])

3. For all examples so far discussed it is known that $\sigma_{ac}(H) \cap (-\infty, E_g + m_0)$ is void, hence by proposition 2 one has $\sigma_{ac}(H) = [E_g + m_0, \infty)$. (In fact for all examples it is known that the spectrum of H in $[E_g, E_g + m_0)$ is purely discrete so that $[E_g + m_0, \infty)$ is the essential spectrum).

4. Does $V_+^{(0)}$ coincide with the closed linear span $\mathcal{F}_b$ of all eigenvectors of H? In this case we would have $\mathcal{F} = \oplus_i \mathcal{F}(\Omega^i)$, where Ω^i are the orthonormal eigenvectors of H, and thus, in particular, absence of singular continuous spectrum. Nothing seems to be known of this question, besides models without vacuum polarization ([Ho-K 1-4]) or the explicitely solvable model $:\phi^2:(g)$. ([Ros 3]).

5. Does $V_+^{(0)} = V_-^{(0)}$ hold? We shall see that this question is related to the unitarity of a scattering operator.

6. The convergence of the $a_t^{\#}$ to asymptotic creation-annihilation operators extends to the case where $a^{\#}$ are replaced by any operators in the norm closed algebra generated by the $e^{i\phi(h)}$. For a physical interpretation of the constructions in terms of asymptotic creation-annihilation operators and their implications in terms of observables in certain fermion models, see [Pru], [Pr-M 1,2].

II.5 The Scattering Operator.

II.5.1 Introduction of the scattering operator. Suppose the Hamiltonian H of the preceding sections has a simple eigenvalue E^0 (ground state energy), sitting at the infimum of the spectrum. Call Ω the corresponding normalized eigenstate. We have $W_+(\Omega)\, a^{\#}(h) \subset a_+^{\#}(h) W_+(\Omega)$, as seen using the definition of W_+ given in

Section II.4. Hence

$$\text{s-}\lim_{t\to\infty} a_t^{\#}(h_1)\ldots a_t^{\#}(h_n)\Omega = a_+^{\#}(h_1)\ldots a_+^{\#}(h_n)\Omega , \tag{1}$$

for all $h_i \in \mathscr{S}(\hat{R}^s)$, $i=1,\ldots,n$. Define $S \equiv W_-^*(\Omega)W_+(\Omega)$. (2)

Then S maps $\mathcal{F}$ into $\mathcal{F}$ and is a contraction. The commutation relation $SH_0 \subset H_0S$ follows easily from the intertwining relations between H, H_0 given by the generalized wave operators $W_+(\Omega)$ (Section II.4).

In the following we shall have to consider products of operators A_i, $i = 1,\ldots,n$ out of a certain set of operators on $\mathcal{F}$. We shall make the convention of writing $A_1\ldots A_n$ not only for the cases $n = 1, 2\ldots$ but also for the case $n = 0$, i.e. for the case where none of the factors are present: by convention we set $A_1\ldots A_n$ equal the identity operator on $\mathcal{F}$ for $n = 0$. Then we have for all $n, m = 0, 1, 2, \ldots$ $(a^*(h_1)\ldots a^*(h_n)\Omega_0,\ S\ a^*(f_1)\ldots a^*(f_m)\Omega_0) = (a_-^*(h_1)\ldots a_-^*(h_n)\Omega,\ a_+^*(f_1)\ldots a_+^*(f_m)\Omega)$ for all $h_r, f_{r'} \in \mathscr{S}(\hat{R}^s)$ $r = 1,\ldots n$; $r' = 1,\ldots, m$. These are the S-matrix elements (transition amplitudes) for scattering of n incoming "physical particles" with distribution of momenta given by $h_1,\ldots, h_n$, created from the physical vacuum Ω by repeated application of the in-creation operators a_-^* (thus represented by a state in $\mathcal{F}_-(\Omega)$), to m outgoing physical particles with distribution of momenta given by $f_1,\ldots,f_m$ and represented by a state in $\mathcal{F}_+(\Omega)$.[1]

Remark 1: In the next Section we shall see that the above definition corresponds to the formal one used usually in physics, in terms of powers series in the coupling constant.

Remark 2: If the Hamiltonian H of the preceding Sections has, besides E^0, other eigenvalues $E^i (\geq E^0)$, with correspondent orthonormal eigenvectors Ω^i, then one nas formulae like (1),with $W_+(\Omega^i)$ replacing $W_+(\Omega)$ and one can define an operator S' from $\mathcal{F}_+ \equiv \bigoplus_i \mathcal{F}_+(\Omega^i)$ to $\mathcal{F}_- \equiv \bigoplus \mathcal{F}_-(\Omega^i)$, as $(S'\psi)^{(i)} = S^{ij}\psi^{(j)}$,

with $S^{ij} \equiv W_-^*(\Omega^i)\, W_+(\Omega^j)$ and where i, j denote the components in $\mathcal{F}_-(\Omega^i)$, $\mathcal{F}_+(\Omega^i)$. S' would be unitary from $\mathcal{F}_+$ to $\mathcal{F}_-$ if and only if $\mathcal{F}_- = \mathcal{F}_+$ and from $\mathcal{F}$ to $\mathcal{F}$ if and only if moreover $V_+^{(0)} = V_-^{(0)} = \mathcal{F}_b$, where $\mathcal{F}_b$ is the closed linear span of all eigenvectors of H. However there are some problems with the physical interpretation of such operators S^{ij}. (See e.g. [Pr-M 2]).

II.5.2 Strong asymptotic and analytic expansions in powers of the coupling constant for the asymptotic creation and annihilation operators and the S operator. In order to give a connection between the above definition of the scattering matrix in terms of asymptotic creation-annihilation operators and the traditional one of quantum fields physics given in terms of formal perturbation series, we shall now briefly discuss the dependence on the coupling constant λ, for small values of $|\lambda|$, of the asymptotic-creation-annihilation operators and the relative S operator, constructed from the Hamiltonian $H = (H_0 + \lambda V)^{**}$.

This study will also give us an asymptotic series analogue of the known result of non-relativistic quantum mechanics on the unitarity of the S-matrix for sufficiently weak forces.[2] For this discussion we shall abandon for simplicity the general formulations of the preceding sections and stick rather to the case of the examples 1), 2), and 3) discussed in Section II.2.[3]

<u>Proposition 1:</u> Let λV be any of the space cut-off interactions of the examples 1), 2), 3) discussed in Section II.2 (polynomial interactions for $s = 1$, ultraviolet cut-off polynomial interactions for $s \geq 2$, ultraviolet cut-off bounded interactions). Let $H = (H_0 + \lambda V)^{**}$ be the correspondent Hamiltonian, $a_{\pm}^{\#}$ the correspondent asymptotic creation-annihilation operators. Set $V(t) \equiv \exp(+itH_0)\, V \exp(-itH_0)$. Then the following asymptotic series for small value of $|\lambda|$ hold for all ψ in the dense domain $C^{\infty}(H)$ of the Fock space and for all h in the one particle space $\mathcal{H}$ with constants $C_{N+1}^{(1)}$, $C_{N+1}^{(2)}$ independent of λ:

1) $$a_{+}^{\#}(h)\psi = \sum_{\ell=0}^{N} (i\lambda)^{\ell} \int_{t_\ell \leq \dots \leq t_1} \{[V(t_\ell),\dots,[V(t_1),a^{\#}(h)]\dots]\}\psi \, dt_1 \dots dt_\ell + R_{N+1}^{(1)} (a_{+}^{\#}(h))\psi,$$

where $||R_{N+1}^{(1)} (a_{+}^{\#}(h))\psi|| \leq C_{N+1}^{(1)} |\lambda|^{N+1}$.

2) $$a_{+}^{\#}(h)\psi = \sum_{\ell=0}^{n} (i\lambda)^{\ell} \int_{t_\ell \leq \dots \leq t_1} \{[V(t_\ell),\dots,[V(t_1),a^{\#}(h)]\dots]\}_{-} \psi + R_{N+1}^{(2)}(a_{+}^{\#}(h))\psi,$$

where $\{\dots\}_{-}\psi \equiv \underset{t\to-\infty}{\text{s-lim}} \exp(itH) \exp(-itH_0) \{\dots\} \exp(itH_0)\exp(-itH)\psi$ (which exists) and $||R_{N+1}^{(2)}(a_{+}^{\#}(h))\psi|| \leq C_{N+1}^{(2)} |\lambda|^{N+1}$.

3) Similar expansions hold also when $a_{+}^{\#}$ and $\{\dots\}_{-}\psi$ are replaced by $a_{-}^{\#}$ and
$\{\dots\}_{+}\psi = \underset{t\to+\infty}{\text{s-lim}} \exp(itH) \exp(-itH_0)\{\dots\} \exp(itH_0) \exp(-itH)\psi$.

<u>Remark:</u> 1) gives the action of the asymptotic creation-annihilation operators on $C^{\infty}(H)$ in terms of the first N terms (N arbitrary, finite) of a power series in λ, with operator coefficients, which can be computed since they depend only on the explicitly given quantities V, $a^{\#}(h)$ and H_0, plus a remainder which is $O(|\lambda|^{N+1})$ in norm.

2) gives the $t = +\infty$ creation-annihilation operators in terms of the first N terms of a power series in λ involving only quantities at $t = -\infty$ plus again a remainder $O(|\lambda|^{N+1})$. The proof of existence of the strong limit giving $\{\dots\}_{-}\psi$ is made by methods similar to the ones explained in II.2 which yield asymptotic creation-annihilation operators.

1), 2) prove that certain formal power series expansions of

the Dyson-Källen-Schwinger type, well known in physics, are actually asymptotic power series expansions for the models considered.

Proof: The central idea is to iterate the relation

$$a_+(h)\psi = a\psi + i\lambda \int_0^\infty e^{isH} [V, a(h_s)] e^{-isH} \psi \, ds,$$

using higher order estimates of the type (5) Sec. II.2 to control the domain and estimate the remainders. (For details see [Al-H.K.4]). ∎

Proposition 2: For the same models as in proposition 1, one has the following asymptotic power series expansions of the scattering operator defined by (2):

$$S B \Omega_0 = \sum_{\ell=0}^{N} (i\lambda)^\ell \int\dots\int_{t_\ell \leq \dots \leq t_1} \{[V(t_\ell), \dots [V(t_1), B]\dots]\}\Omega_0 + R_{N+1}(SB)\Omega,$$

for all N, where B is any operator of the form $B = \prod_{i=1}^{n} a^*(h_i)$, and with $||R_{N+1}(SB)\Omega|| \leq C_{N+1} |\lambda|^{N+1}$, C_{N+1} being a constant independent of λ.

Corollary: For $|\lambda| \neq 0$ sufficiently small the scattering operator is non-trivial in the sense that $S\psi \neq 0$, $S\psi \neq \psi$ for all $\psi \in C^\infty(H)$.

Proof: The steps are: a) Prove first $W_+(\Omega)\, B\Omega_0 = B_+\Omega$, where $B_+ =$ s-lim $\exp(itH) \exp(-itH_0)\, B \exp(itH_0) \exp(-itH)$ on $C^\infty(H)$. b) Prove for B_+ an expansion in terms of-quantities of the type of the expansion 2) in proposition 1. c) Insert this expansion in $W_-{}^*(\Omega)W_+(\Omega)B\Omega_0 = W_-{}^*(\Omega)\, B_+\, \Omega$ and use the fact that $W_-{}^*(\Omega)\, \{\cdot\}_- = W_-{}^*(\Omega)\, W_-(\Omega)\, \{\cdot\} = \{\cdot\}$. For details see [Al-H.K.4]. ∎

Remark: Proposition 2 permits the computation, modulo an error $O(|\lambda|^{N+1})$, of the scattering operator on a dense domain, since the N+1 terms on the right-hand side are all expressed through explicitly given quantities V, H_0, B.

Proposition 3: For the case of the space cut-off bounded interactions of Example 3) in Section II.2, the matrix elements of S between dense sets of vectors in $\mathcal{F}$ are analytic functions of λ for all $|\lambda| < \lambda_0$ and some $\lambda_0 > 0$. The "linked cluster expansion" of these S-matrix elements (expansion as sum of "externally connected graphs"), along the lines of the formal power series expansions of the S-matrix used in physics, is actually convergent to the S-matrix elements, for all $|\lambda|$ sufficiently small.

Proof: The proof uses an expression of the S-matrix elements in terms of correlation functions of a classical statistical mechanics gas. The expression is derived using constructions of the so-called Euclidean-Markoff approach to quantum field theory and we shall return to this in Chapter IV. For details see [Al-H.K.2]. ∎

II.5.3 Unitarity of the S-operator in the sense of asymptotic series:

Using the asymptotic series expansion for the S-operator in powers of the coupling constant one can prove unitarity in the sense of asymptotic series (i.e. in the language common to physical literature, "unitarity in all orders of perturbation theory", where however in our case, perturbation theory is to be understood in the stronger sense of asymptotic series). We have namely, restricting ourselves, for the sake of simplicity, to a statement concerning the examples 1), 2), 3) of Section II.2 (polynomial and bounded space cut-off interactions):

Proposition 1: For the models of examples 1), 2), 3) of Section II.2, the S-operator is unitary in the sense of asymptotic expansions in powers of the coupling constant, so that, for all $n = 0, 1, 2, \ldots$ $S^* S \psi = 1\,\psi + R_{n+1}(\psi)$ and $SS^* \psi = 1\,\psi + R'_{n+1}(\psi)$, where $\|R_{n+1}(\psi)\| \leq C_{n+1} |\lambda|^{n+1}$, $\|R'_{n+1}(\psi)\| \leq C'_{n+1} |\lambda|^{n+1}$, for all $\psi \in C^\infty(H)$, where C_{n+1}, C'_{n+1} are independent of λ. In other words, S^*S and SS^* are two self-adjoint contraction operators which are strongly infinitely differentiable on $C^\infty(H)$ at $\lambda = 0$, with all their derivatives equal to the null operator (multiplication by zero) on $C^\infty(H)$, for $\lambda = 0$.

Remark: If S^*S and SS^* could be proved to be not only strongly C^∞ at $\lambda = 0$ but also strongly analytic at $\lambda = 0$, then we would have unitarity not only in the sense of asymptotic series but also in the usual sense that $S^*S=SS^*=1$ for all $|\lambda|$ sufficiently small. For the bounded space cut-off interactions of example 3) in Section II.2, we know by proposition 3, II.5.2 that suitable matrix elements of S between dense domains of $\mathfrak{F}$ are analytic in λ in a neighborhood of $\lambda = 0$. It can be proven moreover that the same is true of S^*S and SS^*. Since however the domains themselves depend on λ, some more information is needed to use this and the above to conclude $S^*S = SS^* = 1$. [Al-H.K. 6]

Proof: One computes $S^*SB\Omega_0$ (and $SS^*B\Omega_0$, interchanging + with - in the former) as $W_+^*(\Omega)W_-(\Omega)SB\Omega_0$, inserting the asymptotic expansion for $SB\Omega_0$ given by proposition 2. Then the - quantities $W_-(\Omega)SB\Omega_0$ are expressed through + quantities, similarly as in proposition 1. Finally one obtains an expansion for $W_+^*(\Omega)W_-(\Omega)SB\Omega_0$, all terms of which, the first excluded, vanish identically as a consequence of a simply algebraic identity. See [Al-H.K.4]

Remark: Stronger results on unitarity can be obtained in the case of interactions which have, in addition to a space cut-off, also the property of non-polarizing the vacuum. More precisely consider interactions of the form (3), I.3 with $M<\infty, W(w_{io})=W(w_{oi})=0$, for all i, and with kernels which are "suitably smooth" (for precise conditions see[Ho.K1-4]). Suppose moreover V is such that $H_0 + \lambda V$ is essentially self-adjoint and bounded from below (for all $\lambda \in R$). In this case all results of this chapter apply. In addition one has (due to lack of vacuum polarization) that the strong limits of $\exp(itH)\exp(-itH_0)$ for $t \to \pm\infty$ exist [Ci],[Ho.K4]and define wave operators $U_\pm$ with the usual intertwining relations, giving unitary equivalence of H_0 in $\mathfrak{F}$ and the parts of H in $U_\pm U_\pm^* \mathfrak{F}$. Moreover for $|\lambda|$ sufficiently small, $E(\Delta)H$ operating in $E(\Delta)\mathfrak{F}$ and $E_0(\Delta)H_0$ in

$E_0(\Delta)\mathcal{F}$ are unitarity equivalent, where E, E_0 are the spectral measures corresponding to H, H_0 and Δ is any bounded interval. For any real α the restriction of the scattering operator $S = U_-^* U_+$ to $E_0(\alpha)\mathcal{F}$ is unitary and strongly analytic in $|\lambda|<\lambda_0(\alpha)$. These results have been proved by Høegh-Krohn by a very nice adaptation and extension of Friedrichs-Rejto's method of gentle perturbations ([Re]). Further results hold for the Fermi case (or the mixed fermi-boson case, with interaction linear in the boson fields). Of course, it would be extremely worthwhile extending these methods both to cases with non-smooth kernels (translation invariant) and/or vacuum polarization.

III. Remarks About Translation Invariant Models Without Vacuum Polarization.

In the preceding section we have analyzed models with interaction V such that $V\Omega_0 \neq 0$ (presence of vacuum polarization) but with smooth kernels, which (cfr. the discussions of Section I.6) corresponds to space cut-off (and an ultraviolet cut-off, for local relativistic interactions of the form (2) of Section I.6 in more than one space dimension). In this chapter we shall make some remarks on a different type of models which present difficulties that are in a sense complementary to those of the preceding section. Namely we shall consider models for which $V\Omega_0 = 0$ but the kernels of V contain δ- distributions due to the translation invariance of V (no space cut-off). This gives problems of defining one-particle states, which are absent in space cut-off models. To have a first look into the kind of problems involved, let us take advantage of the translation invariance of both H_0 and V and introduce the spectral decomposition $\mathcal{F} = \int^{\oplus} \mathcal{F}_p \, dP$ of $\mathcal{F}$ with respect to the self-adjoint operator P, the "total momentum operator" (generator of space translations), defined on $\mathcal{F}$ by $d\Gamma(k)$, where k is thought of as a multiplication operator on $L_2(\hat{R}^s)$. Call $H_0|_p$, $H|_p$ the restrictions of H_0, H to $\mathcal{F}_p$. $H_0|_p$ has the eigenvalues zero and $\mu(P) \equiv \sqrt{m_0^2 + P^2}$, with eigenvectors $^P\Omega_0 \equiv \{1, 0, \ldots\}$ and $\{0, \psi^{(1)}(P), 0, \ldots\}$. Whenever in V there are terms that create or destroy exactly one particle (i.e. terms of the form $W(w_{1j})$, $W(w_{j1})$) we have that $\mu(P)$ is not eigenvalue of $H|_p$ and $\{0, \psi^{(1)}(P), 0, \ldots\}$ is not an eigenvector of $H|_p$. In this case it is easy to see that not only there do not exist the strong limits of $\exp(itH)\exp(-itH_0)$ but also the strong limits of creation-annihilation operators of the type of those of Section II do not exist.

Remark: In the case where all terms of the form $W(w_{1j})$, $W(w_{j1})$ are absent from V, then the spectrum of $H|_p$ in $\mathcal{F}_p$ has again the eigenvalues 0, $\mu(P)$ and a continuum in $[2\mu(P/2), \infty)$. In this case one has, despite translation invariance, again a situation similar to the one of the Remark of Section II.5.3: both $\underset{t\to\mp\infty}{\text{s-lim}} \exp(itH)\exp(-itH_0)$ and $\underset{t\to\mp\infty}{\text{s-lim}}\, a_t^{\#}(h)$ exist in this case. See [Ho-K 6].

We now return to the general case in which the lowest eigenvalue of $H|_p$ above zero (it it exists) is in general different from $\mu(P)$. There is then a so called "one-particle energy shift" (which corresponds in the relativistic invariant case to a so called mass shift). A well studied particular kind of models which show this phenomenon are the so called Lee models.[1] These models have no true "number of particles divergences" and, in this sense, are more similar to quantum mechanical many-body systems than to systems with infinitely many particles like the one we are discussing. The scattering problems for Lee models are a simpler version of the ones for the so called "persistent models" which we are going to describe briefly and have number of particle divergences and in fact all features of local relativistic interactions besides the fact that they lack the vacuum polarizing terms (those causing $V\Omega_0 \neq 0$). The model is described as follows. One assumes two different sorts of indefinitely many particles, a particles and b particles. The states of the system are elements of the Hilbert tensor product $\mathcal{F} \equiv \mathcal{F}_a \otimes \mathcal{F}_b$, where $\mathcal{F}_n$ for n = a, b, is the symmetric Fock space for particles n (if they were alone). It will be useful to write $\mathcal{F} = \bigoplus_{m,n=0}^{\infty} \mathcal{F}^{(m,n)} = \bigoplus_{n=0}^{\infty} \mathcal{F}^{(n)}$, with $\mathcal{F}^{(m,n)} \equiv \mathcal{F}_a^{(m)} \otimes \mathcal{F}_b^{(n)}$, $\mathcal{F}^{(n)} \equiv \bigoplus_{m=0}^{\infty} \mathcal{F}^{(m,n)}$. The free Hamiltonian of the system is $H = H^a \otimes 1 + 1 \otimes H^b$, where H_0^n are the free Hamiltonians in $\mathcal{F}_n$, n = a, b. Similarly the number operator is $N = N^a \otimes 1 + 1 \otimes N^b$, where N^n is the number operator in $\mathcal{F}_n$. The creation-annihilation operators for n particles, $n^{\#}(h)$ in $\mathcal{F}_n$, defined in Chapter I, have natural extensions to maps on $\mathcal{F}$: $a^{\#}(h) \otimes 1$ and $1 \otimes b^{\#}(h)$. For simplicity we shall write $n^{\#}(h)$, n = a,b instead of $a^{\#}(h) \otimes 1$ and $1 \otimes b^{\#}(h)$. Let $\mathring{D}^{(n)}$ be the set of all elements of $\mathcal{F}(n)$ of the form $\psi = \{\psi^{(m,n)} \mid \psi^{(m,n)} = 0$ except for finitely many $m; \psi^{(m,n)} \varepsilon \mathcal{S}$ in all variables$\}$ Consider an ultraviolet cut-off interaction of the form $\lambda V_\varepsilon = \lambda(V_\varepsilon^c + V_\varepsilon^a)$, where $\lambda \varepsilon R$ is the coupling constant and V_ε^c is the bilinear form

$$\psi \times \psi' \in D_0^{(n)} \times D_0^{(m)} \to V_\varepsilon^c\{\psi, \psi'\} = \int_{R^{3s}} (b(k_1)a(k_2)\,\psi,\; b(-k_3)\psi')$$

$$\frac{\hat{\chi}_\varepsilon(k_1)}{\sqrt{\mu(k_1)}} \quad \frac{\hat{\chi}_\varepsilon(k_2)}{\sqrt{\mu(k_2)}} \quad \frac{\hat{\chi}_\varepsilon(-k_3)}{\sqrt{\mu(k_3)}} \quad \delta\,(k_1 + k_2 + k_3)\, dk_1\, dk_2\, dk_3\ , \quad \text{and}$$

V_ε^a is the adjoint form, where $\hat{\chi}_\varepsilon(\cdot)$ is a smooth ultraviolet cut-off (e.g. in $C_0^\infty(\hat{R}^s)$), tending to 1 as $\varepsilon \to 0$. It is easily seen that for $\varepsilon > 0$, λV_ε is actually a symmetric operator, bounded from $\mathcal{F}^{(m,n)}$ into $\mathcal{F}^{(m-1,n)} \oplus \mathcal{F}^{(m+1,n)}$, thus in particular leaving $\mathcal{F}^{(n)}$ invariant. Moreover λV_ε satisfies in each $\mathcal{F}^{(n)}$, the Rellich-Kato relative boundedness with respect to H_0, so that $H_\varepsilon = H_0 + \lambda V_\varepsilon$ is self-adjoint on $D(H_0)$ and bounded from below on each $\mathcal{F}^{(n)}$. One has $V_\varepsilon^c \Omega_0 = 0$, hence the absence of vacuum polarization. Moreover $V_\varepsilon b^*(h)\Omega_0 \notin \mathcal{F}^{(0,1)}$ and one has the problem of a shift of the one b-particle energies by a finite amount, depending on ε (and going to infinity as $\varepsilon \to 0$). Also $\exp(itH_\varepsilon)\, b^*\,(\exp(-it\mu)h)\,\exp(-itH_\varepsilon)$ does not converge strongly. To cope with these problems one does a so-called "mass renormalization" i.e. one changes $H_0 + \lambda V_\varepsilon$ to $\hat{H}_\varepsilon = H_0 + \lambda V_\varepsilon + M_\varepsilon$, where the symmetric operator M_ε, chosen to map

$\mathcal{F}^{(n)}$ into $\mathcal{F}^{(n)}$, is such that $H_0 + \lambda V_\varepsilon + M_\varepsilon$ is self-adjoint on $D(H_0)$ and $H_0 + \lambda V_\varepsilon + M_\varepsilon \upharpoonright \mathcal{F}_p$ has both zero and $\mu(P)$ as eigenvalues. Moreover M_ε is chosen in such a way as to solve the ultraviolet problem ($H_0 + V_\varepsilon$ undefined as self-adjoint operator for $\varepsilon \to 0$) in the sense that $\hat{H}_\varepsilon = H_0 + \lambda V_\varepsilon + M_\varepsilon$ converges in the generalized (resolvent) sense on each $\mathcal{F}^{(n)}$ to a self-adjoint operator, bounded from below on each $\mathcal{F}^{(n)}$. Since by construction the spectra of H_0 and $\hat{H}_\varepsilon$ are identical in the whole subspace $\mathcal{F}^{(0,1)} \oplus \mathcal{F}^{(0)}$ there exists an intertwining operator T_ε such that $\hat{H}_\varepsilon T_\varepsilon = T_\varepsilon H_0$ on $\mathcal{F}^{(0,1)} \oplus \mathcal{F}^{(0)}$. T_ε and M_ε can be constructed, following essentially lines indicated by Friedrichs [Fr 2], as perturbation series in λ [Ec 2] with operator coefficients, which converge strongly on dense domains of $\mathcal{F}$. It turns out that T_ε is unbounded on $\mathcal{F}$ but such that $T_\varepsilon \; e^{-(1/2\ \ln 2)N}$ is a bounded operator. T_ε is an example of the so called dressing transformations, which play an important role also in the construction of certain approximations to local relativistic interactions. T_ε is an isometry from $\mathcal{F}^{(0,1)} \oplus \mathcal{F}^{(0)}$ to its range. It turns out that the strong limits of $\exp(it\hat{H}_\varepsilon)\, T_\varepsilon \exp(-itH_0)$ taken on a dense subset of $\mathcal{F}$, exist as $t \to \pm\infty$ and are partial isometric generalized wave operators $W_\pm$ on $\mathcal{F}$. Moreover there exists densely defined linear operators $\hat{b}^*(h)$, linear also with respect to $h \in L_2(\hat{R}^s)$, also given as convergent perturbation series in λ, such that: $\hat{b}^*(h)\Omega_0 = T_\varepsilon\, b^*(h)\Omega_0$ for all $|\lambda|$ sufficiently small. Furthermore

$$\prod_{i=1}^{n} \hat{b}_t^*(h_i)\Omega_0 = e^{it\hat{H}_\varepsilon}\, T_\varepsilon\, e^{-itH_0} \prod_{i=1}^{n} b^*(h_i)\Omega_0, \quad \text{where}$$

$\hat{b}_t^*(h) \equiv e^{it\hat{H}_\varepsilon}\; \hat{b}^*\,(e^{-it\mu}h)\, e^{-it\hat{H}_\varepsilon}$. One can prove that $\underset{t\to\pm\infty}{\text{s-lim}}\, \hat{b}_t^*(h)$ exist on a dense domain of $\mathcal{F}$ and define asymptotic creation-annihilation operators $\hat{b}_\pm^\#$. These, together with $\hat{a}_\pm^\#(h) = \underset{t\to\pm\infty}{\text{s-lim}}\, e^{it\hat{H}_\varepsilon}\, a^\#(e^{\mp it\mu}h)\, e^{-it\hat{H}_\varepsilon}$ have the correct (renormalized) canonical commutation relations on a dense domain independent of h,f e.g. $[\hat{C}_+^{n}(h),\ \hat{C}_+^{n'}{}^*(f)] = \delta_{nn'}(\bar{h}, Z_n^{-1}f)$, where $n, n' =$ a, b and where $(\cdot,\cdot)$ is the L_2-scalar product and $Z_n(k)$ is the constant one for $n =$ a and it is a certain (finite, $\neq 0$) "field strength" renormalization function for $n =$ b, given as a power series in λ (with $Z_b(k)$ equal 1 for $\lambda = 0$). $\hat{C}_+^{n\,\#}$ stands for $\hat{a}_+^\#$ if $n =$ a and for $\hat{b}_+^\#$ if $n =$ b. Moreover e.g. $e^{-itH}\ \hat{C}_+^*(h)\ e^{+itH} = \hat{C}_+^*\ (e^{-it\mu}h)$, on the same domain. ([Al 1,2]).

Having these results the discussion of the construction of the asymptotic Fock space and of the S-matrix can then be pursued along similar lines as in Section II. No results on unitarity are, however, known. If a space cut-off is introduced in V_ε, then Høegh-Krohn's method, as mentioned in Remark of Section II, 5.3 gives unitarity of the S-operator at fixed free energy for $|\lambda|$ small enough. Alternative ways of constructing scattering quantities in this model are also known. Whereas the above were in a sense based on techniques essentially adapted from non-relativistic quantum mechanics ("extension of Cook's method") the alternative ways are adaptations of techniques developed in the context of 'axiomatic quantum field theory' (Haag-Ruelle theory, LSZ theory). This latter approach will be mentioned in the next chapter. It is a happy fact thus that these translation invariant models are in the intersection of the mentioned two "zones of influences". Other scattering problems and large classes of persistent models have been discussed particularly by Fröhlich.[2] In a class of persistent models, stronger results than the one mentioned above on the "solution of the one-body problem" (determination of M_ε and T_ε $b^*(h)\Omega_0$), independent of restrictions on the coupling constant (also for $\varepsilon \to 0$) are available. Models more singular than the one mentioned above as $\varepsilon \to 0$ have also been studied.[3] Finally we would like to mention that results have been obtained also for the mass zero case (for the a particles) (infrared problem).[4] This has particular relevance since persistent models seem to be, from a physical point of view, particularly good approximations for the study of quantum electrodynamics.[5]

IV. Local Relativistic Models.

In this chapter we shall discuss local relativistic models and their scattering quantities. From the discussions of Chapter II we know that their interactions cannot be defined as operators on Fock space. One can hope at most, and we shall see that this hope will be fulfilled, to be able to use some information gained through the study of the space cut-off (or, for that matter, any other suitable) approximation to relativistic interactions, approximation which is defined on Fock space, as a tool to work oneself eventually out of Fock space. In fact the same situation is met with any translation invariant interaction which gives rise to vacuum polarization. The infinite volume problem has been solved for such interactions only recently and in few cases. We shall essentially concentrate on the case of local relativistic perturbations, for which the desiderata giving guidelines on "where to go" when leaving Fock space have been written up in careful mathematical terms long ago, after a thorough physical discussion of the principles.

IV.1 What one would like to have eventually (Wightman axioms + scattering theory).

The mentioned basic desiderata for a satisfactory local relativistic theory of quantized fields are traditionally formulated in terms of axioms, which we shall now describe for the case for which we shall discuss in the next section a constructive verification in a class of models. The axioms we shall give are the so called "Wightman axioms for a single, real, scalar field $A(\underline{x})$" (in s space dimensions). We shall write the axioms somewhat in "slogan style", but we hope the meaning is clear. For more precise statements and discussion of the axioms, we refer to any of a series of excellent accounts. (See bibl. 4 to Intr.). Our present exposition follows [St 2]

IV.1.1 Wightman axioms. Ax1: "Quantum mechanics". One should have a separable Hilbert space $\mathcal{H}$, the unit rays of which are the states of the physical system and certain linear operators in $\mathcal{H}$ are the observables of the system. Ax2: "Special relativity". In $\mathcal{H}$ acts the (orthochronous)inhomogeneous Lorentz group by a continuous unitary representation $U(\underline{a},\Lambda)$, where $\underline{a} \in R^{s+1}$, Λ is a homogeneous Lorentz transformation. Ax3: "Spectrum". The infinitesimal generators of time translations H and of space translations P are such that H and $H^2 - P^2 \equiv M^2$ are non-negative operators on $\mathcal{H}$ and zero is a simple eigenvalue of H, P, M^2, $U(\underline{a}, \Lambda)$, with eigenvector Ω (the physical vacuum, unique up to a phase). Ax4: "Field theory (the basic quantities are expressed by a field operator $A(\underline{x})$)".
a) Field operator: For any $f \in \mathcal{S}(R^{s+1})$ there exists a linear densely defined operator A(f), linear in f, defined on a dense linear domain in $\mathcal{H}$, independent of f and containing Ω and such that $A(f)D \subset D$ and $U(\underline{a},\Lambda) D \subset D$. A(f) is symmetric for f real-valued and $(\chi, A(f)\psi)$ is a tempered distribution as functional of f, for all ψ, $\chi \in D$ Moreover the linear hull D_0 of Ω, $A(f_1)\ldots A(f_n)\Omega$ is dense in $\mathcal{H}$.
b) Covariance: $U(\underline{a},\Lambda)\, A(f)\, U(\underline{a},\Lambda)^{-1} = A(f_{(\underline{a},\Lambda)})$, where $f_{(\underline{a},\Lambda)}(\underline{x}) \equiv f(\Lambda^{-1}(\underline{x} - \underline{a}))$ for all $f \in \mathcal{S}(R^{s+1})$.

c) Locality: $[A(f), A(g)] = 0$ on D if $(x^0 - y^0)^2 - (x-y)^2 < 0$ for all $\underline{x} \equiv (x^0, x)$ in the support of f and all y in the support of g; $\bar{f}$, $g \in C_0^\infty(R^{s+1})$. This concludes the list of the Wightman axioms. These are trivially verified by the choice $\mathcal{H}$ = Fock space, $A(f)=\phi(f)$, where $\phi(\underline{x})$, $\underline{x} \equiv (t,x)$ is the free field of Section I. It is a non-trivial task to find models which are not of this trivial type (or of closely related ones, leading to trivial scattering), and satisfy all Wightman axioms. This has been solved up to now only for s = 1. (see next section).

The practical verification of the Wightman axioms, written above in a form stressing the "extrapolation" from the known trivial model of free fields and Fock space, is by using an equivalent set of axioms, written in terms of a set of distributions ("Wightman functions") instead of operators. The new set of axioms and its equivalence with the previous ones is the content of the following: Wightman Reconstruction Theorem: Let $W^{(n)}$, n = 1, 2, ... be a sequence of tempered distributions, where $W^{(n)}$ depends on n variables $\underline{x}_1 \equiv (x_1{}^0, x_1), \ldots, \underline{x}_n \equiv (x_n{}^0, x_n)$, $x_i{}^0 \in R$, $x_i \in R^s$, i = 1,..., n. Suppose the $W^{(n)}$ have the following properties:
1) Positive definiteness:

$$\sum_{j,k} \int\ldots\int \bar{f}_j(\underline{x}_1,\ldots,\underline{x}_j)\, f_k(\underline{y}_1,\ldots,\underline{y}_k)\, W^{(j+k)}(\underline{x}_j,\ldots,\underline{x}_1,\underline{y}_1,\ldots,\underline{y}_k)$$

for all finite sequences f_0, $f_1(\underline{x}_1)$, $f_2(\underline{x}_1, \underline{x}_2)$,... of test functions $f_n \in \mathcal{S}(R^{(s+1)n})$.

2) Relativistic transformation:
$W^{(n)}(\underline{x}_1,\ldots,\underline{x}_n) = W^{(n)}(\Lambda\,\underline{x}_1 + \underline{a},\ldots,\Lambda\,\underline{x}_n + \underline{a})$, $\underline{a} \in R^{s+1}$, Λ = homogeneous orthochronous Lorentz transformation.

3) Spectrum condition:
$\hat{W}^{(n)}(\underline{P}_1,\ldots,\underline{P}_n) = \delta(\sum_{j=1}^{n} \underline{P}_j)\, \hat{W}'^{(n)}(\underline{P}_1, \underline{P}_1 + \underline{P}_2,\ldots, \underline{P}_1 +\ldots+ \underline{P}_n)$
and $\hat{W}'^{(n)}(\underline{q}_1,\ldots,\underline{q}_{n-1}) = 0$ if any $\underline{q}_i^0 \equiv (q_i^0,q_i)$ does not satisfy $(q_i{}^0)^2 - (q_i)^2 > 0$. ($\hat{W} \equiv$ Fourier transform of W).

4) Cluster decomposition: $\lim_{\lambda\to\infty} [W^{(n)}(\underline{x}_1,\ldots,\underline{x}_j, \underline{x}_{j+1}+\lambda\underline{a},\ldots,\underline{x}_n + \lambda\underline{a}) - W^{(j)}(\underline{x}_1,\ldots, \underline{x}_j)\, W^{(n-j)}(\underline{x}_{j+1},\ldots, \underline{x}_n)] = 0$ whenever $(a^0)^2 - a^2 < 0$.

5) Hermiticity: $W^{(n)}(\underline{x}_1,\ldots, \underline{x}_n) = \bar{W}^{(n)}(\underline{x}_n,\ldots, \underline{x}_1)$, where - means complex conjugate.

6) Locality: $W^{(n)}(\underline{x}_1,\ldots,\underline{x}_j, \underline{x}_{j+1},\ldots,\underline{x}_n) = W^{(n)}(\underline{x}_1,\ldots,\underline{x}_{j+1},\underline{x}_j,\ldots\underline{x}_n)$, whenever $(x_j^0 - x_{j+1}^0)^2 - (x_j - x_{j+1})^2 < 0$, $j = 1, 2,\ldots, n-1$.

Then there exists (uniquely up to unitary equivalence) a separable Hilbert space $\mathcal{H}$, a continuous unitary representation $U(\underline{a},\Lambda)$ of the inhomogeneous orthochronous Lorentz group on $\mathcal{H}$, a unique state Ω invariant under $U(\underline{a},\Lambda)$, and a symmetric scalar field $A(f)$ satisfying Axioms 1) to 4) above such that $(\Omega,A(f_1)\ldots A(f_n)\Omega) = W^{(n)}(f_1,\ldots,f_n)$, for any $f_i \in \mathcal{S}(R^{s+1})$. For a proof see e.g. [Str-W].

IV.1.2 Haag-Ruelle scattering theory. In order to associate to a theory, which satisfies all Wightman axioms, a scattering theory, additional postulates are needed. The following are sufficient, by theorems of Haag-Ruelle, to yield a scattering theory.

I. Zero is an isolated eigenvalue of H and of the mass operator $M^2 = H^2 - P^2$ i.e. the distance between zero and the rest of the spectrum of H and M^2 is strictly positive. This postulate is called the "mass gap" postulate. It excludes the presence of any zero mass "asymptotic particles."

II. The infimum of the spectrum of M^2, zero excluded, is an isolated eigenvalue m^2 and the correspondent eigenvectors span a subspace $\mathcal{H}_1$ of $\mathcal{H}$ (1 particle subspace) and the restriction of $U(\underline{a},\Lambda)$ to $\mathcal{H}_1$ is an irreducible representation of Wigner type [m, 0] of the inhomogeneous orthochronous Lorentz group.

Note that II is stronger than I. I and II are assumptions on the spectrum of M^2 and are "natural" for obtaining a particle structure and a scattering theory. There is a third assumption of a more technical nature, which simplifies matters but is not indispensable.

III. Let $A(f)$ be the field which appears in the Wightman axioms. Write $A(\underline{x})$ for the correspondent operator-valued tempered distribution. Then the assumption is that $(\Omega, A(\underline{x})\psi) \neq 0$ for every ψ belonging to the one particle space $\mathcal{H}_1$ of assumption II.

Under the Wightman axioms and assumptions I to III, n-particle scattering states can be constructed using a theorem by Haag-Ruelle. In order to state this theorem we need some notation. Choose a function $\tilde{f}(\underline{p})$ of the form $\tilde{f}(\underline{p}) = \hat{g}(p)\ \chi\ ((p^0)^2 - p^2)\,\theta(p^0)$, where $\hat{g}$ is any function in $C_0^\infty\ (\hat{R}^s)$ and $\chi(\alpha)$ is a $C_0^\infty(R)$ function, equal to a suitable constant c in a neighborhood of $\alpha = m^2$ contained in the interval $(0, 4m^2)$, $0 \leq \chi \leq c$ everywhere and with support in $(0, 4m^2)$ (m is the one particle mass of assumption II). c is chosen so that $\|\tilde{A}(\tilde{f})\Omega\| = 1$, where $\tilde{A}$ is the Fourier transform field, defined by $\tilde{A}(\tilde{f}) = A(f)$. Note that in this case $\tilde{A}(\tilde{f})\Omega$ belongs to $\mathcal{H}_1$ i.e. is a one-particle state. Form now (note the analogy with the method of Chapters II, III) $\tilde{A}_t(\tilde{f}) \equiv \exp(itH)\ \tilde{A}(\exp(-it\mu)\tilde{f})\ \exp(-itH)$, where μ is the function of $p \in \hat{R}^s$ given by $\mu(p) \equiv \sqrt{m^2 + p^2}$ and m is the one particle mass of assumption II.

Remark: $\tilde{A}_t(\tilde{f})$ exists as a densely defined operator on the domain D of Ax. 3: in fact $\tilde{A}_t(\tilde{f}) = \tilde{A}(\tilde{f}_t')$, where

$$\tilde{f}_t'\ (p^0, p) \equiv e^{it(p^0 - \mu(p))}\ \tilde{f}(p^0, p).$$

Haag-Ruelle theorem: Let $\tilde{f}_i$ be test functions with the same properties as the $\tilde{f}$ described above.* Then the vectors $\Phi_t(\tilde{f}_1,..,\tilde{f}_n) \equiv \prod_{i=1}^{n} \tilde{A}_t(\tilde{f}_i)\Omega = e^{itH} \prod_{i=1}^{n} \tilde{A}\ (e^{-it\mu}\tilde{f}_i)\Omega$ converge strongly in $\mathcal{H}$ as $t\to\pm\infty$. Call $\mathcal{H}_\pm$ the closed linear span of Ω and of all the vectors of the form $\Phi_\pm(\tilde{f}_1,...,\tilde{f}_n) = \text{s-}\lim_{t\to\pm\infty} \Phi_t(\tilde{f}_1,...,\tilde{f}_n)$. Asymptotic creation operators are then defined by

$a_\pm^*(\tilde{f})\ \Phi_\pm\ (\tilde{f}_1,...,\tilde{f}_n) = \lim_{t\to\pm\infty} \tilde{A}_t(\tilde{f})\ \Phi_t\ (\tilde{f}_1,...,\tilde{f}_m)$, they are closable and their adjoints are called $a_\pm(\tilde{f})$. $a_\pm^{\#}(\tilde{f})$ satisfy canonical commutation relations among themselves and have the usual free commutation relations with e^{itH}:

$$e^{-itH}\ a_\pm^{\#}(\tilde{f})\ e^{+itH} = a_\pm^{\#}(e^{\mp it\mu}\tilde{f}).$$

Moreover $U(\underline{a},\Lambda)\ a_\pm(\tilde{f})\ U(\underline{a},\Lambda)^{-1} = a_\pm(U(\underline{a},\Lambda)\tilde{f})$, where $(U(\underline{a},\Lambda)\tilde{f})(\underline{p}) = \tilde{f}(\Lambda^{-1}\underline{p})\ e^{i(a^0 p^0 \pm \underline{a}p)}$. For a proof see e.g. [He 2], [Jo].

Remark: Note that above theorem only gives the strong convergence of states, not of the operators $\tilde{A}_t(\tilde{f})$ themselves. For the models of Chapters II, III strong convergence of the operators themselves was obtained. In the general Wightman Haag-Ruelle framework only partial results of this kind are proved. However, above convergence of states to asymptotic states is enough to define an S-operator as the linear map from $\mathcal{H}_-$ to $\mathcal{H}_+$ defined by $S\Phi_-(\tilde{f}_1,...,\tilde{f}_n) = \Phi_+(\tilde{f}_1,...,\tilde{f}_n)$. Moreover many interesting general properties of scattering amplitudes (e.g. cluster, one particle properties, dispersion relations, off

* and nonoverlapping in case s = 1,2 [He 2].

mass-shell crossing symmetry etc.) have been derived with no essential new assumption. However, questions like the unitarity of S and detailed spectral properties of H can hardly be answered in this general framework. They certainly acquire a more concrete and perhaps approachable aspect when asked in the non-trivial models which have been constructed and give substance to the above theory. In the next section we shall give some results on such models and indicate briefly in an example how one goes about proving the Wightman axioms and part of the assumptions for scattering theory.

IV.2 Local Relativistic Quantum Field Theoretical Models Satisfying the Wightman Axioms with Scattering Theory. Non-trivial models satisfying all Wightman axioms exist up to date only for the number of space dimensions s equal to one. There are 2 types of such models which have been constructed: 1) "Weak coupling": $P(\phi)_2$ models. These are models constructed from interactions of the form

$$V = \lambda \sum_{j=0}^{2n} c_j \int_R :\phi(x)^j: dx, \qquad ([\text{G1-J-S}])$$

with $c_{2n} > 0$ and for which there exists a strictly positive number Λ_0 such that for $|\lambda|/m_0^2 < \Lambda_0$ all the Wightman axioms are verified.* 2) "Models with exponential interactions". These are models constructed from interactions of the form $V = \iint :e^{\alpha\phi(x)}: dx\, d\nu(\alpha)$ where $\alpha \in R$, $d\nu(\alpha)$ is any positive finite measure with support in the interval

$$\left(-\frac{4}{\sqrt{\pi}}, +\frac{4}{\sqrt{\pi}}\right)$$

of the real line and such that $d\nu(-\alpha) = d\nu(\alpha)$.([Ho.K.15][Al-H.K.5]).

Remark: In Chapter II we have already discussed to some extent the space cut-off approximation for models of type 1). The space cut-off approximation to models of type 2) was also mentioned in a couple of remarks at the end of Section II.4. The complete verification of the Wightman axioms requires a solution of the problem of how to remove the space cut-off. In particular a good solution of this problem is crucial for proving results on the spectrum of H and M^2. Part of the procedure for the removal of the space cut-off is common to both models of type 1) and 2) and involve the introduction of the so called Schwinger functions for the space cut-off models in question and the exploitation of a representation of these as expectations with respect to a suitable measure incorporating the interaction. We shall do this below in some details only for the case of the models of type 2) but the construction in the other case goes essentially in the same way. However, the actual removal of the space cut-off from the Schwinger functions is different in models of type 1) and type 2). In models of type 1) the procedure is of a highly refined perturbative nature, involving inductively defined expansions in asymptotic power series in λ/m_0^2 for small values of $|\lambda|/m_0^2$ and a suitable step by step exploitation of the good properties of a certain approximation to the interaction free theory and of the locality of the interaction. In models of

* They are also satisfied for all λ for $P(\phi)=a\phi^4+b\phi^2+\mu\phi$. $a>0$, $\mu\neq 0$ [Si-G].

type 2) the solution of the infinite volume limit is much simpler, no expansion procedure is used and instead an extensive use of certain monotonicity properties of the interaction is made, as we shall see below. As for verification of the further assumptions for Haag-Ruelle scattering theory, i.e. the spectral assumptions I, II and the technical assumption III described above, the situation is presently the following: Assumption I is verified both for models of type 1) and 2). Assumption II is verified for models of type 1) but not yet completely for models of type 2) (but is very likely to hold). Assumption III is verified both for models of type 1) and 2), the latter being understood in the sense that, given the existence of a one particle subspace $\mathcal{H}_1$ (which would be the case if II were completely verified), then one can prove, using the particularities of the interaction, that III holds. Again the methods of proof differ in the two types of models in much the same way as for the construction of the infinite volume limit.

Remark: Further results are available for both models. E.g. for models of type 1) it is proved that the one particle mass m converges to the bare mass m_0 as the coupling constant λ goes to zero. Moreover the spectrum is void in an interval (m,m^*), with $m^* > m$ and $m^* \to 2m_0$ as $\lambda \to 0$. An "n particle cluster expansion" has also been proved, which in a sense is a beginning of a complete particle description. For models of type 2 strong information on the mass gap is available. In particular one knows that it has size larger or equal to the bare mass ("repulsiveness") and has monotonic dependence on the parameters m_0 and λ, if we write $\lambda d\nu(\alpha)$ instead of $d\nu(\alpha)$ in the definition of the interaction (no restriction on $|\lambda|$).

We shall now indicate a sketch of proof of the quoted results for models of type 2), to give the flavor of some of the methods which have found application in the solution of the problem of constructing models for nontrivial local relativistic quantum fields, at least in two space-time dimensions. The proof uses extensively the representation of the free fields $\phi(g)$ and of their Wick powers $:\phi^n:(g)$, defined in Section I.2 respectively I.5, as operators multiplication by measurable functions. In Section I.5 we have seen that in the unitary map of the Fock space to an $L_2(Q, \Sigma, \mu_0)$ space, $:\phi^n:(g)$ goes over to the function $:\xi^n:(g)$ in $L_p(Q)$ for all $1 \leq p < \infty$. In order to define interactions given formally as $\iint :e^{\alpha\phi(x)}: g(x)\,dx\,d\nu(\alpha)$, with say $g \in C_0^\infty(R)$, $g \geq 0$, we will show that $\int :e^{\alpha\xi}:(g)\,d\nu(\alpha)$ is a function in $L_2(Q)$, so that the Fock space operator $\iint :e^{\alpha\phi(x)}: g(x)dx\,d\nu(\alpha)$ is then defined as a symmetric operator by the preimage under U of $\int :e^{\alpha\xi}:(g)\,d\nu(\alpha)$. Consider the sum

$$\sum_{n=0}^{M} \int \frac{\alpha^n}{n!} :\xi^n:(g)\,d\nu(\alpha).$$

For M finite it is the image under U of the corresponding sum with ϕ replacing ξ. The limit $M \to \infty$ exists in $L_2(Q)$, as we will show by an explicit computation, using the restriction on the support of $d\nu$.[3] In fact the restrictions enter in the following way. We have to show that $\sum_{n=0}^{M} \frac{\alpha^n}{n!} : \xi^n : (g)$ converges in $L_2(Q)$ as $M \to \infty$. But $:\xi^n: (g) \in L_2^{(n)}(Q)$, (where $L_2^{(n)}(Q)$ was defined in Section I.4) hence it is enough to prove that

$$\sum_{n=0}^{\infty} \frac{\alpha^{2n}}{n!} \iint g(x)\, G(x-y)\, g(y)\, dx\, dy < \infty,$$

where $G(z)$ is the kernel of $(-\Delta + m_0^2)^{-1}$ and since it is a positive function this is equivalent with

$$\iint g(x)\, e^{\alpha^2 G(x-y)}\, g(y)\, dx\, dy < \infty . \qquad (1)$$

But $G(x)$ (which is a modified Bessel function) is smooth for $x \neq 0$, decreasing like $e^{-m_0|x|}$ as $|x| \to \infty$ and at the origin it as a logarithmic singularity of the form $-1/2\pi \,\mathrm{Ln}\, |x|$. This introduced in above condition (1) shows that it is sufficient to have the support of $d\nu(\alpha)$ in $(-\sqrt{2\pi}, \sqrt{2\pi})$, which is satisfied in particular when the assumption stated under 2) above is made. Call now $\int : e^{\alpha\xi}: (g)\, d\nu(\alpha) \equiv V_g^W$ the limit function[4], which is in $L_2(Q)$. We call correspondingly $\int : e^{\alpha\phi}: (g)\, d\nu(\alpha) \equiv V_g$ the image under U* of V_g^W: this defines the space cut-off interaction V_g as the strong limit of $\sum_{n=0}^{M} \int \frac{\alpha^n}{n!} : \phi^n : (g)\, d\nu(\alpha)$ as $M \to \infty$.

Having so defined the space cut-off interaction V_g, we would like now to associate to $H_0^W + V_g^W$ a self-adjoint operator on $L_2(Q)$ (which then would give a self-adjoint operator, the Hamiltonian for the space cut-off interaction, on Fock space $\mathcal{F}$). A first step in the proof of essential self-adjointness is to see that V_g^W is a non-negative function. The idea is to make rigorous the formal computation which tells

$$:e^{\alpha\xi}: (g) = e^{\alpha\xi}(g)\, e^{-1/2\ \alpha^2 \langle g,g\rangle} \geq 0,$$

where $\langle\cdot,\cdot\rangle$ is the scalar product in the basic Sobolev space K, over which the stochastic process ξ is constructed. If we replace everywhere ξ by ξ_ε, where $\xi_\varepsilon(x)$ is the ultraviolet cut-off field $\xi_\varepsilon(x) = \xi(\chi_\varepsilon(x-\bullet))$, $\varepsilon > 0$, $\chi_\varepsilon(y) \equiv \varepsilon^{-1}\chi(y/\varepsilon)$, with $\chi(\cdot)\varepsilon C_0^\infty(R), \chi(-\cdot)=\chi(\cdot)$ $\int \chi(x)\, dx = 1$, $\chi \geq 0$, then the above equation and inequality hold rigorously and therefore $V^W_{g,\varepsilon} \geq 0$, where $V^W_{g,\varepsilon}$ is defined as V^W_g with ξ_ε replacing ξ. But $V^W_{g,\varepsilon} \to V^W_g$ as $\varepsilon \to 0$, as seen by an explicit computation. This then proves $V^W_g > 0$. Hence the Friedrichs self-adjoint extension H^F of $H_0^W + V_g^W$ exists.

The strong approximation of e^{-tH^F} with $H_{g,k} \equiv H_0^W + V^W_{g,k}$, $V^W_{g,k} \equiv V^W_g$ for $V^W_g \leq k$ and equal zero for $V^W_g > k$, proves also that

e^{-tH^F} is a contraction semigroup on $L_p(Q)$, for all $1 \leq p < \infty$. From this one can show rather directly that $(z-H^F)^{-1}\,\overline{L}_\infty(Q)$, which is contained in $D(H_0) \cap D(V)$, is a domain of essential self-adjointness for $H_0^W + V_g^W$. Hence the closure of $H_0^W + V_g^W$ is equal to H^F and we call H_g^W this non-negative operator, which is the Hamiltonian of the space cut-off interaction. It has also been shown by Høegh-Krohn that the infimum of the spectrum of H_g^W is a non-negative simple eigenvalue E_g, isolated from the rest of the spectrum.[5,6] Call Ω_g the corresponding eigenvector. In the limit $g \to 1$ we know from the general remarks of section II, that H_g^W will not be an operator any longer on vectors of $\mathcal{F}$. So in order to construct the limit for $g \to 1$ of the whole structure $(\mathcal{F}, H_g)$, and finding possibly a structure $(\mathcal{H}, H)$, as the one described by the Wightman axioms, we look at the Wightman Reconstruction Theorem of IV.1.1. This tells us that a Wightman theory is completely characterized by the set of functions $W^{(n)}(\cdot)$, or, if we prefer, by a certain positive linear functional on a space of test functions, described by the set of all $W^{(n)}(\cdot)$.[7] It is quite natural to introduce correspondingly a set of functions $W_g^{(n)}(\cdot)$, defined in analogy with the desired Wightman functions ${}^gW^{(n)}(\cdot)$, but in terms of the quantities $(\mathcal{F}, H_g)$ for the space cut-off model. Then if the $W_g^{(n)}(\cdot)$ converge as $g \to 1$, they might hopefully give limiting functions $W^{(n)}(\cdot)$ satisfing the desired properties enumerated in the Wightman reconstruction theorem. Now $W^{(n)}(t_1, x_1; \ldots; t_n, x_n)$, with $(t_i, x_i) \equiv \underline{x}_i$, were defined as tempered distributions $(\Omega, A(t_1, x_1)\ldots A(t_n, x_n)\Omega)$ (to be tested with test functions $f_i \in \mathcal{S}(R^2)$, $i = 1,\ldots, n$, to yield $(\Omega, A(f_1)\ldots A(f_n)\Omega)$. By analogy we define formally

$$W_g^{(n)}(t_1,x_1;\ldots;t_n,x_n) \equiv (\Omega_g, \phi_g(t_1, x_1)\ldots \phi_g(t_n, x_n)\Omega_g), \text{ where}$$

$\phi_g(t,x) \equiv e^{itH_g}\, \phi(x)\, e^{-itH_g}$ is the time t field, obtained by letting evolve the free time zero field $\phi(x)$ under the action of the (normalized) Hamiltonian $\bar{H}_g \equiv H_g - E_g$ of the space cut-off interaction. The definition, as it stands, is formal, but we do not need to make it precise here since in fact we shall not study the $W_g^{(n)}$ functions themselves but rather their analytic continuations to imaginary times, the so-called Schwinger functions for the space cut-off interactions, which are, for reasons which will become apparent below, much more suited for the study of infinite volume limits. These Schwinger functions are defined, for $t_1 \leq t_2 \leq \ldots \leq t_n$, as

$$S_g^{(n)}(t_1x_1,\ldots,t_nx_n) \equiv (\Omega_g, \phi(x_1)e^{(t_1-t_2)\bar{H}_g}\phi(x_2)e^{(t_2-t_3)\bar{H}_g}\ldots\phi(x_n)\Omega_g)$$

(i.e. formally as analytic continuations of $W_g^{(n)}(\tau_1x_1,\ldots,\tau_nx_n)$ to purely imaginary values of τ_i, $i = 1,\ldots,n$, with $\mathrm{Im}\,\tau_1 \leq \ldots \leq \mathrm{Im}\,\tau_n$). The right hand side should be understood as distribution to be tested e.g. with test functions $f_1,\ldots,f_n$ in $\mathcal{S}(R)$, yielding $S_g^{(n)}(t_1f_1,\ldots,t_nf_n) \equiv (\Omega_g, \phi(f_1)\exp((t_1-t_2)\bar{H}_g)\phi(f_2)\exp((t_2-t_3)\bar{H}_g\ldots \phi(f_{n-1})\exp((t_{n-1}-t_n)\bar{H}_g)\,\phi(f_n)\Omega_g)$. It will be shown below that these are indeed finite quantities. The study of the infinite volume limit $g \to 1$ on the $S_g^{(n)}(\cdot)$, as compared with the $W_g^{(n)}(\cdot)$, is, roughly speaking, facilitated by having the contraction semigroup $\exp(-t\bar{H}_g)$ instead of the unitary group $\exp(itH_g)$, by the fact that the $S_g^{(n)}(\cdot)$ have a representation as moments of a measure, to be shown below, and their limit for $g \to 1$, $S^{(n)}$, should be euclidean invariant (and euclidean group is simpler than Lorentz group!) Having the limit $S^{(n)}(\cdot)$ of the $S_g^{(n)}(\cdot)$ as $g \to 1$ we will get, by continuing to imaginary times, functions $W^{(n)}(\cdot)$, the limits of the $W_g^{(n)}(\cdot)$, which satisfy all the assumptions of the Wightman reconstruction theorem and by this yield a theory satisfying all Wightman axioms. To get started with the study of the infinite volume limit, it is useful to go back a step and consider the simpler quantities defined similarly to $S_g^{(n)}(\cdot)$ but with Ω_g replaced by Ω_0:

$$(\Omega_0, e^{(-t-t_1)\bar{H}_g}\,\phi(f_1)e^{(t_1-t_2)\bar{H}_g}\,\phi(f_2)\ldots e^{(t_{n-1}-t_n)\bar{H}_g}\,(f_n)e^{(t_n-t)\bar{H}_g}\Omega_0)$$

$$\text{for } -t \leq t_1 \leq t_2 \leq \ldots \leq t_n \leq t. \tag{2}$$

Let now t go to infinity. We have that $e^{-t\bar{H}_g}\,\Omega_0$ goes strongly to $(\Omega_0, \Omega_g)\Omega_g$, since zero is an isolated simple eigenvalue of $\bar{H}_g$. Thus we expect the above expression, divided by the normalization $(\Omega_0, \exp(-2t\bar{H}_g)\Omega_0)$ to converge as $t \to \infty$ to the Schwinger functions $S_g^{(n)}(t_1f_1,\ldots,t_nf_n)$. That this is indeed so and that the $S_g^{(n)}$ are finite follows from the $L_2(Q)$ representation of the Fock space, whereby Ω_0 is replaced by 1, $\phi(f_i)$ by $\xi(f_i)$ and the semigroup $\exp(-t\bar{H}_g)$ by the operator $\exp(-t(H_g^W - E_g))$. Using essentially the Lie-Trotter-Kato product formula together with the fact, mentioned before, that $\exp(-tH_0^W)$ is a contraction from $L_p(Q)$ to $L_q(Q)$ for any given $p, q > 1$ and all $t > T(p,q)$ and, on the other hand, $V_g^W \geq 0$, we have that $\exp(-t(H_g^W-E_g))$ maps strong convergence in $L_2(Q)$ into strong convergence in $L_p(Q)$, is a contraction on $L_p(Q)$ and since the $\xi(f_i) \in L_p(Q)$ for all $1 \leq p < \infty$ we have finally that

$$(\Omega_0,\exp(-2tH_g)\Omega_0)^{-1}(\Omega_0,\exp(-(t-t_1)\bar{H}_g)\phi(f_1)\ldots\phi(f_n)\exp(t_n-t)\bar{H}_g)\Omega_0)\to$$

$$S_g^{(n)}(t_1f_1,\ldots,t_nf_n) \text{ as } t \to \infty\,. \tag{3}$$

So let us study the expression (2). For passing then to the study of the limit $g \to 1$ it is extremely useful to rewrite (3) in a suitable way by mapping $L_2(Q)$ isometrically onto a "fixed time" subspace of a larger space $L_2(\check{Q})$ in such a way that $L_2(\check{Q})$ plays a role similar to a path space (but with euclidean invariance properties! see below). In this procedure the basic space K, over which the stochastic process ξ was constructed, is mapped into a time τ subspace of a larger space $\check{K}$ in the following sense. Let j_τ be the map defined by $(j_\tau f)(t,x) = \delta(t-\tau)f(x)$, for every $f \in K$. It is easily seen that j_τ is a partial isometry from K to a subspace of $\check{K}$, where $\check{K}$ is the Sobolev space $H_{-1}(R^2)$ defined as the closure of the real $\mathcal{S}(R^2)$ functions with respect to the scalar product defined by $\langle F,G\rangle_{\check{K}} = (F,(-\Delta+m_0^2)^{-1}G)$, $(\cdot,\cdot)$ being the $L_2(R^2)$ scalar product and Δ the Laplacian on R^2.[8] Let now $\check{\xi}(\cdot)$ be the real generalized unit Gaussian process indexed by $\check{K}$ (with mean zero and covariance given by the scalar product $\langle\cdot,\cdot\rangle_{\check{K}}$). Consider the corresponding measure space $(\check{Q}, \check{\Sigma}, \check{\mu}_0)$ and $L_2(\check{Q}, \check{\Sigma}, \check{\mu}_0)$ space, defined correspondently as the spaces (Q, Σ, μ_0) and $L_2(Q, \Sigma, \mu_0)$ of Section I.4. We now carry over the map j_τ to a map J_τ from $L_2(Q)$ into $L_2(\check{Q})$, defined by its action on the multiplication operator $\xi(h)$, $h \in K$ by

$$J_\tau \, \xi(h) \, J_\tau^* = \check{\xi}(\delta_\tau \otimes h),$$

where δ_τ is the δ-function $\delta_\tau(t) = \delta(t-\tau)$. J_τ is a partial isometry with initial domain $L_2(Q)$ and final domain equal to the projection of $L_2(\check{Q})$ onto the subspace generated by the functions which are measurable with respect to the σ-algebra $\check{\Sigma} \cap \pi_\tau$, where π_τ is the hyperplane $\{(t,x) \in R^2,\ t = \tau\} \subset R^2$. The action of $\exp(-tH_0^W)$, $t > 0$, on $L_2(Q)$ is given by $J_0 \exp(-tH_0^W) J_0^* = (J_0 J_0^*) U_t (J_0 J_0^*)$, where U_t is the unitary map of $L_2(\check{Q})$ into itself defined by

$$(U_t f)(\tau,x) = f(\tau-t,x).$$

Using then essentially the Lie-Trotter-Kato product formula for $\exp(-tH_g^W)$ in terms of $\exp(-tH_0^W)$ and $\exp(-tV_g^W)$, one can prove the following Feynman-Kac-Nelson formula:[10]

$$J_0 \exp(-tH_g^W) J_0^* = (J_0 J_0^*) \exp\left(-\int_0^t V_g(\tau)\, d\tau\right) U_t \, (J_0 J_0^*), \tag{4}$$

where $V_g(\tau) \equiv \int :e^{\alpha\check{\xi}}: (\delta_\tau \otimes g) d\nu(\alpha)$ is defined as a function in $L_2(\check{Q})$, equal to the $L_2(\check{Q})$ strong limit of

$$\sum_{n=0}^{M} \frac{\alpha^n}{n!} \int :\check{\xi}^n: (\delta_\tau \otimes g)\, d\nu(\alpha)$$

as $M \to \infty$, in the same way as V_g^W was defined above, using the restriction on the support of the measure $d\nu(\alpha)$. (The Wick powers $:\check{\xi}^n:$ on $L_2(\check{Q})$ are defined in the same way as the Wick powers $:\xi^n:$ on $L_2(Q)$). Since $V_g(\tau) \geq 0$, one has $\exp(-\int_0^t V_g(\tau)\, d\tau) \in L_\infty(Q)$. The fundamental formula (4) expresses the action of $\exp(-tH_g^W)$ on $L_2(Q)$ (and thus of the semigroup generated by the space cut-off Hamiltonian on Fock space $\mathcal{F}$) through measurable functions defined over the path space $L_2(\check{Q})$. This has as a consequence that (2) is equal to $\int \check{\xi}(\delta_{t_1} \otimes f_1) \ldots \check{\xi}(\delta_{t_n} \otimes f_n) e^{(-\int_{-t}^{t} V_g(\tau) d\tau)} d\check{\mu}_0$. In particular $(\Omega_0, \exp(-2tH_g)\Omega_0) = \int \exp(-\int_{-t}^{t} V_g(\tau) d\tau)\, d\check{\mu}_0$. Thus

$$(\Omega_0,\ \exp(-2tH_g)\Omega_0)^{-1} (\Omega_0, \exp(-t-t_1)\bar{H}_g)\phi(f_1)\exp(t_1-t_2)\bar{H}_g)\ldots$$
$$\phi(f_n)\exp(t_n-t)\bar{H}_g)\Omega_0) = \int \check{\xi}(\delta_{t_1} \otimes f_1) \ldots \check{\xi}(\delta_{t_n} \otimes f_n)\, d\check{\mu}_{g,t} \tag{5}$$

where $d\check{\mu}_{g,t}$ is a new measure, absolutely continuous with respect to the Gaussian measure $d\check{\mu}_0$, and defined by

$$d\check{\mu}_{g,t} \equiv \left(\int \exp\left(-\int_{-t}^{t} V_g(\tau)\right) d\check{\mu}_0\right)^{-1} \exp\left(-\int_{-t}^{t} V_g(\tau)\, d\tau\right) d\check{\mu}_0.$$

Finally we have fulfilled our first purpose, namely to express the

Schwinger functions of the space cut-off interaction, $S_g^{(n)}$, through expectations of measurable functions: in fact from (3) and (5) we have $S_g^{(n)}(t_1 f_1,\dots,t_n f_n) = \lim_{t\to\infty} S_{g,t}^{(n)}(t_1 f_1,\dots,t_n f_n)$, for all $t_1 \leq \dots \leq t_n$, $f_i \in \mathcal{S}(R)$, $i = 1,\dots,n$; $n = 1, 2, \dots$, where $S_{g,t}^{(n)}(t_1 f_1,\dots,t_n f_n) \equiv \int \check{\xi}(\delta_{t_1} \otimes f_1)\dots\check{\xi}(\delta_{t_n} \otimes f_n)\, d\check{\mu}_{g,t}$. In these formulae the non-negative space cut-off function g can be taken to be e.g. the characteristic function of an interval $[-\ell, +\ell] \subset R$. Our original program was to find the limit for $g \to 1$ of the $S_g^{(n)}(\cdot)$. In terms of the new set of functions $S_{g,t}^{(n)}(\cdot)$ the problem is to control the double limit $\lim_{g\to 1} \lim_{t\to\infty} S_{g,t}^{(n)}(\cdot)$.

Take now g to be the characteristic function of the interval $[-\ell, +\ell]$ and set, for such g, $S_B^{(n)}(\cdot) \equiv S_{g,t}^{(n)}(\cdot)$, with $B = [-\ell, +\ell] \times [-t, t] \subset R^2$. Then above double limit appears as the limit of $S_B^{(n)}(\cdot)$ as $B \to R^2$ in a particular way. It is very convenient to generalize the problem and introduce for any bounded compact subset B in R^2 quantities $S_B^{(n)}(\cdot)$ defined by $S_B^{(n)}(f_1,\dots,f_n)=\int\check{\xi}(f_1)\dots\check{\xi}(f_n)d\check{\mu}_B$ with $f_i \in \mathcal{S}(R^2)$ and $d\check{\mu}_B = (\int \exp(-V(\chi_B))d\check{\mu}_0)^{-1} \exp(-V(\chi_B))d\check{\mu}_0)$, with $V(\chi_B) \equiv \int :e^{\alpha\check{\xi}}:(\chi_B)\, d\nu(\alpha)$. $V(\chi_B)$ is a $L_2(\check{Q})$ function, due to the restrictions of the support of the measure $d\nu(\alpha)$ (as seen analogously as we did for V_g^W and $V_g(\tau)$). It turns out that one can construct the limit, uniformly on compact sets, of these quantities $S_B^{(n)}(\cdot)$ as $B \to R^2$, in the sense that the distance of any point in any compact fixed subset of R^2 to the boundary of B increases to ∞, and an $\varepsilon/2$ -argument then yields the identification

$$\lim_{B\to R^2} S_B^{(n)}(\cdot) = \lim_{g\to 1} \lim_{t\to\infty} S_{g,t}^{(n)}(\cdot),$$

and thus the wanted limit Schwinger functions $S^{(n)}(\cdot)$. One advantage of this construction using the $S_B^{(n)}(\cdot)$ for general $B \subset R^2$ is that the $S^{(n)}(\cdot)$ will be invariant with respect to the euclidean group of R^2 and thus, once a suitable analytic continuation of the $S^{(n)}(\cdot)$ is made, the Wightman functions will be inhomogenous Lorentz-invariant. Before however to come to these points we shall sketch the proof of the existence of the limit of $S_B^{(n)}$ as $B \to R^2$. The basic idea is to look at the variation of the $S_B^{(n)}(\cdot)$ when the compact bounded subset B is made a little bigger (in all directions). It turns out that the $S_B^{(n)}$ are positive and non-decreasing and that $\inf_{B\subset R^2} S_B^{(n)}$ is their limit when $B \to R^2$. The proof goes as follows: Let $\gamma \geq 0$, $\gamma \in C_0^\infty(R^2)$ and define $S^{(n)}_{\chi_B+\eta\gamma}$, $\eta > 0$, in the same was as $S_B^{(n)}$, with χ_B replaced by $\chi_B + \eta\gamma$. Consider the directional derivative of $S_B^{(n)}$ at the point χ_B in the direction γ: defined as $D_\gamma S_B^{(n)} = \lim_{\eta\downarrow 0} \eta^{-1} [S^{(n)}_{\chi_B+\eta\gamma} - S_B^{(n)}]$. This can be proved to exist and to be equal to the expression obtained by formal differentiation with respect to χ_B i.e. to be equal to

$$C(\check{\xi}(f_1)\dots\check{\xi}(f_n);V(\chi_B)) \equiv \langle V(\chi_B)\rangle\langle\check{\xi}(f_1)\dots\check{\xi}(f_n)\rangle - \langle\check{\xi}(f_1)\dots\check{\xi}(f_n)V(\chi_B)\rangle, \quad (6)$$

where $\langle\cdot\rangle$ is the expectation $\int \cdot \, d\check{\mu}_B$ with respect to the measure $d\check{\mu}_B$. Such a difference measures the amount of "correlation" between the random variables $\check{\xi}(f_1)\ldots\check{\xi}(f_n)$ and $V(\chi_B)$ in the distribution given by the measure $d\check{\mu}_B$. Estimates of correlations of a similar type have been used extensively in the study of lattice models in statistical mechanics (e.g. well known "Griffiths inequalities" have been obtained).[11] The difference is here that one has R^2 instead of a lattice. This suggests the introduction[12] of a "lattice approximation" to the field model, in which R^2 is replaced by the lattice of points $L_\delta \equiv \{ n\delta \mid n \in Z^2\}$, δ being a fixed positive number (lattice spacing, to be set eventually equal to zero) and the stochast process $\check{\xi}(f)$ is replaced by its lattice approximation given, for $f \in C_0^\infty(R^2)$, by $\check{\xi}_\delta(f) = \sum_{n\in Z^2} \delta^2 \, \check{\xi}(\eta_{\delta,n}) \, g(n\delta)$, with

$$\eta_{\delta,n} \equiv (2\pi)^{-2} \int_{T_\delta^2} e^{ik(x-n\delta)} \check{\mu}(k) \, \check{\mu}_\delta(k)^{-1} \, dk, \quad T_\delta{}^2 \equiv [-\pi/\delta, \pi/\delta]^2$$

being the dual torus to L_δ and $\check{\mu}_\delta(k)^2$ the discrete approximation $\delta^{-2}[4-2\cos(\delta k_1) - 2\cos(\delta k_2)] + m_0{}^2$ of $\check{\mu}(k) = \sqrt{m_0{}^2+k_1{}^2+k_2{}^2}$, $k \equiv (k_1, k_2) \in R^2$. Note that the $\xi(\eta_{\delta,n})$ are Gaussian (proper) random variables and the family of all such random variables constitutes a so-called discrete Markovian field (in the sense of Dobrushin and Spitzer). Before showing the usefulness of the lattice approximation, let us remark that by the limit $\delta \to 0$ we really get back the continuous theory. In fact $\check{\xi}_\delta(f) \to \check{\xi}(f)$ as $\delta \to 0$ in $L_p(\check{Q})$ for all $1 \le p < \infty$ ([Guer-R-S]) and moreover $V_\delta(\chi_B) \to V(\chi_B)$ in $L_2(\check{Q})$ ([A1-HK5]), where $V_\delta(\chi_B)$ is defined as $V(\chi_B)$ with $\check{\xi}_\delta$ replacing $\check{\xi}$. Thus $S_{B,\delta}{}^{(n)}(\cdot) \to S_B{}^{(n)}(\cdot)$, where $S_{B,\delta}{}^{(n)}(\cdot)$ are the lattice approximation of the Schwinger functions, defined in the same way as the $S_B{}^{(n)}(\cdot)$, but with $\check{\xi}$ replaced by $\check{\xi}_\delta$ (and consequently $V(\chi_B)$ by $V_\delta(\chi_B)$). Moreover we shall study the correlation quantities (6) in the lattice approximation, i.e. with $\check{\xi}$ replaced by $\check{\xi}_\delta$ and V by V_δ, and prove that they are ≤ 0. This then carries over to (6) in the limit $\delta \to 0$, by the convergence results we have just mentioned. In the lattice approximation and with f_i of compact support, the correlation quantities (6) are expressed in terms of only finitely many random variables, so that the expectations actually reduce to expectations on a finite dimensional measure space $(R^n, \mathcal{B}(R^n), \mu_B{}^{(n)})$, $\mu_B{}^{(n)}$ being the restriction of μ_B to $(R^n, \mathcal{B}(R^n))$ and $\mathcal{B}(R^n)$ being the Borel sets of R^n. The measure $\mu_B{}^{(n)}$ has the special form of being the product of $G(x_1,\ldots,x_n) \equiv \prod_{i=1}^n F(x_i)$ times the Gaussian measure proportional to

$$e^{-\frac{1}{2}\sum_{i,j=1}^n x_i A_{ij} x_j} \, d^n x,$$

where the part $G(x_1,\ldots,x_n)$ comes from the interaction, namely from $e^{-V(\chi_B)}$ and the Gaussian measure comes from the interaction-free part in the lattice approximation and has the special property that the matrix $((A_{ij}))$, besides being symmetric and positive definite, has only diagonal and next to the diagonal non-vanishing elements, and the off diagonal elements are ≤ 0. It turns out that measures of above form fall into a class for which "correlation inequalities" of following form can be proved: $<(x_1)^{j_1}\ldots(x_n)^{j_n}> \geq 0$, $<(x_1)^{j_1+\ell_1}\ldots(x_n)^{j_n+\ell_n}> - <(x_1)^{j_1}\ldots(x_n)^{j_n}> <(x_1)^{\ell_1}\ldots(x_n)^{\ell_n}> \geq 0$ for any non-negative integers j_i, ℓ_i, $i = 1,\ldots, n$, where $(x)^j$ stands for the j-th power of the random variable x and $<\cdot>$ for expectation with respect to the measure in question.[11] This then yields immediately the "correlation inequalities" $S_{B,\delta}^{(n)}(f_1,\ldots,f_n) \geq 0$, $S_{B,\delta}^{(j+\ell)}(f_1,\ldots,f_{j+\ell}) \geq S_{B,\delta}^{(j)}(f_1,\ldots,f_j)\, S_{B,\delta}^{(\ell)}(f_{j+1},\ldots f_{j+\ell})$ for all n, j, ℓ. We would like to apply this to the lattice approximation $C(\check{\xi}_\delta(f_1)\ldots\check{\xi}_\delta(f_n);\, V_\delta)$ of $C(\check{\xi}(f_1)\ldots\check{\xi}(f_n);\, V)$, but we cannot yet do so because of the presence of $V_\delta(\chi_B)$ instead of a product of $\check{\xi}_\delta$'s. Formally however

$$V_\delta(\chi_B) = \int :e^{\alpha\check{\xi}_\delta}:(\chi_B)\, d\nu(\alpha) =$$
$$\int e^{\alpha\check{\xi}_\delta}(\chi_B)\, e^{(-1/2)\alpha^2 <\chi_B,\, \chi_B> \check{K}}\, d\nu(\alpha) =$$
$$\sum_{j=0} \int dx\ \chi_B(x) \left(\int \frac{\alpha^{2j}}{2j!}(\check{\xi}_\delta(x))^{2j} e^{-\alpha^2/2 <\chi_B,\, \chi_B> \check{K}}\, d\nu(\alpha)\right),$$

where in the expansion only even powers survive, since by assumption $d\nu(\alpha) = d\nu(-\alpha)$. Then $V_\delta(\chi_B)$ appears as a sum and integrals of terms of the form $(\check{\xi}_\delta(x))^{2j}$, with positive coefficients, and hence $C(\check{\xi}_\delta(f_1)\ldots\check{\xi}_\delta(f_n);V_\delta)$ as such a sum and integrals of terms

$$<(\check{\xi}_\delta(x))^j><\check{\xi}(f_1)\ldots\check{\xi}(f_n)> - <\check{\xi}(f_1)\ldots\check{\xi}(f_n)(\check{\xi}_\delta(x))^j>,$$

each of which is formally

$$S_{B,\delta}^{(j)}(x\ldots x)\, S_{B,\delta}^{(n)}(f_1\ldots f_n) - S_{B,\delta}^{(n+j)}(f_1\ldots f_n\, x\ldots x) \leq 0$$

and hence (formally) $C(\check{\xi}_\delta(f_1)\ldots\check{\xi}_\delta(f_n);\, V_\delta(\chi_B)) \leq 0$. To make this whole argument rigorous, we have only to go through it with $\check{\xi}_{\delta,\varepsilon}$ everywhere instead of $\check{\xi}_\delta$, where $\check{\xi}_{\delta,\varepsilon}$ is the regularization of $\check{\xi}_\delta$ defined by $\check{\xi}_{\delta,\varepsilon}(x) \equiv \check{\xi}_\delta(\chi_\varepsilon(x-\cdot))$, where $\chi_\varepsilon(x)$ is for $\varepsilon > 0$ a smooth approximation of $\delta(x)$ converging for $\varepsilon \to 0$ to $\delta(x)$. Then $(\xi_{\delta,\varepsilon}(x))^{2j}$ are measurable functions on $L_p(\check{Q})$ and one can prove that all steps in above formal computation are allowed, obtaining finally that $C(\check{\xi}_\delta(f_1)\ldots\check{\xi}_\delta(f_n);\, V_\delta(\chi_B))$ is ≤ 0. But the limit $\varepsilon \to 0$ can be done and yields the result of above formal computation and thus, after removing the lattice approximation ($\delta \to 0$, which we know we can by the above quoted convergence results),

$$D_B\, S_B^{(n)} \leq 0.$$

From the construction we have moreover $S_B^{(n)}(\cdot) > 0$. Thus the $S_B^{(n)}(\cdot)$ are positive, non-increasing functions of χ_B, bounded above by the so-called free Schwinger functions $S_0^{(n)}(\cdot)$, defined as $S_B^{(n)}(\cdot)$ but with $d\check{\mu}_0$ instead of $d\check{\mu}_B$. It is then easily seen that $\lim_{j\to\infty} S_{B_j}^{(n)}(\cdot)$ exists and is equal to $\inf_{B\subset R^2} S_B^{(n)}(\cdot)$, where $\{B_j\}$ is any sequence of bounded compact sets converging to R^2 in such a way that the distance from the boundary ∂B_j to any point of any fixed compact set of R^2 goes to ∞ as $j \to \infty$ and the infimum is taken over all bounded compact subsets of R^2. The Schwinger functions $S^{(n)}(\cdot)$ of the infinite volume theory are then defined as $\inf_{B\subset R^2} S_B$. They satisfy $0 < S^{(n)}(\cdot) < S_0{}^{(n)}(\cdot)$. By the construction one can see that these functions have properties corresponding to those of the Wightman functions $W^{(n)}(\cdot)$ entering the Wightman reconstruction theorem. In particular they are invariant under the euclidean group in R^2 and symmetric under interchanges of their arguments. They have cluster properties with respect to translations in R^2 (these are even exponential, so as to yield a mass gap of size $\geq m_0$). As limits of the $S_g{}^{(n)}(\cdot)$, they are easily shown to possess a "positive-definiteness property" and to belong to a theory with the correct "spectrum property." Briefly all assumptions of a theorem ([O-S]) of Osterwalder and Schrader giving the connections between axiomatically defined euclidean Schwinger functions and Wightman functions are satisfied, and this then yields a set of Wightman functions $W^{(n)}(\cdot)$, defined as analytic continuation of the $S^{(n)}(\cdot)$, and satisfying all Wightman axioms. Alternatively one can use the fact that all Nelson's axioms for Markoff euclidean field theories ([Ne 4]) are satisfied ([Fro 5]) and a theorem of Nelson yields again a constructio of the Wightman functions satisfying all Wightman axioms.

Remark: If we were just interested in constructing a set of Wightman functions satisfying all Wightman axioms, without wanting to have explicitly a connection with a Hamiltonian space cut-off interaction, we could have started directly from a set of euclidean Schwinger functions $S_B{}^{(n)}(\cdot)$, the perturbation being then defined purely as a perturbation of the Gaussian measure $d\check{\mu}_0$ (corresponding to the case of no interaction) by the "Gibbsian factor" $(\int e^{-V(X_B)}d\check{\mu}_0)^{-1}\, e^{-V(X_B)}$. This is a special case of a Euclidean Markoff approach to constructive quantum field theory advocated recently especially by Nelson.

As far as the verification of the assumptions I and III for scattering theory (in addition to the already verified Wightman axioms) let us just mention that they follow from the domination of $S^{(n)}(\cdot)$ by the free Schwinger functions and the above correlation inequalities, yielding exponential cluster properties of the Schwingeı functions. See [Al-H.K5];[Si3]. In fact the measure $d\check{\mu}_B$ converges weak to a strongly clustering measure on $\mathcal{S}'(R^2)$ as $B \to R^2$.

Remark: Besides the local relativistic invariant models of type 1) and 2), there is a third type of non-trivial translation and rotation invariant models, with vacuum polarization, in which the infinite volume problem has been solved. ([Al-H.K1]). These models are constructed starting from the bounded interaction models of example 3) in Section II.2. Also in this case the construction of the infinite volume limit is made through the study of Schwinger functions $S_B{}^{(n)}(\cdot)$, defined as before, with $d\check{\mu}_B$ now given by $(\int e^{-V(X_B)}d\check{\mu}_0)^{-1}\, e^{-V(X_B)}$, with $V(X_B) \equiv \int e^{i\alpha\xi}(X_B)d\mu(\alpha)$.[14]

However, the technique used for proving the convergence of $S_B{}^{(n)}$ as

$B \to R^{s+1}$ is in this case different[15] from the one described above for models of type 2). It relies namely on a relation which exists between the generating functional $\int e^{i\xi(f)} d\check{\mu}_B$, $f \in C_0^\infty(R^{s+1})$ of the Schwinger functions and the correlation functions $\rho_B^{(n)}$ "classical gas of variably changed particles". The latter satisfy Kirkwood-Salzburg integral equations and can be shown to have limits as $B \to R^{s+1}$ for all values of the coupling constant $|\lambda| \leq \Lambda_0$, where $\Lambda_0 > 0$ is explicitly given and depends on the ultraviolet cut-off. The technique for the convergence proof is an adaptation of the usual Banach space technique for dilute classical gases [Ru 2]. The convergence of $\rho_B^{(n)}$ to limit functions $\rho^{(n)}$ then implies the convergence of $\int e^{i\xi(f)} d\check{\mu}_B$, yielding a set $\{S^{(n)}\}$ of Schwinger functions for the infinite volume limit of the models. These are analytic in the coupling constant, non-trivial (different from zero and from the free Schwinger functions $S_0^{(n)}$), have cluster properties with respect to space and time translations and yield by analytic continuation in the time variables the analogues of Wightman function $W^{(n)}$, with cluster properties with respect to translations in R^{s+1}. A reconstruction theorem, analogue to the one given in Section IV.1, yields then a physical Hilbert space $\mathcal{H}$ and field theory, as in IV.1, with the only difference that only translations (in R^{s+1}) and rotations (in R^s) are unitarily represented but not the whole Lorentz group. Especially from the point of view of physics, a nice feature of these models is that $S^{(n)}$ and $W^{(n)}$ possess indeed those "linked cluster expansions", discussed often in the physical literature as formal expansions. In these cases the expansions actually converge, for $|\lambda|$ sufficiently small. As far as the specific scattering quantities are concerned, we have already mentioned in Chapter II results in the case of a space cut-off interaction. One finds that certain "truncated off-shell scattering amplitudes" of the space cut-off interaction converge, as the space cut-off is removed, to "truncated off-shell scattering amplitudes" for the infinite volume limit model. These are expressed through (non-trivial) spectral density functions (Fourier-Laplace transforms of the $\rho^{(n)}(\cdot)$), analytic in λ and which exhibit explicitly analyticity in the complex energy variables, outside the union of certain real hyperplanes. The implications of these facts include a proof of (off-shell) "crossing symmetry". The missing step for having a completely satisfactory scattering theory in this case is the proof that the poles of the spectral density functions, which are isolated for $\lambda = 0$ and correspond to the one particle energies $\mu(k)$, remain isolated also for $\lambda \neq 0$, $|\lambda|$ sufficiently small. ([Al-H.K. 2,3]).

In conclusion let us summarize by saying that in the models discussed in this section existence of scattering quantities has been either completely or partly proven and one is on the verge of being able to study specific questions of scattering, and thereby substantiating and enriching, by exploiting the specific dynamic features of the models, general results obtained axiomatically in the late 50's and in the 60's. There is certainly no lack of open problems in these 2-dimensional models: study of scattering amplitudes, of unitarity, completeness, location of eigenvalues, study of the continuous spectrum of M^2 and H, etc.

Finally sooner or later the problem has to be solved of how to lift the number of space dimensions from s = 1 to the realistic case s = 3. Here certainly problems will be more difficult but it seems now eventually very likely that some of the methods already proved to be powerful for s = 1 will survive, suitably adapted, the limit $s \to 3$.

Acknowledgement: It has been a pleasure to be at this conference and I would like to thank Professors James LaVita and Jean-Paul Marchand for the invitation. These lectures have been strongly influenced by innumerable most illuminating discussions with Dr. R. Høegh-Krohn, and I am happy to use the opportunity to give him my heartily thanks.

Bibliographical Notes

Introduction

1. For a very concise introduction to the physical problems of quantum field theory see e.g. [Ma].
2. For the physical principles and formalism of traditional quantum field theory see e.g. [Bj-D], [Bo-S], [Fi 2], [Ka], [Ro], [Schw], [We].
3. For the development of quantum electrodynamics see e.g. [Schwi].
4. For the axiomatic approach see e.g. [Brandeis 65], [Jo], [Lille 57], [St 2], [Str-W], [Varenna 58], [Wi-G].
5. For the historical roots of the constructive approach see e.g. [Fo], [Schwi], [We] and [Fr 1,2], [Se 1]. For earlier work in the systematic development of the constructive approach see [Gl 3], [Ja], [La], [Ne 1],[wi]. For general reading and further references see e.g. [Gl 1,2],[Gl-J 2,3], [He 1,2], [Ja 3.4], [Se 3,4], [Si 1], [Si-H.K], [Wi 1]. Most expository are [Gl 2], [Ja 2]. For the systematic construction of $(\phi^4)_2$ models see e.g. [Gl-J 5-8], [Di-G], [Gl-J-S], [Ne 3], for the $P(\phi)_2$ models [Ros 1,2], [Se 3,4], [Si-H.K.], for the bounded interactions [Al-H.K. 1-4], [Ho-K 8,9,10], [Str-Wi], for the exponential interactions [Ho-K 15], [Al-H.K. 5]. More specific references

are given in the course of the lectures. A new appraoch, the so called Markoff-Euclidean approach, has been developed recently, inspired by earlier work described in [Sy] and starting with [Ne 3]. See e.g. [Al-H.K. 1], [Al-H.K. 5] [Guer 2,3], [Guer-R-S 1,2], [Ho-K 13], [Ne 3,4,5], [New], [O-S], [Si 3,5], [Si-G].

Chapter I.1

For the historical origins of Fock space see e.g. the short comments and references in [Schw], part two, p. 121. Fock's paper [Fo] gives the basic physical formalism. The mathematical foundations were given by Cook's paper [Co 1]. Fock space is sometimes called, for good reason, Fock-Cook space; however, for simplicity we have accorded to the more common shorter denomination. A mathematical foundation of the so-called "Schrödinger or wave representation" and its equivalence with the particle representation has been given by Segal. See references to section I.4. Other references for the Fock space are e.g. [Fi 2], [Fr 1], [Gui], [Jo], [Wi-G].

Chapter I.2

The main references for this section are [Co 1] and [Gl-J3]. See also e.g. the references of the bibliographical note to section I.1 and [He 1].

Chapter I.3

General and/or introductory references are e.g. [Fr 2], [Gl 1,2], [Gl-J 2,3], [He 1], [Ja 1]. For an introduction to the problem of the lower bound on $H_0 + V$ see e.g. [Ga], [Gl-J1].

Chapter I.4

The Schrödinger space representation was first discussed thoroughly by Segal [Se 1]. In our lectures we have used also [Gl-J3], [Ho-K 13], [Ne 5]. See also e.g. [Ge-V], [Hi], [Se 4], [Si-H.K.], [Xia].

Chapter I.5

See the references of chapter I.3, especially [Fr 2], [Gl 1], [He 1], [Ja 1].

Chapter II

Basic original general references for this chapter are [Ho-K 11] and [Ka-M 1]. Constructions of asymptotic creation-annihilation operators along lines related to those of this chapter are used e.g. in [Al 1,2], [Ao-K-M], [Ao-M], [Di], [Fro 2], (Ho-K 5-8], 11, 12, 14], [Ka-M 1,2], [Mu-K], [Pru], [Pr-M 1,2]. For related original physical discussions see e.g. [Eck], [Fr 2], [He 1]. For the corresponding formalism for non-relativistic n particle systems (second quantization) see e.g. [He 2], [Kl-Z], [Sa]. More specific references to the different sections are given in the main text and II.1: [Fr 2]; II.2: [Ho-K 11, 12], [Ka-M 1]. See also [Al 1,2]. II.3: [Di], [Ka-M 1], [Ho-K 5, 8, 10-12]; II.4: [Ho-K 5, 8, 10-12], [Si 1]; II.5: [Ho-K 1-4, 5, 8, 10-12]. Also [Al-H.K 4].

Chapter III.

The persistent model described in this chapter is a modification, with relativistic kinematics, of a model studied originally by Nelson ([Ne 1]) and Cannon ([Ca]). Its study was suggested in [He 1] and was carried through, as far as the construction of the basic quantities is concerned, by Eckmann ([Ec 1,2]), adapting and extending methods developed by Hepp's and Schrader's study of a Y_2-Lee model ([He 1], [Schr 1]). For more details on the model described and for the study of scattering theory in the model see [Al 1,2], [Fro 1]. See also the references given in the footnotes to this chapter.

Chapter IV.

The references to Chapter IV.1 are mentioned already in the main text. For Chapter IV.2 see the references mentioned in the bibliographical note to Introduction, point 5. Statistical mechanical methods for the study of the Markoff-Euclidean formulation of quantum field theoretical models are first used in [Al-H.K. 1] and [Guer-R-S 1]. The lattice approximation was developed in [Guer-R-S 2] and [Ne 6]; applications are also given in [Al-H.K. 5], [Si 3,5], [Si-G].

Footnotes.

Introduction.

1. See e.g. the historical notes in the introduction of [Jo].
2. See e.g. the references 2 of the bibliographical note for more details.
3. See e.g. the references and the discussion given in [Wil], ch. II, §6.
4. See e.g. [v.N]; [Pu]; Ch. IV, p. 63.
5. To be described in chapter I below.
6. See also the bibliographical notes to the introduction.
7. Other axiomatic approaches were also developed, from different points of view. E.g. the Araki - Haag - Kastler approach to local quantum theory, which has not entirely clarified connections with Wightman's approach, and Bogoliubov's approach, based on an S-operator.
For the former see e.g. [Ar],[Ha-K], for the latter the references given below, in footnote 12).
8. This had been inbetween also analyzed to a certain extent. See e.g. [Fr1],[Fr2],[Ga].
9. References will be given in the course of the lectures. See also the bibliographical notes to this introduction.
10. For the problems in higher dimensions see e.g. [Gℓ1], [He1].
11. See the bibliographical note 4.
12. We regret however not having time to say something on some recent developments in related areas, like the work done along the lines of Bogoliubov S-matrix formulation of quantum field theory (see e.g. [Ep-G] and references given there) on analytic properties of scattering amplitudes, on formal power series([He 3]), on formulation of LSZ - perturbation theory([St 1]) and on the infrared problems (see references in Ch. III).
13. The models of chapter II (space cut-off models), chapter III (translation invariant models without vacuum polarization) and of the quoted remark to chapter IV (euclidean - invariant models with vacuum polarization) are all interesting quantum field theoretical models, but lack full invariance with respect to the inhomogeneous Lorentz group. In this sense they are intermediate between the non-relativistic n-particle theory and the full local relativistic models with non-trivial scattering matrix with which the rest of chapter IV is concerned. We would like to mention here that, for reasons of time, we will not be able to say anything either on the second quantization formalism for n particle non-relativistic systems, which has similarities with the formalism of chapter II (and partly III, IV) nor on "intermediate models" other than the quoted ones, and which can be

divided into 2 types, characterized in a physical terminology which will become clear in the course of the lectures:
I) explicitly solvable local covariant models, with physically trivial scattering (e.g. [De], [Wi 1]).
II) external field models, with Lorentz covariance but replacing the interacting field by a classical field, e.g. [Sei], [Wi 2

Section I.1.

1. The symmetric Fock space is also called boson Fock space. It is the Hilbert space of states of a system which consists of zero, one, two,...boson (i.e. integer spin) particles. Fock space was introduced quite early in quantum field theory: see the references to this section. Later on in this section we shall only consider a particular kind of boson Fock space, namely the one associated to scalar (i.e. spin zero) and neutral (uncharged) boson particles.
2. It is well known, and we shall come back to this many times, that in the most interesting physical cases of interactions which are translation invariant and cause "vacuum polarization" the perturbations are always so singular as to be undefinable as operators on Fock space (at most, in some favorable cases, they can be bilinear forms on Fock space). In this case the perturbation problem is only formulated in Fock space for suitable ("space cut-off" and moreover possibly "ultraviolet cut-off") approximations of the interaction and for the removal of the approximation one has eventually to go outside Fock space (do the removal on suitable functionals defined on Fock space) and construct a new space called "physical Hilbert space" (as opposed to the "bare" Fock space) on which the Hamiltonian ("physical Hamiltonian") acts. However also in this case we can look upon the physical Hilbert space and Hamiltonian as limits in a suitable sense of objects defined on Fock space and in this sense we still speak of being confronted with a "singular" perturbation problem in Fock space.
3. Both realizations are useful. Roughly speaking the "particle representation" is a spectral representation for the free Hamiltonian and thus suitable as a reference point when discussing existence of scattering quantities. The "wave representation" is a spectral representation of the fields and thus suitable for discussing interactions (given in terms of fields) and determining the Hamiltonian. In pratical problems an exploitation of both is often useful.
4. Then n-fold Hilbert tensor product $\mathcal{H}^{\otimes n} \equiv \mathcal{H} \otimes \ldots \otimes \mathcal{H}$ of $\mathcal{H}$ with itself is defined as the closure of the algebraic tensor product $\mathcal{H} \otimes \ldots \otimes \mathcal{H}$ in the natural norm given by the scalar product defined by $(\psi_1 \otimes \ldots \otimes \psi_n, \chi_1 \otimes \ldots \otimes \chi_n) = \prod_{i=1}^{n} (\psi_i, \chi_i)_{\mathcal{H}}$,

where $(\cdot,\cdot)_H$ is the scalar product in H. The symmetric group $\mathcal{S}_n$ (group of all permutations of n objects) has a unitary representation on $H^{\otimes n}$, defined by $\pi \in \mathcal{S}_n \to U_\pi \; \psi_1 \otimes \ldots \otimes \psi_n = \psi_{\pi_1} \otimes \ldots \otimes \psi_{\pi_n}$, where π sends $(1,\ldots,n)$ into $(\pi_1,\ldots,\pi_n)$. $S=(n!)^{-1} \sum_\pi U_\pi$ is the projector onto the symmetric subspace.

5. I.e. the Hilbert space closure (with respect to the scalar product naturally defined on $\mathcal{F}$ from the one present on H, see below) of the symmetric tensor algebra over H.

6. Of course the interesting physical case is s = 3. However dynamically interesting local relativistic perturbations can be handled up to now, only for the case s = 1 (and partially for s = 2).

7. Whenever we want to distinguish explicitly momentum space and functions on it from the conjugate position space and functions on it, we shall put a "$\wedge$" on momentum space quantities. Variables in momentum space are denoted k, p, q; those in position space by x,y,z.

Section I.2.

1. As opposite to the ordinary non-relativistic quantum mechanical processes in which the total number of particles is conserved (i.e. it is a constant of the motion).

2. Cfr. the introduction.

3. In fact the free energy operator $\frac{1}{n^2}\left(-\sum_{\ell=1}^{n} \frac{d^2}{dx_\ell^2} + x_\ell^2 - 1\right)$ can be written as $\sum_{\ell=1} a_\ell^* a_\ell$.

4. a(k) is not closable as an operator. Hence a*(k), which should be formal adjoint to a(k), can certainly not be defined as the operator adjoint to a(k). Instead we define it as the linear form adjoint to the one defined by a(k). Bilinear forms are often called sesquilinear forms: e.g. [Kat], Ch. VI, p. 308.

5. This is first shown for all $\psi \in \mathcal{F}^{(n-1)} \cap D_0$, $\psi' \in \mathcal{F}^{(n)} \cap D_0$ and then extended to all D_0 observing that $a(h) \upharpoonright \mathcal{F}^{(n)}$ extends to a bounded operator from $\mathcal{F}^{(n)}$ to $\mathcal{F}^{(n-1)}$.

6. This is done similarly as in footnote 5), observing that $a^*(h) \upharpoonright \mathcal{F}^{(n)}$ extends naturally to a bounded operator from $\mathcal{F}^{(n)}$ to $\mathcal{F}^{(n+1)}$.

7. This follows from the canonical commutation relations given in point 3).

8. This is in analogy with the expression in terms of a_ℓ, a_ℓ^* of the harmonic oscillator position operator: $x_\ell = \frac{1}{\sqrt{2}}(a_\ell + a_\ell^*)$. Cfr. Section I.4.

9. We shall now put a hat ^ on momentum space functions.

10. As seen e.g. from the fact that every vector D_0 is an analytic vector for $\phi(g)$.

11. We find this formulation useful for later reference. In terms of operator valued distributions one has $\Gamma(U_0(\underline{a},\Lambda))\phi(F)\Gamma(U_0(\underline{a},\Lambda))^{-1} = \phi(F_{\underline{a},\Lambda})$, where F is any test function in $\mathcal{S}(R^{s+1})$ for the distribution $\phi(\underline{x})$, with $\underline{x} \equiv (t,x)$ and $F_{\underline{a},\Lambda}(\underline{x}) \equiv F(\Lambda^{-1}(\underline{x}-\underline{a}))$

12. For later reference (cfr. Chapter IV) let us also remark that the locality property can be written covariantly as $[\phi(G_1), \phi(G_2)] = 0$ on a dense domain, if $(x^0-y^0)^2 - (x-y)^2 < 0$ for all $\underline{x} \equiv (x^0,x)$ in the support of G_1 and $\underline{y} \equiv (y^0,y)$ in the support of G_2, with $G_i \in C_0^\infty(R^{s+1})$, $i = 1, 2$.

Section I.3.

1. Non-symmetric in general. Linear combinations of such perturbations however can be chosen to be symmetric and these are in fact the only perturbations we shall consider later (see also footnote 2) below).

2. As in ordinary quantum mechanics only symmetric perturbations V have direct physical meaning, since the Hamiltonian constructed from H_0 and V should be self-adjoint.

3. Arising e.g. for approximations of local relativistic interactions, but never for the local relativistic interactions themselves. See the discussion in Sections I.5, I.6.

4. E.g. if V is a polynomial interaction (to be defined below) of odd degree and as smooth a kernel as we like, then one has always $H_0 + V$ unbounded from below. See [Ga],[Gl-J 1],[0].

5. This is put in evidence by the physically transparent discussion of [Eck] as well as by perturbation theoretical considerations ([Fa], [Fr2]). For mathematical results under the assumptions $w_{0i} = w_{i0} = 0$ see Chapter III and the Remark to Section II 5.3.

Section I.4.

1. This is essentially the same connection which exists between the energy representation (spectral representation for the energy) and the Schrödinger representation (spectral representation for the position operator x) in the theory of the non-relativistic quantum mechanical harmonic oscillator.

2. For any wave function $\hat{g}(k)$ like the ones considered in Section I.2 when introducing the free field $\phi(\hat{g})$, with the property $\hat{g}(-k) = \bar{\hat{g}}(k)$ and $\hat{g}/\sqrt{\mu} \in L_2(\hat{R}^s)$, we have that the Fourier transform of $\hat{g}/\sqrt{\mu}$ is in $L_2(\hat{R}^s)$. Viceversa every $g \in K$ has a Fourier transform in $L_2(\hat{R}^s)$. The Sobolev space K can be characterized also as the closure of the real $C_0^\infty(R^s)$ functions $g(x)$ with respect to the scalar product $\langle f,g\rangle \overset{\Omega}{=} (f,(-\Delta+m_0^2)^{-\frac{1}{2}} g)$, where $(\cdot,\cdot)$ is the $L_2(R^s)$ scalar product and Δ is the Laplacian

on R^s.

3. I.e. a real-valued measurable function.

4. Other choices of Q, which of course lead to the same Gaussian process (modulo equivalences) and have been found convenient are the choice $Q = \mathcal{S}'(R^s)$ (connected with Minlos theorem) and Q=spectrum of the von Neumann algebra generated by the spectral projections of the time zero free fields $\phi(g)$, $g \in C_0^\infty(R^s)$. As a matter of fact however the particular choice of (Q,Σ,μ_0) space does not enter in an essential way e.g. in the applications of Chapter IV.

5. We shall often write shortly $L_p(Q)$ for $L_p(Q, \Sigma, \mu_0)$

6. Particularly in connection with the study of "white noise" (Wiener-Ito construction). See e.g. [Hi] part III, p. 94

7. See [Se1]; also e.g. [Hi], part IV, p 94; [Xi], Ch. V, p 332.

8. So that $S_n(g_1 \otimes \ldots \otimes g_n) = \frac{1}{n!} \sum_\pi g_{\pi_1} \otimes \ldots \otimes g_{\pi_n}$ the sum going over all permutations $\pi = (\pi_1,\ldots,\pi_n)$ of $(1,\ldots,n)$.

Section I.5.

1. Below we shall see that $:\phi(x)^n:$ has a natural expression (random-variable distribution) on $L_2(Q)$.

2. The interactions we study (boson self-interactions) should only be restricted by the requirements to yield eventually models satisfying certain general postulates (like e.g. those of Chapter IV). They are not a priori suggested or dictated by nature, like it is the case e.g. in quantum electrodynamics. As a matter of fact very little is known on the form of the true "strong interactions" between bosons.

3. Let $\chi(\cdot)$ be a real valued even $C_0^\infty(R^s)$ function with $\int \chi(x)\,dx = 1$ and define, for every $\varepsilon > 0$, $\chi_\varepsilon(x) \equiv \varepsilon^{-s} \chi(x/\varepsilon)$. Then $\chi_\varepsilon(x)$ is a smooth approximation of the δ-function, converging to $\delta(x)$ as $\varepsilon \to 0$. Define the regularized approximations ("ultraviolet cut-off approximations": cfr Section I.6) to $\phi(x)$ and $\xi(x)$ as $\phi(\chi_\varepsilon(x-\bullet)) \equiv \phi_\varepsilon(x)$ and $\xi(\chi_\varepsilon(x-\bullet)) \equiv \xi_\varepsilon(x)$. Note $\xi_\varepsilon(x) = U\, \phi_\varepsilon(x) U^*$. Define $:\phi^n:(g)$ as the sum (7) and $:\phi_\varepsilon{}^n:(g)$ likewise, with g replaced by its convolution with χ_ε. Define $:\xi_\varepsilon{}^n:(g)$ as we defined $:\xi^n:(g)$, but with ξ_ε replacing ϕ resp. ξ. On one hand $||(:\phi_\varepsilon^n:(g) - :\phi^n:(g))\Omega_0|| \to 0$ as $\varepsilon \to 0$, as seen using the fact that the kernels of $:\phi_\varepsilon^n:(g)$ converge in L_2 - norm to those of $:\phi^n:(g)$, together with Proposition 1 of I.3 (and the fact that Ω_0 is in the domain of any power of N). On the other hand $:\xi_\varepsilon^n:(g) \to :\xi^n:(g)$ as $\varepsilon \to 0$, in $L_2(Q,\Sigma,\mu_0)$, as proved by an explicit computation. But $:\xi_\varepsilon^n:(g) = U :\phi_\varepsilon^n:(g) U^*$ (as seen e.g. from the fact that $:(\phi_\varepsilon(g))^j:$ is a linear combination of powers $(\phi_\varepsilon(g))^j$ of

$\phi_\varepsilon(g)$, $j \leq n$ which are mapped by U into the L_p-functions $(\xi_\varepsilon(g))^j$. Since $U :\phi_\varepsilon{}^n: (g) U^*$ converges strongly to $:\xi^n:(g)$ and $:\phi^n_\varepsilon: (g)\, \Omega_0$ to $:\phi^n: (g)\, \Omega_0$, we have then by the unitarity of U^ε as a map from $\mathcal{F}$ to $L_2(Q)$, $U :\phi^n: (g)\Omega_0 = :\xi^n:(g)$ hence $U :\phi^n: (g) U^* = :\xi^n:(g)$.

Section I.6.

1. $(\psi, H_0\, \Omega_0) = 0$ for all ψ but $V\{\psi,\Omega_0\} \neq 0$ for some ψ. If V is not only a bilinear form but even a symmetric operator such that Ω_0 is in the domain of V (see e.g. Section II for examples), then $H_0\Omega_0 = 0$ but $V\Omega_0 \neq 0$, thus $(H_0 + V)\Omega_0 \neq 0$. In this case the Fock vacuum is an eigenvector of the free Hamiltonian but not of the total Hamiltonian, hence the name of being "polarized" by the interaction.
2. E.g. Lee models : see e.g. [He 1], [Schr 1].
3. If $w_{j0} = 0 = w_{0j}$ for all j for the general perturbations of section I.3 (absence of vacuum polarization) then the presence of the δ-distribution does not hinder the L_2-character of the kernels, provided they are smooth and L_2 at infinity in momentum space (and at the origin if $m_0 = 0$).
4. It arises e.g. in the euclidean invariant models of [Al-H.K 1,2,3].
5. It transforms Ω_0 in a vector with norm $\int|\delta(k_1 +\ldots+k_n)\; f\;(k_1,\ldots,k_n)|^2\, dk_1\ldots dk_n = \infty$. This is a manifestation of Haag's theorem and holds whenever the interactions can be written as $V = \int v(x)\, dx$, where $v(x)$ transforms under translation by $a \in R^s$ to $v(x + a)$. This is an easy consequence of the translation invariance of Ω_0.
6. We have already introduced this procedure shortly in (7) of Section I.5.
7. See also footnote 3 in I.5.
8. The case s = 1 however, as remarked above, does not need the introduction of an ultraviolet cut-off. For s=2 a change of Hilbert space is necessary.
9. E.g. for $:P(\phi,g):_2$ and $:P(\phi_\varepsilon,g):_{s+1}$, $s \geq 0$, $\varepsilon > 0$ interactions, H_0 and V can be shown to have a common dense domain and $H_0 + V$ is essentially self-adjoint on this domain (see e.g. [Gl-J 2,3], [Ros1,2],[Se3,4],[Si-H.K]). In more singular models it can happen that $D(V) \cap D(H_0) = \{0\}$. Then of course "$H_0 + V$" must be defined in some more involved way. See e.g. [Gl-J 4], [0].
10. Of course 3) and 2) are related and e.g. the localization of the absolutely continuous spectrum as given in models depends on constructions of 3).

11. This is (at least partly) the case for the interactions studied in [Al-H.K 1,2,3] (See also Remark in Chapter IV and [Gl-J-S].

Section II.1.

1. At least in the traditional approach.
2. Some more singular situations can be treated essentially by the same methods however: [Di], [Ho-K 14], [Mu-K].
3. For the study of a suitable limit of W(t)(in fermion models) see [Pru], [Pr-M]. See also Lemma 4.1 in [Al-H.K. 2].
4. For scattering quantities for Yukawa interactions see e.g. [Ho-K 11], [Ya], and, without ultraviolet cut-off [Di]. For first results on existence of quantities in the infinite volume limit see e.g. [Gl-J 3], [Schr 2].

Section II.2.

1. All functions h,f appearing as arguments of operators $a^{\#}(h)$ in this chapter will be functions on momentum space $\hat{R}^s$. We do not write the "momentum space" label $\wedge$ on them, since no confusion can arise, no position space functions being used.
2. μ^{θ} is the operator defined on $L_2(\hat{R}^s)$ by multiplication by the θ-th power of $\mu(k) \equiv \sqrt{m_0^2 + k^2}$. The symbol $d\Gamma(\cdot)$ was defined in Section I.1.
3. [Ne 2] See also e.g. [Gl-J 2] and references given therein.
4. References to these models are e.g. [Al-H.K. 1,2,3,4], (Ho-K 8,9,10], [Str-Wi]. Other references are given in [Al-H.K. 1]. See also e.g. [Go].

Section II.4.

1. H can even have, in certain cases, eigenvalues embedded in the continuum ([Ros 3], [Si 4]; see also [Si 1]). This is a phenomenon which is expected to disappear (unless selection rules are present) when the space cut-off is removed. For the correspondent problem in n particle systems see e.g. [Al 3], [Ba], [Si 2].

Section II.5.

1. Mind that we have thus a transition amplitude from incoming asymptotic particles to out-going asymptotic particles defined directly in terms of the asymptotic creation operators $a_{\pm}^{\#}$. Above formula shows that the original creation-annihilation operators $a^{\#}$ play now a role only in the computation of the transition amplitudes, but are not associated directly with the asymptotic n-particle states (they were associated to the n-particle spaces $\mathcal{F}^{(n)}$ of $\mathcal{F}$, in the sense that $\mathcal{F}^{(n)}$ is the closed linear span of $a^*(h_1)\ldots a^*(h_n)\Omega_0$, for all $h_i \in H$. For this reason the operators $a^{\#}(h)$ are sometimes called "bare"

creation-annihilation operators and said to create, when applied to the Fock vacuum Ω_0 (called also "bare vacuum"), "bare particle vectors (states)" and bare particle subspaces $\mathcal{F}^{(n)}$. N, the number operator we defined in $\mathcal{F}$, is then called "number of bare particles operator."

2. See e.g. [Io-Oć], [Pr] and references in [Al 3].
3. A more abstract version of these results could also be given, adapting the proof of [Al-H.K. 4]. Expansions of the same type were first considered for smooth fermi interactions in [Ho-K 11]. See also [Al 1,2].

Section III.

1. See e.g. [Guer 1], [He 1], [Schr 1].
2. [Fro 1, 2].
3. See e.g. [Fro 1], [Gr 1,2], [Sl].
4. [Fro 2, 3, 4].
5. See e.g. [Bl], [Ku-F], [Pa-F].

Section IV.2

1. For the original proofs see [Ha], [Ru 1]. See also [Str].
2. See also e.g. [Gl-J 2, 3], [Gl-J 5-8].
3. As a matter of fact the condition on $d\nu$ to be finite can be replaced by $\int d\nu(s)d\nu(t)/(2\pi - st) < \infty$.
4. The label w reminds us that we have to do with a quantity defined on the "wave picture" space $L_2(Q, \Sigma, \mu_0)$.
5. The proof of the existence of the lowest eigenvalue (and in fact of the discreteness of the spectrum in $[Eg, Eg + m_0]$ (see footnote 6 for improvements on this point) is based on a norm convergence result of $\exp(-t(H_0^w + V_{g,k}^w))$ to $\exp(-tH_g^w)$, for $|t|$ sufficiently big, together with the information on the lower part of the spectrum of $H_0^w + V_{g,k}^w$ (available, since $V_{g,k}^w$ is bounded). The simplicity of E_g is based on the positivity of the kernel of $\exp(-tH_g^w)$ on $L_2(Q)$, which itself follows from the strong convergence of $\exp(-tH_{g,k}^w)$ to $\exp(-tH_g^w)$, together with the fact that $\exp(-tH_{g,k}^w)$ has a positive kernel (this is proved using that $\exp(-tH_0^w)$ has a positive kernel, that $V_{g,k}^w \geq 0$, and the Lie-Kato-Trotter product formula).
6. It follows as a corollary from the results on the spectrum of the infinite volume Hamiltonian, in the physical Hilbert space, to be discussed below, that the gap between E_g and the rest of the spectrum is exactly m_0, for all $g \in L_1(R) \cap L_2(R)$.
7. See e.g. [Wy].

8. The construction of path space ($L_2(\check{Q}, \check{\Sigma}, \check{\mu})$ and $\check{\xi}$) goes of course through also for an arbitrary number of space dimensions s, with $K = H_{-1/2}(R^{s+1})$ and $\check{K} = H_{-1}(R^{s+1})$. Since however in the present case the interaction is only defined for s = 1, we shall write the construction only for s = 1.

9. Of course, since $\mathcal{F}$ and $L_2(Q)$ are unitarily equivalent, to any J_τ there is associated an isometry J_τ' from $\mathcal{F}$ to $L_2(Q)$.

10. By the observation of the previous footnote, the formula (4) can be replaced by a similar one in which H_g^w is replaced by H_g and J_0 by J_0'.

11. See e.g. [Gi].

12. [Guer-R-s 1,2], [Ne 6]. For the model under discussion [Al-H.K. 5].

13. Mind that, as remarked in footnote 4 to Section I.4, μ_B can be realized as a measure on $\mathcal{S}'(R^2)$.

14. $\check{\xi}$ and the corresponding path space are defined similarly as we did above for s = 1: cfr. footnote 8.

15. But still inspired by statistical mechanics, which is a natural tool in the study of Euclidean-Markoff field theories.

<u>References</u>:

[Al 1] S. Albeverio, Scattering theory in some models of quantum fields, I, J. Math. Phys. <u>14</u>, (1973) 1800-1816.

[Al 2] S. Albeverio, Scattering theory in some models of quantum fields, II, Helv. Phys, Acta. <u>45</u>, (1972) 303-321.

[Al 3] S. Albeverio, On bound states in the continuum of N-body systems and the Virial Theorem, Annals. of Phys. <u>71</u>, (1972), 167-276.

[Al-H.K 1] S. Albeverio, R. Hoegh-Krohn, Uniqueness of the physical vacuum and the Wightman functions in the infinite volume limit for some non-polynomial interactions, Commun. Math. Phys. <u>30</u>, (1973) 171-200.

[Al-H.K 2] S. Albeverio, R. Hoegh-Krohn, The scattering matrix for some non-polynomial interactions; I, Oslo Univ. Preprint, Helv. Phys. Acta <u>46</u>, (1973) 504-534.

[Al-H.K 3] S. Albeverio, R. Hoegh-Krohn, The scattering matrix for some non-polynomial interactions, II, Oslo Univ. Preprint, Helv. Phys. Acta <u>46</u>, (1973) 535-545.

[Al-H.K 4] S. Albeverio, R. Hoegh-Krohn, Asymptotic series for the scattering operator and asymptotic unitarity of the space cut-off interactions, Oslo Univ. Preprint, Nuovo Cimento 18A, (1973) 285-307.

[Al-H.K 5] S. Albeverio, R. Hoegh-Krohn, The Wightman axioms and the mass gap for strong interactions of exponential type in two-dimentional space-time, Oslo Univ Preprint Mathematics No. 12, May, 1973. To appear in J. Funct. Analys., 1974.

[Al-H.K 6] S. Albeverio, R. Hoegh-Krohn, Unpublished.

[Ao-K-M] M. Aoki, Y. Kato, N. Mugibayashi, Asymptotic fields in model field theories, II, Cut-off Yukawa interaction, Kagawa-Kobe Preprint, May, 1971.

[Ao-M] M. Aoki, N. Mugibayashi, Hamiltonian defined as a graph limit in a simple system with an infinite renormalization, Kagawa-Kobe Preprint.

[Ar] H. Araki, Local quantum theory, I (in [Varenna 68]).

[Ba] E. Balslev, Absence of positive eigenvalues of Schrödinger operators, UCLA Preprint, 1973.

[Bj-D] J. Bjorken, S. Drell, Relativistic quantum fields, McGraw-Hill, New York, 1965.

[Bl 1] P. Blanchard, Discussion mathématique du modèle de Pauli et Fierz relatif à la catastrophe infrarouge, Comm. Math. Phys. 15 (1969) 156-172.

[Bo-S] N.N. Bogoliubov, D.V. Shirkov, Introduction to the theory of quantized fields, Interscience, New York 1959.

[Bon] P.J.M. Bongaarts, Linear fields according to I.E. Segal, in [London 71].

[Brandeis 65] Axiomatic field theory, 1965, Brandeis University Summer Institute in Theoretical Physics, Vol. 1, Gordon and Breach, 1966.

[Ca] J.T. Cannon, Quantum field theoretic properties of a model of Nelson: domain and eigenvector stability for perturbed linear operators, J. Funct. An. 8, (1971), 101-152.

[Ci] A.L. Cistjakov, On the scattering operator in the space of second quantization, Sov. Math. Dokl, 5, 1199-1202 (1964) (transl.)

[Co 1] J.M. Cook, The mathematics of second quantization, Trans. Amer. Math. Soc. 74, (1953) 222-245.

[Co 2] J.M. Cook, Asymptotic properties of a boson field with given source, Journ. Math. Phys. 2, (1961) 33-45.

[De] G.F. Dell'Antonio, A model field theory: the Thirring model, Lectures given at Schladming (1973), to appear in Acta Phys Austr.

[Di] J. Dimock, Spectrum of local Hamiltonians in the Yukawa field theory, J. Math. Phys. 13, (1972) 477-481.

[Di-G] J. Dimock, J. Glimm, Measures on Schwartz distribution space and applications to $p(\phi)_2$ field theories, Courant Preprint (1973).

[Dir] P.A.M. Dirac, The quantum theory of the emission and absorption of radiation, Proc. Roy. Soc. London, Ser. A, 114, (1927) 243-265.

[Ec 1] J.P. Eckmann, Hamiltonians of persistent interactions, Thesis, University of Geneva, (1970).

[Ec 2] J.P. Eckmann, A model with persistent vacuum, Commun. Mathe. Phys. 18, (1970) 247-264.

[Eck] H. Eckstein, Scattering in field theory, Nuovo Cimento 4, (1956) 1017-1058.

[Ep] H. Epstein, Some analytic properties of scattering amplitudes in quantum field theory, in [Brandeis 65].

[Ep G] H. Epstein, V. Glaser, Le rôle de la localité dans la renormalization perturbative en theorie quantique des champs, in [Les Houches 70].

[Ez] H. Ezawa, Some examples of the asymptotic field in the sense of weak convergence, Ann. Phys. 24, (1963) 46-62.

[Fa] L.D. Faddeev, Sov. Phys. Dokl. 8, (1964) 881.

[Fi 1] M. Fierz, Beispiel einer lösbaren Feldtheorie, Zeitschr. f. Phys. 171, (1963) 262-264.

[Fi 2] M. Fierz, Einführung in die relativistische Quantumfeld theorie und in die Physik der Elementarteilchen (ausgearbeitet von S. Albeverio und Dr. M. Kummer) VMP Verlag, Zürich, 1965.

[Fo] V. Fock, Konfigurationsraum und zweite Quantelung, Zeitschr. f. Phys. 75, 1932, 622-647.

[Fr 1] K.O. Friedrichs, Mathematical aspects of the quantum theory of fields, Interscience, New York, 1953.

[Fr 2] K.O. Friedrichs, Perturbation of spectra in Hilbert space, Am. Math. Soc., 1965.

[Fro 1] J. Fröhlich, Mathematical discussion of models with persistent vacuum, preprint ETH 1971 and note in preparation.

[Fro 2] J. Fröhlich, On the infrared problem in a model of scalar bosons I, ETH doctoral thesis, Zürich.

[Fro 3] J. Fröhlich, On the infrared problem in a model of scalar electrons and massless scalar bosons, II, 1972.

[Fro 4] J. Fröhlich, LSZ asymptotic condition and structure of scattering states in theories without a mass gap, in preparation.

[Fro 5] J. Fröhlich, An exercise on the exponential models of S. Albeverio and R. Høegh-Krohn, in preparation.

[Fro 6] J. Fröhlich, Quantum field theory and the theory of Markov processes, Seminars given at CERN, 1973, to appear.

[Ga] A Galindo, On a class of perturbations in quantum field theory, Proc. Nat. Acad. Sci. USA, 48, (1962), 1128-1134.

[Ge-V] I.M. Gelfand, N. Ya Vilenkin, Generalized functions, Vol. 4, Academic Press, New York 1964.

[Gi] J. Ginibre, General formulation of Griffiths' Inequalities, Commun. Math. Phys. 16, (1970) 310-328.

[Gl 1] J. Glimm, Models for quantum field theory, in [Varenna 1968].

[Gl 2] J. Glimm, The foundations of quantum field theory, Adv. Math. 3, (1969) 101-125.

[Gl 3] J. Glimm, Yukawa coupling of quantum fields in two dimensions I, Commun. Math. Phys. 5 (1967) 343-386.

[Gl-J 1] J. Glimm, A. Jaffe, Infinite renormalization of the Hamiltonian is necessary, I. Math. Phys. 10, (1969), 2213-2214.

[Gl-J 2] J. Glimm, A. Jaffe, Quantum field theory models in [Les Houches 1970].

[Gl-J 3] J. Glimm, A. Jaffe, Boson quantum field models in [London, 1971]

[Gl-J 4] J. Glimm, A. Jaffe, Positivity of the $(\phi^4)_3$ Hamiltonian Courant-Harvard Preprint, 1972.

[Gl-J 5] J. Glimm, A. Jaffe, A $\lambda\ \phi^4$ quantum field theory without cutoffs, I, Phys. Rev. 176, 1945-1951 (1968).

[Gl-J 6] J. Glimm, A Jaffe, A $\lambda\ \phi^4$ quantum field theory without cutoffs II, The field operators and the approximate vacuum; Ann. Math. 91, (1970) 362-401.

[Gl-J 7] J. Glimm, A. Jaffe, A $\lambda\ \phi^4$ quantum field theory without cutoffs III, The physical vacuum, Acta. Math 125, (1970) 203-261.

[Gl-J 8] J. Glimm, A. Jaffe, A $\lambda\ \phi^4$ quantum field theory without cutoffs IV, Perturbations of the Hamiltonian, J. Math. Phys. 13, (1972) 1568-1584.

[Gl-J-S] J. Glimm, A. Jaffe, T. Spencer, The Wightman axioms and particle structure in the $P(\phi)_2$ quantum field model, New York Univ. Preprint, 1973.

[Go] N.S. Gonchar, Green's functions in the euclidean domain without space-time cutoffs, Kiev Preprint, 1972.

[Gr-S] O.W. Greenberg, S.S. Schweber, Clothed operators in simple models, Nuovo Cimento 8, (1958) 378-405.

[Gro 1] L. Gross, Existence and uniqueness of physical ground states, J. Funct. Analys. 10, (1972) 52-109.

[Gro 2] L. Gross, The relativistic polaron without cutoffs, Commun. Math. Phys. 31 (1973), 25-73.

[Gue-V] M. Guenin, G. Velo, On the scalar field model, Helv. Phys. Acta. 42, 102-116 (1969).

[Guer 1] F. Guerra, Equivalence problems in models with infinite renormalization, Nuovo Cimento, 68A, (1970) 258-280.

[Guer 2] F. Guerra, Uniqueness of the vacuum energy density and van Hove phenomenon in the infinite volume limit for two-dimensional self-coupled Bose fields, Phys. Rev. Letts. 28, (1972) 1213-1215.

[Guer 3] F. Guerra, Introduction to euclidean-Markov methods in constructive quantum field theory, Lectures given at the Xth Winter School of Theoretical Physics, Karpacz, 1973.

[Guer-R-S 1] F. Guerra, L. Rosen, B. Simon, Statistical mechanics results in the $P(\phi)_2$ quantum field theory, Physics Letters B, 44, (1973) 102-104.

[Guer-R-S 2] F. Guerra, L. Rosen, B. Simon, The $P(\phi)_2$ quantum field theory as classical statistical mechanics, Princeton University Preprint, to appear in Ann. of Math.

[Gui] A. Guichardet, Symmetric Hilbert spaces and related topics, Springer, Berlin, 1972.

[Ha 1] R. Haag, On quantum field theories, Mat-Fys. Medd, Danske Vid. Selsk., 29, No. 12, (1955) p. 37.

[Ha 2] R. Haag, Quantum field theories with composite particles and asymptotic conditions, Phys. Rev. 112, (1958) 669-673.

[Ha-K] R. Haag, D. Kastler, An algebraic approach to quantum field theory, Jour. Math. Phys. 5, (1964) 848-861.

[He 1] K. Hepp, Théorie de la renormalization, Springer Verlag, Heidelberg (1969).

[He 2] K. Hepp, On the connection between Wightman and LSZ quantum field theory, in [Brandeis 65].

[He 3] K. Hepp, Renormalization theory, in [Les Houches 70].

[Hi] T. Hida, Stationary stochastic processes, Princeton University Press, Princeton, 1970.

[Ho-K 1] R. Høegh-Krohn, Partly gentle perturbations with application to perturbation by annihilation-creation operators, Thesis N.Y.U. 1966.

[Ho-K 2] R. Høegh-Krohn, Partly gentle perturbations with application to perturbation by annihilation-creation operators, Nat. Acad. of Science, Vol. 58, No. 6, (1967) 2187-2192.

[Ho-K 3] R. Høegh-Krohn, Partly gentle perturbation with application to perturbation by annihilation-creation operators. Comm. Pure and Applied Math. Vol. 21, (1968) 313-342.

[Ho-K 4] R. Høegh-Krohn, Gentle perturbations by annihilation-creation operators, Com. Pure and Applied Math. Vol.21 (1968) 343-357.

[Ho-K 5] R.Høegh-Krohn, Asymptotic fields in some models of quantum field theory I, Jour. Math. Phys. Vol. 9, No. 12 (1968) 2075-2080.

[Ho-K 6] R. Høegh-Krohn, Asymptotic fields in some models of quantum field theory II. Jour. Math. Phys. Vol. 10 No. 4, (1969) 639-643.

[Ho-K 7] R. Høegh-Krohn, Asymptotic fields in some models of quantum field theory III, Jour. Math. Phys. Vol. 11, No. 1, (1970) 185-189.

[Ho-K 8] R. Høegh-Krohn, Boson fields under a general class of cutoff interactions, Comm. Math. Phys. 12, (1969) 216-225.

[Ho-K 9] R. Høegh-Krohn, Boson fields under a general class of local relativistic invariant interactions., Commun. Math. Phys. 14, (1969) 171-184.

[Ho-K 10] R. Høegh-Krohn, Boson fields with bounded interaction densities, Commun. Math. Phys. 17, (1970) 179-193.

[Ho-K 11] R. Høegh-Krohn, On the scattering operator for quantum fields, Comm. Phys. 18, (1970) 109-126.

[Ho-K 12] R. Høegh-Krohn, On the spectrum of the space cutoff $:P(\phi):$ Hamiltonian in two space-time dimensions, Comm. Math. Phys. 21, (1971) 256-260.

[Ho-K 13] R. Høegh-Krohn, Infinite dimensional analysis with applications to self-interacting boson fields in two space-time dimensions, to appear in Proc. Functional Analysis Meeting, Aarhus, Spring, 1972.

[Ho-K 14] R. Høegh-Krohn, The scattering operator for boson fields with quadratic interactions, Unpublished.

[Ho-K 15] R. Høegh-Krohn, A general class of quantum fields without cutoffs in two space-time dimensions, Commun. Math. Phys. 21, (1971) 244-255.

[Ho-K 16] R. Høegh-Krohn, Private communication.

[Io-O'c] R.J. Iorio, M. O'Carroll, Asymptotic completeness for multiparticle Schrödinger Hamiltonians with weak potentials, Commun. Math. Phys. 27, (1972) 137-145.

[Ja 1] A. Jaffe, Constructing the $\lambda(\phi^4)_2$ theory, in [Varenna 68].

[Ja 2] A. Jaffe, Whither axiomatic field theory?, Rev. Mod. Phys. 41, (1969) 576-580.

[Ja 3] A. Jaffe, Wick polynomials at a fixed time, J. Math. Phys. 7, (1966) 1251-1255.

[Ja 4] A. Jaffe, Dynamics of a cutoff $\lambda\ \phi^4$ field theory, Princeton University thesis, 1965.

[Jo] R. Jost, The general theory of quantized fields, Ann. Math. Soc. Providence, 1965.

[Ka] D. Kastler, Introduction à L'electrodynamique quantique, Dunod, Paris, 1961

[Kat] T. Kato, Perturbation theory for linear operators, Springer, New York, 1966.

[Ka-M 1] Y. Kato, N. Mugibayashi, Regular perturbations and asymptotic limits of operators in quantum field theory, Progr. Theor. Phys. 30, (1963) 103-133.

[Ka-M 2] Y. Kato, N. Mugibayashi, Asymptotic fields in model space theories, I, Progr. Theor. Phys. 45, (1971) 628.

[Kl-Z] J. Klein, I.I. Zinnes, On nonrelativistic field theory: interpolating fields and long range forces, Iona-Fordham Preprint, 1973.

[Ku-F] P.P. Kulish, L.D. Faddeev, Asymptotic condition and infared divergence in quantum electrodynamics, Theor. and Math. Phys. 2, (1970) 153-

[La] O. Lanford, Construction of quantum fields interacting by a cutoff Yukawa coupling, Princeton University thesis (1966).

[Les Houches 70] Statistical mechanics and quantum field theory, Les Houches 1970; Ed. C. DeWitt, R. Stora, Gordon and Breach, New York, 1971.

[Lille 57] Les problèmes mathématiques de la théorie quantique des champs, Lille 1957, CNRS, Paris, 1959.

[London 71] Mathematics of contemporary physics, London Mathematical Society Symposia, 1971, Ed. R. Streater, Academic Press, London and New York, 1972.

[Ma] P. Mandl, Introduction to quantum field theory, Interscience, New York, 1959.

[Mu] N. Mugibayashi, Asymptotic fields in quantum field theory, Report presented to the International Seminar on Statistical Mechanics and Field Theory, Haifa, (1971).

[Mu-K] N. Mugibayashi, Y. Kato, Regular perturbation and asymptotic limits of operators in fixed source theory, Progr. Theor. Phys. 31, (1964) 300-310.

[Ne 1] E. Nelson, Interaction of Nonrelativistic Particles with a quantized scalar field, Journ. Math. Phys. 5, (1964) 1190-1197.

[Ne 2] E. Nelson, A quartic interaction in two dimensions, in Mathematical theory of elementary particles, ed. R. Goodman and I. Segal, MIT Press, Cambridge, (1966).

[Ne 3] E. Nelson, Quantum fields and Markoff fields, in Proceedings of Summer Institute of Partial Differential Equations, Berkeley, (1971), Ann. Math. Soc,. Providence, (1973).

[Ne 4] E. Nelson, Construction of quantum fields from Markoff fields, J. Funct. Anal. 12, (1973), 97-112.

[Ne 5] E. Nelson, The free Markoff field, J. Funct. Anal. 12, (1973) 211-227.

[Ne 6] E. Nelson, In preparation.

[New] C. Newman, The construction of stationary two-dimensional Markoff fields with an application to quantum field theory, J. Funct. Anal. 14 (1973), 44-61.

[O] K. Osterwalder, On the Hamiltonian of the cubic boson selfinteraction in four dimensional space-time, Fortschr. d. Phys. 19, (1971) 43-113.

[O-S] K. Osterwalder, R. Schrader, Axioms for euclidean green's functions, Commun. Math. Phys. 31, (1973), 83-112.

See also corrections given at the Erice Summer School on Constructive Quantum Field Theory, (1973).

[Pa-F] W. Pauli, M. Fierz, Zur Theorie der Emission langwelliger Lichtquanten, Nuovo Cimento 15, (1938), 167-188.

[Pr] R. Prosser, Convergent perturbation expansions for certain wave operators, J. Math. Phys. 5, (1964) 708-

[Pru] E. Prugovecki, Scattering theory in Fock space, J. Math. Phys. 13, (1972) 969-976.

[Pr-M 1] E. Prugovecki, E.B. Manoukian, Spin-momentum distribtuion observables and asymptotic fields for fermion fields with cutoff self interaction, Nuovo Cimento, 10B, (1972) 421-446.

[Pr-M 2] E. Prugovecki, E.B. Manoukian, Asymptotic behavior of spin-momentum distribution observables for fermion fields with cutoff self-coupling, Commun. Math. Phys. 24, (1972) 133-150.

[Pu] C.R. Putnam, Commutation Properties of Hilbert Space operators and related topics, Springer Verlag, New York (1967), Ch. IV.

[Re] P.A. Rejto, On gentle perturbations, I, II, Comm. Pure Appl. Math. 16, (1963), 279-303; 17, (1964) 257-292.

[Ro] P. Roman, Introduction to quantum field theory, J. Wiley, New York, (1969).

[Ros 1] L. Rosen, A $\lambda\ \phi^{2n}$ field theory without cutoffs, Comm. Math. Phys. 16, (1970) 157-183.

[Ros 2] L. Rosen, A $(\phi^{2n})_2$ quantum field theory: higher order estimates, Comm. Pure and Appl. Math. 24, (1971) 417-457.

[Ros 3] L. Rosen, Renormalization of the Hilbert Space in the mass shift model, J. Math. Phys. 13, (1972) 918-927.

[Ru 1] D. Ruelle, On the asymptotic condition in quantum field theory, Helv. Phys. Acta. 35, (1962) 147-163.

[Ru 2] D. Ruelle, Statistical mechanics, Rigorous results, W.A. Benjamin, New York (1969).

[Sa] W. Sandhas, Commun. Math. Phys. 3, (1966) 358.

[Schn] W. Schneider, S-Matrix und interpolierende Felder, Helv. Phys. Acta, 39, (1966) 81-106.

[Schr 1] R. Schrader, On the existence of a local Hamiltonian in the Galilean invariant Lee model, Comm. Math. Phys. 10, (1968) 155-178).

[Schr 2] R. Schrader, Yukawa quantum field theory in two space-time dimensions without cutoffs, Ann. of Phys. 70, (1972) 412-457.

[Schw] S.S. Schweber, An introduction to relativistic quantum field theory, Row, Perterson & Co., Evanston (1961).

[Schwi] J. Schwinger (Editor), Selected papers on Quantumelectrodynamics, Dover, New York, (1958).

[Se 1] I.E. Segal, Tensor algebras over Hilbert spaces, I, Trans. Amer. Math. Soc. 81, (1956) 106-134.

[Se 2] I.E. Segal, Mathematical problems of relativistic physics, Am. Math. Soc., Providence, (1963).

[Se 3] I.E. Segal, Construction of nonlinear local quantum processes: I, Ann. Math. 92, (1970) 462-481.

[Se 4] I.E. Segal, Nonlinear functions of weak processes, I, J. Funct. Anal. 4, (1969), 404-456.

[Sei] R. Seiler, Quantum theory of particles with spin zero and one half in external fields, Comm. Math. Phys. 25, (1972) 127-151.

[Sem] Yu.A. Semenov, On the construction of the Bogoliubov scattering operator in the $(P^{2m}(\phi))_2$ theory, Kiev Preprint, (1972).

[Si 1] B. Simon, Studying spatially cutoff $(\phi^{2n})_2$ Hamiltonians, pp. 197-221 of book in Ref. [St 2].

[Si 2] B. Simon, Absence of positive eigenvalues in a class of multiparticle quantum systems, Preprint Marseille.

[Si 3] B. Simon, Correlation inequalities and the mass gap in $P(\phi)_2$, I:Domination by the two-point function, Comm. Math. Phys. 31, (1973), 127.

[Si 4] B. Simon,Continuum embedded eigenvalues in a spatially cutoff $P(\phi)_2$ field theory, Proc. AMS 35 (1972) 223-226.

[Si 5] B. Simon, Correlation inequalities and the mass gap in $P(\phi)_2$, II:Uniqueness of the vacuum for a class of strongly coupled theories, Toulon Preprint (1973).

[Si-G] B. Simon, R.B. Griffiths, The $(\phi^4)_2$ field theory as a classical Ising model, Commun. Math. Phys. 33 (1973), 145-164.

[Si-H.K] B. Simon, R. Høegh-Krohn, Hypercontractive semi-groups and two dimensional self-coupled Bose fields, J. Funct. Analys, 9, (1972) 121-180.

[Sl] A. Sloan, The relativistic polaron without cutoffs in two space-dimensions, Cornell University Thesis (1971).

[St 1] O. Steinmann, Perturbation expansions in axiomatics field theory, Springer Verlag, Heidelberg (1971).

[St 2] O. Steinmann, Connection between Wightman and LSZ Field Theory, in Statistical Mechanics and Field Theory, Lectures 1971 Haifa Summer School, Ed. R.N. Sen, C. Weil, Halsted Press, Jerusalem (1972) 269-291.

[Str] R.F. Streater, Uniqueness of the Haag-Ruelle scattering states, J. Math. Phys. 8, (1967) 1685-1693.

[Str-W] R.F. Streater, A.S. Wightman, PCT, Spin and Statistics and All That, Benjamin, New York, (1964).

[Str-Wi] R.F. Streater, I.F. Wilde, The time evolution of quantized fields with bounded quasi-local interaction density, Commun. Math. Phys. 17, (1970) 21-32.

[Strei] L. Streit, The construction of physical states in quantum field theory, Acta Phys. Austr., Suppl. VII, (1970) 355-391.

[Sy] K. Symanzik, Euclidean quantum field theory, in [Varenna 68].

[Varenna 58] Problemi matematici della teoria quantistica delle particelle elemeniari e dei campi, Scuola Internationale di Fisica, Varenna (1958).

[Varenna 68] Teoria quantistica locale, Ed. R. Jost, Rendiconti Scuola Internazionale Fisica, Varenna (1968), Academic Press, New York, (1969).

[v.N] J. v.Neumann, Die Eindeutigkeit der Schrödingerschen Operatoren, Math. Ann. 104, (1928) 570-578.

[We] G. Wentzel, Quantum theory of fields, Interscience, New York, (1949).

[Wi 1] A.S. Wightman, Introduction to some aspects of the relativistic dynamics of quantized fields, in "1964 Cargese Summer School Lectures" (M. Levy, Ed), Gordon and Breach, New York, (1967).

[Wi 2] A.S. Wightman, Relativistic wave equations as singular hyperbolic systems, Princeton University, (1971).

[Wi-G] A.S. Wightman, L. Gårding, Fields as operator-valued distributions in relativistic quantum theory, Ark. f. Phys. 28, (1964) 129-184.

[Wy] W. Wyss, On Wightman's theory of quantized fields, pp. 533-575 in Lectures in Theoretical Physics, XIth, Ed. K. Mahanthappa, W. Brittin, Boulder Summer School 1968, Gordon and Breach, New York (1969).

[Xia] Xia-Dao-Xing, Measure and integration theory on infinite-dimensional spaces, Academic Press, New York, (1972).

[Ya] Ya Yakimiv, Asymptotic fields for approximated renormalized two-dimensional Yukawa model, Theor. Math. Phys. 9, (1971) 1169.

HISTORY AND PERSPECTIVES IN SCATTERING THEORY.

A STRICTLY PERSONAL VIEW.

Hans Ekstein

Argonne National Laboratory

This summer institute was mostly concerned with a variety of scattering theory that originated in the middle fifties. It may be interesting to recall the circumstances and motivation for the revision and reconstruction that was begun in this period.

The main unsolved problem, then as well as now, was relativistic quantum field theory. Despite the dazzling quantitative successes of renormalization prescriptions in quantum electrodynamics, no consistent theory capable of guiding research into other interactions was found.

After the failure of many clever attempts to eliminate the inconsistencies of the Lagrangian field theories, some physicists felt that the trouble might be found in the then accepted starting formula for the calculation of cross sections. There were two suspects in this matter: the definition of the S-matrix for a well-defined Hamiltonian and the meaning of the ill-defined functions of field operators which were accepted as Hamiltonians. The first suspicion motivated the reexamination of general scattering theory, the second the axiomatic approach of the Wightman field theory. The basis of scattering theory was in a poor state, to be sure, and there seemed to be a hope that by a correct mathematical statement of the physical scattering problem at least some of the obscurities and inconsistencies of relativistic field theories could be removed. This hope was supported by the empirical discovery that even non-relativistic theories with well-defined Hamiltonians (e.g. phonon scattering) led to absurd results if the time-dependent Dirac scattering theory was applied to them literally. In the current language of the day, the discovery was that "you gave to renormalize even finite theories".

J. A. LaVita and J.-P. Marchand (eds.), Scattering Theory in Mathematical Physics, 383–389.

The two approaches to scattering theory known then as now were the stationary and the time-dependent ones. The stationary statement requires the solution of a partial differential equation and has apparently nothing to do with the Hilbert-space formulation of quantum mechanics.

The wave functions were originally introduced by Schrödinger [1] as analogs of the wave fields in optics, related to particles by the Hamilton-Jacobi formalism. Since Schrödinger's analogy had come to be considered as a heuristic rather than a postulational basis of quantum theory, these non-L^2-functions had no clear physical interpretation. For the single-channel case, Dirac's [2] additional interpretative rule regarding $|\psi|^2$ as relative probability is plausible, but what about multi-channel scattering?

The present consensus is that the stationary wave-functions have no direct physical interpretation, and that they are mathematical auxiliaries of a time-dependent theory. The problem of making this connection precise is still with us. At any rate, the stationary approach was not at the center of attention because it requires a configuration space of finite dimensionality while the problem at hand required the inclusion of infinitely many particles. Also, there was an attempt to make the statement of a relativistic problem manifestly covariant, and this excludes singling out a definite time-direction.

Therefore, the basic formula for the scattering cross section used by Tomonaga, Schwinger and Dyson was, except for "covariant" typography and special features, the result of Dirac's time-dependent scattering theory [2].

In this theory, the division of the Hamiltonian into an unperturbed H_0 and an interaction is essential. But what is H_0? According to Dirac or (more explicitly) Heitler [3], it is that self-adjoint operator of which the initial states are eigenfunctions. We have to ask the experimenter: which operator are your initial states eigenfunctions of? Since most operators proposed for H_0 have no eigenstates, he may have some trouble answering the question.

Another answer, accepted by the practitioners of the black scattering magic of the day, was this: H_0 is the kinetic energy of the particles or asymptotic particles. But what particles? Dressed or bare? What state of undress would the Supreme Court of Scattering condone?

For nonrelativistic single-channel theory, the answer was, of course, unambiguous, but even for nonrelativistic N-body scattering with indistinguishable particles, one cannot properly define a self-adjoint H_0 as the kinetic energy of distant fragments, nor even a channel Hamiltonian H_α defined on a supspace. Clearly, a non-arbitrary basic formula had to be intrinsic and not dependent on the choice of H_0. Hence arose the challenge to develop a scattering theory without the use of an unperturbed Hamiltonian H_0 or channel Hamiltonian H_α.

The use of channel Hamiltonians H_α, useful as it may be in nonrelativistic theory, was of no avail in the field theoretic case, because of the identity of the particles in different clusters. This difficulty can, in principle, be circumvented in nonrelativistic quantum mechanics by using first only distinguishable particles, and then projecting the result into a symmetric or antisymmetric subspace of Hilbert space, but for infinitely many particles - e.g. two bare nucleons surrounded by their meson clouds of infinitely many mesons - this device is impracticable.

The wish to state the scattering problem without an unperturbed or channel Hamiltonian became almost an obsession with me in the late 40's, but it was not until 1956 that I was "sort of" successful. About the same time others, G.C. Wick, Nishijima and R. Haag, pursued similar ideas. The resultant version of the basic formula of scattering theory was well summarized in the review by Brenig and Haag [4]. The then adopted basic formula had to be corrected in two respects: by an energy- and a wave function-renormalization. Another clarification implied in the new approach was pointed out later by F. Coester [5]: the sound feeling that the kinetic energy of distant "dressed" particles must play a primary rôle in the theory finds its expression in then existence of a "free" Hamiltonian H_F. It is *not* a part of H, but it is a self-adjoint operator on a different Hilbert space $\mathcal{H}_F$.

In retrospect, the vagaries of the uncritical scattering theory appear as a comedy of errors; it is as though a man started from the assumption 2 × 2 = 5 and corrected the resulting absurdities by a clever scheme, called renormalization. Unfortunately, this conceptual clarification of scattering did not lead to effective algorithms for relativistic cross sections, as some of us had hoped, because the specific Hamiltonians or Lagrangians of the theory were ill-defined.

A pragmatist may well argue that the moral of the story is that conceptual clarity in physics is really not important, but we did obtain a deeper and clearer view of the problem and, as a byproduct, obtain a systematic nonrelativistic multichannel theory, and, if locality is disregarded, a systematic N-body relativistic scattering theory [5].

How is it possible that it took so long before silly mistakes in the basic statement were found?

To understand this, one must remember the then existent relation between physics and mathematics, profoundly different from the present. The physical basis of the time-dependent Dirac formalism, even where H_0 was given, was very shaky. If one follows the accepted interpretative rules, the formula gives the theoretical prediction for the outcome of the following sequence of physical operations: first, switch off the interaction Hamiltonian, H_I, prepare the initial state, switch on H_I, let a long time elapse, switch off H_I, and determine the final eigenstate of H_0. This enormity, quite literally, is the operational counterpart of the scattering formalism, according to one of the most careful writers on the subject [3]. The gap between the actual experiment and this imagined sequence of manipulations is appalling. Other authors introduced an adiabatic switching-on and off of the interaction, in order to obtain asymptotic convergence of

$$e^{i\,Ht}\,e^{-i\,H_0 t}\,f,$$

where f is a plane wave. As Barry Simon [6] says: One can't help picturing a pair of experimenters, one shouting to the other: "I have the beam ready, turn on the target, but do it very slowly."

The basic mistake in the time-dependent formulation was the assumption that the initial state of the system is an eigenstate of H_0. These are plane waves, and they never diffuse out of the range of a potential, no matter how short this range. Hence it was necessary to invoke an adiabatic exponential switching off of the interaction which is wholly fictitious.

It is incomprehensible to most participants of this conference that physicists should have created these difficulties for themselves. Everybody knows that H_0 has no eigenfunctions, and as soon as one uses L^2 functions, the necessity for switching off disappears. The wave packet leaves the interaction region all by itself, and it effectively becomes blind to the interaction. However, for single-channel scattering, all the extravagant idealizations of what one may call the pre-scientific scattering theory were quite justifyable, and led to correct results. To this day, most experimenters know only this kind of scattering theory and most quantitative results are due to it.

Physicists being pragmatic by nature, it is not surprising that they were impatient with quibblers (like me), since they saw no reasonable alternative to the way they did things.

I remember discussing my work with M.L. Goldberger. Of course, he said, scattering theory should properly be done with wave packets, but he was pessimistic about the success of my attempts to "work with wave packets of the most general type" because he thought the result would depend strongly on the shape of the wave packet. I was too naive at the time to appreciate the truth of his warning. I disregarded domain questions, and anybody could have reduced my paper ad absurdum by letting H act on a function outside its domain. Later, when I was shocked by this discovery, Karl Menger said that I reminded him of the rider over the Bodensee - the antihero of the poem in which a man rides inadvertently over the thinly frozen Bodensee, arrives safely in Switzerland, but upon learning what he had done, dies from fright. Indeed, ignoring all pitfalls, I had stumbled on something that was almost right.

For multichannel scattering theory, working with L^2 functions became imperative because the existence of time-limits was essential. What pragmatists may consider as only cosmetic in single-channel theory becomes indispensable for the definition of in- and out states in multichannel theory.

Moral: don't let success spoil you. If you get away with being sloppy in one field, you may not be as lucky in another field.

The amateurish attitude towards mathematics was the rule, not the exception among theorists in the early 50's. A mathematical physicist was, typically, a man who solved electrostatic boundary value problems. As to the value of rigorous mathematics for quantum theory, many theorists quoted approvingly a sentence attributed to Pauli: "Herr Neumann, wenn es in der Physik aufs Beweisen ankame, waren Sie ein grosser Physiker." (If it was proving that mattered in physics, you would be a great physicist.)

One of the unexpected and fortunate by-products of the reconstruction of scattering theory in the middle 50's was the recruitment of mathematicians so that now mathematical physics is one again a vital and indispensable branch of physics. I think that this accomplishment was largely the merit of R. Haag, who took important steps toward the rigorization of scattering theory and inspired others to complete the job. At the same time a similar rejuvenation was undertaken by A.S. Wightman in Princeton. The result is the vigorous type of mathematical physics that characterized this summer school.

Is advanced mathematics really good for physics? According to Niels Bohr, a good physicist can get along by using only addition and multiplication. I think he is right: once a matter is physically understood, calculations on the back of a used envelope are enough to obtain fair numerical estimates and the rest is engineering. The trouble is only that so few of us have sufficiently good physical intuition. We need mathematics as a crutch for our stumbling efforts, and proofs as a reassurance that we are not going astray.

But mathematical physicists have done more for physics: just to take our particular field, at the Flagstaff meeting R. Lavine opened the door to a conceptual revolution in scattering theory by an approach without S-matrix, and at this meeting J.M. Combes gave a tantalizing glimpse on a new concept of resonant states.

On the other hand, there may be an excess of mathematical technicalities in the field. It would be absurd to require that everything in physics should be proved. There is a famous story about a physics seminar chairman who said after a mathematician's talk on hydrodynamics: most of us have always thought that water flows downhill, but I am happy to hear it confirmed by such rigorous mathematics. Proofs are necessary for physics when intuition is uncertain.

We are, I think, entering a new territory, as I mentioned above in connection with Lavine's Flagstaff talk. The S-matrix that was the target of the theory for many years does not exist in general. In the presence of massless particles, each S-matrix element that corresponds to a finite number of particles vanishes because of the infrared catastrophe. To be sure, one can try to consider an infinite sum of such matrix elements (or their squares) and argue that $0 \cdot \infty$ gives a finite result, if the limit is cleverly defined. There is, I think, a danger in trying obstinately to patch up an old theory when the conditions call for a new one. To take an example from political science: if a theory based on the assumption that Chief Executives are law-abiding becomes untenable in the light of new information, the political scientist can patch up the old theory by redefining law-abidingness, but he would do better to start a new theory in which power and virtue are not necessarily linked.

And now I can give the reason for my display of recollections. It was a cautionary tale: vestigia terrent!

Just as it was tempting to persevere in the use of H_0 or H_α by superimposing corrective prescriptions (renormalization) on an inadequate basic formalism, so it is tempting to define all kinds of substitutes for the inadequate strict S-matrix. I hope that we will squarely meet the new challenge. It is true that mathematical enormities will not be in the way as they were in the 40's and 50's, but it is perfectly possible to commit mathematics that only purports to be physical theory. There is, I think, a tendency to write papers on mathematically well-defined objects that have quite misleading physical names attached to them. As we did away with H_0 in the fifties, so the challenge now is to do away with S. In the next few years, the physical interpretation of quantum theory will become important for scattering theory. This is where the second kind of vestigia terrent: cursory and sloppy treatment of the connection between operators and laboratory experiments was sufficient, and nobody bothered about thorough axiomatization of this connection, just as sloppy mathematics gave correct results in single-channel theory. But beware! For the new field of research, it will not be sufficient to begin a paper by <u>defining</u> a mathematical object as, say, an infrared-catastrophe system. The art of finding the mathematical counterpart of a piece of hardware will have to be raised from black magic to science.

<u>References</u>:

1. E. Schrödinger, Ann. of Phys. (4), <u>79</u>, (1926) 489.
2. P.A.M. Dirac, *The Principles of Quantum Mechanics*, 2nd Ed. Oxford Univ. Press, (1935).

3. W. Heitler, *The Quantum Theory of Radiation*, Oxford University Press, (1949) p. 88.

4. Reprinted in *Quantum Scattering Theory*, Ed. M. Ross, Indiana U. Press., Bloomington, (1963), p. 47.

5. F. Coester, Helv. Phys. Acta 38 (1965), p.7.

6. B. Simon, *Quantum Mechanics for Hamiltonians Defined as Quadratic Forms*, Princeton University Press, Princeton, (1971).